T0213607

Quantum Field Theory and its
Macroscopic Manifestations

Boson Condensation, Ordered Patterns
and Topological Defects

Quantum Field Theory and its Macroscopic Manifestations

Boson Condensation, Ordered Patterns and Topological Defects

Massimo Blasone & Giuseppe Vitiello
Università di Salerno & INFN, Italy

Petr Jizba
Czech Technical University, Prague, Czech Republic

Imperial College Press
ICP

Published by

Imperial College Press
57 Shelton Street
Covent Garden
London WC2H 9HE

Distributed by

World Scientific Publishing Co. Pte. Ltd.
5 Toh Tuck Link, Singapore 596224
USA office: 27 Warren Street, Suite 401-402, Hackensack, NJ 07601
UK office: 57 Shelton Street, Covent Garden, London WC2H 9HE

Library of Congress Cataloging-in-Publication Data
Blasone, Massimo.
    Quantum field theory and its macroscopic manifestations : Boson condensation, ordered patterns, and topological defects / by Massimo Blasone, Petr Jizba & Giuseppe Vitiello.
        p. cm.
    Includes bibliographical references and index.
    ISBN-13: 978-1-84816-280-8 (hardcover : alk. paper)
    ISBN-10: 1-84816-280-4 (hardcover : alk. paper)
    1. Quantum field theory. 2. Quantum theory. 3. Crystals. 4. Ferromagnetism.
5. Superconductors. I. Jizba, Petr. II. Vitiello, Giuseppe. III. Title.
    QC174.45.B5575 2011
    530.14'3--dc22

                                                            2010034113

British Library Cataloguing-in-Publication Data
A catalogue record for this book is available from the British Library.

First published 2011 (Hardcover)
Reprinted 2016 (in paperback edition)
ISBN 978-1-911299-72-1

Cover image:
"Landscape", 1961 – Oil on wooden panel, 20 × 20 cm, by Pasquale Vitiello (1912–1962)
La Pittura in Italia – II Novecento (1900/1990) (C. Pirovano Ed., Electa, Milano, 1992)

Printed in Singapore

To Kateřina, Hana, and Marina.

Preface

In 1951 and in 1952 Van Hove observed that there are theories where no normalizable state vectors belong to the common domain of both the free Hamiltonian H_0 and the perturbed (full) Hamiltonian H. Faith in the general applicability and validity of perturbation theory was then deeply shaken and a whole conception of the physical world was put in doubt. Perturbation theory rests indeed on the assumption that perturbed and unperturbed state vectors belong to the same Hilbert space. Friedrichs' book, where the existence of a host of unitarily inequivalent irreducible representations of the canonical commutation relations was discussed at length, came out in 1953 and not much later the Haag theorem was formulated. Actually, it was in those years that the discussion on basic principles, such as Lorentz invariance, spectral conditions, locality, etc., on which a reliable quantum field theory should be founded, led to the programme of the Axiomatic Formulation of Quantum Field Theory (QFT), starting indeed from the works by Haag, Gärding, Wightman, Schweber and others. The papers by Lehmann, Symanzik and Zimmermann (LSZ), published in "il Nuovo Cimento" between 1954 and 1958, laid solid bases for future developments of QFT. The LSZ formalism is founded on the so-called asymptotic condition which requires that a field theory must have an interpretation in terms of asymptotic particles with definite quantum numbers. Such a condition has been the guiding criterion underlying most of the work done in QFT in subsequent years, from renormalization theory to the search for a unified theory of the basic interactions among the constituents of matter. The crucial problem, which has been attracting the attention of many physicists, is indeed how to map Heisenberg fields, in terms of which the dynamics is given, to the asymptotic fields, in terms of which observables are constructed. In an early unpublished note, consistent with the LSZ formalism and written in Naples by Dell'Antonio and Umezawa in 1964, it is stressed that this map can only be a weak map, i.e., a map among matrix elements computed in the Hilbert space for the asymptotic fields. In subsequent years the study of such a mapping, called the dynamical map or the Haag expansion, has been central in the research activity of Umezawa and has revealed many subtle mechanisms through which the basic dynamics manifests itself at the level of the observables. One of these mechanisms, through which the consistency between the dynamical and the phenomenological level of the theory is expressed, is the dynamical rearrangement of symmetry in spontaneously broken symmetry theories.

A very important development occurred when the dynamical generation of long-range correlations, mediated by the Nambu–Goldstone (NG) boson quanta, was discovered in the early '60s, with subsequent implications in local gauge theories, such as the Higgs mechanism, which is one of the pillars of the standard model of elementary particles. It is interesting to remark that exactly the discovery of these collective modes gave strength to non-perturbative approaches, which could then establish themselves as complementary, or even, in some cases, alternative to the perturbation theory paradigm based on the ontological postulate of the asymptotic condition. The discovery of the existence in QFT of the unitarily inequivalent representations of the canonical commutation relations, which was in some sense shocking in the previous decade, could be better appreciated. The many inequivalent representations appeared to be a richness of QFT, which was thus recognized to be, due to such a specific feature indeed, the proper frame where systems endowed with many physically different phases could be described. QFT turns out not to be simply the "extension" of Quantum Mechanics (QM) to systems with an infinite number of degrees of freedom. Instead, QFT appears to be drastically different from QM. The von Neumann theorem, known for a long time and stating the unitary equivalence of the irreducible representations of the canonical commutation relations in QM, makes QM intrinsically not adequate to describe the variety of physically (unitarily) inequivalent phases of a given system. The crucial point is that such a theorem fails to hold in QFT, indeed, due to the infinite number of degrees of freedom. Spontaneous breakdown of symmetry, thermal field theory, phase transitions in a variety of problems, the process of defect formation during the process of non-equilibrium symmetry breaking phase transitions characterized by an order parameter, could then be studied by exploiting the whole manifold of the inequivalent representations in QFT.

In these studies, the prominent role played by coherent states was recognized, and attention was more and more focused on this, especially after the discovery of laser beams in quantum optics. It appeared that the "physical differences" among inequivalent representations are the differences in the degree of coherence of the boson condensates in the respective vacua.

The developments of QFT very briefly depicted above constitute the basis on which this book rests. The existence of the unitarily inequivalent representations is, indeed, a recurrent theme in our presentation. It is explored in several Chapters and shown to be especially related with finite temperature and dissipation in QFT, to the point that QFT can be recognized to be an *intrinsically* thermal quantum theory. The possibility of defining operators such as entropy and free energy in QFT and the role played there by them has been explored. The emerging picture is that no microscopic physical system may be considered completely isolated (closed) since it is always in interaction with the background fluctuations. From a different perspective, dissipation is discussed in relation to the proposal put forward by 't Hooft, according to which classical deterministic systems with information loss at high energy (Planck scale) may exhibit quantum behavior at low energy.

Quantum dynamics underlies macroscopic systems exhibiting some kind of ordering, such as superconductors, ferromagnets or crystals. Even the large-scale structures in the Universe, as well as the ordering in the biological systems, appear to be the manifestation of the microscopic dynamics ruling the elementary components of these systems. Therefore, in our discussion of the spontaneous breakdown of symmetry and collective modes, we stress that one crucial achievement has been recognizing that quantum field dynamics is not confined to the microscopic world: crystals, ferromagnets, superconductors, etc. are *macroscopic quantum systems*. They are quantum systems not in the trivial sense that they are made by quantum components (like any physical system), but in the sense that their macroscopic properties, accounted for by the order parameter field, cannot be explained without recourse to the underlying quantum dynamics. The problem is then to understand how the observed macroscopic properties are generated out of the quantum dynamics; how the *macroscopic* scale characterizing those systems is dynamically generated out of the *microscopic* scale of the quantum elementary components. Such a *change of scale* is understood to occur through the condensation of the NG boson quanta in the system ground state. Even in the presence of a gauge field, the NG boson fields do not disappear from the theory; they do not appear in

the spectrum of physical particles, as the Higgs mechanism predicts; however, they do condense in the system vacuum state, thus creating a number
of physically detectable properties originating from the vacuum structure
so generated. Many of the physical examples we study in this book are
characterized by the phenomenon of NG boson condensation. In this connection, we also consider the question of the dynamical generation of the
macroscopic stability out of fluctuating quantum fields.

Moreover, a variety of phenomena are also observed where quantum particles coexist and interact with *extended macroscopic objects* which show a
classical behavior, e.g., vortices in superconductors and superfluids, magnetic domains in ferromagnets, dislocations in crystals and other *topological
defects*, fractal structures and so on. One is thus also faced with the question of the quantum origin of topological defects and of their interaction
with quanta. This is a crucial issue for the understanding of symmetry
breaking phase transitions and structure formation in a wide range of systems, from condensed matter to cosmology. We are thus led to discuss how
the generation of ordered structures and of extended objects is explained in
QFT. We show that topological defects are originated by non-homogeneous
(localized) coherent condensation of quanta. The approach we follow is thus
in some sense alternative to the one in which one starts from the classical
soliton solutions and then quantizes them. Along the same line of thought,
also oscillations of mixed particles, with particular reference to neutrinos,
which manifest themselves on large (macroscopic) space distances appear
to be connected to a (microscopic) condensation mechanism in the vacuum
state.

As a general result stemming out of our discussion in this book we could
say that recognizing the existence of the collective NG boson modes in spontaneously broken symmetry theories has produced a *shift of paradigm* (*á
la* Kuhn): the former purely atomistic vision of the world, although necessary, turns out to be *not sufficient* to explain many physical phenomena.
One needs to integrate such an atomistic vision with the inclusion of the
dynamical generation of *collective* modes.

Throughout the book we have not specifically considered many important computational and conceptual questions and problems that have
marked in a significant way the historical development of QFT, among these
primarily renormalization problems. Neither have we discussed string theory, inflationary scenarios in cosmology and some recent theoretical and
experimental achievements, such as, for example, the ones in the Bose–
Einstein condensation of atoms in magnetic traps or other kinds of potential

wells, and related developments in quantum optics and quantum computing. Our choice is motivated by the fact that the present book is not meant to be one on the general formalism of QFT and the whole spectrum of its applications. In any case, we apologize to the reader for neglecting many important topics and for many holes in our presentation.

We use both the operator formalism and the functional integration formalism. In the operator formalism the particle and wave-packet physical picture is more transparent, while in the functional formalism the general mathematical structure underlying the symmetry properties and the correlation functions appears more evident. From a formal point of view, the price we pay for the apparent non-homogeneous treatment is compensated by the multi-faceted understanding of the theoretical structure under study. Another price we pay for the variety of arguments treated is a non-uniform notation: our preference has been to adopt the general criterion of keeping contact with the notation of the original works.

The level of the presentation has been finalized to a readership of graduate students with a basic knowledge of quantum mechanics and QFT. Some of the presented material grew from graduate courses on elementary particle physics and/or condensed matter physics which the authors taught at the University of Salerno and Czech Technical University in Prague. The matter is organized as shown in the following table of contents and purposely several arguments and notions have been repeated in different Sections and Chapters for the reader's convenience. Much formalism is confined to the Appendices, where, however, the reader can find short discussions of conceptually and computationally important topics, such as Glauber coherent states and generalized coherent states. Some material on classical soliton theory, homotopy theory and defect classification is confined to Chapter 10, which may be skipped by the reader who is familiar with such topics.

Summarizing, the book contains an overview of many QFT results obtained by many research groups and by ourselves. It is therefore imperative to warmly thank all those colleagues and collaborators with whom we have had the good fortune to work or to discuss some of the problems considered in this book. This is certainly not a complete list, and we apologize for that. It includes T. Arimitsu, V. Srinivasan, H. Matsumoto, S. Kamefuchi, Y. Takahashi, H. Ezawa, E. Del Giudice, T. Evans, R. Rivers, J. Klauder, E. C. Sudarshan, H. Kleinert, J. Tolar, J. Niederle, E. Celeghini, A. Widom, Y. Srivastava, R. Mańka, E. Alfinito, O. Romei, A. Iorio, A. Capolupo, G. Lambiase, A. Kurcz, F. C. Khanna, P. A. Henning, E. Graziano, A. Beige, R. Jackiw, R. Haag, P. L. Knight, G. Vilasi, G. Scarpetta, F. Mancini, D.

Steer, O. Pashaev, P. Sodano, Y. X. Gui, K. Fujii, T. Yabuki, F. Buccella, S. De Martino, S. De Siena, F. Illuminati, F. Dell'Anno, M. Di Mauro, A. Stabile, I. Rabuffo, M. Tarlini, M. de Montigny, M. Piattelli-Palmarini, A. Plotnitsky, M. Milani, S. Doglia, T.H. Elze, R.C. Ji, N. E. Mavromatos, E. Pessa, G. Minati, K. Yasue, M. Jibu, G. G. Globus, G. L. Sewell, G. 't Hooft, G. E. Volovik, J. Swain, W. J. Freeman, K. H. Pribram, W. H. Zurek, H. Haken, the late M. Marinaro, E. R. Caianiello, G. Preparata, A. O. Barut, L. O' Raifeartaigh, and of course H. Umezawa. A special thank you goes to Francesco Guerra, Tom Kibble and Mario Rasetti for their constant encouragement. We also thank Francesco Guerra for calling our attention to the unpublished 1964 paper by Gianfausto Dell'Antonio and Hiroomi Umezawa on the dynamical map. Without the patient efforts, the advice and the assistance of Katie Lydon, Lizzie Bennett and Jacqueline Downs of Imperial College Press and Ms E. H. Chionh of World Scientific Publishing Company, we would never have been able to finish this book. To them also all our warm thanks.

Salerno, Prague, July 2010

Massimo Blasone
Petr Jizba
Giuseppe Vitiello

Contents

Preface   vii

1. The structure of the space of the physical states   1

    1.1   Introduction   1

    1.2   The space of the states of physical particles   2

    1.3   The Weyl–Heisenberg algebra and the Fock space   7

    1.4   Irreducible representations of the canonical commutation
            relations 10

    1.5   Unitarily equivalent representations 15

    1.6   The Stone–von Neumann theorem 17

    1.7   Unitarily inequivalent representations 19

    1.8   The deformation of Weyl–Heisenberg algebra 21

            1.8.1   Self-similarity, fractals and the Fock–Bargmann
                      representation 26

    1.9   The physical particle energy and momentum operator . . 30

    1.10 The physical Fock space and the physical fields 32

Appendix A   Strong limit and weak limit   35

Appendix B   Glauber coherent states   37

Appendix C   Generalized coherent states   41

Appendix D q-WH algebra, coherent states and theta func-
                  tions   49

2.   Inequivalent representations of the canonical commuta-
     tion relations   53

     2.1   Introduction   53
     2.2   Heisenberg fields, physical fields and the dynamical map   54
     2.3   Examples of inequivalent representations   58
     2.4   The Haag theorem and non-perturbative physics   66
     2.5   The momentum operator   67
     2.6   Time evolution and asymptotic limits   68
     2.7   Inequivalent representations in flavor mixing   71

Appendix E   Computation of $\langle 0|\psi_i(x)|\alpha_n\alpha_m\rangle$ 85

Appendix F   Computation of $|0(\theta)\rangle$ 87

Appendix G   Orthogonality of flavor vacua at different times   89

Appendix H   Entanglement in neutrino oscillations   91

3.   Spontaneous breakdown of symmetry and the Goldstone
     theorem   97

     3.1   Introduction 97
     3.2   Invariance and symmetry   99
     3.3   Irreducible representations of the symmetry group 101
     3.4   Symmetry and the vacuum manifold   102
     3.5   Boson transformation and inequivalent representations .   106
     3.6   Spontaneous symmetry breaking and functional integrals .   108
     3.7   The Goldstone theorem   112
           3.7.1 $U(1)$ symmetry   112
           3.7.2 $SU(2)$ symmetry 115
     3.8   Spontaneous symmetry breaking in local gauge theories .   119
           3.8.1   The $U(1)$ local gauge model 119
           3.8.2   The chiral gauge model   124
     3.9   Finite volume effects   126
     3.10 Space-time dimensionality   128

Appendix I   The order parameter space   133

Appendix J   The Mermin–Wagner–Coleman theorem   135

4. Dynamical rearrangement of symmetry and macroscopic manifestations of QFT   137

    4.1   Introduction 137
    4.2   Dynamical rearrangement of symmetry 139
        4.2.1 $SU(2)$ symmetry 139
        4.2.2 Global $U(1)$ symmetry 143
        4.2.3 Local $U(1)$ symmetry and the emergence of classical Maxwell equations 145
    4.3   The boson transformation theorem and the non-homogeneous boson condensation 153
        4.3.1 Topological singularities, gapless modes and macroscopic observables 156
        4.3.2 Defect formation in the process of symmetry breaking phase transitions 158
    4.4   Group contraction and spontaneous symmetry breaking . 159
        4.4.1 The infrared effect 159
        4.4.2 Group contraction, boson condensation and macroscopic quantum systems 165
        4.4.3 The collective behavior of quantum components and group contraction 166
    4.5   Quantum fluctuations and macroscopic stability 169
        4.5.1 Quantum mechanical decoherence and stability of macroscopic quantum systems 174

Appendix K   Group contraction and Virasoro algebra   179

Appendix L   Phase locking in the N atom system   183

5. Thermal field theory and trajectories in the space of the representations   185

    5.1   Introduction 185
    5.2   Doubling the degrees of freedom 188
        5.2.1 The two-slit experiment 189
    5.3   Thermo Field Dynamics: A brief introduction 191
        5.3.1 The propagator structure in TFD 199
        5.3.2 Non-hermitian representation of TFD 201
        5.3.3 TFD for fields with continuous mass spectrum . . 202

5.4   The q-deformed Hopf algebra and the doubling of the field
      degrees of freedom . . . . '.   206
5.5   Free energy, entropy and the arrow of time. Intrinsic ther-
      mal nature of QFT   212
      5.5.1   Entropy and system-environment entanglement .   214
5.6   Thermal field theory and the gauge field   215
5.7   Boson condensation at finite temperature   221
      5.7.1   Free energy and classical energy   225
5.8   Trajectories in the space of representations   231

6. Selected topics in thermal field theory   235

6.1   Introduction   235
6.2   The Gell-Mann–Low formula and the closed time-path for-
      malism   236
6.3   The functional integral approach   240
      6.3.1   Generating functionals for Green's functions . . .   240
      6.3.2   The Feynman–Matthews–Salam formula   246
      6.3.3   More on generating functionals   249
6.4   The effective action and the Schwinger–Dyson equations .   252
6.5   Imaginary-time formalism   257
6.6   Geometric background for thermal field theories   260
      6.6.1   The η-ξ spacetime   261
      6.6.2   Fields in η-ξ spacetime   264

Appendix M   Thermal Wick theorem   271

Appendix N   Coherent state functional integrals   275

N.1   Glauber coherent states   275
N.2   Generalized coherent states   279

Appendix O   Imaginary-time formalism and phase transitions   287

O.1   Landau–Ginzburg treatment   290

Appendix P   Proof of Bogoliubov inequality   293

7. Topological defects as non-homogeneous condensates. I   297

7.1   Introduction   297
7.2   Quantum field dynamics and classical soliton solutions   .   298

7.2.1   The dynamical map and the boson transformation   299

7.2.2   The quantum coordinate   302

7.3   The $\lambda\phi^4$ kink solution   304

7.3.1   The kink solution and temperature effects   309

7.3.2   The kink solution: closed time-path approach . .   312

7.4   The sine-Gordon solution   319

7.4.1   The quantum image of the Bäcklund transformations   324

7.5   Soliton solutions of the non-linear Schrödinger equation .   327

7.5.1   The ferromagnetic chain   328

7.5.2   Non-linear Schrödinger equation with Toda lattice back-reaction potential   334

7.5.3   Ring solitons in the Scheibe aggregates   336

7.6   Fermions in topologically non-trivial background fields .   339

7.7   Superfluid vortices   344

8.   Topological defects as non-homogeneous condensates. II   349

8.1   Introduction   349

8.2   Vortices in $U(1)$ local gauge theory   350

8.3   Topological solitons in gauge theories   353

8.3.1   Homogeneous boson condensation   357

8.3.2   The vortex of scalar electrodynamics   359

8.3.3   The 't Hooft–Polyakov monopole   362

8.3.4   The sphaleron   365

8.4   The $SU(2)$ instanton   367

9.   Dissipation and quantization   373

9.1   Introduction   373

9.2   The exact action for damped motion   374

9.2.1   Quantum Brownian motion   378

9.3   Quantum dissipation and unitarily inequivalent representations in QFT   379

9.3.1   The arrow of time and squeezed coherent states .   383

9.4   Dissipative non-commutative plane   385

9.4.1   The dissipative quantum phase interference . . .   388

9.5   Gauge structure and thermal features in particle mixing .   390

9.6   Dissipation and the many-body model of the brain   396

9.7   Quantization and dissipation   404

Appendix Q   Entropy and geometrical phases in neutrino mixing   413

Appendix R   Trajectories in the memory space   417

10. Elements of soliton theory and related concepts   423

    10.1   Introduction   423
    10.2   The Korteweg–de Vries soliton   424
    10.3   Topological solitons in $(1 + 1)$-d relativistic field theories .   428
        10.3.1   The sine-Gordon soliton   429
        10.3.2   The $\lambda\phi^4$ kink   431
    10.4   Topological solitons in gauge theories   434
        10.4.1   The Nielsen–Olesen vortex   434
        10.4.2   The 't Hooft–Polyakov monopole   438
    10.5   Topological defect classification and the Kibble–Zurek
        mechanism for defect formation   442
        10.5.1   Exact homotopy sequences   445
        10.5.2   Topological defects in theories with SSB   447
    10.6   Derrick theorem and defect stability   453
    10.7   Bogomol'nyi bounds   458
    10.8   Non-topological solitons   463
    10.9   Instantons and their manifestations   466
        10.9.1   Collective coordinates and fermionic zero modes .   480

Appendix S   Bäcklund transformation for the sine-Gordon
              system   483

Appendix T   Elements of homotopy theory   487

Bibliography   497

Index   519

Chapter 1

The structure of the space of the physical states

1.1 Introduction

Symmetry principles play a central role in the understanding of natural phenomena. However, it is not always easy to recognize symmetries in physical observations since at a phenomenological level they can manifest as distorted, "rearranged" symmetries. For example, the fundamental symmetry between protons and neutrons, the nucleons, does not manifest as an exact symmetry, but as a "broken symmetry": the charge independence of nuclear interaction is indeed violated by the electromagnetic interaction. In general, various symmetry schemes, which are quite successful, also appear to be in some way approximate symmetry schemes [79, 343, 443, 476, 617], i.e., one has to disregard some phenomenological aspects, e.g., mass differences, which violate certain symmetry requirements. A way of looking at this situation is to interpret the observed deviations from the exact symmetry as a phenomenological distortion or rearrangement of the basic symmetry. Examples of rearranged symmetries are easily found in solid state physics: crystals manifest a periodic structure, but do not possess the continuous translational invariance of the Hamiltonian of molecular gas. Ferromagnets present rotational invariance around the magnetization axis, but not the original $SU(2)$ invariance of the Lagrangian. In superconductivity and superfluidity the phase invariance is the one that disappears.

The crucial problem one has to face in the recognition of a symmetry is, then, the intrinsic two-level description of Nature: one aspect of this duality concerns original symmetries ascribed to "basic" entities, the other aspect concerns the corresponding rearranged symmetries of observable phenomena. This two-level description of Nature was soon recognized in Quantum Field Theory (QFT) as the duality between fields and particles. Without

1

going into the historical developments of this concept, which are outside the purpose of this book, we only recall, as an example, how fundamental this duality is in the renormalization theory, where the distinction is crucial between "bare" and "observed" particles, namely the distinction between basic fields and their physical "manifestation".

In the following Sections we will focus our attention on some structural aspects of QFT in order to prepare the tools to be used in the study of the mechanisms through which the dynamics of the basic fields leads to their observable physical manifestation. Thus, the core of our interest will be the structure of the space of the physical fields, which will bring us to study that peculiar nature of QFT consisting in the existence of infinitely many unitarily inequivalent representations of the canonical (anti-)commutation relations, and thus to the analysis of the von Neumann theorem, of the Weyl–Heisenberg algebra, the characterization of the physical fields and the coherent states. Our discussion will include in a unified view, topics such as the squeezing and self-similarity transformations, fractals and quantum deformation of the Weyl–Heisenberg algebra. A glance at the table of contents shows how these subjects are distributed in the various Sections and Appendices.

1.2   The space of the states of physical particles

Let us consider a typical scattering process between two or more particles. By convenient measurements we can identify the kind, the number, the energy, etc., of the particles before they interact (incoming particles); there is then an interaction region which is precluded to observations and finally we can again measure the kind, the number, the energy, etc., of the particles after the interaction (outgoing particles). The sum of the energies of the incoming particles is observed to be equal to the sum of the energies of the outgoing particles. Incoming particles and outgoing particles are referred to as "physical particles", or else as "observed" or "free" particles, where the word "free" does not exclude the possibility of interaction among them; it means that the interaction among the particles can be considered to be negligible far away, in space and time, from the interaction region. The total energy of the system of free particles is given in a good approximation by the sum of the energies of the single particles. We require that the energy of the physical particles is determined as a certain function of their momenta. In solid state physics the physical particles are usually called quasiparticles.

Among the quasiparticles, a special role is played by those, such as phonons, magnons, etc., called "collective modes", which are responsible for long-range correlation among the elementary components of the system.

The possibility of identifying outgoing and/or incoming particles resides in the possibility of setting our particle detector far away from the interaction region (at a space distance $\mathbf{x} = \pm\infty$ from the interaction region) and to let it be active well before and/or well after the interaction time (at a time $t = \pm\infty$ with respect to the interaction time). In other words, we assume that the interaction forces among the particles go to zero at spacetime regions far away from the spacetime interaction region.[1] This is indeed possible in most cases. However, there are important cases in which this "switching off" of the interaction is not possible due to intrinsic properties of the interaction. When this latter situation occurs, we cannot apply the usual methods of perturbation theory. The validity of the perturbative methods relies indeed on the possibility of correctly defining "asymptotic" states for the system under study, namely states properly defined in spacetime regions where the interaction effects are negligible.

We now briefly summarize the main steps in the construction of the Hilbert space for the physical particles. Among several possible strategies [115, 343, 466, 558, 599, 666], we mostly follow [617, 619, 621].

The state of a single particle is classified by the suffices (i, s), where i specifies the spatial distribution of the state, while s specifies other freedoms (e.g., spin, charges, etc.). For simplicity, we assume we are dealing with only one kind of particle (e.g., only electrons, or only protons, etc.). We must use wave packets to specify spatial distributions, because plane waves like $\exp(i\mathbf{k} \cdot \mathbf{x})$ are not normalizable and do not form a countable set. On the other hand, it is well known that an orthonormalized complete set of square-integrable functions $\{f_i(\mathbf{x}), i = 1, 2, \dots\}$ is a countable set. Thus we introduce the creation operators $\alpha_i^{s\dagger}$ and $\beta_i^{s\dagger}$ for particles and their antiparticles, respectively, with spatial distribution $f_i(\mathbf{x})$, i.e., in wave-packet states, as

$$\alpha_i^{s\dagger} = \frac{1}{(2\pi)^{3/2}} \int d^3k \, f_i(\mathbf{k}) \alpha_{\mathbf{k}}^{s\dagger}, \tag{1.1a}$$

$$\beta_i^{s\dagger} = \frac{1}{(2\pi)^{3/2}} \int d^3k \, f_i(\mathbf{k}) \beta_{\mathbf{k}}^{s\dagger}. \tag{1.1b}$$

In these equations we have also introduced the creation operators $\alpha_{\mathbf{k}}^{s\dagger}$ and

---

[1] Self-interaction, which dresses each particle and is responsible for wave function renormalization, is implicitly taken into account in the concept of physical particle.

$\beta_{\mathbf{k}}^{s\dagger}$ for particles and their anti-particles of momentum $\mathbf{k}$. Their hermitian conjugates α_i^s, β_i^s and $\alpha_{\mathbf{k}}^s$, $\beta_{\mathbf{k}}^s$, respectively, denote the annihilation operators. In Eqs. (1.1) $f_i(\mathbf{k})$ are the Fourier amplitudes of $f_i(\mathbf{x})$

$$f_i(\mathbf{x}) = \int \frac{d^3k}{(2\pi)^3} f_i(\mathbf{k}) e^{i\mathbf{k}\cdot\mathbf{x}} . \qquad (1.2)$$

The possibility of expressing the spatial distribution of the state by a discrete index follows from the fact that we use the square-integrable functions (wave-packets) $f_i(\mathbf{x})$ which form a countable set. The space of states so constructed is a separable Hilbert space, whose states can be expressed as superposition of the countable set of basis vectors $f_i(\mathbf{k})$.

The inner product is:

$$(f_i, g_j) = \int d^3x\, f_i^*(\mathbf{x}) g_j(\mathbf{x}) = \int \frac{d^3k}{(2\pi)^3} f_i^*(\mathbf{k}) g_j(\mathbf{k}) . \qquad (1.3)$$

The orthonormality condition is:

$$(f_i, f_j) = \int \frac{d^3k}{(2\pi)^3} f_i^*(\mathbf{k}) f_j(\mathbf{k}) = \delta_{ij} . \qquad (1.4)$$

For the norm we will use the notation $|f_i| = (f_i, f_i)^{1/2} = 1$. In general, any square-integrable normalized function $f(x)$ is expressed as

$$f(x) = \sum_i a_i f_i(\mathbf{x}) , \qquad (1.5)$$

where a_i (and $f(x)$) may depend on time and $\sum_i |a_i|^2 = 1$ due to the normalization $(f, f) = 1$:

$$(f, f) = \sum_{i,j} a_i^* a_j (f_i, f_j) = \sum_{i,j} \delta_{ij} a_i^* a_j = 1 . \qquad (1.6)$$

We also introduce, in analogy with Eqs. (1.1), the wave-packet operators associated to the spatial distribution $f(\mathbf{x})$:

$$\alpha_f^{s\dagger} = \frac{1}{(2\pi)^{3/2}} \int d^3k\, f(\mathbf{k}) \alpha_{\mathbf{k}}^{s\dagger} , \qquad (1.7a)$$

$$\beta_f^{s\dagger} = \frac{1}{(2\pi)^{3/2}} \int d^3k\, f(\mathbf{k}) \beta_{\mathbf{k}}^{s\dagger} , \qquad (1.7b)$$

where $f(\mathbf{k})$ are the Fourier amplitudes of $f(\mathbf{x})$.

For brevity, we will omit the suffix s when no confusion arises. We next assume the existence of the physical vacuum state $|0\rangle$ defined as

$$\alpha_i|0\rangle = 0 = \beta_i|0\rangle . \qquad (1.8)$$

The conjugate state $\langle 0|$ is such that

$$\langle 0|\alpha_i^\dagger = 0 = \langle 0|\beta_i^\dagger , \qquad (1.9)$$

and $\langle 0|0\rangle = 1$. From Eqs. (1.1), it is then, for any $\mathbf{k}$,

$$\alpha_\mathbf{k}|0\rangle = 0 = \beta_\mathbf{k}|0\rangle , \qquad (1.10a)$$

$$\langle 0|\alpha_\mathbf{k}^\dagger = 0 = \langle 0|\beta_\mathbf{k}^\dagger . \qquad (1.10b)$$

The Hilbert space of physical particle states is cyclically constructed by repeated applications of $\alpha_i^\dagger$ and $\beta_i^\dagger$ on $|0\rangle$, e.g., the one-particle state is $\alpha_i^\dagger|0\rangle = |\alpha_i\rangle$. Let us denote by n_i the number of particles with spatial distribution given by $f_i(\mathbf{x})$. The n-particle state of the system is specified by the state vector $|n_1, n_2, \ldots\rangle$, where the numbers n_i are assigned for each i, and is given by

$$|n_1, n_2 \ldots\rangle = \prod_i \frac{1}{\sqrt{n_i!}}(\alpha_i^\dagger)^{n_i}|0\rangle , \qquad (1.11)$$

in the case of boson particles and by

$$|n_1, n_2 \ldots\rangle = \widetilde{\prod_i}(\alpha_i^\dagger)^{n_i}|0\rangle , \qquad (1.12)$$

in the case of fermion particles. In Eq. (1.12) $\widetilde{\prod}_i$ denotes product with n_i restricted to the values 0 or 1 for any i. The conjugate vectors are introduced in the usual way, e.g., for bosons

$$\langle n_1, n_2 \ldots| = \langle 0| \prod_i \frac{1}{\sqrt{n_i!}}(\alpha_i)^{n_i} . \qquad (1.13)$$

The action of α_i and $\alpha_i^\dagger$ on the states of the Hilbert space is given by

$$\alpha_i|n_1, \ldots, n_i, \ldots\rangle = \sqrt{n_i}|n_1, \ldots, n_i - 1, \ldots\rangle , \qquad (1.14a)$$

$$\alpha_i^\dagger|n_1, \ldots, n_i, \ldots\rangle = \sqrt{n_i + 1}|n_1, \ldots, n_i + 1, \ldots\rangle , \qquad (1.14b)$$

for bosons and by

$$\alpha_i|n_1, \ldots, n_i, \ldots\rangle = n_i\eta(n_1, \ldots n_{i-1})|n_1, \ldots, n_i - 1, \ldots\rangle , \qquad (1.15a)$$

$$\alpha_i^\dagger|n_1, \ldots, n_i, \ldots\rangle = (1 - n_i)\eta(n_1, \ldots n_{i-1})|n_1, \ldots, n_i + 1, \ldots\rangle , \qquad (1.15b)$$

for fermions, with $n_i = 0, 1$ and

$$\eta(n_1, \ldots n_{i-1}) = (-1)^{\sum_{j<i} n_j} . \qquad (1.16)$$

In a standard fashion [619] one can show that the set $\{|n_1, n_2, \ldots\rangle\}$ is an orthonormalized set of vectors in the Hilbert space:

$$\langle n_1', n_2', \ldots|n_1, n_2, \ldots\rangle = \prod_i \delta_{n_i' n_i} . \qquad (1.17)$$

We are using the simplified notation $|n_1, n_2, \ldots\rangle \equiv |n_1\rangle \otimes |n_2\rangle \otimes \ldots$, where n_i can be any non-negative integer for bosons; it is 0 or 1 for fermions. In this notation, the vacuum $|0\rangle$ is $|0, 0, \ldots\rangle \equiv |0\rangle \otimes |0\rangle \otimes \ldots$. Eqs. (1.14) show that

$$N_i |n_1, \ldots, n_i, \ldots\rangle = n_i |n_1, \ldots, n_i, \ldots\rangle, \qquad (1.18)$$

where $N_i = \alpha_i^\dagger \alpha_i$ is the number operator, or, by restoring the suffix s,

$$N_i^s = \alpha_i^{s\dagger} \alpha_i^s. \qquad (1.19)$$

The total number is given by $N = \sum_{i,s} N_i^s$. Let us consider particles which are bosons. Eqs. (1.14) and (1.17) give for any n_i's

$$\langle n_1, n_2, \ldots | [\alpha_i^s, \alpha_j^{r\dagger}] | n_1, n_2, \ldots\rangle = \mathbb{1}\delta_{ij}\delta_{rs}, \qquad (1.20a)$$

$$\langle n_1, n_2, \ldots | [\alpha_i^s, \alpha_j^r] | n_1, n_2, \ldots\rangle = 0 = \langle n_1, n_2, \ldots | [\alpha_i^{s\dagger}, \alpha_j^{r\dagger}] | n_1, n_2, \ldots\rangle, \qquad (1.20b)$$

$$\langle n_1, n_2, \ldots | [\alpha_i^s, \mathbb{1}] | n_1, n_2, \ldots\rangle = 0 = \langle n_1, n_2, \ldots | [\alpha_i^{s\dagger}, \mathbb{1}] | n_1, n_2, \ldots\rangle, \qquad (1.20c)$$

and similarly for β_i^s and $\beta_i^{s\dagger}$. Here $\mathbb{1}$ denotes the identity operator. Consistency with Eqs. (1.1) then requires that

$$[\alpha_{\mathbf{k}}^s, \alpha_{\mathbf{l}}^{r\dagger}] = \mathbb{1}\, \delta(\mathbf{k} - \mathbf{l})\delta_{rs}, \qquad (1.21a)$$

$$[\alpha_{\mathbf{k}}^s, \alpha_{\mathbf{l}}^r] = 0 = [\alpha_{\mathbf{k}}^{s\dagger}, \alpha_{\mathbf{l}}^{r\dagger}], \qquad (1.21b)$$

$$[\alpha_{\mathbf{k}}^s, \mathbb{1}] = 0 = [\alpha_{\mathbf{k}}^{s\dagger}, \mathbb{1}], \qquad (1.21c)$$

and similarly for $\beta_{\mathbf{k}}^s$ and $\beta_{\mathbf{k}}^{s\dagger}$. Due to the delta function $\delta(\mathbf{k}-\mathbf{l})$ appearing in Eqs. (1.21), these equations are to be understood in the sense of distribution theory, namely

$$\int \frac{d^3k\, d^3l}{(2\pi)^3}\, f^*(\mathbf{k})g(\mathbf{l})[\alpha_{\mathbf{k}}^s, \alpha_{\mathbf{l}}^{r\dagger}] = \int \frac{d^3k}{(2\pi)^3} f^*(\mathbf{k})g(\mathbf{k})\, \delta_{rs} = (f, g)\, \delta_{rs}, \qquad (1.22)$$

with $f(\mathbf{k})$ and $g(\mathbf{k})$ being suitable test functions. Then, for the operators introduced in Eq. (1.7) we have

$$[\alpha_f^s, \alpha_g^{r\dagger}] = (f, g)\delta_{rs}, \qquad (1.23a)$$

$$[\alpha_f^s, \alpha_g^r] = 0 = [\alpha_f^{s\dagger}, \alpha_g^{r\dagger}]. \qquad (1.23b)$$

In the case of fermions we obtain relations similar to (1.20)–(1.23) with anticommutators replacing the commutators. The vectors of the Hilbert space can also be proven to be, under particle permutations, fully symmetrical states for bosons and fully antisymmetrical states for fermions.

By repeated applications of creation and annihilation operators one can move from one member to another member in the set $\{|n_1, n_2, \ldots\rangle\}$; however, the operators $\alpha_{\mathbf{k}}^s$ (and $\beta_{\mathbf{k}}^s$) are not bounded operators: due to

Eqs. (1.21), they do not map normalizable vectors on normalizable ones; indeed, $\langle 0|\alpha_{\mathbf{k}}^s \alpha_{\mathbf{k}}^{s\dagger}|0\rangle = \delta(\mathbf{0})$, which is not finite. However, operators associated with spatial distribution $f_i(\mathbf{x})$ (or $f(\mathbf{x})$) give $|\alpha_i^{s\dagger}|0\rangle|^2 = 1$ (or $|\alpha_f^{s\dagger}|0\rangle|^2 = 1$) due to Eqs. (1.1) and Eq. (1.4) (or, for $\alpha_f^{s\dagger}$, due to Eqs. (1.7) and Eq. (1.6)).

We consider in more detail the mathematical nature of the Hilbert space in the following Section.

1.3   The Weyl–Heisenberg algebra and the Fock space

We require that it must be possible to express any vector in the Hilbert space, i.e., any physical state of the system, as a superposition of the vectors of the basis. This implies that the Hilbert space must be separable, i.e., it must contain a countable basis of vectors, say $\{\boldsymbol{\xi}_n\}$. In such a case, for any vector $\boldsymbol{\xi}$ of the space and any arbitrary $\epsilon > 0$ there exist a sequence $\{c_n\}$ such that $|\boldsymbol{\xi} - \sum_n c_n \boldsymbol{\xi}_n| < \epsilon$, which means that $\boldsymbol{\xi}$ can be approximated by the linear superposition $\sum_n c_n \boldsymbol{\xi}_n$ to any accuracy.

If we thus require that our space must be a separable Hilbert space, it is not correct to use the set $\{|n_1, n_2, \ldots\rangle\}$ as a basis, because this is not a countable set. To prove this, let us consider for simplicity a fermion system where n_i can assume only the values 0 or 1. Then, we consider the set of positive numbers

$$\{0.n_1 n_2 \ldots\},\tag{1.24}$$

with n_i assuming only the values 0 or 1. Using the binary system, we see that the set (1.24) covers all the real values in the interval $[0, 1]$, i.e., it is a non-countable set. On the other hand, there is a one-to-one correspondence between the set (1.24) and the set $\{|n_1, n_2, \ldots\rangle\}$, and thus we conclude that the latter is a non-countable one. Since the set $\{|n_1, n_2, \ldots\rangle\}$ for bosons is larger than the one for fermions, also in the boson case it is a non-countable set. To remedy this situation, we observe that physical states do not really contain an infinite number of particles. The number of particles can be as large as we want, but does not need to be infinite. We then select a countable subset $\mathcal{S}$ from the set $\{|n_1, n_2, \ldots\rangle\}$ as follows:

$$\mathcal{S} = \left\{ |n_1, n_2, \ldots\rangle, \; \sum_i n_i = \text{finite} \right\}.\tag{1.25}$$

This set contains the vacuum $|0, 0, \ldots\rangle$ but does not contain states like $|1, 1, 1, \ldots\rangle$ where $n_i = 1$ for all i. Actually, one can extract infinitely many countable subsets from the set $\{|n_1, n_2, \ldots\rangle\}$, each of them representing a different possible representation of the (anti-)commutation relations of the operators α_i, $\alpha_i^\dagger$, $i = 1, 2, \ldots$. Two of these representations will be said to be unitarily inequivalent representations when one arbitrary vector of one of them cannot be expressed as a superposition of base vectors of the other representations. We will discuss this point below and we will see that the existence of unitarily inequivalent representations is a characterizing feature of QFT, not present in Quantum Mechanics [123, 569, 621, 648, 649].

We now prove that the set $\mathcal{S}$ is a countable set.

Consider the state $|n_1, n_2, \ldots\rangle$ belonging to $\mathcal{S}$. Since this state contains only a finite number of particles, there is an integer number, say p, for which $n_p \neq 0$ and $n_i = 0$ for $i > p$. Then, to each vector in $\mathcal{S}$ we can associate two numbers, i.e., p and $N = \sum_i n_i$. For each product pN there exists only a finite number of vectors in $\mathcal{S}$ because for each pN we can distribute a finite number of particles only in a finite number of states. This means that we can label the vectors in $\mathcal{S}$ as $\boldsymbol{\xi}_a$ with $a = 1, 2, \ldots$ in such a way that pN is not decreasing for a increasing. The set $\mathcal{S}$ is thus countable and $\boldsymbol{\xi}_a$ are orthonormal vectors: $(\boldsymbol{\xi}_a, \boldsymbol{\xi}_b) = \delta_{ab}$.

We consider now the linear space $\mathcal{H}_F$ defined by

$$\mathcal{H}_F = \left\{ \boldsymbol{\xi} = \sum_{a=1}^{\infty} c_a \boldsymbol{\xi}_a, \quad \sum_{a=1}^{\infty} |c_a|^2 = \text{finite} \right\}. \tag{1.26}$$

$\mathcal{H}_F$ is separable because the set $\{\boldsymbol{\xi}_a\}$ is countable. If $\boldsymbol{\zeta}$ and $\boldsymbol{\eta}$ are vectors of $\mathcal{H}_F$,

$$\boldsymbol{\zeta} = \sum_{a=1}^{\infty} b_a \boldsymbol{\xi}_a, \quad \sum_{a=1}^{\infty} |b_a|^2 = \text{finite}, \tag{1.27a}$$

$$\boldsymbol{\eta} = \sum_{a=1}^{\infty} d_a \boldsymbol{\xi}_a, \quad \sum_{a=1}^{\infty} |d_a|^2 = \text{finite}, \tag{1.27b}$$

the inner product is defined as

$$(\boldsymbol{\zeta}, \boldsymbol{\eta}) = \sum_a b_a^* d_a. \tag{1.28}$$

The vectors in $\mathcal{H}_F$ thus have finite norm:

$$|\boldsymbol{\xi}|^2 = (\boldsymbol{\xi}, \boldsymbol{\xi}) = \sum_a |c_a|^2 = \text{finite}. \tag{1.29}$$

Note that a vector $\boldsymbol{\zeta}$ of $\mathcal{H}_F$ is the null vector, i.e., $\boldsymbol{\zeta} = 0$, if and only if all the coefficients $b_a = (\boldsymbol{\xi}_a, \boldsymbol{\zeta})$ (cf. (1.28)) are zero.

The linear space $\mathcal{H}_F$ is called the *Fock space* of physical particles. In this space one considers the set D of all the finite summations of the basis vectors of $\mathcal{H}_F$:

$$D \doteq \left\{ \xi_N^{(D)} = \sum_{a=1}^{N} c_a \xi_a, \ N \text{ finite} \right\}. \tag{1.30}$$

The set D can be proven to be dense in $\mathcal{H}_F$, i.e., every vector of D belongs to $\mathcal{H}_F$ and every vector of $\mathcal{H}_F$ is either a member of D or the limit of a Cauchy sequence of vectors in D. This last property can be expressed as

$$\xi = \lim_{N \to \infty} \xi_N^{(D)}, \tag{1.31}$$

where ξ is a vector of $\mathcal{H}_F$. Eq. (1.31) has to be understood in the sense of the strong limit (see Appendix A for a definition of strong and weak limit). When the relations (1.20) and (1.21) are computed using vectors belonging to D, then they hold for any vector in $\mathcal{H}_F$ and we can write the commutation relations in the Hilbert space as

$$[\alpha_i^s, \alpha_j^{r\dagger}] = \mathbb{1}\delta_{ij}\delta_{rs}, \tag{1.32a}$$

$$[\alpha_i^s, \alpha_j^r] = 0 = [\alpha_i^{s\dagger}, \alpha_j^{r\dagger}], \tag{1.32b}$$

$$[\alpha_i^s, \mathbb{1}] = 0 = [\alpha_i^{s\dagger}, \mathbb{1}], \tag{1.32c}$$

and similar relations for $\beta_{\mathbf{k}}^s$ and $\beta_{\mathbf{k}}^{s\dagger}$. The algebra (1.32) generated by $\alpha_i^s, \alpha_i^{s\dagger}$ and $\mathbb{1}$, for any s, and similarly the algebra (1.21), is a Lie algebra and is called the Heisenberg algebra or also the Weyl–Heisenberg (WH) algebra. Instead of Eqs. (1.32) and (1.21) we have anti-commutation relations in the case of fermions. The algebra (1.32) (and (1.21)) is also referred to as the canonical commutation relation (or anti-commutation relation, in the case of fermions) algebra.

As a final remark we observe that assigning the algebra (1.32) (or (1.21)) is not enough to specify the particular countable subset $\mathcal{S}$ one may select out of the set $\{|n_1, n_2, \ldots\rangle\}$. Since the states are obtained by cyclic operation of $\alpha_i^\dagger$ on the vacuum state, in order to specify $\mathcal{S}$ one needs to assign also the vacuum state annihilated by the α_i and on which the $\alpha_i^\dagger$ operators are defined as shown in Eq. (1.11).

1.4   Irreducible representations of the canonical commutation relations

We here assume that the particles under consideration are bosons. We use $\hbar = c = 1$. We introduce the dimensionless operators q_i and p_i defined by

$$q_i = \frac{1}{\sqrt{2}}(\alpha_i + \alpha_i^\dagger)\,, \tag{1.33a}$$

$$p_i = \frac{1}{i\sqrt{2}}(\alpha_i - \alpha_i^\dagger)\,. \tag{1.33b}$$

They satisfy the canonical commutation relations

$$[q_i, p_j] = i\mathbb{1}\delta_{ij}\,, \tag{1.34a}$$

$$[q_i, q_j] = 0 = [p_i, p_j]\,, \tag{1.34b}$$

$$[q_i, \mathbb{1}] = 0 = [p_i, \mathbb{1}]\,. \tag{1.34c}$$

This is also called the Weyl–Heisenberg algebra (generated by the operators $q_i, p_i, \mathbb{1}$, for any i). We note that by a convenient transformation leaving invariant the canonical commutation relations (1.34) the q_i and p_i operators can be given the dimensions appropriate to the usual phase-space coordinates. Since we are considering a finite number of particles (cf. Eq. (1.25)) the present situation is very similar to that in Quantum Mechanics (QM). In particular, the Hilbert space under consideration is the oscillator realization of the canonical variables q_i and p_i. It is known to be a complete space and we can use the well-known [660, 661] "unitarization" or "extension" procedure, in which one considers the operators

$$U_i(\sigma) = \exp(i\sigma p_i)\,, \tag{1.35a}$$

$$V_i(\tau) = \exp(i\tau q_i)\,, \tag{1.35b}$$

instead of p_i and q_i, with σ and τ real parameters. However, it should be stressed that in QM the p_i and q_i operators are not bounded operators [621, 649, 660, 661]. In the present QFT case, as noted in the previous Sections, the unbounded operators $\alpha_{\mathbf{k}}$ and $\alpha_{\mathbf{k}}^\dagger$ are smeared out by use of the square-integrable test functions $f_i(\mathbf{k})$ (cf. Eqs. (1.1)), so that in (1.33) we are using wave-packets operators α_i and $\alpha_i^\dagger$. One can show that the operators $U_i(\sigma)$ and $V_i(\sigma)$ are bounded operators and therefore their definition can be "extended" on the whole $\mathcal{H}_F$ (see below). The conclusion is that the Fock space of the physical particles is a representation of the unitary operators $U_i(\sigma)$ and $V_i(\tau)$, with $i = 1, 2, \ldots$.

We also introduce $U(\boldsymbol{\sigma})$ and $V(\boldsymbol{\tau})$ as

$$U(\boldsymbol{\sigma}) = \exp\left(i\sum_{i=1}^{\infty}\sigma_i p_i\right), \tag{1.36a}$$

$$V(\boldsymbol{\tau}) = \exp\left(i\sum_{i=1}^{\infty}\tau_i q_i\right), \tag{1.36b}$$

where we assume that only a finite number of σ_i and τ_i are not zero. The operators $U(\boldsymbol{\sigma})$ and $V(\boldsymbol{\tau})$ satisfy the so-called Weyl algebra:

$$U(\boldsymbol{\sigma})U(\boldsymbol{\zeta}) = U(\boldsymbol{\sigma}+\boldsymbol{\zeta}), \tag{1.37a}$$
$$V(\boldsymbol{\tau})V(\boldsymbol{\eta}) = V(\boldsymbol{\tau}+\boldsymbol{\eta}), \tag{1.37b}$$
$$U(\boldsymbol{\sigma})V(\boldsymbol{\tau}) = \exp(i\boldsymbol{\sigma}\cdot\boldsymbol{\tau})V(\boldsymbol{\tau})U(\boldsymbol{\sigma}). \tag{1.37c}$$

The relations (1.37) reflect the canonical commutation relations (1.34) (or (1.32)). It is also customary to introduce the so-called Weyl operator $W(\boldsymbol{z})$:

$$W(\boldsymbol{z}) \equiv \exp(i\boldsymbol{\sigma}\cdot\boldsymbol{\tau})U(\sqrt{2}\boldsymbol{\sigma})V(\sqrt{2}\boldsymbol{\tau}), \tag{1.38}$$

with $\boldsymbol{z} \equiv \boldsymbol{\sigma} + i\boldsymbol{\tau}$. Eqs. (1.37) then lead to

$$W(\boldsymbol{z}_1)W(\boldsymbol{z}_2) = \exp[-i\Im m(\boldsymbol{z}_1^*\cdot\boldsymbol{z}_2)]W(\boldsymbol{z}_1+\boldsymbol{z}_2). \tag{1.39}$$

The knowledge of $U(\boldsymbol{\sigma})$ and $V(\boldsymbol{\tau})$ can tell us about $p_i\boldsymbol{\xi}$ and $q_i\boldsymbol{\xi}$, respectively, whenever such vectors belong to $\mathcal{H}_F$. Indeed,

$$p_i\boldsymbol{\xi} = -i\left(\frac{d}{d\sigma_i}U(\boldsymbol{\sigma})\right)_{\sigma=0}\boldsymbol{\xi}, \tag{1.40a}$$

$$q_i\boldsymbol{\xi} = -i\left(\frac{d}{d\tau_i}V(\boldsymbol{\tau})\right)_{\tau=0}\boldsymbol{\xi}. \tag{1.40b}$$

Below we show that any operator which commutes with $U(\boldsymbol{\sigma})$ and $V(\boldsymbol{\tau})$ is a multiple of the identity operator, which means that the Fock space is an irreducible representation of the canonical variables q_i and p_i, i.e., of the annihilation and creation operators of physical particles, or, equivalently, of the Weyl operator introduced above. Sometimes we refer to the Fock space irreducible representation of the Weyl operator as the Weyl system.

In conclusion, the description of the system in terms of physical (boson) particles naturally leads to canonical variables $\{q_i, p_i, i = 1, 2, \ldots\}$ whose irreducible representation is the Fock space defined above.

Extension of the Weyl operators on $\mathcal{H}_F$

We introduce

$$U_i^M(\sigma) = \sum_{n=0}^{M} \frac{1}{n!}(i\sigma p_i)^n \,, \tag{1.41}$$

where M is a positive and finite integer. Since the action of any power of α_i and $\alpha_i^\dagger$ on a vector of the basis gives another vector of the basis, the action of any positive power of p_i and of q_i on vectors $\boldsymbol{\xi}_N$ in the dense set D creates a superposition of finite number of vectors in D, which is still a vector of D. Then, the sequence of vectors $U_i^M(\sigma)\boldsymbol{\xi}_N^{(D)}$ has a limit for $M \to \infty$; thus we define the operation of $U_i(\sigma)$ on D as

$$U_i(\sigma)\boldsymbol{\xi}_N^{(D)} = \lim_{M\to\infty} U_i^M(\sigma)\boldsymbol{\xi}_N^{(D)} \,.$$

Due to the unitarity of $U_i(\sigma)$,

$$|U_i(\sigma)\boldsymbol{\xi}_N^{(D)}| = |\boldsymbol{\xi}_N^{(D)}| \,, \tag{1.42}$$

from which we conclude that the operator $U_i(\sigma)$ is a bounded operator and therefore its definition can be "extended" on the whole $\mathcal{H}_F$ in the following way: let $\boldsymbol{\xi}$ be a vector of $\mathcal{H}_F$; if it is a vector of D, the action of $U_i(\sigma)$ on $\boldsymbol{\xi}$ is well defined. If $\boldsymbol{\xi}$ is not a vector belonging to D, we can find in D a Cauchy sequence $\{\boldsymbol{\xi}_N^{(D)}\}$ whose limit is $\boldsymbol{\xi}$; then we define the action of $U_i(\sigma)$ on $\boldsymbol{\xi}$ as

$$U_i(\sigma)\boldsymbol{\xi} = \lim_{N\to\infty} U_i(\sigma)\boldsymbol{\xi}_N^{(D)} \,. \tag{1.43}$$

In a similar way, we can define the action of $V_i(\tau)$ on $\mathcal{H}_F$.

We now show that any operator which commutes with $U(\boldsymbol{\sigma})$ and $V(\boldsymbol{\tau})$ is a multiple of the identity operator, which implies that the Fock space is an irreducible representation of the canonical variables q_i and p_i, i.e., of the annihilation and creation operators of physical particles. To see this we note that if $\boldsymbol{\xi}$ is a vector of $\mathcal{H}_F$, $\alpha_i \boldsymbol{\xi} = 0$ for all i, when and only when $\boldsymbol{\xi} = c|0,0,\ldots\rangle$, due to (1.8), with c an ordinary number. If A is an operator commuting with q_i and p_i, for all i, i.e., with α_i and $\alpha_i^\dagger$, for all i, then $\alpha_i A|0,0,\ldots\rangle = A\alpha_i|0,0,\ldots\rangle = 0$, i.e., $A|0,0,\ldots\rangle = c|0,0,\ldots\rangle$, with c an ordinary number. Since any vector of the basis (1.25) is constructed by repeated operations of $\alpha_i^\dagger$, e.g.,

$$|n_1, n_2, \ldots\rangle = f(n_1, n_2, \ldots)(\alpha_1^\dagger)^{n_1}(\alpha_2^\dagger)^{n_2}\ldots|0,0,\ldots\rangle \,, \tag{1.44}$$

with $f(n_1, n_2, \ldots)$ some function of n_i (consistent with Eq. (1.14)), we have

$$A|n_1, n_2, \ldots\rangle = c|n_1, n_2, \ldots\rangle \,,$$

i.e., using Eq. (1.26) $A\boldsymbol{\xi} = c\boldsymbol{\xi}$ for any $\boldsymbol{\xi}$ in $\mathcal{H}_F$. This means that $A = c\mathbb{1}$, with $\mathbb{1}$ the identity operator.

Labeling the irreducible representations

Since, as already noted in Section 1.3, the choice of the countable basis is not unique, we now consider the problem of labeling the Weyl operators and the Weyl systems.

We consider the transformation

$$\boldsymbol{\sigma} \to \frac{1}{\rho}\boldsymbol{\sigma}, \qquad \boldsymbol{\tau} \to \rho\boldsymbol{\tau}, \qquad (1.45)$$

with ρ a non-zero real c-number, $\rho \neq 0$. This is a canonical transformation, since $\Im m(z_1^* \cdot z_2)$ in Eq. (1.39) is left invariant under it and the Weyl algebra (1.39) is therefore preserved:

$$W^\rho(z_1)W^\rho(z_2) = \exp[-i\Im m(z_1^* \cdot z_2)]W^\rho(z_1 + z_2), \quad \rho \neq 0, \qquad (1.46)$$

where $W^\rho(z) \equiv W(\frac{1}{\rho}\boldsymbol{\sigma} + i\rho\boldsymbol{\tau})$. We thus see that the transformation parameter ρ acts as a label for the Weyl systems.

We observe [341] that the transformation (1.45) can be equivalently thought as applied to p_i and to q_i instead of σ_i and τ_i. Let us consider for simplicity one specific value of i (extension to many values of i is straightforward). Therefore, we will omit the index i in the following:

$$p \to p(\rho) \equiv \frac{1}{\rho}p, \qquad q \to q(\rho) = \rho q, \quad \rho \neq 0. \qquad (1.47)$$

The *action* variable $J = \int p\,dq$ is invariant under (1.47); this clarifies the physical meaning of the invariance of the area $\Im m(z_1^* \cdot z_2)$ under (1.45). Of course, the transformation (1.47) is a canonical transformation: $([q,p] = i) \to ([q(\rho),p(\rho)] = i)$. By inverting Eqs. (1.33) we have

$$\alpha(\rho) = \frac{1}{\sqrt{2}}(q(\rho) + ip(\rho)) = \frac{1}{2}(u(\rho)\alpha + v(\rho)\alpha^\dagger), \qquad (1.48a)$$

$$\alpha^\dagger(\rho) = \frac{1}{\sqrt{2}}(q(\rho) - ip(\rho)) = \frac{1}{2}(u(\rho)\alpha^\dagger + v(\rho)\alpha), \qquad (1.48b)$$

where

$$u(\rho) \equiv \left(\rho + \frac{1}{\rho}\right), \qquad v(\rho) \equiv \left(\rho - \frac{1}{\rho}\right), \qquad (1.49)$$

so that $u^2 - v^2 = 1$. Eqs. (1.48) are then recognized to be nothing but Bogoliubov transformations; specifically, they are the squeezing transformations occurring in solid state physics and in quantum optics [280] and in elementary particle physics, see, e.g., [13, 108]. Let $\rho \equiv e^{-\varsigma} = \rho(\varsigma)$, $\varsigma \neq \infty$ and real. Eqs. (1.48) are then put in the form

$$\alpha(\varsigma) = \alpha\cosh\varsigma - \alpha^\dagger\sinh\varsigma, \qquad (1.50a)$$

$$\alpha^\dagger(\varsigma) = \alpha^\dagger\cosh\varsigma - \alpha\sinh\varsigma, \qquad (1.50b)$$

where we have used $\alpha(\zeta) \equiv \alpha(\rho(\zeta))$. The ρ-labeling or parametrization is called the *Bogoliubov parametrization* of the Weyl algebra or Weyl systems [341]. The generator of the Bogoliubov transformations (1.50) is:

$$\hat{S}(\zeta) \equiv \exp\left(\frac{\zeta}{2}(\alpha^2 - \alpha^{\dagger 2})\right), \qquad (1.51)$$

$$\alpha(\zeta) = \hat{S}^{-1}(\zeta)\alpha\hat{S}(\zeta), \qquad (1.52a)$$
$$\alpha^{\dagger}(\zeta) = \hat{S}^{-1}(\zeta)\alpha^{\dagger}\hat{S}(\zeta). \qquad (1.52b)$$

In quantum optics $\hat{S}(\zeta)$ is called the squeezing operator [279, 664], ζ being the squeezing parameter. We note that the r.h.s. of Eq. (1.51) is an $SU(1,1)$ group element. In fact, by defining $K_- = \frac{1}{2}\alpha^2$, $K_+ = \frac{1}{2}\alpha^{\dagger 2}$, $K_z = \frac{1}{2}(\alpha^{\dagger}\alpha + \frac{1}{2})$, one easily checks they close the algebra $su(1,1)$ (cf. Appendix C).

In the transition from QM to QFT, namely from finite to infinite number of degrees of freedom, one must operate in the complex linear space $\mathcal{E}_C = \mathcal{E} + i\mathcal{E}$ instead of working in C^M, where M denotes the (finite) number of degrees of freedom ($i = 1, 2, \ldots, M$). Here $\mathcal{E}$ denotes a real linear space of square-integrable functions f; we shall denote by $F = f + ig$, $f, g \in \mathcal{E}$, the elements in $\mathcal{E}_C$. The scalar product $\langle F_1, F_2 \rangle$ in $\mathcal{E}_C$ is defined through the the scalar product (f, g) in $\mathcal{E}$:

$$\langle F_1, F_2 \rangle = (f_1, f_2) + (g_1, g_2) + i[(f_1, g_2) - (f_2, g_1)]. \qquad (1.53)$$

In QFT the Weyl operators and their algebra become

$$W(F) = \exp[i(f, g)]U(\sqrt{2}f)V(\sqrt{2}g), \qquad (1.54a)$$
$$W(F_1)W(F_2) = \exp(-i\Im m\langle F_1, F_2 \rangle)W(F_1 + F_2). \qquad (1.54b)$$

It must be stressed that the use of the complex linear space $\mathcal{E}_C$ in QFT is required to smear out spatial integrations of field operators by means of test functions f.

Our discussion in this Section has been confined to the case of boson operators. We will see that, although fermion operators cannot be traced back to the canonical variables $\{q_i, p_i, i = 1, 2, \ldots\}$, nevertheless there exist also for them infinitely many unitarily inequivalent Fock spaces which are irreducible representations of the anti-commutation relations.

1.5   Unitarily equivalent representations

We now go back to the Weyl–Heisenberg algebra (1.34) (or (1.32), (1.21)). In the following, we will omit for simplicity the suffix i. As customary, we introduce for each i the notation $e_1 = ip$, $e_2 = iq$, $e_3 = i\mathbb{1}$ [519]. By regarding these as elements of an abstract Lie algebra, we recognize the WH algebra introduced above to be, for each i, a real three-dimensional Lie algebra given by the commutation relations

$$[e_1, e_2] = e_3, \qquad [e_1, e_3] = 0 = [e_2, e_3]. \tag{1.55}$$

The generic element x of the algebra is written as $x = (s; x_1, x_2) = x_1 e_1 + x_2 e_2 + s e_3$, with s, x_1 and x_2 real numbers; or,

$$x = is\mathbb{1} + i(\tau q - \sigma p) = is\mathbb{1} + (g\alpha^\dagger - g^*\alpha), \tag{1.56}$$

where we have used Eq. (1.33) and we have put

$$x_1 \equiv -\sigma, \quad x_2 \equiv \tau, \quad g = \frac{1}{\sqrt{2}}(\sigma + i\tau), \tag{1.57}$$

and g^* is the complex conjugate of g. The commutator of the elements $x = (s; x_1, x_2)$ and $y = (t; y_1, y_2)$ is

$$[x, y] = B(x, y)e_3, \qquad B(x, y) = x_1 y_2 - x_2 y_1, \tag{1.58}$$

where $B(x, y)$ is recognized to be the standard symplectic form on the (x_1, x_2) plane.

It is now possible to construct the Lie group corresponding to the Lie algebra by exponentiation:

$$\exp(x) = \exp(is\mathbb{1})D(g), \qquad D(g) = \exp(g\alpha^\dagger - g^*\alpha). \tag{1.59}$$

By use of the formula

$$\exp(A)\exp(B) = \exp\left(\frac{1}{2}[A, B]\right)\exp(A + B), \tag{1.60}$$

which holds provided $[A[A, B]] = 0 = [B[A, B]]$, one obtains the multiplication law

$$D(f)D(g) = \exp(i\Im m(fg^*))D(f + g), \tag{1.61}$$

from which

$$D(f)D(g) = \exp(2i\Im m(fg^*))D(g)D(f). \tag{1.62}$$

This last relation has to be compared with the last one of Eqs. (1.37), of which it provides another realization. Like the operators U and V, the operators $D(g)$ are bounded operators and defined on the whole $\mathcal{H}_F$. In conclusion, the operators $\exp(is\mathbb{1})D(g)$ form a representation of the group whose elements are specified by three real numbers $\gamma = (s; x_1, x_2)$, or by a real number s and a complex number g, $\gamma = (s; g)$. This group is called the Weyl–Heisenberg group and denoted by W_1. The multiplication rule is:

$$(s; x_1, x_2)(t; y_1, y_2) = (s + t + B(x, y); x_1 + y_1, x_2 + y_2). \qquad (1.63)$$

The center of the group W_1, i.e., the set of all the elements commuting with every element of W_1, is given by the elements $(s; 0)$. Let us denote by $T(\gamma)$ any unitary irreducible representation of W_1. $T(\gamma)$ is also said to be a unitary irreducible representation of the canonical commutation relations ((1.32) or (1.21), or (1.34)). Then the operators $T(s; 0)$ form a unitary representation of the subgroup $\{(s; 0)\}$. They are specified by a real number λ:

$$T^\lambda(s; 0) = \exp(i\lambda s)\mathbb{1}. \qquad (1.64)$$

Furthermore, there are representations for which $\lambda = 0$ and are specified by a pair of real numbers, say μ and ν: $T^{\mu\nu}(\gamma) = \exp\{i(\mu x_1 + \nu x_2)\}\mathbb{1}$.

By generalizing $D(g)$ in Eq. (1.59), we can also introduce the operator

$$G(g) = \exp\left(\int d^3k \left[g_\mathbf{k}\alpha_\mathbf{k}^\dagger - g_\mathbf{k}^*\alpha_\mathbf{k}\right]\right), \qquad (1.65)$$

which, acting on $\alpha_\mathbf{k}$, generates the transformation

$$G^{-1}(g)\alpha_\mathbf{k}G(g) = \alpha_\mathbf{k} + g_\mathbf{k} \equiv \alpha_\mathbf{k}(g). \qquad (1.66)$$

The commutation rules for the $\alpha_\mathbf{k}(g)$ and $\alpha_\mathbf{k}^\dagger(g)$ operators are the same as the ones in Eq. (1.21). The transformation (1.66) is thus a canonical transformation since it preserves the commutation rules. Relations similar to Eqs. (1.61) and (1.62) hold for the operator $G(g)$.

Clearly, $\alpha_\mathbf{k}(g)$ acts as the annihilation operator on the state

$$|0(g)\rangle \equiv G^{-1}(g)|0\rangle, \qquad (1.67)$$

but it does not annihilate $|0\rangle$. Note that

$$\langle 0(g)|0(g)\rangle = 1. \qquad (1.68)$$

Then we can construct the "new" Fock space $\mathcal{H}_F(g)$ by using $|0(g)\rangle$ as the vacuum and by repeating the construction followed for obtaining $\mathcal{H}_F$. In this way we get another representation of the canonical commutation relations (1.21). They are unitarily equivalent representations provided $G^{-1}(g)$ is a unitary operator. In the next Section we show that in QFT there exist infinitely many unitarily inequivalent representations, and we discuss the conditions under which this happens and the related physical meaning.

1.6   The Stone–von Neumann theorem

By use of Eq. (1.60) we can write Eq. (1.65) as

$$G^{-1}(g) = \exp\left(-\frac{1}{2}\int d^3k d^3q\, g_{\mathbf{k}} g_{\mathbf{q}}^* \delta(\mathbf{k} - \mathbf{q})\right)$$
$$\times \exp\left(-\int d^3k\, g_{\mathbf{k}} \alpha_{\mathbf{k}}^\dagger\right) \exp\left(\int d^3q\, g_{\mathbf{q}}^* \alpha_{\mathbf{q}}\right), \qquad (1.69)$$

and thus

$$|0(g)\rangle = \exp\left(-\frac{1}{2}\int d^3k |g_{\mathbf{k}}|^2\right) \exp\left(-\int d^3k\, g_{\mathbf{k}} \alpha_{\mathbf{k}}^\dagger\right)|0\rangle, \qquad (1.70)$$

which shows that the inner product of the vacuum state $|0(g)\rangle$ for the "new" operators $\alpha_{\mathbf{k}}(g)$ with the "old" vacuum $|0\rangle$ is

$$\langle 0|0(g)\rangle = \exp\left(-\frac{1}{2}\int d^3k |g_{\mathbf{k}}|^2\right), \qquad (1.71)$$

which is zero provided

$$\frac{1}{2}\int d^3k |g_{\mathbf{k}}|^2 = \infty. \qquad (1.72)$$

For instance, this happens when $g_{\mathbf{k}} = c\delta(\mathbf{k})$, with c a real constant. In such a case, by using the delta function representation

$$\delta(\mathbf{k}) = \frac{1}{(2\pi)^3}\int d^3x\, e^{i\mathbf{k}\cdot\mathbf{x}}, \qquad (1.73)$$

and denoting the volume by $V = \int d^3x$, we can formally write

$$\exp\left(-\frac{1}{2}\int d^3k d^3q\, g_{\mathbf{k}} g_{\mathbf{q}}^* \delta(\mathbf{k} - \mathbf{q})\right)$$
$$= \exp\left(-\frac{c^2}{2(2\pi)^3}\int d^3x \int d^3k d^3q\, e^{i(\mathbf{k}-\mathbf{q})\cdot\mathbf{x}}\delta(\mathbf{k})\delta(\mathbf{q})\right)$$
$$= \exp\left(-\frac{1}{2}\frac{Vc^2}{(2\pi)^3}\right) \to 0, \qquad \text{for } V \to \infty. \qquad (1.74)$$

Similarly, for $g_{\mathbf{k}} = c\delta(\mathbf{k})$ and $g'_{\mathbf{k}} = c'\delta(\mathbf{k})$, $c \neq c'$, we obtain

$$\langle 0(g')|0(g)\rangle \to 0, \qquad \text{for } g' \neq g \quad \text{and } V \to \infty. \qquad (1.75)$$

Thus Eqs. (1.71) and (1.75) are zero in the infinite volume limit, and in that limit the representations $\mathcal{H}_F(g)$ and $\mathcal{H}_F(g')$ are unitarily inequivalent for each set of c-numbers $g = \{g_{\mathbf{k}} = c\delta(\mathbf{k}), \forall\ \mathbf{k}\}$ and $g' = \{g_{\mathbf{k}} = c'\delta(\mathbf{k}), \forall\ \mathbf{k}\}$, with $\{g_{\mathbf{k}}\} \neq \{g'_{\mathbf{k}}\}$, $\forall\ \mathbf{k}$. In other words, there is no unitary generator $G^{-1}(g)$ which maps $\mathcal{H}_F$ onto itself in the infinite volume limit. If, on the contrary, the volume is finite, i.e., the number of degrees of freedom is finite, Eqs. (1.71) and (1.75) are not zero and the representations $\mathcal{H}_F(g)$ and $\mathcal{H}_F(g')$ are unitarily equivalent (and therefore physically equivalent): they are related by a unitary transformation. This is what happens in Quantum Mechanics where only systems with a finite number of degrees of freedom are considered. In the case of infinite volume, Eq. (1.70) is instead only a formal relation: the vacuum $|0(g)\rangle$ cannot be expressed as a superposition of states belonging to $\mathcal{H}_F$ (or to $\mathcal{H}_F(g')$, $g' \neq g$) in the infinite volume limit. Since the c-numbers $g = \{g_{\mathbf{k}} = c\delta(\mathbf{k}), \forall\ \mathbf{k}\}$ span a continuous domain, in the infinite volume limit we have infinitely many unitarily inequivalent representations $\{\mathcal{H}_F(g), \forall\ g = \{g_{\mathbf{k}} = c\delta(\mathbf{k}), \forall\ \mathbf{k}\}\}$ labeled by g. Due to Eq. (1.66), g is called the shift parameters. In concrete cases one needs to operate at finite volume and the infinite volume limit has to be performed only at the end of the computations.

In conclusion, we thus have arrived at the so-called Stone–von Neumann theorem [580, 648, 649] (or, simply, von Neumann theorem), which states that for systems with a finite number of degrees of freedom, which is always the case with Quantum Mechanics, the representations of the canonical commutation relations are all unitarily equivalent to each other. In QFT, the number of degrees of freedom is infinite and the von Neumann theorem does not hold: infinitely many unitarily inequivalent representations of the canonical (anti-)commutation relations exist.

Our discussion is not confined to the relativistic domain. QFT applies also to non-relativistic many-body systems in condensed matter physics. In this last case, one considers the so-called thermodynamic limit in which the infinite volume limit is understood in such a way that the density N/V is kept constant, with N denoting the particle number. One way to visualize this is to consider that at the boundary surfaces of the system the potential barrier is not infinite. Thus wave-packets can spread outside and a continuous distribution of momentum is allowed.

We will comment in the following Section on the physical meaning of the existence of infinitely many unitarily inequivalent representations in QFT.

It is finally necessary to comment on the case of fermions. We have discussed the von Neumann theorem for the bosonic case where the creation and annihilation operators may be introduced through the operators $\{q_i, p_i\}$ as in (1.33). This cannot be the case for the fermion creation and annihilation operators, which need to be directly introduced as in (1.1), without reference to the operators $\{q_i, p_i\}$. However, the result of the von Neumann theorem also holds true for the fermionic case, namely infinitely many unitarily inequivalent representations of the anti-commutation relations also exist in the infinite volume limit. We will give an explicit example of this in Chapter 2, Example 2b, which may be adopted as an explicit proof of the von Neumann theorem for the fermionic case.

A final remark is that the operator $D(g)$ (or $G(g)$) introduced above is the generator of coherent states related with the Weyl–Heisenberg group [519]. Essential notions on single mode coherent states (Glauber coherent states) are presented in Appendix B (for their functional integral representation see Appendix N). In Appendix C we discuss how to extract a complete set of coherent states from an over-complete set.

1.7 Unitarily inequivalent representations

In Section 1.3 we have seen that the set $\{|n_1, n_2, \ldots\rangle\}$ is not a countable set and thus it cannot be used as a basis for the space of states if we require such a space to be a separable one. Then, we have extracted from $\{|n_1, n_2, \ldots\rangle\}$ the subset $\mathcal{S}$ as in Eq. (1.25):

$$\mathcal{S} = \left\{ |n_1, n_2, \ldots\rangle, \qquad \sum_i n_i = \text{finite} \right\}, \tag{1.76}$$

and shown that this is a countable set. The root of the existence of the infinitely many unitarily inequivalent representations in QFT is in the fact that there are infinitely many ways of choosing a separable subspace out of the original non-separable one. To different countable subsets there correspond different, i.e., unitarily inequivalent, representations of the commutation relations. The meaning of this is that a state vector of a given representation cannot be expressed as a superposition of vectors belonging to another inequivalent representation. We therefore must be careful in selecting the representation describing the physical states of our system under given boundary conditions. Consider, for instance, a ferromagnetic system at a temperature below the Curie temperature. The fact that the ferromagnetic state cannot be expressed as a superposition of non-ferromagnetic (paramagnetic) states means that there is no unitary operator connecting the ferromagnetic phase with the non-ferromagnetic one. Indeed, if such an operator existed, its unitarity would imply that characterizing observables would be left unaltered under its action connecting the ferromagnetic phase to the non-ferromagnetic one. However, for example, the observable magnetization does change from non-zero to zero in the process of transition from the ferromagnetic to the non-ferromagnetic phase. These phases are thus physically different in their observable properties and they are therefore to be described by unitarily inequivalent representations.

In conclusion, the existence of many unitarily inequivalent representations allows the description of systems which may be in physically different phases under different boundary conditions. Such a situation is excluded in Quantum Mechanics (QM) since there, as we have seen, the von Neumann theorem [648] guarantees that all the representations are unitarily, and therefore physically, equivalent. In this sense, QM can only describe systems in a single specified physical phase. From such a perspective we may say that QFT is drastically different from QM and it provides a much richer framework than Quantum Mechanics. In the course of this book we will see in more detail how the description of physical phases is carried on and what its relation is with the mechanism of the spontaneous breakdown of symmetry.

We have seen that in the case of the shift transformation, $\alpha_{\mathbf{k}} \rightarrow \alpha_{\mathbf{k}}(g) = \alpha_{\mathbf{k}} + g_{\mathbf{k}}$, with $g_{\mathbf{k}} = c\delta(\mathbf{k})$, the vacua $|0\rangle$ and $|0(g)\rangle$ turn out to be orthogonal (the corresponding representations are unitarily inequivalent). Let us now see the physical meaning of this.

The number $N_{\mathbf{k}} = \alpha_{\mathbf{k}}^\dagger(g)\alpha_{\mathbf{k}}(g)$ of particles $\alpha_{\mathbf{k}}(g)$ in the vacuum $|0\rangle$ is given by

$$\langle 0|\alpha_{\mathbf{k}}^\dagger(g)\alpha_{\mathbf{k}}(g)|0\rangle = |g_{\mathbf{k}}|^2 . \tag{1.77}$$

We then say that there are $|g_{\mathbf{k}}|^2$ bosons of momentum $\mathbf{k}$ condensed in the state $|0\rangle$. The total number of condensed bosons is

$$\int d^3k \langle 0|\alpha_{\mathbf{k}}^\dagger(g)\alpha_{\mathbf{k}}(g)|0\rangle = \int d^3k |g_{\mathbf{k}}|^2 = c^2\delta(0) = c^2\frac{V}{(2\pi)^3} , \tag{1.78}$$

i.e., it is proportional to the system volume V. Thus, an infinite number of bosons is condensed in the vacuum $|0\rangle$ in the infinite volume limit. However, the density of these condensed bosons is everywhere finite, even in the infinite volume limit:

$$\rho = \frac{1}{V}\int d^3k |g_{\mathbf{k}}|^2 = \frac{1}{(2\pi)^3}c^2 . \tag{1.79}$$

The meaning of $g_{\mathbf{k}} = c\delta(\mathbf{k})$ is therefore that the density of the boson condensation in the vacuum state $|0\rangle$ is spatially homogeneous, i.e., everywhere the same, and finite. This also means that when $g_{\mathbf{k}} = c\delta(\mathbf{k})$, the transformation (1.66) does not violate the translational invariance of the vacuum state. On the other hand, since $\langle 0(g)|\alpha_{\mathbf{k}}^\dagger(g)\alpha_{\mathbf{k}}(g)|0(g)\rangle = 0$ everywhere, there are no $\alpha_{\mathbf{k}}(g)$ bosons condensed in $|0(g)\rangle$. We thus see that the two vacua $|0\rangle$ and $|0(g)\rangle$ are different because of their different content in the condensation of $\alpha_{\mathbf{k}}(g)$ bosons, being this infinite in one of them in the infinite volume limit. This depicts the physical meaning of the unitary inequivalence between the representations associated to the two vacua. Physical, local observables may thus turn out to be different in the two vacua since the boson condensation density is different in each of them.

In each representation $\mathcal{H}_F(g)$, for any set $g = \{g_{\mathbf{k}} = c\delta(\mathbf{k}); \forall \mathbf{k}\}$ (including $g = 0 = \{g_{\mathbf{k}} = 0; \forall \mathbf{k}\}$, $\langle 0|\alpha_{\mathbf{k}}^{\dagger}\alpha_{\mathbf{k}}|0\rangle = 0$), we have a set of creation and annihilation operators $\{\alpha_{\mathbf{k}}^{\dagger}(g), \alpha_{\mathbf{k}}(g); \forall \mathbf{k}\}$. For each g, i.e., for each representation, we may assume that the associated set $\{\alpha_{\mathbf{k}}^{\dagger}(g), \alpha_{\mathbf{k}}(g); \forall \mathbf{k}\}$ forms an irreducible set of operators. This means that there are, depending on the physical phase in which the system sits, different sets of physical particles appropriate to the system description in that phase. On the other hand, one may always define the action of one set of operators for a given set g on the representation labeled by a different g' ($g' \neq g$). For example, the action of $\alpha_{\mathbf{k}}(g)$ on $\mathcal{H}_F(g = 0)$ is well defined, as shown in the discussion above, through the mapping $\alpha_{\mathbf{k}} \to \alpha_{\mathbf{k}}(g) \equiv \alpha_{\mathbf{k}} + g_{\mathbf{k}}$, $g_{\mathbf{k}} = c\delta(\mathbf{k})$.

Finally, a comment on the operation of normal ordering, by which a given product of a number of operator factors is rearranged in such a way that all the annihilation operators are on the right and all the creation operators on the left. Such a normal ordering is usually denoted by $: \cdots :$, where dots between the colon denote the operator factors. Expectation value in the vacuum state of normal ordered products is thus zero. However, the discussion above implies that normal ordering is representation dependent, since the annihilation operator in one representation is not such in another unitarily inequivalent representation, and thus normal ordered products have non-zero expectation value in the vacuum of the last representation. A better notation for normal ordering could be $: \cdots :_g$, the label g specifying the representation [90, 438] (see also Section 5.3).

1.8 The deformation of Weyl–Heisenberg algebra

As we shall see in the present Section and in Chapter 5, the quantum-deformed Hopf algebra is a characterizing structural feature of QFT, intimately related with the existence of the unitarily inequivalent representations of the canonical commutation relations CCR [141–143, 152, 337, 338, 341, 342, 632, 633]. Quantum deformed algebras, usually denoted as q-algebras, are deformations in the enveloping algebras of Lie algebras whose structure appears to be an essential tool for the description of composed systems. The general properties of q-algebras are better known than those of q-groups. The interest in q-groups arose almost simultaneously in statistical mechanics, in conformal theories, in solid state physics as well as in the study of topologically non-trivial solutions to non-linear equations [207, 359, 437]. The WH algebra admits two inequivalent deformations: one which is properly a q-algebra [144, 145], the other (on which we shall focus our attention here), denoted as q-WH and often referred to as $osp_q(2|1)$, was originated by the seminal work of Biedenharn [75] and Mac-Farlane [433]. The q-WH algebra is characterized by the property that its intrinsic nature of superalgebra, proper also to the WH algebra itself, plays a non-trivial role, in view of the form of the coproduct. It can therefore be referred to as a Hopf superalgebra [150, 393].

The q-WH algebra has been shown [142, 143] to be related to coherent states, to squeezed coherent states, to the Bloch functions in periodic potentials, to lattice QM and in general to the physics of discretized (periodic) systems. For completeness, we briefly discuss in this Section the q-deformed WH algebra and its relation with the Fock–Bargmann representation (FBR) in QM. For a more detailed account see [142] and [143]. The relation with coherent states and the theta functions is briefly presented in Appendix D.

We consider for simplicity the operators for one single mode. The WH algebra is generated by the operators $\{a, a^\dagger, \mathbb{1}\}$ with commutation relations

$$[a, a^\dagger] = \mathbb{1}, \quad [N, a] = -a, \quad [N, a^\dagger] = a^\dagger, \tag{1.80}$$

and the other commutators vanishing. Here $N \equiv a^\dagger a$. The representation of (1.80), which here we denote by $\mathcal{K}$, is the Fock space generated by the eigenkets of N with integer (positive and zero) eigenvalues. Any state vector $|\psi\rangle$ in $\mathcal{K}$ is thus described by the set $\{c_n; \ c_n \in \mathcal{C}\}$ defined by

$$|\psi\rangle = \sum_{n=0}^{\infty} c_n |n\rangle \,, \text{ i.e., by its expansion in the complete orthonormal set of}$$

eigenkets $\{|n\rangle\}$ of N.

Upon defining $H \equiv N + \frac{1}{2}$, the three operators $\{a, a^\dagger, H\}$ close on $\mathcal{K}$ the relations

$$\{a, a^\dagger\} = 2H \,, \quad [H, a] = -a \,, \quad [H, a^\dagger] = a^\dagger \,, \tag{1.81}$$

and the other (anti-)commutators vanishing. These relations are equivalent to (1.80) on $\mathcal{K}$ and show the intrinsic nature of superalgebra of such a scheme.

In terms of the operators $\{a_q, \bar{a}_q, H; \ q \in \mathbb{C}\}$ the q-deformed version of (1.81), the q-WH algebra, is [150, 393]:

$$\{a_q, \bar{a}_q\} = [2H]_{\sqrt{q}} \,, \quad [H, a_q] = -a_q \,, \quad [H, \bar{a}_q] = \bar{a}_q \,, \tag{1.82}$$

where we utilized the customary notation

$$[x]_q \equiv \frac{q^{\frac{1}{2}x} - q^{-\frac{1}{2}x}}{q^{\frac{1}{2}} - q^{-\frac{1}{2}}} \ . \tag{1.83}$$

The q-WH structure defined by (1.82) together with the related coproduct

$$\Delta(H) = H \otimes \mathbb{1} + \mathbb{1} \otimes H \ \Rightarrow \ \Delta(H) = N \otimes \mathbb{1} + \mathbb{1} \otimes N + \tfrac{1}{2}\mathbb{1} \otimes \mathbb{1}, \tag{1.84a}$$

$$\Delta(a_q) = a_q \otimes q^{\frac{1}{4}H} + q^{-\frac{1}{4}H} \otimes a_q \,, \tag{1.84b}$$

$$\Delta(\bar{a}_q) = \bar{a}_q \otimes q^{\frac{1}{4}H} + q^{-\frac{1}{4}H} \otimes \bar{a}_q \,, \tag{1.84c}$$

is a quantum superalgebra (graded Hopf algebra) and, consequently, all relations (1.82) are preserved under the coproduct map.

In the space $\mathcal{K}$ (i.e., in the space spanned by the vectors $\{|n\rangle; \ n \in \mathcal{N}\}$), Eqs. (1.82) can be rewritten in the equivalent form [75, 150, 433], which makes them more explicitly analogous to the un-deformed case:

$$a_q \bar{a}_q - q^{-\frac{1}{2}} \bar{a}_q a_q = q^{\frac{1}{2}N}, \quad [N, a_q] = -a_q \,, \quad [N, \bar{a}_q] = \bar{a}_q \ ; \tag{1.85}$$

or, by introducing $\hat{a}_q \equiv \bar{a}_q q^{N/2}$,

$$[a_q, \hat{a}_q] \equiv a_q \hat{a}_q - \hat{a}_q a_q = q^N, \quad [N, a_q] = -a_q \,, \quad [N, \hat{a}_q] = \hat{a}_q \ . \tag{1.86}$$

Eqs. (1.85) and (1.86) are deformations only at the algebra level of (1.80). Thus (1.82)–(1.84) is the relevant mathematical structure. However, we prefer to resort henceforth to (1.86), even though the whole discussion could be based on (1.82), since it is perfectly correct as far as we remain in $\mathcal{K}$ and it is the most similar to the usual form (1.80) of the WH algebra.

The q-WH algebra and the Fock–Bargmann representation

In the following, let q be any complex number. The notion of hermiticity for the generators of q-WH algebra associated with complex q is non-trivial and has been studied in [142] and [151] in connection with the squeezing of the generalized coherent states $(GCS)_q$ over $\mathcal{K}$.

We now discuss the functional realization of Eqs. (1.86) by means of finite difference operators in the complex plane, in the Fock–Bargmann representation (FBR) of QM [142, 143, 151, 152].

In the FBR, state vectors are described by entire analytic functions, i.e., uniformly converging in any compact domain of the complex z-plane (see also Appendix D), contrary to the usual coordinate or momentum representation where no condition of analyticity is imposed.

The FBR of the operators with commutation relations (1.80) is [519]:

$$N \to z\frac{d}{dz} , \quad a^{\dagger} \to z , \quad a \to \frac{d}{dz} . \tag{1.87}$$

The corresponding eigenkets of N (orthonormal under the Gaussian measure $d\mu(z) = \frac{1}{\pi}e^{-|z|^2}dzd\bar{z}$) are:

$$u_n(z) = \frac{z^n}{\sqrt{n!}} , \quad u_0(z) = 1 \quad (n \in \mathcal{N}_+) . \tag{1.88}$$

The FBR is the Hilbert space generated by the $u_n(z)$, i.e., the whole space $\mathcal{F}$ of entire analytic functions. Each state vector $|\psi\rangle$ is associated, in a one-to-one way, with a function $\psi(z) \in \mathcal{F}$ by:

$$|\psi\rangle = \sum_{n=0}^{\infty} c_n|n\rangle \to \psi(z) = \sum_{n=0}^{\infty} c_n u_n(z) . \tag{1.89}$$

Note that, as expected in view of the correspondence $\mathcal{K} \to \mathcal{F}$ (induced by $|n\rangle \to u_n(z)$),

$$a^{\dagger} u_n(z) = \sqrt{n+1}\, u_{n+1}(z) , \quad a\, u_n(z) = \sqrt{n}\, u_{n-1}(z) , \tag{1.90}$$

$$N\, u_n(z) = a^{\dagger}a\, u_n(z) = z\frac{d}{dz}\, u_n(z) = n\, u_n(z) . \tag{1.91}$$

Eqs. (1.90) and (1.91) establish the mutual conjugation of a and $a^{\dagger}$ in the FBR, with respect to the measure $d\mu(z)$.

We now consider the finite difference operator D_q defined by:

$$D_q f(z) = \frac{f(qz) - f(z)}{(q-1)z} , \tag{1.92}$$

with $f(z) \in \mathcal{F}$, $q = e^{\zeta}$, $\zeta \in \mathbb{C}$. D_q is the so-called q-derivative operator [76], which, for $q \to 1$ ($\zeta \to 0$), reduces to the standard derivative. By using Eqs. (1.88) and (1.90), it may be written on $\mathcal{F}$ as

$$D_q = \frac{1}{(q-1)z}\left(q^{z\frac{d}{dz}} - 1\right) = q^{\frac{z}{2}\frac{d}{dz}}\frac{1}{z}\left[z\frac{d}{dz}\right]_q . \tag{1.93}$$

Consistency between (1.92) and (1.93) can be proven by first "normal ordering" the operator $\left(z\frac{d}{dz}\right)^n$ in the form:

$$\left(z\frac{d}{dz}\right)^n = \sum_{m=1}^{n} S_n^{(m)} z^m \frac{d^m}{dz^m} , \tag{1.94}$$

where $S_n^{(m)}$ denotes the Stirling numbers of the second kind, defined by the recursion relations [3]

$$S_{n+1}^{(m)} = m\, S_n^{(m)} + S_n^{(m-1)} , \tag{1.95}$$

and then expanding in formal power series the exponential $\left(q^{z\frac{d}{dz}} - 1\right)$, and considering the identity:

$$\frac{1}{m!}\left(e^{\theta} - 1\right)^m = \sum_{n=m}^{\infty} \mathcal{S}_n^{(m)} \frac{\theta^n}{n!} . \qquad (1.96)$$

D_q satisfies, together with z and $z\frac{d}{dz}$, the commutation relations:

$$\left[D_q, z\right] = q^{z\frac{d}{dz}} , \qquad \left[z\frac{d}{dz}, D_q\right] = -D_q , \qquad \left[z\frac{d}{dz}, z\right] = z , \qquad (1.97)$$

which can be recognized as a realization of relations (1.86) in the space $\mathcal{F}$, with the identification

$$N \to z\frac{d}{dz} , \qquad \hat{a}_q \to z , \qquad a_q \to D_q , \qquad (1.98)$$

where $\hat{a}_q = \hat{a}_{q=1} = a^{\dagger}$ and $\lim_{q\to 1} a_q = a$ on $\mathcal{F}$. We stress that, while (1.97) are restricted to $\mathcal{F}$, the operators (1.98) are related to the true algebraic structure (1.82)–(1.84).

The relations analogous to (1.90) for the q-deformed case are

$$\hat{a}_q u_n(z) = \sqrt{n+1}\, u_{n+1}(z) , \qquad a_q u_n(z) = q^{\frac{n-1}{2}} \frac{[n]_q}{\sqrt{n}}\, u_{n-1}(z) . \qquad (1.99)$$

The q-commutator $[a_q, \hat{a}_q]$ is thus defined on the whole $\mathcal{F}$ and acts as

$$[a_q, \hat{a}_q]f(z) = q^N f(z) = f(qz) . \qquad (1.100)$$

Eq. (1.100) provides a remarkable result since it shows that the action of the q-WH algebra commutator $[a_q, \hat{a}_q]$, which is a *linear* form in a_q and $\hat{a}_q$, may be represented in the FBR as the action of the operator q^N which is *non-linear* in the FBR operators a and $a^{\dagger}$.

Finally, we show that the q-WH algebra is related with the squeezing generator. In the Hilbert space of states identified with the space $\mathcal{F}$ of entire analytic functions $\psi(z)$, the identity

$$2z\frac{d}{dz}\psi(z) = \left\{\frac{1}{2}\left[\left(z + \frac{d}{dz}\right)^2 - \left(z - \frac{d}{dz}\right)^2\right] - 1\right\}\psi(z) , \qquad (1.101)$$

holds. We set $z \equiv x + iy$ and introduce the operators

$$\alpha = \frac{1}{\sqrt{2}}\left(z + \frac{d}{dz}\right) , \qquad \alpha^{\dagger} = \frac{1}{\sqrt{2}}\left(z - \frac{d}{dz}\right) , \qquad [\alpha, \alpha^{\dagger}] = \mathbb{1} , \qquad (1.102)$$

namely, in terms of the FBR operators a and $a^{\dagger}$,

$$z = \frac{1}{\sqrt{2}}(\alpha + \alpha^{\dagger}) \to a , \qquad \frac{d}{dz} = \frac{1}{\sqrt{2}}(\alpha - \alpha^{\dagger}) \to a^{\dagger} . \qquad (1.103)$$

In $\mathcal{F}$, $\alpha^\dagger$ is indeed the conjugate of α, as discussed in [150] and [519]. In the limit $y \to 0$, α and $\alpha^\dagger$ turn into the conventional annihilation and creator operators a and $a^\dagger$ associated with x and p_x in the canonical configuration representation, respectively. We then realize that the operator

$$[a_q, \hat{a}_q] = \frac{1}{\sqrt{q}} \, \exp\left(\frac{\zeta}{2}\left(\alpha^2 - {\alpha^\dagger}^2\right)\right) \equiv \frac{1}{\sqrt{q}} \hat{S}(\zeta), \qquad (1.104)$$

where, for simplicity, $q = e^\zeta$ is assumed to be real, is, in the limit $y \to 0$, the squeezing operator [142, 143] in $\mathcal{F}$, well known in quantum optics [664]. A detailed analysis of the relation between the q-WH algebra and the generator of squeezed coherent states is presented in [142, 143, 151, 152]. As shown in Appendix D, the q-WH algebra is related also to the theta functions, which provide an essential tool in the treatment of coherent states [519].

Because the q-algebra has been essentially obtained by replacing the customary derivative with the finite difference operator, the above discussion suggests [142, 143] that whenever one deals with some finite scale (e.g., with some discrete structure, lattice or periodic system, lattice QM) which cannot be reduced to the continuum by a limiting procedure, then a deformation of the operator algebra acting in $\mathcal{F}$ should arise. Deformation of the operator algebra is also expected whenever the system under study involves periodic (analytic) functions, since periodicity is nothing but a special invariance under finite difference operators. The special case of the Bloch functions for periodic potentials in QM is studied in [142, 143]. See [142, 143] for applications to several cases of physical interest. In Chapter 5 we will discuss the q-deformation of the Hopf algebra in connection with thermal field theory and the general algebraic structure of QFT.

1.8.1 *Self-similarity, fractals and the Fock–Bargmann representation*

It is interesting to consider the FBR and the q-deformation of the WH algebra discussed in the previous Section in connection with self-similarity. We follow closely [634] and [635] in the following. In fractal studies, the self-similarity property is referred to as the *most important property* of fractals (p.150 in [516]). In fact, a connection will emerge between fractals and q-deformed coherent states.

Let us consider indeed the fractal example provided by the *Koch curve* (Fig. 1.1). One starts with the step, or stage, of order $n = 0$: the one-dimensional ($d = 1$) segment u_0 of unit length L_0, called the *initiator* [127], is divided by the reducing factor $s = 3$, and the unit length $L_1 = \frac{1}{3} L_0$ is adopted to construct the new "deformed segment" u_1, called the *generator* [127], made of $\alpha = 4$ units L_1 (step of order $n = 1$). The "deformation" of the u_0 segment is only possible provided the one-dimensional constraint $d = 1$ is relaxed. The u_1 segment "shape" lives in some $d \neq 1$ dimensions and thus we write $u_{1,q}(\alpha) \equiv q\,\alpha\,u_0$, $q = \frac{1}{3^d}$, $d \neq 1$ to be determined. The index q is introduced in the notation of the deformed segment u_1.

In general, denoting by $\mathcal{H}(L_0)$ lengths, surfaces or volumes, one has

$$\mathcal{H}(\lambda L_0) = \lambda^d \mathcal{H}(L_0), \tag{1.105}$$

under the scale transformation: $L_0 \to \lambda L_0$. A square S of side L_0 scales to $\frac{1}{2^2} S$ when $L_0 \to \lambda L_0$ with $\lambda = \frac{1}{2}$. A cube V of same side with same rescaling of L_0 scales to $\frac{1}{2^3} V$. Thus $d = 2$ and $d = 3$ for surfaces and volumes, respectively. Note that $\frac{S(\frac{1}{2} L_0)}{S(L_0)} = p = \frac{1}{4}$ and $\frac{V(\frac{1}{2} L_0)}{V(L_0)} = p = \frac{1}{8}$, respectively, so that in both cases $p = \lambda^d$. For the length L_0 it is $p = \frac{1}{2}$; $\frac{1}{2^d} = \lambda^d$ and $p = \lambda^d$ gives $d = 1$.

In the case of any other "hypervolume" $\mathcal{H}$ one considers the ratio

$$\frac{\mathcal{H}(\lambda L_0)}{\mathcal{H}(L_0)} = p, \tag{1.106}$$

and Eq. (1.105) is assumed to be still valid. So,

$$p\, \mathcal{H}(L_0) = \lambda^d \mathcal{H}(L_0), \tag{1.107}$$

i.e., $p = \lambda^d$. For the Koch curve, setting $\alpha = \frac{1}{p} = 4$ and $q = \lambda^d = \frac{1}{3^d}$, the relation $p = \lambda^d$ gives

$$q\alpha = 1, \quad \text{where} \quad \alpha = 4, \quad q = \frac{1}{3^d}, \tag{1.108}$$

i.e.,

$$d = \frac{\ln 4}{\ln 3} \approx 1.2619. \tag{1.109}$$

The non-integer d is called the *fractal dimension*, or the *self-similarity dimension* [516]. The meaning of Eq. (1.108) is that the measure of the deformed segment $u_{1,q}$, with respect to the undeformed segment u_0, is 1: $\frac{u_{1,q}}{u_0} = 1$, i.e., $\alpha q = \frac{4}{3^d} = 1$. In the following we will set $u_0 = 1$.

Steps of higher order n, $n = 2, 3, 4, \ldots, \infty$, can be obtained by iteration of the deformation process. In the step $n = 2$, $u_{2,q}(\alpha) \equiv q\,\alpha\,u_{1,q}(\alpha) = (q\,\alpha)^2 u_0$, and so on. For the nth order deformation:

$$u_{n,q}(\alpha) \equiv (q\,\alpha)\,u_{n-1,q}(\alpha), \quad n = 1, 2, 3, \ldots, \tag{1.110}$$

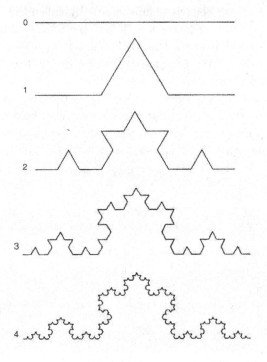

Fig. 1.1   The first five stages of Koch curve.

i.e., for any n

$$u_{n,q}(\alpha) = (q\,\alpha)^n \, u_0 \ .\tag{1.111}$$

By requiring that $\frac{u_{n,q}(\alpha)}{u_0}$ be 1 for any n, this gives $(q\,\alpha)^n = 1$ and Eq. (1.109) is again obtained. We stress that the fractal is mathematically defined in the limit of infinite iterations of the deformation process, $n \to \infty$. The fractal is the limit of the deformation process for $n \to \infty$. The definition of fractal dimension is indeed more rigorously given starting from $(q\alpha)^n = 1$ in the $n \to \infty$ limit [47, 127]. Self-similarity is defined only in the $n \to \infty$ limit. Since $L_n \to 0$ for $n \to \infty$, the Koch fractal is a curve which is non-differentiable everywhere [516].

Eqs. (1.110) and (1.111) express, *in the $n \to \infty$ limit*, the *self-similarity* property of a large class of fractals (the Sierpinski gasket and carpet, the Cantor set, etc.) [47, 127]. Our discussion can be extended to self-affine fractals (invariance under anisotropic magnification is called self-affinity).

Summarizing the discussion of [634], we consider the complex α-plane and note that applying Eq. (1.100) to the basis provided by the functions $u_n(\alpha)$ (cf. Eq. (1.88) where now we have changed z into α) we have:

$$q^N u_n(\alpha) = \frac{(q\alpha)^n}{\sqrt{n!}}, \qquad u_0(\alpha) = 1, \qquad (n \in \mathbb{N}_+). \qquad (1.112)$$

We recall that the FBR is the Hilbert space generated by the $u_n(\alpha)$, i.e., the space $\mathcal{F}$ of entire analytic functions. Eq. (1.112) applied to the coherent state functional (D.2) (cf. Appendix D), gives

$$q^N |\alpha\rangle = |q\alpha\rangle = \exp\left(-\frac{|q\alpha|^2}{2}\right) \sum_{n=0}^{\infty} \frac{(q\alpha)^n}{\sqrt{n!}} |n\rangle. \qquad (1.113)$$

From Eq. (D.1) we obtain

$$a\,|q\alpha\rangle = q\alpha\,|q\alpha\rangle, \qquad q\alpha \in \mathbb{C}. \qquad (1.114)$$

Eq. (1.111), with u_0 set equal to 1, is then obtained by projecting out the nth component of $|q\alpha\rangle$ and restricting to real $q\alpha$, $q\alpha \to \Re e(q\alpha)$:

$$u_{n,q}(\alpha) = (q\alpha)^n = \sqrt{n!}\, \exp\left(\frac{|q\alpha|^2}{2}\right)\langle n|q\alpha\rangle, \qquad \forall\, n,\ q\alpha \to \Re e(q\alpha), \quad (1.115)$$

which, taking into account that $\langle n| = \langle 0| \frac{(a)^n}{\sqrt{n!}}$, gives

$$u_{n,q}(\alpha) = (q\alpha)^n = \exp\left(\frac{|q\alpha|^2}{2}\right)\langle 0|(a)^n|q\alpha\rangle, \qquad \forall\, n,\ q\alpha \to \Re e(q\alpha). \quad (1.116)$$

The operator $(a)^n$ thus acts as a "magnifying" lens [127]. The nth iteration can be "seen" by applying $(a)^n$ to $|q\alpha\rangle$ and restricting to real $q\alpha$:

$$\langle q\alpha|(a)^n|q\alpha\rangle = (q\alpha)^n = u_{n,q}(\alpha), \qquad q\alpha \to \Re e(q\alpha). \qquad (1.117)$$

In conclusion, the nth fractal stage of iteration, with $n = 0, 1, 2, \ldots, \infty$, is represented, in a one-to-one correspondence, by the nth term in the coherent state series Eq. (1.113). The operator q^N applied to $|\alpha\rangle$ (Eq. (1.113)) "produces" the fractal in the functional form of the coherent state $|q\alpha\rangle$. q^N is also called *the fractal operator* [634].

The study of the fractal properties may thus be carried on in the space $\mathcal{F}$ of the entire analytic functions, by restricting, at the end, the conclusions to real α, $\alpha \to \Re e(\alpha)$. Since in Eq. (1.111) it is $q \neq 1$ ($q < 1$), actually one needs to consider the "q-deformed" algebraic structure of which the space $\mathcal{F}$ provides a representation.

Eq. (1.114) expresses the invariance of the coherent state under the action of the operator $\frac{1}{q\alpha}a$ and allows us to consider the coherent functional

$\psi(q\alpha)$ as an "attractor" in $\mathcal{F}$. This reminds us of the fixed point equation $W(A) = A$, where W is the Hutchinson operator [127], characterizing the iteration process for the fractal A in the $n \to \infty$ limit.

The connection between fractals and the (q-deformed) algebra of the coherent states is formally established by Eqs. (1.115), (1.116) and (1.117).

Moreover, the fractal operator q^N is associated with the squeezing transformation (cf. the previous Section). This establishes the relation between the fractal generator process and squeezed coherent states (see also [634]).

In conclusion, for the case of fractals generated iteratively according to a prescribed recipe (deterministic fractals), the functional realization of fractal self-similarity has been obtained in terms of the q-deformed algebra of coherent states. Fractal study can thus be incorporated into the theory of entire analytical functions. From the discussion it appears that the reverse is also true: under a convenient choice of the q-deformation parameter and by a suitable restriction to real α, coherent states exhibit fractal properties in the q-deformed space of the entire analytical functions.

The relation between fractals and coherent states, originally conjectured in [636], introduces dynamical considerations in the study of fractals and of their origin, as well as geometrical insight into coherent state properties. Fractals appear to be global systems arising from local deformation processes.

1.9   The physical particle energy and momentum operator

Let us now make a more specific statement about the energy and the momentum operator of physical particles in the language of the Fock space. We consider a one particle (boson or fermion) wave-packet state $|\alpha_i^\dagger\rangle = \alpha_i^\dagger|0\rangle$,

$$\alpha_i^\dagger|0\rangle = \int \frac{d^3k}{(2\pi)^{3/2}} f_i(\mathbf{k})\alpha_{\mathbf{k}}^\dagger|0\rangle \,. \tag{1.118}$$

The energy operator H_0 is introduced by requiring

$$H_0\,\alpha_i^\dagger|0\rangle = \int \frac{d^3k}{(2\pi)^{3/2}} E_k f_i(\mathbf{k})\alpha_{\mathbf{k}}^\dagger|0\rangle \,, \tag{1.119}$$

with real E_k. Since this should be true for any square-integrable function $f_i(\mathbf{k})$, we have

$$H_0\alpha_{\mathbf{k}}^\dagger|0\rangle = E_k\alpha_{\mathbf{k}}^\dagger|0\rangle \,. \tag{1.120}$$

Such a relation must be understood in the sense of distributions, i.e., in the sense of (1.119), since the states involved in (1.120) are not elements of the Fock space due to the unboundness of the $\alpha_{\mathbf{k}}^\dagger$ operators.

We now consider the scattering of many particles. Before the collision, the system energy is the sum of the energies of the particles entering in the collision region (the incoming particles), say the particles A_{in}, B_{in}, C_{in}, etc. After the collision, the total energy is the sum of the energies of the particles outgoing from the collision region, say the particles A_{out}, B_{out}, C_{out}, etc. If the experimental setup is such that any exchange of energy between our particle system and the environment is negligible, then as a result of our measurement we find that the total energy after the collision equals the total energy before the collision (the principle of conservation of energy). Thus, for many particles we require that

$$H_0 \alpha^\dagger_{\mathbf{k}_1} \ldots \alpha^\dagger_{\mathbf{k}_n} |0\rangle = (E_{k_1} + \cdots + E_{k_n}) \, \alpha^\dagger_{\mathbf{k}_1} \ldots \alpha^\dagger_{\mathbf{k}_n} |0\rangle \,, \tag{1.121}$$

and for $n = 0$

$$H_0 |0\rangle = 0 \,. \tag{1.122}$$

Since E_k is real, we require that $H_0 = H_0^\dagger$. From (1.121) we can derive

$$[H_0, \alpha^\dagger_{\mathbf{k}}] = E_k \alpha^\dagger_{\mathbf{k}} \,, \qquad [H_0, \alpha_{\mathbf{k}}] = -E_k \alpha_{\mathbf{k}} \,. \tag{1.123}$$

These commutators imply that the form of H_0 has to be

$$H_0 = \sum_s \int d^3k \; E_k^s \, \alpha^{s\dagger}_{\mathbf{k}} \alpha^s_{\mathbf{k}} + H_1 \,, \tag{1.124}$$

where the suffix s (cf. Section 1.2) has been restored and H_1 commutes with $\alpha^{s\dagger}_{\mathbf{k}}$ and $\alpha^s_{\mathbf{k}}$. Since the Fock space is an irreducible representation of $\alpha^{s\dagger}_{\mathbf{k}}$ and $\alpha^s_{\mathbf{k}}$, H_1 must be a c-number multiple of the identity operator. On the other hand, its value is determined by the vacuum expectation value of (1.124) (cf. Eq. (1.122)):

$$H_1 = \langle 0|H_0|0\rangle = 0 \,. \tag{1.125}$$

Thus, by considering also the operators $\beta^s_{\mathbf{k}}$ and $\beta^{s\dagger}_{\mathbf{k}}$, we have

$$H_0 = \sum_s \int d^3k \, E_k^s \left(\alpha^{s\dagger}_{\mathbf{k}} \alpha^s_{\mathbf{k}} + \beta^{s\dagger}_{\mathbf{k}} \beta^s_{\mathbf{k}} \right) \,. \tag{1.126}$$

In a similar way we can introduce the momentum operator $\mathbf{P}_0$ as

$$\mathbf{P}_0 = \sum_s \int d^3k \, \mathbf{k} \left(\alpha^{s\dagger}_{\mathbf{k}} \alpha^s_{\mathbf{k}} + \beta^{s\dagger}_{\mathbf{k}} \beta^s_{\mathbf{k}} \right) \,, \tag{1.127}$$

with

$$[\mathbf{P}_0, \alpha^{s\dagger}_{\mathbf{k}}] = \mathbf{k} \, \alpha^{s\dagger}_{\mathbf{k}} \,, \qquad [\mathbf{P}_0, \alpha^s_{\mathbf{k}}] = -\mathbf{k} \, \alpha^s_{\mathbf{k}} \,; \tag{1.128}$$

and

$$\mathbf{P}_0|0\rangle = 0 . \tag{1.129}$$

Although the operators H_0 and $\mathbf{P}_0$ are well defined only on the dense set D, the operators $e^{iH_0 t}$ and $e^{i\mathbf{P}_0 \cdot \mathbf{x}}$, with real t and $\mathbf{x}$, are well defined on the whole $\mathcal{H}_F$. In terms of these last operators, the above commutation relations should be replaced by

$$e^{iH_0 t} \alpha_{\mathbf{k}}^{s\dagger} e^{-iH_0 t} = e^{iE_k^s t} \alpha_{\mathbf{k}}^{s\dagger} , \tag{1.130a}$$

$$e^{-i\mathbf{P}_0 \cdot \mathbf{x}} \alpha_{\mathbf{k}}^{s\dagger} e^{i\mathbf{P}_0 \cdot \mathbf{x}} = e^{-i\mathbf{k}\cdot\mathbf{x}} \alpha_{\mathbf{k}}^{s\dagger} , \tag{1.130b}$$

and their hermitian conjugates.

1.10   The physical Fock space and the physical fields

Eqs. (1.121), (1.126) and (1.127) give an exact meaning to the statement that the total energy and the total momentum of a system of free particles is given by the sum of the energies and of the momenta, respectively, of each particle (cf. Eq. (1.19) for the definition of the number operator). In particular, we take this to be the definition of the free or physical particle state: it is the state where the total energy and the total momentum are given by the sum of the energy and of the momentum, respectively, of each constituent particle. The Fock space of the free particle states is thus the one where the Hamiltonian operator and the momentum operator assume the form (1.126) and (1.127), respectively. We will call it the physical Fock space.

We define the free or physical field $\phi(x)$, with x denoting $\mathbf{x}, t$, by

$$\phi(x) = \int d^3 k \left[u(\mathbf{k}) \alpha_{\mathbf{k}} e^{i\mathbf{k}\cdot\mathbf{x} - iE_k t} + v(\mathbf{k}) \beta_{\mathbf{k}}^{\dagger} e^{-i\mathbf{k}\cdot\mathbf{x} + iE_k t} \right] . \tag{1.131}$$

In general, $\phi(x)$ is a one-column matrix. The fact that the energy E_k of a physical particle is a certain function of its momentum means that the physical field $\phi(x)$ must solve a linear homogeneous equation:

$$\Lambda(\partial)\phi(x) = 0 . \tag{1.132}$$

The differential operator $\Lambda(\partial)$ is in general a square matrix. Its operation is defined on the Fourier transform as $\Lambda(\partial)e^{-ik\cdot x} = \Lambda(ik)e^{-ik\cdot x}$, $k \cdot x \equiv k_\mu x^\mu = E_k t - \mathbf{k}\cdot\mathbf{x}$. The "wave functions" $u(\mathbf{k})$ and $v(\mathbf{k})$ are solutions of

$$\Lambda(ik)u(\mathbf{k}) = 0 , \quad \text{and} \quad \Lambda(-ik)v(\mathbf{k}) = 0 , \tag{1.133}$$

respectively.

Physical particles are thus ingoing and outgoing particles far from the region of interaction. In solid state physics, as mentioned in Section 1.2, the physical particles are called quasiparticles. We will call in-fields or out-fields the fields referring to ingoing or outgoing physical particles, respectively, and denote them by ϕ_{in} and/or ϕ_{out}. In the following Chapters, whenever no misunderstanding arises we will drop the 'in' and/or 'out' indexes. In-fields and out-fields will also be generically called asymptotic fields since they describe particles in spacetime regions where interactions are not felt. The free field equations of type (1.132), in fact, do not contain any information about the interactions. Although the physical particles undergo interaction processes, the language we have set up till now cannot describe such dynamical processes; thus we need another source of information to describe the dynamics of a physical system. The concept of free field is pertinent to one of the aspects of the two-level description of Nature. We thus assume the existence of basic entities, the Heisenberg or interacting fields, in order to account for interactions, the other aspect of this duality. Heisenberg fields satisfy basic relations characterizing the dynamics, the Heisenberg equations. We will come back to this point in the following Chapter.

Appendix A

Strong limit and weak limit

Consider in full generality a linear metric vector space $\mathcal{F}$ (namely a vector space endowed with addition of its elements, multiplication by a scalar and inner product). Let $\boldsymbol{\xi}$ be an element of $\mathcal{F}$ and $|\boldsymbol{\xi}| \equiv (\boldsymbol{\xi}, \boldsymbol{\xi})^{1/2}$ denote the norm of $\boldsymbol{\xi}$. A sequence of elements $\{\boldsymbol{\xi}_n\}$ of $\mathcal{F}$ is said to be a Cauchy sequence if for every $\epsilon > 0$ one can find an $N > 0$ such that

$$|\boldsymbol{\xi}_n - \boldsymbol{\xi}_m| < \epsilon, \tag{A.1}$$

whenever $n, m > N$. A sequence $\{\boldsymbol{\xi}_n\}$ converges to an element $\{\boldsymbol{\xi}\}$ of $\mathcal{F}$ if for every $\epsilon > 0$ there exist an $N > 0$ such that

$$|\boldsymbol{\xi}_n - \boldsymbol{\xi}| < \epsilon, \qquad \text{for } n > N, \tag{A.2}$$

and we write

$$\lim_{n \to \infty} |\boldsymbol{\xi}_n - \boldsymbol{\xi}| = 0. \tag{A.3}$$

A space $\mathcal{F}$ with the property that all Cauchy sequences of elements of $\mathcal{F}$ have a limit that also belongs to $\mathcal{F}$ is called complete. A Hilbert space is a linear metric vector space that is also complete.

A subset $\mathcal{D}$ of elements $\boldsymbol{\xi}$ of $\mathcal{F}$ is said to be dense in $\mathcal{F}$ if, for any element $\boldsymbol{\zeta}$ of $\mathcal{F}$, one can construct a sequence of elements of $\mathcal{D}$ that has $\boldsymbol{\zeta}$ as its limit. If this dense set has a countable basis, the space $\mathcal{F}$ is called separable. In other words, this means that one can find a countable set of orthogonal elements $\{\boldsymbol{\xi}_i\}$ $(i = 1, 2, \dots)$ such that any vector $\boldsymbol{\zeta}$ can be written as

$$\boldsymbol{\zeta} = \sum_{i=0}^{\infty} c_i \boldsymbol{\xi}_i, \tag{A.4}$$

where c_i are complex constants. The equality in (A.4) is understood in the sense of

$$\lim_{N \to \infty} |\boldsymbol{\zeta} - \sum_{i=0}^{N} c_i \boldsymbol{\xi}_i| = 0. \tag{A.5}$$

In a complete space $\mathcal{F}$, the vector $\boldsymbol{\xi}$ is said to be the strong limit [7,351,537] of the sequence $\{\boldsymbol{\xi}_n;\ \boldsymbol{\xi}_n \in \mathcal{F}\}$ if (A.3) is satisfied.

On the other hand, the sequence $\boldsymbol{\xi}_n$ is said to be weakly convergent to $\boldsymbol{\xi}$ if for any arbitrary vector $\boldsymbol{\eta}$ in $\mathcal{F}$ it is

$$\lim_{n\to\infty} (\boldsymbol{\eta},\boldsymbol{\xi}_n) = (\boldsymbol{\eta},\boldsymbol{\xi})\,. \tag{A.6}$$

Moreover, if the sequence $\boldsymbol{\xi}_n$ is a bounded operator sequence, i.e., $|\boldsymbol{\xi}_n| \le M$ for any n and for some constant M independent of n, as we generally assume for the states of the Fock space, $\boldsymbol{\xi}_n$ is weakly convergent to $\boldsymbol{\xi}$ if

$$\lim_{n\to\infty} (\boldsymbol{\eta}_i,\boldsymbol{\xi}_n) = (\boldsymbol{\eta}_i,\boldsymbol{\xi})\,, \tag{A.7}$$

for all the elements $\boldsymbol{\eta}_i$ of a dense set in $\mathcal{F}$ (the Fock space). Sometimes one writes $w - \lim_{n\to\infty} \boldsymbol{\xi}_n = \boldsymbol{\xi}$ to denote the weak limit convergence.

Appendix B

Glauber coherent states

We shall briefly consider here some essentials of single-mode coherent states (CS) commonly called canonical CS, or Glauber coherent states, or Fock–Bargmann coherent states [285, 519]. The functional integrals based on Glauber coherent states will be discussed in Appendix N.1. Group related generalized CS will be considered in Appendices C and N.2.

The *un-normalized* coherent state has the form

$$
\begin{aligned}
|z\rangle &= \sum_n \frac{(za^\dagger)^n}{n!}|0\rangle = \exp(za^\dagger)|0\rangle \\
&= \exp(za^\dagger)\exp(z^*a)|0\rangle = e^{-|z|^2/2}\exp(za^\dagger + z^*a)|0\rangle \\
&= \exp(za^\dagger)\exp(-z^*a)|0\rangle = e^{|z|^2/2}\exp(za^\dagger - z^*a)|0\rangle . \quad \text{(B.1)}
\end{aligned}
$$

Since the state $|n\rangle$ is given by

$$
|n\rangle = \frac{(a^\dagger)^n}{\sqrt{n!}}|0\rangle , \quad \text{(B.2)}
$$

one can alternatively rewrite (B.1) as

$$
|z\rangle = \sum_n \frac{z^n}{\sqrt{n!}}|n\rangle . \quad \text{(B.3)}
$$

It is easy to see that $|z\rangle$ is an eigenstate of a with the eigenvalue z. This is a straightforward implication of the operator formula

$$
e^A B e^{-A} = \sum_{n=1}^{\infty} \frac{1}{n!} C_n, \quad C_0 = B, \quad C_1 = [A, B], \quad C_n = [C_{n-1}, B],
$$

$$
\Rightarrow \quad e^{-za^\dagger} a\, e^{za^\dagger} = a + z . \quad \text{(B.4)}
$$

As a result

$$
a|z\rangle = a\, e^{za^\dagger}|0\rangle = e^{za^\dagger}(a + z)|0\rangle = z|z\rangle ,
$$

$$
\langle z|a^\dagger = z^*\langle z| . \quad \text{(B.5)}
$$

37

The *normalized* CS can be obtained from (B.1) by realizing that

$$\langle z|z\rangle = \sum_{n,m}\langle m|\frac{z^{*m}z^n}{\sqrt{n!\,m!}}\,|n\rangle = \sum_n \frac{|z|^{2n}}{n!} = e^{|z|^2}\,. \qquad (B.6)$$

Thus

$$|z) \equiv |z\rangle_{norm} = e^{-|z|^2/2}\exp(za^\dagger)|0\rangle = \exp(za^\dagger - z^*a)|0\rangle\,. \quad (B.7)$$

The corresponding *completeness relation (or resolution of unity)* for single-mode CS can be easily derived with the aid of (B.3). Indeed,

$$\begin{aligned}
\mathbb{1} &= \sum_n |n\rangle\langle n| = \sum_n \frac{(a^\dagger)^n}{\sqrt{n!}}\,|0\rangle\langle 0|\frac{a^n}{\sqrt{n!}} \\
&= \int \frac{dzdz^*}{2\pi i}\,e^{-zz^*}\sum_{n,m}\frac{(za^\dagger)^n}{n!}\,|0\rangle\langle 0|\frac{(z^*a)^m}{m!} \\
&= \int \frac{dzdz^*}{2\pi i}\,|z\rangle\,e^{-zz^*}\langle z| = \int \frac{dzdz^*}{2\pi i}\,|z)(z|\,. \qquad (B.8)
\end{aligned}$$

The second line results from the identity

$$\begin{aligned}
\int \frac{dzdz^*}{2\pi i}\,e^{-zz^*}z^n(z^*)^m &= \int_0^{2\pi}\frac{d\theta}{\pi}\int_0^\infty dr\,e^{-r^2}r^{n+m+1}\,e^{i\theta(n-m)} \\
&= \delta_{mn}\int_0^\infty dt\,e^{-t}t^n = \delta_{mn}\,\Gamma(n+1) = \delta_{mn}\,n!\,, \qquad (B.9)
\end{aligned}$$

where the polar decomposition $z = re^{i\theta}$ has been used. There is yet another frequently used form of the completeness relation that is particularly useful in the path integral formalism. To obtain it we use

$$\hat{x} = \sqrt{\frac{\hbar}{2\omega m}}\,(a + a^\dagger)\,, \qquad \hat{p} = i\sqrt{\frac{\hbar\omega m}{2}}\,(a^\dagger - a)\,. \qquad (B.10)$$

Clearly $[\hat{x},\hat{p}] = i\hbar\mathbb{1}$. Equations (B.10) imply that

$$a = \sqrt{\frac{m\omega}{2\hbar}}\,\hat{x} + i\sqrt{\frac{1}{2\hbar m\omega}}\,\hat{p}\,. \qquad (B.11)$$

In a similar fashion we decompose z according to the rule

$$z = \sqrt{\frac{m\omega}{2\hbar}}\,x + i\sqrt{\frac{1}{2\hbar m\omega}}\,p\,, \qquad x,p \in \mathbb{R}\,. \qquad (B.12)$$

Direct consequences of (B.10) and (B.12) are

$$\langle \hat{x}\rangle \equiv \frac{\langle z|\hat{x}|z\rangle}{\langle z|z\rangle} = x\,, \qquad \langle \hat{p}\rangle \equiv \frac{\langle z|\hat{p}\,|z\rangle}{\langle z|z\rangle} = p\,, \qquad (B.13)$$

and

$$(\triangle x)^2 (\triangle p)^2 \;\equiv\; \langle (\hat{x} - \langle \hat{x} \rangle)^2 \rangle \langle (\hat{p} - \langle \hat{p} \rangle)^2 \rangle \;=\; \frac{\hbar^2}{4}. \qquad (B.14)$$

Relation (B.13) provides an interpretation for the labels x and p, while (B.14) indicates that states $|z\rangle$ saturate the Heisenberg uncertainty relation. It is customary to utilize an alternative notation for $|z\rangle$, namely $|x, p\rangle \equiv |z\rangle$. In terms of the phase-space variables/operators we can directly write

$$|x,p\rangle \;=\; \exp\left[\frac{i}{4\hbar}\left(\omega m x^2 + \frac{1}{\omega m}p^2\right)\right] e^{i(p\hat{x} - x\hat{p})/\hbar} \,|0\rangle, \qquad (B.15)$$

$$|x,p\rangle \;=\; e^{i(p\hat{x} - x\hat{p})/\hbar} \,|0\rangle, \qquad (B.16)$$

$$\mathbb{1} \;=\; \int \frac{dp\, dx}{2\pi\hbar}\, |x,p\rangle\langle x,p|\, \exp\left[-\frac{1}{2\hbar}\left(\omega m x^2 + \frac{1}{\omega m}p^2\right)\right]$$

$$=\; \int \frac{dp\, dx}{2\pi\hbar}\, |x,p)(x,p|. \qquad (B.17)$$

An important signature of CS is their *over-completeness*. In fact, one should note that relation (B.17) (resp. (B.8)) appears exactly like a resolution of unity used for self-adjoint operators. There is, however, a difference in that the one-dimensional projection operators $|x,p)(x,p|$ are not mutually orthogonal. Indeed,

$$\text{Tr}\Big[|x,p)(x,p|x',p')(x',p'|\Big] \;=\; |(z|z')|^2 \;=\; \frac{e^{2\Re e(z^* z')}}{e^{|z|^2 + |z'|^2}} \;\neq\; \delta_{zz'}. \,(B.18)$$

For this reason it is usually said that the set of CS is over-complete. In fact, the previous result shows that CS are not orthogonal for any $|z\rangle$ and $|z'\rangle$. If, however, the numerical distance $|z - z'|$ is large the states are almost orthogonal. This is because the angle $\theta(\widehat{zz'})$ between the states can be calculated through the relation

$$\cos\theta(\widehat{zz'}) \;=\; \frac{|\langle z|z\rangle|}{\sqrt{\langle z|z\rangle}\sqrt{\langle z'|z'\rangle}}. \qquad (B.19)$$

Using (B.6) we obtain

$$\cos\theta(\widehat{zz'}) \;=\; \exp\left(-\tfrac{1}{2}|z|^2 + \Re e(z^* z') - \tfrac{1}{2}|z'|^2\right)$$

$$=\; \exp\left(-\tfrac{1}{2}|z - z'|^2\right). \qquad (B.20)$$

Inasmuch as $|z - z'| \gg 1$ then vectors $|z\rangle$ and $|z'\rangle$ are close to being orthogonal. In Appendix D we consider the problem of extracting a complete set of CS from an over-complete set. We will see that in order to do that one needs to introduce a regular lattice called the von Neumann lattice.

If we make use of the resolution of the unity (B.8) we can write for a general state $|\psi\rangle$

$$|\psi\rangle = \int \frac{dzdz^*}{2\pi i} \, e^{-|z|^2} \langle z|\psi\rangle \, |z\rangle . \qquad (B.21)$$

Here

$$\langle z|\psi\rangle = \sum_{n=0}^{\infty} \frac{(z^*)^n}{\sqrt{n!}} \, \langle n|\psi\rangle \equiv f_\psi(z^*) . \qquad (B.22)$$

From the fact that $\sum_n |\langle n|\psi\rangle|^2 = 1$ it is clear that the series (B.22) converges for all z^*, and thus it represents a complex function that is holomorphic on the whole complex plane $\mathbb{C}$. Such functions are said to be *entire*. Decomposition (B.21) indicates that the function $f_\psi(z^*)$ is itself a representation of $|\psi\rangle$ (in $\{|z\rangle\}$ basis), and can be regarded as the element of the Hilbert space. The representation $f_\psi(z^*)$ is called a holomorphic representation and the corresponding Hilbert space is known as the Fock–Bargmann, or Segal–Fock–Bargmann space of entire analytical functions [55, 242, 564].

Appendix C

Generalized coherent states

The Glauber coherent states considered in Appendix B, have the three following properties: they are eigenstates of lowering operators [60], they are minimum uncertainty states [500], and they may be generated via translation (or displacement) operators [519].

Various generalizations of the above coherent states have been proposed [379, 519], which maintain only some of the above conditions. Here we consider the generalized coherent states generated via displacement operators [518, 519], related to a Lie group G. Such states have been used in many applications in atomic and nuclear physics and in statistical mechanics [33, 172, 173, 533].

The generalized coherent states related to a Lie group G are constructed in the following way: let $D(g)$, $g \in G$ be an irreducible *unitary* representation of G acting in some Hilbert space $\mathcal{H}$. We choose a normalized fiducial state vector in $\mathcal{H}$ and denote it as $|0\rangle$ (the reason for this notation will be clear shortly). The generalized coherent states corresponding to G are then defined as

$$|0(g)\rangle = D(g)|0\rangle \quad \text{for } \forall g \in G. \tag{C.1}$$

We say that two coherent states $|0(g_1)\rangle$ and $|0(g_2)\rangle$ represent the same state (or are physically equivalent) in $\mathcal{H}$ if

$$D(g_1)|0\rangle = e^{i\alpha(g_1,g_2)}D(g_2)|0\rangle \iff D(g_2^{-1}g_1)|0\rangle = e^{i\alpha(g_1,g_2)}|0\rangle. \tag{C.2}$$

Here the phase factor $\alpha \in \mathbb{R}$ may depend both on g_1 and g_2. As g runs in $|0(g)\rangle$ through G, states $|0(g)\rangle$ travel through the Hilbert space $\mathcal{H}$. In general, however, physically equivalent states will be visited many times during this procedure. Defining the *stability* group $H_{|0\rangle}$ the group of transformations leaving $|0\rangle$ invariant (modulo a phase factor), i.e.,

$$H_{|0\rangle} = \{h \in G : D(h)|0\rangle = e^{i\beta(h)}|0\rangle, \beta(h) \in \mathbb{R}\}, \tag{C.3}$$

we see from (C.2) that $g_2^{-1} g_1 \in H_{|0\rangle}$. Note that $H_{|0\rangle}$ is indeed a subgroup of G, because if h_1 and h_2 belong to $H_{|0\rangle}$ then also $h_1^{-1} h_2$ does.

In this connection and for future reference, we recall that given a subgroup H of a group G, the (left) *coset* of H with respect to $g \in G$, written as gH, is defined as the set of all elements $\{gh; h \in H\}$. An elementary theorem from group theory asserts that two cosets $g_1 H$ and $g_2 H$ for $g_1 \neq g_2$ are either identical or completely disjoint. In this way the group G can be partitioned into disjoint cosets. The collection of cosets of the subgroup H in the group G is usually denoted as G/H and called the coset (or quotient) space of G modulo H. Despite the fact that both G and H are groups, the coset space G/H is generally not a group. Only in cases when H is a *normal* subgroup of G (i.e., when $gH = Hg$ for all $g \in G$) then one can formulate group operations in G/H. For instance, the product law for two cosets $g_1 H$ and $g_2 H$ can be simply defined as the coset $(g_1 g_2)H$: $(g_1 H)(g_2 H) = (g_1 g_2)H$. With this the associativity is obvious, the identity element can be taken as $E \equiv eH = H$, and the inverse of the coset gH is $g^{-1}H$. In such cases the coset space G/H is called the *factor group*.

It is also convenient to recall that the algebra of a d-dimensional Lie algebra is given by the commutators

$$[T_a, T_b] = i C_{ab}{}^c T_c. \tag{C.4}$$

$C_{ab}{}^c$ are the *structure constants* and T_a are hermitian matrices — the group generators ($a = 1, \ldots, d$). The *adjoint* or *regular* representation of the Lie algebra is then defined so that

$$(T_a)_b{}^c = i C_{ba}{}^c \quad \text{or equivalently} \quad (T_a)_{bc} = -i C_{abc}. \tag{C.5}$$

Thus T_a is a $d \times d$ matrix. For instance, the adjoint representation for $SU(2) \cong SO(3)$ has $d = 3$ and hence its representation space is three-dimensional vector space with matrix elements $(T_a)_{bc}$ given explicitly by

$$(T_a)_{bc} = -i\epsilon_{abc}. \tag{C.6}$$

The *fundamental* or *defining* representation of the Lie algebra corresponds to the defining matrix representation. For instance, the fundamental representation of $SO(3)$ corresponds to 3×3 orthogonal matrices of determinant 1. The corresponding representation space is thus three-dimensional. The fundamental representation of $SU(2)$ has two-dimensional representation space.

In future considerations we shall simply denote $H_{|0\rangle}$ as H with the implicit knowledge that H is associated with a fiducial state. We note that both g_1 and g_2 in (C.2) are part of the same stability group $H_{|0\rangle}$ in G.

Let dg be the left-invariant group measure (Haar measure), i.e., for any fixed $g_0 \in G$, $d(g_0 \cdot g) = dg$. Consider now the operator

$$\mathcal{B} = \int_G dg \, |0(g)\rangle\langle 0(g)| \,. \tag{C.7}$$

Due to the invariance of the measure we have for any $g' \in G$

$$D(g')\mathcal{B}D^\dagger(g') = \int_G dg \, |0(g' \cdot g)\rangle\langle 0(g' \cdot g)| = \mathcal{B} \,. \tag{C.8}$$

So $\mathcal{B}$ commutes with all $D(g)$, and hence it must be proportional to the unit operator. This is a result of the fundamental lemma of group theory which asserts that any linear operator commuting with all the operators of an irreducible representation of some group G must be a multiple of the unit operator, i.e.,

$$D(g)\mathcal{B} = \mathcal{B}D(g), \quad \forall g \in G \Rightarrow \mathcal{B} = c^{-1}\mathbb{1}\,. \tag{C.9}$$

This lemma is known as the *first Schur lemma*.

Having measure dg on G, the measure on the coset space G/H is naturally induced by dg. We shall denote this induced measure as dx. With the help of (C.9) the resolution of the unity can be written as

$$\mathbb{1} = c \int_G dg \, |0(g)\rangle\langle 0(g)| = c \int_{G/H} dx \, |0(\mathbf{x})\rangle\langle 0(\mathbf{x})| \,. \tag{C.10}$$

Here c is determined so as to fulfill the consistency condition

$$1 = \langle 0(\mathbf{y})|0(\mathbf{y})\rangle = c \int_{G/H} dx \, |\langle 0(\mathbf{y})|0(\mathbf{x})\rangle|^2, \quad \mathbf{y} \in G/H\,. \tag{C.11}$$

It should be stressed that when

$$\int_G dg \, |\langle 0(g')|0(g)\rangle|^2 = \int_G dg \, |\langle 0|D(g)|0\rangle|^2 = \infty, \tag{C.12}$$

the condition (C.11) cannot be fulfilled with $c \neq 0$. It is thus meaningful to confine only to representations $D(g)$ for which the integral (C.12) is finite, i.e., square integrable representations.

SU(2) coherent states

The $SU(2)$ group has three generators J_1, J_2, J_3. The $SU(2)$ algebra is

$$[J_+, J_-] = 2J_3 \qquad [J_3, J_\pm] = \pm J_\pm. \tag{C.13}$$

Here the ladder operators are defined as $J_\pm = J_1 \pm iJ_2$. The unitary irreducible representations of the $SU(2)$ algebra are finite-dimensional and are spanned by the states $|j, m\rangle$, such that

$$J_3|j, m\rangle = m|j, m\rangle,$$

$$J_\pm|j, m\rangle = \sqrt{(j \mp m)(j \pm m + 1)}\,|j, m \pm 1\rangle, \qquad (|m| \le j). \tag{C.14}$$

The representations of $SU(2)$ are labeled by the eigenvalues of the $SU(2)$ Casimir operator:

$$\mathcal{C} = \mathbf{J}^2 = J_1^2 + J_2^2 + J_3^3$$

$$= \tfrac{1}{2}(J_+J_- + J_-J_+) + J_3^2 = j(j+1)\mathbb{1}, \tag{C.15}$$

i.e.,

$$\mathbf{J}^2|j, m\rangle = j(j+1)|j, m\rangle \quad \text{with} \quad j = 0, \tfrac{1}{2}, 1, \tfrac{3}{2}, \ldots. \tag{C.16}$$

As the fiducial vector we choose the state $|j, -j\rangle$, i.e., the state which is annihilated by the lowering operator: $J_-|j, -j\rangle = 0$. In this way each representation has its unique fiducial state — "vacuum state" $|0\rangle \equiv |j, -j\rangle$. The stability group for this state is the subgroup of rotations around the z-axis, thus $H = U(1)$. According to Eq. (C.10), distinct coherent states are labeled by $\mathbf{x} \in \mathcal{M} = G/H$. By noting that $\mathcal{M} = SU(2)/U(1) \cong \mathcal{S}^2$ we can identify $\mathbf{x}$ with the spherical angular variables θ and φ. The associated state can be written as $|0(\theta, \varphi)\rangle$:

$$|0(\theta, \varphi)\rangle = D(\theta, \varphi)|0\rangle = \exp\left[i\theta(\mathbf{J} \cdot \mathbf{n})\right]|0\rangle, \tag{C.17}$$

with the unit vector $\mathbf{n} = (\sin\varphi, \cos\varphi, 0)$. Using the Gauss decomposition formula

$$D(\theta, \varphi) = e^{\xi J_+}\, e^{\log(1+|\xi|^2)J_3}\, e^{-\xi^* J_-}, \qquad \xi = \tan\frac{\theta}{2}\, e^{i\varphi}, \tag{C.18}$$

one can alternatively use the more economical form

$$|0(\theta, \varphi)\rangle = (1 + |\xi|^2)^{-j} e^{\xi J_+}|0\rangle \equiv |0(\xi)\rangle. \tag{C.19}$$

Relation (C.19) is an analogue of the canonical coherent state relation (B.1). The scalar product of two coherent states $|0(\xi)\rangle$ can be written in the form

$$\langle 0(\xi')|0(\xi)\rangle = \frac{\langle 0|e^{\xi'^* J_-}\, e^{\xi J_+}|0\rangle}{(1 + |\xi'|^2)^j(1 + |\xi|^2)^j} = \frac{(1 + \xi'^*\xi)^{2j}}{(1 + |\xi'|^2)^j(1 + |\xi|^2)^j}. \tag{C.20}$$

In the derivation of (C.20) we used the identity

$$J_+^k |j, -j\rangle = \sqrt{k!} \sqrt{2j(2j-1)\ldots(2j-k+1)} \, |j, -j+k\rangle. \quad (C.21)$$

An implication of Eq. (C.20), that will be relevant later, is that

$$
\begin{aligned}
|\langle 0(\xi') | 0(\xi) \rangle|^2 &= \left(\frac{1 + 2\Re e(\xi'^*\xi) + |\xi'|^2|\xi|^2}{(1+|\xi'|^2)(1+|\xi|^2)} \right)^{2j} \\
&= \left(\frac{1 + \cos\theta' \cos\theta + \sin\theta' \sin\theta \cos(\varphi' - \varphi)}{2} \right)^{2j} \\
&= \left(\frac{1 + \mathbf{m}' \cdot \mathbf{m}}{2} \right)^{2j}.
\end{aligned}
\quad (C.22)
$$

Here $\mathbf{m} = (\sin\theta\cos\varphi, \sin\theta\sin\varphi, \cos\theta)$ is the unit vector parametrizing $\mathcal{M} = \mathcal{S}^2$ (similarly for $\mathbf{m}'$). For this reason it is sometimes convenient to use notation $|0(\theta, \varphi)\rangle = |0(\xi)\rangle = |0(\mathbf{m})\rangle$.

According to Eq. (C.10) the resolution of the unity reads

$$\mathbb{1} = \int_{SU(2)} dg \, |0(g)\rangle\langle 0(g)| = c \int_{\mathcal{S}^2} d\mathbf{m} \, |0(\mathbf{m})\rangle\langle 0(\mathbf{m})|. \quad (C.23)$$

The constant c is determined from the consistency condition

$$
\begin{aligned}
1 &= c \int_{\mathcal{S}^2} d\mathbf{m} \, |\langle 0(\mathbf{m}') | 0(\mathbf{m}) \rangle|^2 = c \int_{\mathcal{S}^2} d\mathbf{m} \left(\frac{1 + \mathbf{m}' \cdot \mathbf{m}}{2} \right)^{2j} \\
&= c \, 4\pi \int_0^1 dx \, x^{2j} = c \frac{4\pi}{2j+1}.
\end{aligned}
\quad (C.24)
$$

So finally the resolution of the unity may be written in one of the following equivalent forms:

$$
\begin{aligned}
\mathbb{1} &= \frac{2j+1}{4\pi} \int_{\mathcal{S}^2} d\mathbf{m} \, |0(\mathbf{m})\rangle\langle 0(\mathbf{m})| \\
&= \frac{2j+1}{4\pi} \int_{\mathcal{S}^2} d\varphi d\theta \, \sin\theta \, |0(\theta, \varphi)\rangle\langle 0(\theta, \varphi)| \\
&= \frac{2j+1}{\pi} \int_{\mathcal{S}^2} \frac{d\xi d\xi^*}{(1+|\xi|^2)^2} \, |0(\xi^*)\rangle\langle 0(\xi)|,
\end{aligned}
\quad (C.25)
$$

where we have used the usual convention

$$d\xi d\xi^* \equiv d\Re e(\xi) \, d\Im m(\xi),$$

where $\Re e$ and $\Im m$ denote the real and the imaginary parts, respectively. This relation will be useful in the construction of the $SU(2)$ coherent state functional integral.

SU(1,1) coherent states

The group $SU(1,1)$ is the group of unitary unimodular matrices of the form

$$g = \begin{pmatrix} \alpha & \beta \\ \beta^* & \alpha^* \end{pmatrix}, \quad \det g = 1, \tag{C.26}$$

i.e., matrices that preserve the quadratic form $|\alpha|^2 - |\beta|^2$. The algebra of the $SU(1,1)$ group is

$$[J_+, J_-] = -2J_3 \qquad [J_3, J_\pm] = \pm J_\pm, \tag{C.27}$$

with the ladder operators $J_\pm = J_1 \pm i J_2$.

The unitary irreducible representations for $SU(1,1)$ are labeled by the eigenvalues of Casimir operators. Because the rank of $SU(1,1)$ is 1, there is only one (quadratic) Casimir operator for the $su(1,1)$ algebra, i.e.,

$$\mathcal{C} = J_3^2 - J_1^2 - J_2^2 = J_3^2 - \tfrac{1}{2}(J_+J_- + J_-J_+) = j(j+1)\mathbb{1}. \tag{C.28}$$

As in the $SU(2)$ case, we consider simultaneous eigenstates of $\mathcal{C}$ and J_3

$$\mathcal{C}|j,m\rangle = j(j+1)|j,m\rangle,$$

$$J_3|j,m\rangle = m|j,m\rangle,$$

$$J_\pm|j,m\rangle = \sqrt{(m \mp j)(m \pm j \pm 1)}\,|j,m\pm 1\rangle. \tag{C.29}$$

However, in contrast with $SU(2)$ there is more than one way in which the spectrum $\{j,m\}$ may be realized. This is because the Casimir operator (C.28) is a semi-definite operator. Owing to this, there are four classes (so-called series) of the unitary irreducible representations of $SU(1,1)$ (see, e.g., [519]). All of these representations are infinite dimensional. It should be emphasized that $SU(1,1)$ has also non-unitary representations (e.g., non-unitary principal series [387]), but we shall refrain from considering non-unitary representations. Analogous algebraic considerations as in $SU(2)$ would lead one to four classes of unitary irreducible representations in the $SU(1,1)$ case. These are:

(1) *The principal continuous series* $C_j(q_0)$:
$$j = -\tfrac{1}{2} + is, \quad m = q_0 + n \quad (s \in \mathbb{R}^+,\ n \in \mathbb{Z},\ q_0 \in \mathbb{R},\ |q_0| \le \tfrac{1}{2})$$

(2) *The principal discrete series* $D_j^+(q_0)$:
$$j = -|q_0| - \tilde{n}, \quad m = -j + n \quad (\tilde{n},\ n \in \mathbb{N}_0,\ q_0 \in \mathbb{R},\ |q_0| \le \tfrac{1}{2})$$

(3) *The principal discrete series* $D_j^-(q_0)$:
$$j = -|q_0| - \tilde{n}, \quad m = j - n \quad (\tilde{n}, \ n \in \mathbb{N}_0, \ q_0 \in \mathbb{R}, \ |q_0| \le \tfrac{1}{2})$$

(4) *The supplementary continuous series* $E_j(q_0)$:
$$-\tfrac{1}{2} < j < -|q_0|, \quad m = q_0 + n \quad (n \in \mathbb{Z}, \ q_0 \in \mathbb{R}, \ |q_0| \le \tfrac{1}{2}) \,.$$

The so-called Bargmann index q_0 cannot be determined from algebraic considerations alone and the representation must be labeled both by value of j and q_0. However, $SU(1,1)$ has the maximal compact subgroup $U(1)$ for whose unitary representations the possible values of Bargmann index is restricted to 0 or 1/2. These representations are known as Bargmann representations.

In the following we confine ourselves to the discussion of D_j^+ only:

The principal discrete series D_j^+: $2j = -\tilde{n}, \quad m = -j + n \quad (\tilde{n}, \ n \in \mathbb{N}_0)$
$$\text{i.e., } j = 0, -\tfrac{1}{2}, -1, -\tfrac{3}{2}, \ldots, \quad m = |j|, |j| + 1, |j| + 2, \ldots.$$

As the fiducial "ground" state $|0\rangle$ we choose the state $|j, -j\rangle$. Similarly as in the $SU(2)$ case, such a state is annihilated by J_-: i.e., $J_- |j, -j\rangle = 0$. Thus each representation in D_j^+ has its unique fiducial vector. The stability group for $|j, -j\rangle$ is the subgroup of rotations around the z-axis. The coherent states $|0(\mathbf{x})\rangle$ are then completely determined by points $\mathbf{x}$ on the coset space $\mathcal{M} = SU(1,1)/U(1) \cong H_+^2$. The manifold H_+^2 represents the upper sheet of the two-sheet hyperboloid: $H_+^2 = \{\mathbf{m}; \ \mathbf{m}^2 = m_3^2 - m_2^2 - m_1^2 = 1, \ m_3 > 0\}$. This two-dimensional surface can be conveniently parametrized by the hyperbolic "angular" variables τ and φ according to prescription

$$\mathbf{m} = (\sinh \tau \cos \varphi, \sinh \tau \sin \varphi, \cosh \tau) \,. \tag{C.30}$$

Parameter $\mathbf{x} \in \mathcal{M}$ can be identified with variables τ and φ. The coherent state $|0(\tau, \varphi)\rangle$ can thus be written as

$$|0(\mathbf{x})\rangle = |0(\tau, \varphi)\rangle = D(\tau, \varphi)|0\rangle = \exp\left[i\tau\,(\mathbf{J} \cdot \mathbf{n})\right]|0\rangle, \tag{C.31}$$

with the unit vector $\mathbf{n} = (\sin \varphi, \cos \varphi, 0)$. The Gauss decomposition allows us to write $D(\tau, \varphi)$ in the ordered form

$$D(\tau, \varphi) = e^{\zeta J_+}\, e^{\log(1-|\zeta|^2) J_3}\, e^{-\zeta^* J_-}, \quad \zeta = \tanh\frac{\tau}{2}\, e^{i\varphi}. \tag{C.32}$$

Consequently

$$|0(\tau, \varphi)\rangle = (1 - |\zeta|^2)^{|j|}\, e^{\zeta J_+}|0\rangle \equiv |0(\zeta)\rangle. \tag{C.33}$$

Overlap of two such coherent states is then

$$\langle 0(\zeta')|0(\zeta)\rangle = (1 - |\zeta'|^2)^{|j|}(1 - |\zeta|^2)^{|j|} \langle 0|e^{\zeta'^* J_-} e^{\zeta J_+}|0\rangle$$

$$= (1 - |\zeta'|^2)^{|j|}(1 - |\zeta|^2)^{|j|} (1 - \zeta'^*\zeta)^{-2|j|} . \quad \text{(C.34)}$$

The transition probability between two coherent states can be written as

$$|\langle 0(\zeta')|0(\zeta)\rangle|^2 = \left(\frac{1 - 2\Re e(\zeta'^*\zeta) + |\zeta'^*|^2|\zeta|^2}{(1 - |\zeta'|^2)(1 - |\zeta|^2)} \right)^{-2|j|}$$

$$= \left(\frac{1 - \sinh\tau' \sinh\tau \cos(\varphi' - \varphi) + \cosh\tau' \cosh\tau}{2} \right)^{-2|j|}$$

$$= \left(\frac{1 + \mathbf{m}' \cdot \mathbf{m}}{2} \right)^{-2|j|} . \quad \text{(C.35)}$$

Here the pseudo-Euclidean scalar product is defined as $\mathbf{m}' \cdot \mathbf{m} = m'_3 m_3 - m'_2 m_2 - m'_3 m_3$. Defining $|0(\tau, \varphi)\rangle = |0(\zeta)\rangle \equiv |0(\mathbf{m})\rangle$ we can write the resolution of the unity as

$$\mathbb{1} = \int_{SU(1,1)} dg \, |0(g)\rangle\langle 0(g)| = c \int_{H^2_+} d\mathbf{m} \, |0(\mathbf{m})\rangle\langle 0(\mathbf{m})| . \quad \text{(C.36)}$$

The constant c follows from the normalization condition

$$1 = c \int_{H^2_+} d\mathbf{m} \, |\langle 0(\mathbf{m})|0(\mathbf{m})\rangle|^2 = c \int_{H^2_+} d\mathbf{m} \left(\frac{1 + \mathbf{m}' \cdot \mathbf{m}}{2} \right)^{-2|j|}$$

$$= c2\pi \int_1^\infty dx \, x^{-2|j|} = c\frac{2\pi}{2|j| - 1} . \quad \text{(C.37)}$$

Note that the integral is convergent (i.e., the representation is square integrable) only when $|j| > 1/2$. Only such representations will concern us here. In the end the resolution of the unity reads

$$\mathbb{1} = \frac{2|j| - 1}{2\pi} \int_{H^2_+} d\mathbf{m} \, |0(\mathbf{m})\rangle\langle 0(\mathbf{m})|$$

$$= \frac{2|j| - 1}{4\pi} \int_0^{2\pi} d\varphi \int_0^\infty d\tau \, \sinh\tau \, |0(\tau, \varphi)\rangle\langle 0(\tau, \varphi)|$$

$$= \frac{2|j| - 1}{\pi} \int_{H^2_+} \frac{d\zeta d\zeta^*}{(1 - |\zeta|^2)^2} \, |0(\zeta^*)\rangle\langle 0(\zeta)| . \quad \text{(C.38)}$$

Here again $d\zeta d\zeta^* \equiv d\Re e(\zeta) \, d\Im m(\zeta)$. The resolution of unity (C.38) will serve as a useful starting point in setting up the $SU(1,1)$ coherent state functional integral in Appendix N.2.

Appendix D

q-WH algebra, coherent states and theta functions

As an application of the result expressed by Eq. (1.100), we shall show that the action of the commutator $[a_q, \hat{a}_q]$ may be related in the Fock–Bargmann representation (FBR) to the action of the coherent states (CS) displacement operator.

The FBR provides a transparent frame to describe the usual CS [379, 380, 519]. In this Appendix we change notation with respect to Appendix B. Here we replace z by α, so that the CS are now written as:

$$|\alpha\rangle = D(\alpha)|0\rangle \; ; \quad a|\alpha\rangle = \alpha|\alpha\rangle \,, \quad a|0\rangle = 0 \,, \quad \alpha \in \mathbb{C} \,, \qquad \text{(D.1)}$$

$$|\alpha\rangle = \exp\left(\frac{-|\alpha|^2}{2}\right) \sum_{n=0}^{\infty} \frac{\alpha^n}{\sqrt{n!}}|n\rangle = \exp\left(\frac{-|\alpha|^2}{2}\right) \sum_{n=0}^{\infty} u_n(\alpha)|n\rangle \,. \qquad \text{(D.2)}$$

The relation between the CS and the basis $\{u_n(z)\}$ (Eq. (1.88)) of the entire analytic function is here made explicit: $u_n(\alpha) = e^{\frac{1}{2}|\alpha|^2}\langle n|\alpha\rangle$. The unitary displacement operator $D(\alpha)$ in (D.1) is given by:

$$D(\alpha) = \exp(\alpha a^\dagger - \bar{\alpha} a) = \exp\left(-\frac{|\alpha|^2}{2}\right) \exp(\alpha a^\dagger) \exp(-\bar{\alpha}\, a) \,, \qquad \text{(D.3)}$$

and the relations hold

$$D(\alpha)D(\beta) = \exp(i\Im m(\alpha\bar{\beta}))D(\alpha + \beta) \,, \qquad \text{(D.4)}$$

$$D(\alpha)D(\beta) = \exp(2i\Im m(\alpha\bar{\beta}))D(\beta)D(\alpha) \,. \qquad \text{(D.5)}$$

Eq. (D.4) is nothing but the WH group law, also referred to as the Weyl integral representation (cf. Eqs. (1.61) and (1.62)).

In order to extract a complete set of CS $\{|\alpha_n\rangle\}$, from the over-complete set $\{|\alpha\rangle\}$, it is necessary to introduce a regular lattice L in the α-complex plane [55, 519]. The points (lattice vectors) α_n of L ($\{\alpha_n \in \mathbb{C}; n =$

(n_1, n_2); $n_j \in \mathcal{Z}\}$) are given by $\alpha_{\mathbf{n}} = n_1 \Omega_1 + n_2 \Omega_2 \equiv \mathbf{n} \cdot \mathbf{\Omega}$, with the two lattice periods Ω_j, $j = 1, 2$ linearly independent, i.e., such that $\Im m(\Omega_1 \bar{\Omega}_2) \neq 0$.

We recall [519] that the set $\{|\alpha_{\mathbf{n}}\rangle\}$ (with the exclusion of the vacuum state $|0\rangle \equiv |\alpha_0\rangle$) can be shown to be complete, invoking square integrability along with analyticity [56], if the lattice elementary cell has area $\Im m(\Omega_1 \bar{\Omega}_2) = \pi$ (L is called, in this case, the von Neumann lattice).

The lattice vectors $\alpha_{\mathbf{n}}$ describe the discrete translational invariance of L: $\alpha_{\mathbf{n+m}} = \alpha_{\mathbf{n}} + \alpha_{\mathbf{m}}$, i.e.,

$$e^{\alpha_{\mathbf{n}} \frac{d}{d\alpha}} |\alpha\rangle\Big|_{\alpha = \alpha_{\mathbf{m}}} = |\alpha_{\mathbf{n+m}}\rangle . \tag{D.6}$$

The denumerable set of points $\alpha_{\mathbf{n}}$ is now mapped onto the set $\{z_{\mathbf{n}}$; $z_{\mathbf{n}} \in \mathbb{C}\}$ with $z_{\mathbf{n}} \equiv e^{\alpha_{\mathbf{n}}}$. Assuming that the two periods Ω_1 and Ω_2 have imaginary parts incommensurate with π and among themselves, such a map is one-to-one and no point $z_{\mathbf{n}}$ lies on the real axis in the z plane (notice that the set $\{z_{\mathbf{n}}\}$ does *not* constitute a lattice in z, but it has the structure of concentric circles).

The function $z = e^{\alpha}$, which interpolates among these points, is analytical in its domain of definition, and, along with the basis functions $\{u_n(\alpha)\}$, the new functions $\{\tilde{u}_n(z) \equiv u_n(\ln z) = u_n(\alpha)\}$, with $\tilde{u}_n(z) \in \mathcal{F}$, may be introduced ($\mathcal{F}$ denotes the space of the entire analytical functions). It is then straightforward to check that, in $\mathcal{F}$,

$$[a_{q_{\mathbf{m}}}, \hat{a}_{q_{\mathbf{m}}}] \tilde{u}_n(z) = q_{\mathbf{m}}{}^{z \frac{d}{dz}} \tilde{u}_n(z) = \tilde{u}_n(q_{\mathbf{m}} z) = u_n(\alpha + \alpha_{\mathbf{m}}) = q_{\mathbf{m}}{}^{\frac{d}{d\alpha}} u_n(\alpha) , \tag{D.7}$$

where the complex parameter $q_{\mathbf{m}} \equiv e^{\alpha_{\mathbf{m}}}$ has been introduced and we used Eq. (1.100). Therefore,

$$[a_{q_{\mathbf{n}}}, \hat{a}_{q_{\mathbf{n}}}] \tilde{f}_m(z)\Big|_{z = z_{\mathbf{r}}} = \tilde{f}_m(q_{\mathbf{n}} z_{\mathbf{r}}) = f_m(\alpha_{\mathbf{r}} + \alpha_{\mathbf{n}}) = q_{\mathbf{n}}{}^{\frac{d}{d\alpha}} f_m(\alpha)\Big|_{\alpha = \alpha_{\mathbf{r}}}$$
$$= \exp\left[-i\Im m(\alpha_{\mathbf{n}} \bar{\alpha}_{\mathbf{r}})\right] \langle m | D(\alpha_{\mathbf{n}}) | \alpha_{\mathbf{r}} \rangle , \tag{D.8}$$

where we utilized Eq. (D.4), and the notation $f_m(\alpha) \equiv \exp\left(-\frac{|\alpha|^2}{2}\right) u_m(\alpha)$. The operator $[a_q, \hat{a}_q]\big|_{q = q_{\mathbf{n}}}$, which realizes the mapping $\tilde{f}(z_{\mathbf{m}}) \to \tilde{f}(q_{\mathbf{n}} z_{\mathbf{m}})$, can thus be thought of as extendible to the map on the α-plane $|\alpha_{\mathbf{m}}\rangle \to |\alpha_{\mathbf{n+m}}\rangle$.

In the FBR we have, for any $f \in \mathcal{F}$

$$D(\beta) f(\alpha) = \exp\left(-\frac{|\beta|^2}{2}\right) \exp(\alpha\beta) f(\alpha - \bar{\beta}) , \quad f \in \mathcal{F} , \tag{D.9}$$

so that, by using Eq. (D.7), we can write, for $q = e^{\zeta}$ and $z = e^{\alpha}$,

$$[a_q, \hat{a}_q] \tilde{f}(z) = e^{\frac{1}{2}|\zeta|^2} \bar{q}^{\alpha} D(-\bar{\zeta}) f(\alpha) . \tag{D.10}$$

Eqs. (D.8) and (D.10) show the relation between the q-WH algebra operator $[a_q, \hat{a}_q]$ and the CS displacement operator in the frame of the theory of entire analytical functions.

We conclude that the existence of a quantum deformed algebra signals the presence of finite lengths in the theory and provides the natural framework for the physics of discretized systems, the q-deformation parameter being related with the lattice constants.

The lattice structure is also of crucial relevance in the relation between theta functions and the complete system of CS. In order to establish such a relation, we look for the common eigenvectors $|\theta\rangle$ of the CS operators $D(\alpha_{\mathbf{n}})$ associated to the regular lattice L [519]. A common set of eigenvectors exists if and only if all the $D(\alpha_{\mathbf{n}})$ commute, i.e., when the $D(\Omega_j)$ commute, as indeed happens on the von Neumann lattice (cf. Eq. (D.5)).

The eigenstates $|\theta\rangle$ are characterized by two real numbers ϵ_1 and ϵ_2, so that we denote them by $|\theta_\epsilon\rangle$, which are eigenvectors of $D(\Omega_i)$:

$$D(\Omega_j)|\theta_\epsilon\rangle = e^{i\pi\epsilon_j}|\theta_\epsilon\rangle \ , \ j = 1, 2 \ , \ 0 \le \epsilon_j \le 2 \ . \tag{D.11}$$

We thus see that $|\theta_\epsilon\rangle$, which belongs to a space which is the extension [519] of the Hilbert space where the operators $D(\alpha)$ act, corresponds to a point on the two-dimensional torus. The action of $D(\alpha)$ on $|\theta_\epsilon\rangle$ generates a set of generalized coherent states. Use of Eqs. (D.11) and (D.4) gives

$$D(\alpha_{\mathbf{m}})|\theta_\epsilon\rangle = e^{i\pi F_\epsilon(\mathbf{m})}|\theta_\epsilon\rangle \ , \tag{D.12}$$

with $\alpha_{\mathbf{m}} = \mathbf{m} \cdot \mathbf{\Omega}$, an arbitrary lattice vector, and

$$F_\epsilon(\mathbf{m}) = m_1 m_2 + m_1 \epsilon_1 + m_2 \epsilon_2 \ . \tag{D.13}$$

On the other hand, the system of CS is associated, in the FBR, with a corresponding set of entire analytic functions, say $\theta_\epsilon(\alpha)$. Eq. (D.9) with $\bar{\alpha} = -\alpha_{\mathbf{m}}$ shows that Eq. (D.12) may be written as

$$\theta_\epsilon(\alpha + \alpha_{\mathbf{m}}) = \exp\big(i\pi F_\epsilon(-\mathbf{m})\big) \exp\left(\frac{|\alpha_{\mathbf{m}}|^2}{2}\right) \exp(\bar{\alpha}_{\mathbf{m}}\alpha)\theta_\epsilon(\alpha) \ , \tag{D.14}$$

which is the functional equation for the theta functions [69, 484, 519]. We emphasize that such relation is obtained by considering the CS system corresponding to the von Neumann lattice L. The relation with the q-WH algebra is established by realizing that in $\mathcal{F}$ the functional equation (D.14) reads

$$[a_{q_{\mathbf{m}}}, \bar{a}_{q_{\mathbf{m}}}]\tilde{\theta}(z) = \tilde{\theta}(q_{\mathbf{m}} z)$$

$$= \exp\big(i\pi F_\epsilon(-\mathbf{m})\big) \exp\left(\frac{|\alpha_{\mathbf{m}}|^2}{2}\right) \exp(\bar{\alpha}_{\mathbf{m}}\alpha)\theta_\epsilon(\alpha) \ . \tag{D.15}$$

The commutator $[a_q, \hat{a}_q]$ acts as shift operator on the von Neumann lattice whereas it acts as z-dilatation operator $(z \to qz)$ in the space of entire analytic functions or, else, as the $U(1)$ generator of phase variations in the z-plane, $\arg(z) \to \arg(z) + \theta$, when $q = e^{i\theta}$.

It is remarkable that Eqs. (D.7), (D.8) and (D.10) represent the action of the q-WH algebra commutator $[a_q, \hat{a}_q]$, (bi-)linear in a_q and $\hat{a}_q$, through the action of the CS displacement operator which is non-linear in the FBR operators a and $a^\dagger$. Conversely, the non-linear operator $D(\alpha)$ is represented by the linear form $[a_q, \hat{a}_q]$ in the q-WH algebra.

The reader is referred to [143] for the relation between q-WH algebra and lattice QM, and q-WH algebra and Bloch functions.

Chapter 2

Inequivalent representations of the canonical commutation relations

2.1 Introduction

In Chapter 1 we introduced the concept of free field as one aspect of the duality in the structure of QFT. In the present Chapter we will introduce the other aspect of this duality, namely, the concept of the Heisenberg field. This will turn out to be crucial for the understanding of the phenomenon of the dynamical rearrangement of symmetry which occurs whenever the spontaneous breakdown of symmetry occurs.

As already mentioned in the previous Chapter, although the physical particles undergo interaction processes, free field equations, such as Eq. (1.132), do not contain any information about the interactions. Thus we assume the existence of basic entities, the Heisenberg fields, which satisfy basic relations characterizing the dynamics, namely, the Heisenberg equations. Let $\psi(x)$ denote generically the Heisenberg fields. We formally write the Heisenberg field equation as

$$\Lambda(\partial)\psi(x) = J(\psi(x)), \tag{2.1}$$

where $\Lambda(\partial)$ is a differential operator and J is some function of $\psi(x)$. Eq. (2.1) can be written in such a way that $\Lambda(\partial)$ can be made equal to the differential operator in the free field equations for ϕ (cf. Eq. (1.132)). The Heisenberg equations describe the dynamical properties of our system and can be derived from the system Lagrangian in the canonical formalism of QFT. We stress, however, that Eq. (2.1) is only a formal relation among the field operators $\psi(x)$, unless we define their operational meaning, i.e., unless we specify the vector space $\mathcal{H}$ on which they operate. This is the Hilbert space for the so-called "bare" particles. On the other hand, since we can observe only physical quantities, any kind of useful description must be related to such physical quantities. Therefore, the only space we will consider is the Fock space of physical particles.

Thus, in order to give a physical meaning to the description in terms of Heisenberg fields we must introduce a mapping between such a description and the description in terms of physical fields. For this reason we require that the operators $\psi(x)$ must have well defined matrix elements among states of the Fock space of physical particles $\mathcal{H}_F$. Then we read equations such as (2.1) as equations among matrix elements in $\mathcal{H}_F$.

In the following Sections we present some remarks about the mapping between Heisenberg fields and physical fields and, by means of explicit examples, we show that there are conditions under which the Hilbert space for the Heisenberg fields is unitarily inequivalent to the Fock space for the physical fields, which is the content of the Haag theorem in QFT [114, 308, 558, 569, 579]. We also discuss inequivalent representations in the context of flavor mixing, where the coherent state structure of the vacuum leads to physical effects. In this framework we finally discuss entanglement in neutrino mixing and oscillations.

2.2   Heisenberg fields, physical fields and the dynamical map

In this Section we discuss some of the properties of the dynamical mapping among Heisenberg fields and physical fields. Although in the following Chapters we will often use functional integration techniques, it is, however, instructive to consider general properties of the operatorial form of the dynamical map. Explicit examples of dynamical maps are given in the following Section and in the following Chapters.

The dynamical map of the Heisenberg fields $\psi_i(x)$ in terms of the physical fields, generically denoted by $\phi_j(x)$, is written as [617, 619]:

$$
\psi_i(x) = \chi_i + \sum_j Z_{ij}^{1/2} \phi_j(x)
$$

$$
+ \sum_{j,k} \int d^4\xi_1 \int d^4\xi_2 f_{ijk}(x, \xi_1, \xi_2) : \phi_j(\xi_1)\phi_k(\xi_2) :
$$

$$
+ \sum_{j,k,l} \int d^4\xi_1 \int d^4\xi_2 \int d^4\xi_3 f_{ijkl}(x, \xi_1, \xi_2, \xi_3) : \phi_j(\xi_1)\phi_k(\xi_2)\phi_l(\xi_3) :
$$

$$
+ \dots , \tag{2.2}
$$

where $j, k, l, \dots$ are indices for different physical fields, χ_i is a c-number constant, which is zero when $\psi_i(x)$ is a fermion (half-integer spin) field

(χ_i can be thought to be related to the square root of the density of the boson condensation), $Z_{ij}^{1/2}$ is a c-number constant called the renormalization-tion factor, $\phi_j(x)$ here stands for both $\phi_j(x)$ and $\phi_j^\dagger(x)$, $f_{ijk}(x, \xi_1, \xi_2)$, etc. are c-number functions and the dots denote terms which contain higher order normal products. We refer to χ_i, $Z_{ij}^{1/2}$, $f_{ijk}(x, \xi_1, \xi_2)$, etc. as to the coefficients of the dynamical map.

We stress that the map (2.2) must be read as an equality among matrix elements between states of the Fock space $\mathcal{H}_F$ of the physical particles. It is not an equality among field operators, but an equality among matrix elements. The presence of normal products is due to the fact that we are indeed interested in the computation of matrix elements. Equalities which have to be understood as equalities among matrix elements are called weak equalities, or weak relations. Since (2.2) is a weak equality, so the Heisenberg equations (2.1) must be read as weak relations.

Since the Heisenberg fields ψ_i given by (2.2) must satisfy the field equations (2.1), which describe the dynamics of the system under consideration, the coefficients of the map are determined by these field equations. For this reason it is called the "dynamical map". Clearly, a different dynamics (i.e., a different set of Heisenberg equations) will determine different coefficients and thus a different mapping among the Heisenberg and the physical fields. On the other hand, since the Heisenberg equations are to be read as weak relations, the same dynamics (i.e., the same set of Heisenberg equations) may lead to different solutions when unitarily inequivalent representations (Hilbert spaces of physical states) are used in computing the matrix elements. The choice of the representation to be used to describe our system is thus of crucial importance in solving the dynamics: the same dynamics may be realized in different ways in different (i.e., unitarily inequivalent) representations. The choice of the representation may thus be considered as a boundary condition under which the Heisenberg equations have to be solved. In QM this problem does not arise, since there all the representations of the canonical commutation relations are unitarily equivalent due to the von Neumann theorem [648].

We observe that there is no problem of convergence of the summation on the r.h.s. of (2.2) because all the normal product terms are linearly independent and only a finite number of terms contribute to a given physical process.

In practical cases it happens that one has to compute matrix elements of products of fields at the same point x. Then one must use the traditional care [115, 343, 558, 599, 618] with limiting procedures such as $\lim_{\epsilon \to 0} \psi(x)\psi(x + \epsilon)$.

In computing the map (2.2) one needs to choose to work with the set of incoming physical fields, $\{\phi_{in,j}(x)\}$, or outgoing physical fields, $\{\phi_{out,j}(x)\}$. Of course, we assume the existence of a unitary operator, the S-matrix, which transforms $\{\phi_{in,j}(x)\} \leftrightarrow \{\phi_{out,j}(x)\}$. Our choice will be that of the incoming fields $\{\phi_{in,j}(x)\}$. As a consequence of this choice the coefficients in (2.2) must be of retarded nature, i.e., the domain of time integration in (2.2) must be from $-\infty$ to t, which means that physical fields affect Heisenberg fields only from the past. The coefficients of the map must have advanced nature if one chooses the outgoing fields $\{\phi_{out,j}(x)\}$.

Note that a unitary transformation leaves the dynamical mapping unaltered, while the coefficients of the map are affected.

With our choice of $\{\phi_{in,j}(x)\}$ the retarded nature of the coefficients implies that time integration in the map is well defined only when contributions from t to $+\infty$ vanish. Below we will see that it is then essential that the physical states are wave-packet states, not plane-wave states.

We assume the translational invariance of the map, i.e., the translation $\xi_\mu \to \xi_\mu + a_\mu$ in the fields $\{\phi_{in,j}(\xi)\}$ induces the translation $x_\mu \to x_\mu + a_\mu$ in $\psi_i(x)$. The coefficients $f_{ijk}(x, \xi_1, \xi_2)$, etc. will then depend on the differences $x - \xi_1$, $x - \xi_2$, etc., and we can write

$$f_{ijk}(x - \xi_1, x - \xi_2) = \theta(t_x - t_{\xi_1})\theta(t_x - t_{\xi_2})F_{ijk}(x - \xi_1, x - \xi_2), \quad (2.3)$$

etc. In Appendix E we consider the computation of the matrix element $\langle 0|\psi_i(x)|\alpha_n\alpha_m\rangle$ as an example. From that computation it clearly appears that the wave-packet nature of the physical states plays a crucial role. It also appears that the time dependence of the Heisenberg fields, namely of their matrix elements, is controlled by the energies of the free fields. The term in (E.9) depending on $\omega_1 + \omega_2$ represents the balance of energies of the initial and final free states. Other very simple matrix elements are, e.g.:

$$\langle 0|\psi_i(x)|0\rangle = \chi_i, \quad (2.4a)$$

$$\langle 0|\psi_i(x)|\alpha_j\rangle = Z_{ij}^{1/2}u_j(x), \quad (2.4b)$$

$$\langle \beta_j|\psi_i(x)|0\rangle = Z_{ij}^{1/2}v_j(x). \quad (2.4c)$$

In Eqs. (2.4) there is no summation on repeated indices and $u_j(x)$ and $v_j(x)$ are related to $u_j(\mathbf{k})$ and $v_j(\mathbf{k})$ (cf. Eq. (1.131)) by Fourier transforms, e.g.,

$$u_j(x) \equiv \int \frac{d^3k}{(2\pi)^{3/2}} f_j(\mathbf{k})u(\mathbf{k})e^{i\mathbf{k}\cdot\mathbf{x} - iE_k t}. \quad (2.5)$$

Eqs. (2.4) show that

$$\lim_{t \to -\infty} (\psi_i(x) - \chi_i - \sum_j Z_{ij}^{1/2} \phi_{in,j}(x)) = 0 \,. \qquad (2.6)$$

Indeed, due to the retarded nature of the coefficients $f_{ijk...}$, the limit $t \to -\infty$ excludes higher order terms. In the Lehmann–Symanzik–Zimmermann (LSZ) [416–420] formalism of QFT one introduces

$$a_{ij}(t) = -(2\pi)^3 \int d^3x \, u_j^\dagger(x) \Gamma_4(\overrightarrow{\partial} - \overleftarrow{\partial})(\psi_i(x) - \chi_i) \,, \qquad (2.7)$$

where $\Gamma_4(\overrightarrow{\partial} - \overleftarrow{\partial})$ is γ_4 for the Dirac field and $-i(\overrightarrow{\partial}_0 - \overleftarrow{\partial}_0)$ for the scalar field. The field $(\psi_i(x) - \chi_i)$ is also called interpolating field. We find

$$\lim_{t \to -\infty} a_{ij}(t) = Z_{ij}^{1/2} \alpha_{in,j} \,, \qquad \text{no summation on } j, \qquad (2.8)$$

which shows that the interpolating field (weakly) approaches the incoming field at $t \to -\infty$. $\alpha_{in,j}$ denotes the wave-packet operator (cf. Eqs. (1.1)). Once more, the use of smeared functions is essential: the limit in Eq. (2.8) would be meaningless otherwise. By using

$$\lim_{t \to +\infty} \frac{e^{iEt}}{i(E - i\epsilon)} = 2\pi\delta(E) \,, \qquad (2.9)$$

one can show that, unless the matrix element $\langle \beta_n | a_{ij}(t) | \alpha_m \rangle$ and the other matrix elements vanish at $t \to +\infty$, it is

$$\lim_{t \to -\infty} a_{ij}(t) \neq \lim_{t \to +\infty} a_{ij}(t) \,. \qquad (2.10)$$

Defining the annihilation (wave-packet) operators of outgoing particles by

$$\lim_{t \to +\infty} a_{ij}(t) = Z_{ij}^{1/2} \alpha_{out,j} \,, \qquad \text{no summation on } j, \qquad (2.11)$$

Eqs. (2.8) and (2.10) then mean

$$\alpha_{in,j} \neq \alpha_{out,j} \,, \qquad (2.12)$$

which shows that in-fields are different from out-fields and therefore that there exists interaction, confirming that our definition of free particles does not mean that there is no interaction. In-fields (out-fields) are the asymptotic limits of the interpolating fields and, as already said, we assume that in-fields and out-fields are related by the unitary S-matrix operator S:

$$S\alpha_{in,j}S^{-1} = \alpha_{out,j} \,, \qquad (2.13a)$$

$$S\phi_{in,j}(x)S^{-1} = \phi_{out,j}(x) \,. \qquad (2.13b)$$

Thus, S is different from the identity operator when and only when Eq. (2.12) holds. Under the S-matrix operation the vacuum $|0\rangle$ is assumed to be stable, and then the one-particle state $|\alpha_{in,j}\rangle$ goes into the one-particle state $|\alpha_{out,j}\rangle$:

$$S|0\rangle = |0\rangle = S^{-1}|0\rangle\,, \quad S|\alpha_{in,j}\rangle = |\alpha_{out,j}\rangle\,, \qquad (2.14)$$

i.e., no reactions occur in vacuum or single particle state. The fact that the S-matrix elements only involve incoming and outgoing fields makes a specific choice of the interpolating fields immaterial in the theory [619].

Let us note that there is not necessarily a one-to-one correspondence between the set $\{\psi_i\}$ and the sets $\{\phi_{in,j}\}$ and/or $\{\phi_{out,j}\}$. Assume for example that the set of Heisenberg fields has n members and the set of out-fields has m members. It could happen that the asymptotic limit $t \to -\infty$ of $(\psi_i(x) - \chi_i)$ gives us a set of p in-field, with $p < m$. This means that for some values of the index i the asymptotic limit $t \to -\infty$ vanishes. In such a circumstance we say that we have p elementary particles and $m - p$ composite particles [617, 619, 644]. We see then that composite particle fields do not appear in the linear part of the map (cf. Eq. (2.6)).

2.3   Examples of inequivalent representations

As already observed, the time dependence of the Heisenberg fields is controlled by the physical fields in each term of the map. This implies that the coefficients of the dynamical map, apart from their retarded or advanced nature discussed in Section 2.1, must be time-independent. By using the free Hamiltonian (1.126) one can prove [617, 619, 644] that:

$$\langle a|[H_0, \psi_i(x)]|b\rangle = \langle a|\frac{1}{i}\frac{\partial}{\partial t}\psi_i(x)|b\rangle\,, \qquad (2.15)$$

where $|a\rangle$ and $|b\rangle$ are any two vectors belonging to a set D dense in $\mathcal{H}_F$. On the other hand, we can always introduce the operator H written in terms of the Heisenberg fields $\psi_i(x)$, called the Heisenberg Hamiltonian operator, such that

$$\langle a|[H, \psi_i(x)]|b\rangle = \langle a|\frac{1}{i}\frac{\partial}{\partial t}\psi_i(x)|b\rangle\,. \qquad (2.16)$$

By comparing (2.15) and (2.16), we can write

$$H = H_0 + W_0\,, \qquad (2.17)$$

with W_0 a c-number constant. In deriving (2.17) we have used the fact that any quantity that commutes with the irreducible set $\{\phi_{in,j}\}$ must be a c-number constant. We stress that Eq. (2.17), where H is written in terms of ψ_i and H_0 written in terms of $\phi_{in,j}$, is a direct consequence of the dynamical map and it is a weak equality. Moreover, due to Eq. (2.12), Eq. (2.17) does not imply that there is no interaction.

In order to clarify the role of the existence of infinitely many unitarily inequivalent representations in connection with the Hamiltonian operator we discuss now a couple of explicit examples.

Example 1

We start with the so-called van Hove model [625, 626]:

$$H = \int d^3k [\omega_k a_\mathbf{k}^\dagger a_\mathbf{k} + \nu_\mathbf{k}(a_\mathbf{k} + a_\mathbf{k}^\dagger)], \qquad (2.18)$$

with $a_\mathbf{k}$ boson operators and $\nu_\mathbf{k}$ c-numbers for any $\mathbf{k}$; ω_k is assumed to depend only on the modulus k of $\mathbf{k}$. The hermiticity of H implies that ω_k and $\nu_\mathbf{k}$ are real. Unless $\nu_\mathbf{k} = c\delta(\mathbf{k})$, with c a real c-number, the Hamiltonian (2.18) is not translational invariant (see the comment after Eq. (1.79)). The commutation relations are

$$[a_\mathbf{k}, a_\mathbf{l}^\dagger] = \delta(\mathbf{k} - \mathbf{l}), \qquad (2.19)$$

and all other commutators are zero. We denote by $|0\rangle$ the vacuum state belonging to the Heisenberg field Hilbert space $\mathcal{H}$ and annihilated by $a_\mathbf{k}$:

$$a_\mathbf{k}|0\rangle = 0. \qquad (2.20)$$

$\mathcal{H}$ is an irreducible representation of the canonical commutation relations (2.19). We now show that there exist a representation of (2.19) where the Hamiltonian (2.18) assumes the form (2.17), namely,

$$H = \int d^3k E_k \alpha_\mathbf{k}^\dagger \alpha_\mathbf{k} + W_0. \qquad (2.21)$$

The commutation relations for the α operators are

$$[\alpha_\mathbf{k}, \alpha_\mathbf{l}^\dagger] = \delta(\mathbf{k} - \mathbf{l}), \qquad [\alpha_\mathbf{k}, \alpha_\mathbf{l}] = 0 = [\alpha_\mathbf{k}^\dagger, \alpha_\mathbf{l}^\dagger], \qquad (2.22)$$

and all other commutators are zero. In other words, we show that there exists the Fock space for the physical states. We will also see that there are conditions under which such a representation belongs to the set of the infinitely many unitarily inequivalent representations of (2.19).

It is easily seen that H given by (2.18) can be expressed as in (2.21) by use of the transformation

$$a_\mathbf{k} \to \alpha_\mathbf{k} = a_\mathbf{k} - g_\mathbf{k}, \qquad (2.23)$$

provided $g_\mathbf{k}$ is chosen to be the real c-number

$$g_\mathbf{k} = -\frac{\nu_\mathbf{k}}{\omega_k}, \tag{2.24}$$

for $\omega_k \neq 0$ for all k. The Hamiltonian (2.18) assumes then the form (2.21) (cf. (2.17)) with E_k and W_0 given by

$$E_k = \omega_k, \qquad W_0 = -\int d^3 k \frac{\nu_\mathbf{k}^2}{\omega_k}, \tag{2.25}$$

respectively, for $\omega_k \neq 0$ for all k.

Eq. (2.23) provides a very simple example of dynamical map: $a_\mathbf{k} = \alpha_\mathbf{k} + g_\mathbf{k}$. It is a canonical transformation since it leaves the canonical commutation relations unchanged (cf. Eqs. (2.19) and (2.22)) and it can also be regarded as an example of Bogoliubov transformation. It can be formally induced by the operator (cf. (1.65))

$$G(g) = \exp\left(\int d^3 k g_\mathbf{k}(a_\mathbf{k} - a_\mathbf{k}^\dagger)\right), \tag{2.26}$$

$$\alpha_\mathbf{k} \equiv \alpha_\mathbf{k}(g) = G^{-1}(g)a_\mathbf{k}G(g) = a_\mathbf{k} - g_\mathbf{k}. \tag{2.27}$$

Eqs. (1.61) and (1.62) hold for the operator $G(g)$ in (2.26).

Clearly, $\alpha_\mathbf{k}(g)$ acts as the annihilation operator on the state

$$|0(g)\rangle \equiv G^{-1}(g)|0\rangle, \qquad \alpha_\mathbf{k}(g)|0(g)\rangle = 0. \tag{2.28}$$

However, it does not annihilate $|0\rangle$. Similarly, $a_\mathbf{k}$ does not annihilate $|0(g)\rangle$. Note that

$$\langle 0(g)|0(g)\rangle = 1. \tag{2.29}$$

We recall that the operators $\alpha_\mathbf{k}$, $\alpha_\mathbf{k}^\dagger$, do not map normalizable vectors into normalizable vectors and thus we need to introduce smeared-out operators like in Eqs. (1.1), in terms of which wave-packet states are constructed in the Fock space $\mathcal{H}_F(g)$ for the physical states by using $|0(g)\rangle$ as the vacuum. In conclusion, we now have another representation, $\mathcal{H}_F(g)$, of the canonical commutation relations (2.19): $\mathcal{H}$ and $\mathcal{H}_F(g)$ are unitarily equivalent representations provided $G^{-1}(g)$ is a unitary operator. We now show that in the infinite volume limit (i.e., in QFT) this is not the case for some choices of $g_\mathbf{k}$.

The discussion is similar to the one presented in Section 1.6. By use of the Baker–Hausdorff formula

$$\exp A \exp B = \exp\left\{A + B + \frac{1}{2}[A, B] + \frac{1}{12}[A[A, B]] + \dots\right\}, \tag{2.30}$$

choosing

$$A = \int d^3k \, g_{\mathbf{k}} a_{\mathbf{k}}^\dagger, \qquad B = -\int d^3k \, g_{\mathbf{k}} a_{\mathbf{k}}, \qquad (2.31)$$

we can write

$$G^{-1}(g) = \exp\left(-\frac{1}{2}\int d^3k \, g_{\mathbf{k}}^2\right) \exp\left(\int d^3k g_{\mathbf{k}} a_{\mathbf{k}}^\dagger\right) \exp\left(-\int d^3k g_{\mathbf{k}} a_{\mathbf{k}}\right),$$

$$(2.32)$$

and thus

$$|0(g)\rangle = \exp\left(-\frac{1}{2}\int d^3k g_{\mathbf{k}}^2\right) \exp\left(\int d^3k g_{\mathbf{k}} a_{\mathbf{k}}^\dagger\right)|0\rangle, \qquad (2.33)$$

which shows that the inner product of the "new" vacuum state $|0(g)\rangle$ with the "old" vacuum $|0\rangle$ is

$$\langle 0|0(g)\rangle = \exp\left(-\frac{1}{2}\int d^3k g_{\mathbf{k}}^2\right). \qquad (2.34)$$

As we know (cf. Section 1.6), this is zero, provided

$$\frac{1}{2}\int d^3k \, g_{\mathbf{k}}^2 = \infty. \qquad (2.35)$$

For instance, this happens for a translationally invariant system, i.e., for $g_{\mathbf{k}} = g\delta(\mathbf{k})$, with g a real c-number constant. Another situation where (2.35) occurs is when the system has spatial periodicity, e.g., a crystal. The wave function then has Fourier components for discrete values of $\mathbf{k}$,

$$\mathbf{k} = \sum_{i=1}^{3} n_i \ell_i, \qquad n_i = 0, 1, 2, \ldots, \qquad (2.36)$$

where the vectors ℓ_i denote the three directions of the periodicity. In this case, $g_{\mathbf{k}}$ must take the form

$$g_{\mathbf{k}} = \sum_{n_i} g_{n_i} \delta(\mathbf{k} - \sum_i n_i \ell_i), \qquad (2.37)$$

which again gives (2.35).

We observe that when (2.35) is satisfied Eq. (2.33) holds only formally. As already observed in the previous Chapter, we can think of $|0(g)\rangle$ as a state where many a-bosons are condensed, namely, where boson condensation occurs. If we want the condensation to have locally observable effects, we need a non-vanishing density of condensed bosons. The quantity

$$\langle 0(g)|a_{\mathbf{k}}^\dagger a_{\mathbf{k}}|0(g)\rangle = g_{\mathbf{k}}^2, \qquad (2.38)$$

gives the number of condensed bosons of momentum $\mathbf{k}$.

Suppose we consider another set of c-numbers $\{g_{\mathbf{k}} = g'\delta(\mathbf{k}), \forall\, \mathbf{k}\}$ with $g' \neq g$. We then have

$$\langle 0(g')|0(g)\rangle \to 0, \qquad g' \neq g \quad \text{and for } V \to \infty. \tag{2.39}$$

Thus, in the infinite volume limit the representations $\mathcal{H}_F(g)$ and $\mathcal{H}_F(g')$ are unitarily inequivalent; moreover, no unitary generator $G^{-1}(g)$ exists which maps $\mathcal{H}$ onto itself in the infinite volume limit.

In conclusion, provided the condition (2.35) is satisfied, the space $\mathcal{H}_F(g)$ is unitarily inequivalent to the original space $\mathcal{H}$ when the system is translationally invariant (infinite volume). The vacuum $|0(g)\rangle$ cannot be then expressed as a superposition of states belonging to $\mathcal{H}$ (or to $\mathcal{H}_F(g')$, $g' \neq g$). Since the set of c-numbers $\{g_{\mathbf{k}}, \forall\, \mathbf{k}\}$ span a continuous domain, by different choices of such a set we can construct infinitely many unitarily inequivalent representations $\mathcal{H}_F(g)$.

We finally remark that different choices of the set of c-numbers g may represent different choices of the values of the quantities $\nu_{\mathbf{k}}$ (cf. Eq. (2.24)) in the model Hamiltonian (2.18). Different sets of physical field operators $\alpha_{\mathbf{k}}(g)$ (cf. Eq. (2.23)) then correspond to these different choices of g. In full generality, we thus realize that the same dynamics (same operatorial form of the Heisenberg Hamiltonian, and thus of the field equations) may be realized into different representations of the physical fields, namely the physical system may exhibit different phases with different physical content (normal or superconducting, ferromagnetic or non-ferromagnetic, etc.).

Example 2a

We now consider another example of bosonic Hamiltonian, namely,

$$H = \int d^3k [\omega_k(a_{\mathbf{k}}^\dagger a_{\mathbf{k}} + b_{\mathbf{k}}^\dagger b_{\mathbf{k}}) + \nu_k(a_{\mathbf{k}} b_{-\mathbf{k}} + b_{-\mathbf{k}}^\dagger a_{\mathbf{k}}^\dagger). \tag{2.40}$$

The hermiticity of H implies that ω_k and ν_k are real. We assume that $\omega_k^2 > \nu_k^2$. The commutation relations are the usual ones for boson operators

$$[a_{\mathbf{k}}, a_{\mathbf{l}}^\dagger] = \delta(\mathbf{k} - \mathbf{l}), \qquad [b_{\mathbf{k}}, b_{\mathbf{l}}^\dagger] = \delta(\mathbf{k} - \mathbf{l}), \tag{2.41}$$

and all other commutators are zero. We proceed as in the previous example. Let $|0\rangle$ denote the vacuum state belonging to $\mathcal{H}$ annihilated by $a_{\mathbf{k}}$ and $b_{\mathbf{k}}$

$$a_{\mathbf{k}}|0\rangle = 0 = b_{\mathbf{k}}|0\rangle. \tag{2.42}$$

$\mathcal{H}$ is an irreducible representation of the canonical commutation relations (2.41). Also in the present case we can show that there exists a representation of (2.41) where the Hamiltonian H assumes the form (2.17), namely,

$$H = \int d^3k\, E_k(\alpha_{\mathbf{k}}^\dagger \alpha_{\mathbf{k}} + \beta_{\mathbf{k}}^\dagger \beta_{\mathbf{k}}) + W_0. \tag{2.43}$$

The commutation relations for the α operators are

$$[\alpha_{\mathbf{k}}, \alpha_{\mathbf{l}}^{\dagger}] = \delta(\mathbf{k} - \mathbf{l}), \qquad [\beta_{\mathbf{k}}, \beta_{\mathbf{l}}^{\dagger}] = \delta(\mathbf{k} - \mathbf{l}), \qquad (2.44)$$

and all other commutators are zero. Such a representation, which we denote by $\mathcal{H}_F(\theta)$, is the Fock space for the physical states.

The Hamiltonian H given by (2.40) can be expressed in the form of (2.43) by use of the relations

$$a_{\mathbf{k}} = \alpha_{\mathbf{k}} \cosh \theta_k + \beta_{-\mathbf{k}}^{\dagger} \sinh \theta_k, \qquad (2.45a)$$

$$b_{\mathbf{k}} = \beta_{\mathbf{k}} \cosh \theta_k + \alpha_{-\mathbf{k}}^{\dagger} \sinh \theta_k, \qquad (2.45b)$$

provided

$$\cosh 2\theta_k = \frac{\omega_k}{\sqrt{\omega_k^2 - \nu_k^2}}, \qquad (2.46a)$$

$$\sinh 2\theta_k = -\frac{\nu_k}{\sqrt{\omega_k^2 - \nu_k^2}}, \qquad (2.46b)$$

and with E_k and W_0 given by

$$E_k = \sqrt{\omega_k^2 - \nu_k^2}, \qquad (2.47a)$$

$$W_0 = \int d^3 k \left(\sqrt{\omega_k^2 - \nu_k^2} - \omega_k \right), \qquad (2.47b)$$

respectively. Eqs. (2.45) provide another simple example of dynamical map. They are Bogoliubov transformations and are canonical transformations since they preserve the commutation relations. From Eqs. (2.45) we have

$$\alpha_{\mathbf{k}} \equiv \alpha_{\mathbf{k}}(\theta) = G^{-1}(\theta) a_{\mathbf{k}} G(\theta) = a_{\mathbf{k}} \cosh \theta_k - b_{-\mathbf{k}}^{\dagger} \sinh \theta_k, \qquad (2.48a)$$

$$\beta_{\mathbf{k}} \equiv \beta_{\mathbf{k}}(\theta) = G^{-1}(\theta) b_{\mathbf{k}} G(\theta) = b_{\mathbf{k}} \cosh \theta_k - a_{-\mathbf{k}}^{\dagger} \sinh \theta_k, \qquad (2.48b)$$

where

$$G(\theta) = \exp \left[\int d^3 k\, \theta_k (a_{\mathbf{k}} b_{-\mathbf{k}} - b_{-\mathbf{k}}^{\dagger} a_{\mathbf{k}}^{\dagger}) \right]. \qquad (2.49)$$

The vacuum state annihilated by $\alpha_{\mathbf{k}}(\theta)$ is

$$|0(\theta)\rangle \equiv G^{-1}(\theta)|0\rangle, \qquad \alpha_{\mathbf{k}}(\theta)|0(\theta)\rangle = 0, \qquad (2.50)$$

and

$$\langle 0(\theta)|0(\theta)\rangle = 1. \qquad (2.51)$$

However, $\alpha_{\mathbf{k}}(\theta)$ does not annihilate $|0\rangle$. In Appendix F we show that

$$|0(\theta)\rangle = \exp \left(-\frac{V}{(2\pi)^3} \int d^3 k \ln \cosh \theta_k \right) \exp \left(\int d^3 k \tanh \theta_k a_{\mathbf{k}}^{\dagger} b_{-\mathbf{k}}^{\dagger} \right) |0\rangle,$$

$$(2.52)$$

where V is the volume enclosing the system. Unless $\theta_k = 0$ for all k, all the states of $\mathcal{H}$ are orthogonal to those of $\mathcal{H}_F(\theta)$ for $V \to \infty$:

$$\langle 0|0(\theta)\rangle \to 0 \quad \text{for} \quad V \to \infty, \tag{2.53}$$

$$\langle 0|a_i b_j|0(\theta)\rangle \to 0 \quad \text{for} \quad V \to \infty, \tag{2.54}$$

etc. Thus we have another example where the physical state space $\mathcal{H}_F(\theta)$, with $\theta = \{\theta_k\}$ given by Eqs. (2.46), is unitarily inequivalent in the infinite volume limit to the space of the Heisenberg fields $\mathcal{H}$. Again, we can adopt different θ_k values, corresponding to different values in the parameter ν_k (cf. Eqs. (2.46)) appearing in the Hamiltonian and thus to different sets of physical field operators (2.48). For $\mathcal{H}_F(\theta)$ and $\mathcal{H}_F(\theta')$ we have

$$\langle 0(\theta')|0(\theta)\rangle \to 0, \qquad \forall \theta \neq \theta' \text{ and for } V \to \infty, \tag{2.55}$$

etc. We see that $\mathcal{H}$, $\mathcal{H}_F(\theta)$ and $\mathcal{H}_F(\theta')$, for any $\theta \neq \theta'$, are unitarily inequivalent representations in the infinite volume limit.

We remark that the result does not depend on the functional dependence of θ_k on k (except for the possibility that $\theta_k = 0$ for all values of k).

The number of $a_{\mathbf{k}}$ and $b_{\mathbf{k}}$ modes condensed in the state $|0(\theta)\rangle$ is:

$$\langle 0(\theta)|a_{\mathbf{k}}^\dagger a_{\mathbf{k}}|0(\theta)\rangle = \langle 0(\theta)|b_{\mathbf{k}}^\dagger b_{\mathbf{k}}|0(\theta)\rangle = \sinh^2 \theta_k. \tag{2.56}$$

The fact that in the infinite volume limit Eq. (2.52) is only a formal expression is rich in physical meaning: it says that $|0(\theta)\rangle$ is not a state of the Fock space to which $|0\rangle$ belongs. Therefore, one cannot expand the vacuum $|0(\theta)\rangle$ in terms of $|0\rangle$ (and, correspondingly, states constructed on $|0(\theta)\rangle$ in terms of those constructed on $|0\rangle$, as Eq. (2.53) shows): Fock spaces labeled by different θ values are unitarily inequivalent; they provide representations of physically different phases of the system. This situation arises from the fact that $G^{-1}(\theta)$ in general transforms vectors in the Fock space into vectors with infinite number of quanta a and b, i.e., elements of the general non-separable Hilbert space examined in Chapter 1. The different Fock spaces obtained by the transformations for different θ values are separable subspaces of the larger non-separable space. They are orthogonal to each other in the infinite volume limit. If, on the contrary, the number of degrees of freedom is finite, we are led to separable Hilbert spaces, which, due to the von Neumann theorem, are unitarily equivalent to each other.

Example 2b

We can treat in a similar way the case of the Hamiltonian (2.40) for fermion operators. In this case, we have the anticommutation relations:

$$\{a_{\mathbf{k}}, a_{\mathbf{l}}^\dagger\} = \delta(\mathbf{k} - \mathbf{l}) = \{\beta_{\mathbf{k}}, \beta_{\mathbf{l}}^\dagger\}, \tag{2.57}$$

and all other anticommutators vanishing.

The Hamiltonian (2.40) can again be written in the form (2.43) characterizing physical states. This is achieved by use of

$$a_{\mathbf{k}} = \alpha_{\mathbf{k}} \cos\theta_k + \beta_{-\mathbf{k}}^{\dagger} \sin\theta_k \,, \tag{2.58a}$$

$$b_{\mathbf{k}} = \beta_{\mathbf{k}} \cos\theta_k - \alpha_{-\mathbf{k}}^{\dagger} \sin\theta_k \,, \tag{2.58b}$$

with

$$\cos 2\theta_k = \frac{\omega_k}{\sqrt{\omega_k^2 + \nu_k^2}} \,, \tag{2.59a}$$

$$\sin 2\theta_k = \frac{\nu_k}{\sqrt{\omega_k^2 + \nu_k^2}} \,. \tag{2.59b}$$

In Eq. (2.43) E_k and W_0 are now given by

$$E_k = \sqrt{\omega_k^2 + \nu_k^2} \,, \tag{2.60a}$$

$$W_0 = \int d^3 k \left(\omega_k - \sqrt{\omega_k^2 + \nu_k^2} \right) . \tag{2.60b}$$

By inverting Eqs. (2.58) we have the Bogoliubov transformations:

$$\alpha_{\mathbf{k}} \equiv \alpha_{\mathbf{k}}(\theta) = G^{-1}(\theta) a_{\mathbf{k}} G(\theta) = a_{\mathbf{k}} \cos\theta_k - b_{-\mathbf{k}}^{\dagger} \sin\theta_k \,, \tag{2.61a}$$

$$\beta_{\mathbf{k}} \equiv \beta_{\mathbf{k}}(\theta) = G^{-1}(\theta) b_{\mathbf{k}} G(\theta) = b_{\mathbf{k}} \cos\theta_k + a_{-\mathbf{k}}^{\dagger} \sin\theta_k \,, \tag{2.61b}$$

with

$$G(\theta) = \exp[- \int d^3 k \theta_k (a_{\mathbf{k}} b_{-\mathbf{k}} - b_{-\mathbf{k}}^{\dagger} a_{\mathbf{k}}^{\dagger})] \,, \tag{2.62}$$

and

$$|0(\theta)\rangle = \prod_{\mathbf{k}} \left(\cos\theta_k + \sin\theta_k a_{\mathbf{k}}^{\dagger} b_{-\mathbf{k}}^{\dagger} \right) |0\rangle \,. \tag{2.63}$$

Eqs. (2.61) are canonical transformations since they preserve the anticommutation relations (2.57). The number or $a_{\mathbf{k}}$ and $b_{\mathbf{k}}$ modes condensed in the state $|0(\theta)\rangle$ is:

$$\langle 0(\theta) | a_{\mathbf{k}}^{\dagger} a_{\mathbf{k}} | 0(\theta) \rangle = \langle 0(\theta) | b_{\mathbf{k}}^{\dagger} b_{\mathbf{k}} | 0(\theta) \rangle = \sin^2\theta_k \,. \tag{2.64}$$

Also in the fermion case we can generalize to any $\theta' \neq \theta$ and we find that $\mathcal{H}$, $\mathcal{H}_F(\theta)$ and $\mathcal{H}_F(\theta')$, are unitarily inequivalent representations in the infinite volume limit. Note that (2.63) is the superconducting ground state in the BCS theory of superconductivity [54, 536].

In Section 2.7 we present a further example of inequivalent representations, which is of great physical relevance, namely the mixing of fields (fermionic or bosonic) with different masses.

We close this Section by observing that the generator $G(\theta)$ in Eq. (2.49) (and Eq. (2.62)) is the same as the counter-rotating term operator appearing in Hamiltonian models in quantum optics. Such a term is neglected in the rotating wave approximation. In our discussion, we have seen, however, that it may play a crucial role since it is related to the condensate structure of the vacuum and to the existence of unitarily inequivalent representations. See [397, 398] for a more detailed discussion on this subject.

2.4 The Haag theorem and non-perturbative physics

In the examples presented above we have seen that, due to the unitary inequivalence among the spaces $\mathcal{H}$, $\mathcal{H}_F(\theta)$ and $\mathcal{H}_F(\theta')$, with $\theta \neq \theta'$ (or $\mathcal{H}$, $\mathcal{H}_F(g)$ and $\mathcal{H}_F(g')$, with $g \neq g'$ for translationally invariant systems), equations such as (2.33), (2.52) and (2.63) are only formal relations: in the limit of infinite volume any given state belonging to one specific representation cannot be expanded, at any order, into a superposition of states of another inequivalent representation. This is the familiar situation occurring, for example, in the study of superconductors [619], where the superconducting state cannot be reached by perturbation techniques starting from the non-superconducting (normal) state: the normal and the superconducting states are orthogonal states belonging to different physical *phases* of the system. Once the limit to the infinite volume has been performed, and thus we fully operate in QFT, then there is no unitary operator connecting the Heisenberg field space $\mathcal{H}$ with representations indexed by different θ values (or g values), neither connecting these among themselves. In particular, this means that the limits $V \to \infty$ and $\theta \to 0$ are non-commuting limits, so that although $\lim_{V \to \infty} \lim_{\theta \to 0} |0(\theta)\rangle = |0\rangle$, it is

$$\lim_{\theta \to 0} \lim_{V \to \infty} |0(\theta)\rangle \neq |0\rangle, \qquad (2.65)$$

where $\theta \to 0$ denotes $\theta_k \to 0$, $\forall k$ (we could also consider $g \to 0$ with $g_{\mathbf{k}} \to 0$, $\forall \mathbf{k}$ in the case of Example 1). Due to (2.46) (or to (2.24)), $\theta \to 0$ (or $g \to 0$) means the limit of the coupling $\nu_k \to 0$, $\forall k$, in the Hamiltonian (2.40) (or in the Hamiltonian (2.18)). Thus we see that *at infinite V* the eigenstates $|0(\theta)\rangle$ ($|0(g)\rangle$) of the Hamiltonian (2.43) (or (2.21)) do not go into the eigenstates of the "free" Hamiltonian obtained from (2.40) (or (2.18)) in the limit of the coupling constant going to zero. This is the statement of the so-called Haag theorem [114, 202, 308, 558, 569, 579].

2.5 The momentum operator

The momentum operator for the physical particles has been given in Eq. (1.127). In all the examples presented above, by feeding into Eq. (1.127) the appropriate Bogoliubov transformation, we find (we are now replacing in Eq. (1.127) the notation $\mathbf{P_0}$ with $\mathbf{P}$)

$$\mathbf{P} = \sum_s \int d^3k \; \mathbf{k}(a_{\mathbf{k}}^{s\dagger}a_{\mathbf{k}}^s + b_{\mathbf{k}}^{s\dagger}b_{\mathbf{k}}^s) \,, \tag{2.66}$$

i.e., the momentum $\mathbf{P}$ is diagonal in both the old and the new space. This is a general feature. We can understand this by considering the wave-packet state characterized by

$$f(\mathbf{x}) = \int \frac{d^3k}{(2\pi)^3} f(\mathbf{k}) e^{i\mathbf{k}\cdot\mathbf{x}} \,, \tag{2.67}$$

and the corresponding creation operator

$$a_f^{s\dagger} = \frac{1}{(2\pi)^{3/2}} \int d^3k f(\mathbf{k}) a_{\mathbf{k}}^{s\dagger} \,. \tag{2.68}$$

If the state is translated in space by a constant quantity ϵ, then

$$f'(\mathbf{x}) \equiv f(\mathbf{x}+\epsilon) = \int \frac{d^3k}{(2\pi)^3} f(\mathbf{k}) e^{i\mathbf{k}\cdot(\mathbf{x}+\epsilon)} \,, \tag{2.69}$$

so that $a_f^{s\dagger}$ goes into

$$a_{f'}^{s\dagger} = \frac{1}{(2\pi)^{3/2}} \int d^3k f(\mathbf{k}) e^{i\mathbf{k}\cdot\epsilon} a_{\mathbf{k}}^{s\dagger} \,. \tag{2.70}$$

But this translation is known to be generated by

$$a_{f'}^{s\dagger} = e^{i\mathbf{P}\cdot\epsilon} a_f^{s\dagger} e^{-i\mathbf{P}\cdot\epsilon} \,. \tag{2.71}$$

Thus we get

$$e^{i\mathbf{P}\cdot\epsilon} a_{\mathbf{k}}^{s\dagger} e^{-i\mathbf{P}\cdot\epsilon} = e^{i\mathbf{k}\cdot\epsilon} a_{\mathbf{k}}^{s\dagger} \,, \tag{2.72}$$

i.e., to the first order in ϵ

$$[\mathbf{P}, a_{\mathbf{k}}^{s\dagger}] = \mathbf{k} a_{\mathbf{k}}^{s\dagger} \,, \qquad\qquad [\mathbf{P}, a_{\mathbf{k}}^s] = -\mathbf{k} a_{\mathbf{k}}^s \,. \tag{2.73}$$

From these we conclude that $\mathbf{P}$ must have the form

$$\mathbf{P} = \sum_s \int d^3k \; \mathbf{k}(a_{\mathbf{k}}^{s\dagger}a_{\mathbf{k}}^s + b_{\mathbf{k}}^{s\dagger}b_{\mathbf{k}}^s) + c \,, \tag{2.74}$$

where c is commuting with $a_{\mathbf{k}}^s$ and $a_{\mathbf{k}}^{s\dagger}$, and therefore is a c-number constant, since $\mathcal{H}_F$ is an irreducible representation of $a_{\mathbf{k}}^s$ and $a_{\mathbf{k}}^{s\dagger}$. Moreover, since the Bogoliubov transformations are linear transformations, c must be a c-number constant also in the representation $\mathcal{H}_F(\theta)$. However, the constant c must be zero since $\mathbf{P}$ must annihilate the vacuum state $|0\rangle$ to give the correct translation properties to the states ($|0\rangle$ has to be translationally invariant). Thus we recover the form (2.66) for the momentum operator.

In conclusion, since the momentum operator is related with the kinematics, we see that the kinematical properties do not depend on the representation where we realize the dynamics, i.e., the Hamiltonian operator.

2.6   Time evolution and asymptotic limits

Time evolution of a generic operator $A(t)$ is controlled by the Hamiltonian H through the Heisenberg equation

$$\frac{d}{dt}A(t) = -i[A(t), H]. \tag{2.75}$$

It is instructive to consider one of the above examples, e.g., the Hamiltonian (2.40) (Example 2a). We have the Heisenberg equations:

$$\frac{d}{dt}a_{\mathbf{k}}(t) = -i[\omega_k a_{\mathbf{k}}(t) + \nu_k b^{\dagger}_{-\mathbf{k}}(t)], \tag{2.76a}$$

$$\frac{d}{dt}b_{\mathbf{k}}(t) = -i[\omega_k b_{\mathbf{k}}(t) + \nu_k a^{\dagger}_{-\mathbf{k}}(t)], \tag{2.76b}$$

where $k = |\mathbf{k}|$. These equations can be solved by considering the Heisenberg equations for $\alpha_{\mathbf{k}}(t)$ and $\beta_{\mathbf{k}}(t)$. Use of the Hamiltonian (2.43) gives

$$\frac{d}{dt}\alpha_{\mathbf{k}}(t) = -iE_k\alpha_{\mathbf{k}}(t), \tag{2.77a}$$

$$\frac{d}{dt}\beta_{\mathbf{k}}(t) = -iE_k\beta_{\mathbf{k}}(t). \tag{2.77b}$$

Integration under the conditions $\alpha_{\mathbf{k}} = \alpha_{\mathbf{k}}(0)$, $\beta_{\mathbf{k}} = \beta_{\mathbf{k}}(0)$ leads to

$$\alpha_{\mathbf{k}}(t) = e^{-iE_k t}\alpha_{\mathbf{k}}, \quad \beta_{\mathbf{k}}(t) = e^{-iE_k t}\beta_{\mathbf{k}}. \tag{2.78}$$

Use of the Bogoliubov transformations (2.45) then gives

$$a_{\mathbf{k}}(t) = e^{-iE_k t}\alpha_{\mathbf{k}}\cosh\theta_k + e^{iE_k t}\beta^{\dagger}_{-\mathbf{k}}\sinh\theta_k, \tag{2.79a}$$

$$b_{\mathbf{k}}(t) = e^{-iE_k t}\beta_{\mathbf{k}}\cosh\theta_k + e^{iE_k t}\alpha^{\dagger}_{-\mathbf{k}}\sinh\theta_k, \tag{2.79b}$$

which are the solutions of (2.76), with the choice (2.46) for θ_k.

The procedure outlined above provides a very simple example of a general method of solution of the Heisenberg equations for the interacting fields, called the self-consistent method [619]. Summarizing: the dynamics is defined by a certain Hamiltonian (in the previous case Eq. (2.40)) expressed in terms of the Heisenberg operators ($a_{\mathbf{k}}$ and $b_{\mathbf{k}}$). These operators satisfy the Heisenberg equations of motion (the Heisenberg or dynamical field equations). Solving the dynamics means to find solutions to these equations, namely the time dependence of the Heisenberg operators and their realization in the physical Hilbert space, i.e., in terms of the physical field operators $\alpha_{\mathbf{k}}$ and $\beta_{\mathbf{k}}$ such that the Hamiltonian takes the diagonal form (2.43). To do this, one writes a trial relation for the dynamical map to be determined by using the Heisenberg equations. In the example considered above this is the relation (2.79). It depends on the parameters θ_k. We may have many candidate sets of operators $\alpha_{\mathbf{k}}$ and $\beta_{\mathbf{k}}$, each set being specified by a particular choice of θ_k's. The choice (2.46) of θ_k's is the one determined by the requirement that the Hamiltonian acquires the diagonal form (2.43). The corresponding set of operators $\alpha_{\mathbf{k}}$ and $\beta_{\mathbf{k}}$ is the set of physical operators.

Eq. (2.79) is a linear mapping. In general, the dynamical map involves higher order products of the physical field operators. Time dependence of the Heisenberg field equations appears in general only through the time dependence of the physical field operators. The central idea is that the Hamiltonian must provide the total energy of the system as measured in the experiments, which, as we have seen (cf. Section 1.9), is obtained by adding the energies of the incoming (or outgoing) particles which are described by the physical fields.

Proceeding as above also in Example 2b with fermion fields, one obtains

$$a_{\mathbf{k}}(t) = e^{-iE_k t}\alpha_{\mathbf{k}}\cos\theta_k + e^{iE_k t}\beta_{-\mathbf{k}}^{\dagger}\sin\theta_k\,, \qquad (2.80a)$$

$$b_{\mathbf{k}}(t) = e^{-iE_k t}\beta_{\mathbf{k}}\cos\theta_k - e^{iE_k t}\alpha_{-\mathbf{k}}^{\dagger}\sin\theta_k\,. \qquad (2.80b)$$

We leave its derivation as an exercise for the reader. In the case of Example 1, use of the Hamiltonian (2.18) gives the Heisenberg equation

$$\frac{d}{dt}a_{\mathbf{k}}(t) = -i[\omega_k a_{\mathbf{k}}(t) + \nu_{\mathbf{k}}]\,, \qquad (2.81)$$

and, for $\omega_k \neq 0$ for all k, the dynamical map is given by (2.23):

$$a_{\mathbf{k}}(t) = \alpha_{\mathbf{k}}e^{-i\omega_k t} - \frac{\nu_{\mathbf{k}}}{\omega_k}\,. \qquad (2.82)$$

Let us now consider the limit for $t \to -\infty$ of the Heisenberg operators. We expect that the physical fields are recovered when $t \to -\infty$, i.e., at a time well before (or well after, when $t \to +\infty$) the interaction has occurred. Thus, we seek a Heisenberg operator $\tilde{a}_{\mathbf{k}}(t)$ such that

$$\lim_{t\to-\infty}\left(\tilde{a}_{\mathbf{k}}(t) - \alpha_{\mathbf{k}}\right) = 0 = \lim_{t\to-\infty}\left(\tilde{b}_{\mathbf{k}}(t) - \beta_{\mathbf{k}}\right)\,. \qquad (2.83)$$

Eq. (2.83) here denotes the so-called strong limit [7, 351, 537], i.e.,

$$\lim_{t \to -\infty} |(\tilde{a}_{\mathbf{k}}(t) - \alpha_{\mathbf{k}})\boldsymbol{\xi}| = \lim_{t \to -\infty} |(\tilde{b}_{\mathbf{k}}(t) - \beta_{\mathbf{k}})\boldsymbol{\xi}| = 0, \qquad (2.84)$$

where $\boldsymbol{\xi}$ is any vector of the Fock space with norm $|\boldsymbol{\xi}| \equiv (\boldsymbol{\xi}, \boldsymbol{\xi})^{1/2}$ (cf. Section 1.3). However, since $\alpha_{\mathbf{k}}$ and $\beta_{\mathbf{k}}$ are not bounded operators in the Fock space, we must consider wave-packet operators such as those of Eqs. (1.1). Then we consider the limit

$$\lim_{t \to -\infty} (\tilde{a}_f(t) - \alpha_f) = \lim_{t \to -\infty} (\tilde{b}_f(t) - \beta_f) = 0, \qquad (2.85)$$

which is now the weak limit (cf. Section 1.3).

The Heisenberg operator which weakly converges to the physical operator in the $t \to \pm\infty$ limit is called the interpolating operator (cf. Section 2.2).

The choice of the interpolating operator is in general not unique. For example, in the case of the Hamiltonian (2.40) one can show that

$$\tilde{a}_f(t) = i \int \frac{d^3k}{(2\pi)^{\frac{3}{2}} 2E_k} f(\mathbf{k}) e^{iE_k t} \overset{\leftrightarrow}{\frac{\partial}{\partial t}} \left[a_{\mathbf{k}}(t) \cosh\theta_k - b^\dagger_{-\mathbf{k}}(t) \sinh\theta_k \right], \quad (2.86a)$$

$$\tilde{b}_f(t) = i \int \frac{d^3k}{(2\pi)^{\frac{3}{2}} 2E_k} f(\mathbf{k}) e^{iE_k t} \overset{\leftrightarrow}{\frac{\partial}{\partial t}} \left[b_{\mathbf{k}}(t) \cosh\theta_k - a^\dagger_{-\mathbf{k}}(t) \sinh\theta_k \right], \quad (2.86b)$$

as well as

$$\tilde{a}_f(t) = i \int \frac{d^3k}{(2\pi)^{\frac{3}{2}} 2E_k \cosh\theta_k} f(\mathbf{k}) e^{iE_k t} \overset{\leftrightarrow}{\frac{\partial}{\partial t}} a_{\mathbf{k}}(t), \qquad (2.87a)$$

$$\tilde{b}_f(t) = i \int \frac{d^3k}{(2\pi)^{\frac{3}{2}} 2E_k \cosh\theta_k} f(\mathbf{k}) e^{iE_k t} \overset{\leftrightarrow}{\frac{\partial}{\partial t}} b_{\mathbf{k}}(t), \qquad (2.87b)$$

both have α_f and β_f, respectively, as weak limit. In the equations above we have used the notation:

$$g(t) \overset{\leftrightarrow}{\frac{\partial}{\partial t}} f(t) \equiv g(t) \frac{\partial f(t)}{\partial t} - \frac{\partial g(t)}{\partial t} f(t). \qquad (2.88)$$

One example in which the choice of the interpolating operator must be very carefully performed is the one of the van Hove model (2.18). The map

(2.23) suggests that we can use

$$\tilde{a}_f(t) = \int \frac{d^3k}{(2\pi)^{\frac{3}{2}}} f(\mathbf{k}) e^{i\omega_k t} a_\mathbf{k}(t)$$

$$= \int \frac{d^3k}{(2\pi)^{\frac{3}{2}}} f(\mathbf{k}) \alpha_\mathbf{k}(t) - \int \frac{d^3k}{(2\pi)^{\frac{3}{2}}} f(\mathbf{k}) \frac{\nu_\mathbf{k}}{\omega_k} e^{i\omega_k t}, \qquad (2.89)$$

since the second integral on the r.h.s. vanishes due to the Riemann–
Lebesgue lemma, provided $\nu_\mathbf{k}$ is a well-behaved function of k (recall that
$f(\mathbf{k})$ is a square-integrable function). However, when $\nu_\mathbf{k} = c\delta(\mathbf{k})$, with c a
real constant, Eq. (2.89) gives

$$\tilde{a}_f(t) = \alpha_f - \frac{c}{(2\pi)^{\frac{3}{2}}} \frac{f(0)}{\omega_0} e^{i\omega_0 t}, \qquad (2.90)$$

from which we see that the weak limit of $\tilde{a}_f(t)$ does not even exist because of
the oscillating factor $e^{i\omega_0 t}$. Therefore, $\tilde{a}_f(t)$ is not an interpolating operator.
In this case, the correct interpolating operator is

$$\tilde{a}_f(t) = \int \frac{d^3k}{(2\pi)^{\frac{3}{2}}} f(\mathbf{k}) \left[a_\mathbf{k}(t) + \frac{\nu_\mathbf{k}}{\omega_k} \right] e^{i\omega_k t}. \qquad (2.91)$$

2.7 Inequivalent representations in flavor mixing

As a further example of unitarily inequivalent representations, we discuss
here the quantization of mixed fields for both Dirac fermions and charged
bosons [83, 108]. Flavor mixing is an important phenomenon occurring in
particle physics; examples include neutrino mixing and oscillations, quark
mixing and meson mixing [156].

We will see how a consistent treatment of flavor mixing and oscillations
for relativistic quantum fields can be only achieved if the multiple Hilbert
space structure of QFT is properly taken into account. The usual treatment
of particle mixing and oscillations [156, 284] generally neglects this point,
thus leading necessarily to approximate results. Other discussions of flavor
oscillations in QFT can be found in [71, 74, 669].

In the following we limit ourselves to the case of two generations
(flavors), although the main results presented below have general validity [84, 313, 355]. See also [105] for the case of Majorana neutrinos and
neutral boson fields.

Some results of this Section are used in Appendix H for the study of entanglement in flavor neutrino states, again taking into account inequivalent
representations.

Quantization of mixed fields

Fermion mixing

For definiteness, let us consider Dirac neutrino fields, although the discussion is clearly valid for any Dirac fields. We denote by ν_e, ν_μ the neutrino fields with definite flavors and by ν_1, ν_2 the neutrino fields with definite masses m_1, m_2, respectively. The mixing relations are [77]

$$\nu_e(x) = \cos\theta\, \nu_1(x) + \sin\theta\, \nu_2(x)\,, \tag{2.92a}$$

$$\nu_\mu(x) = -\sin\theta\, \nu_1(x) + \cos\theta\, \nu_2(x)\,. \tag{2.92b}$$

Here θ is the mixing angle. The free fields ν_1 and ν_2 are expanded as [108]

$$\nu_i(x) = \frac{1}{\sqrt{V}} \sum_{\mathbf{k},r} \left[u_{\mathbf{k},i}^r(t)\alpha_{\mathbf{k},i}^r + v_{-\mathbf{k},i}^r(t)\beta_{-\mathbf{k},i}^{r\dagger} \right] e^{i\mathbf{k}\cdot\mathbf{x}}\,, \quad i = 1, 2\,, \tag{2.93}$$

where $u_{\mathbf{k},i}^r(t) = e^{-i\omega_{k,i}t}u_{\mathbf{k},i}^r$ and $v_{\mathbf{k},i}^r(t) = e^{i\omega_{k,i}t}v_{\mathbf{k},i}^r$, with $\omega_{k,i} = \sqrt{\mathbf{k}^2 + m_i^2}$. The $\alpha_{\mathbf{k},i}^r$ and the $\beta_{\mathbf{k},i}^r$ ($r = 1, 2$), are the annihilation operators for the vacuum state $|0\rangle_{1,2} \equiv |0\rangle_1 \otimes |0\rangle_2$: $\alpha_{\mathbf{k},i}^r|0\rangle_{1,2} = \beta_{\mathbf{k},i}^r|0\rangle_{1,2} = 0$. The anticommutation relations are:

$$\{\nu_i^\alpha(x), \nu_j^{\beta\dagger}(y)\}_{t=t'} = \delta^3(\mathbf{x} - \mathbf{y})\delta_{\alpha\beta}\delta_{ij}\,, \quad \alpha, \beta = 1, \ldots, 4\,, \tag{2.94a}$$

$$\{\alpha_{\mathbf{k},i}^r, \alpha_{\mathbf{q},j}^{s\dagger}\} = \delta_{\mathbf{k}\mathbf{q}}\delta_{rs}\delta_{ij}; \quad \{\beta_{\mathbf{k},i}^r, \beta_{\mathbf{q},j}^{s\dagger}\} = \delta_{\mathbf{k}\mathbf{q}}\delta_{rs}\delta_{ij}\,, \quad i, j = 1, 2\,. \tag{2.94b}$$

All other anticommutators are zero. The orthonormality and completeness relations are:

$$u_{\mathbf{k},i}^{r\dagger}u_{\mathbf{k},i}^s = v_{\mathbf{k},i}^{r\dagger}v_{\mathbf{k},i}^s = \delta_{rs}\,, \quad u_{\mathbf{k},i}^{r\dagger}v_{-\mathbf{k},i}^s = v_{-\mathbf{k},i}^{r\dagger}u_{\mathbf{k},i}^s = 0\,, \tag{2.95a}$$

$$\sum_r (u_{\mathbf{k},i}^r u_{\mathbf{k},i}^{r\dagger} + v_{-\mathbf{k},i}^r v_{-\mathbf{k},i}^{r\dagger}) = \mathbb{1}\,. \tag{2.95b}$$

We remark that mixing relations such as the relations (2.92) deserve a careful analysis, since they actually represent a dynamical mapping [108]. Let us therefore investigate the structure of the Fock spaces $\mathcal{H}_{1,2}$ and $\mathcal{H}_{e,\mu}$ relative to ν_1, ν_2 and ν_e, ν_μ, respectively. In particular we want to study the relation among these spaces in the infinite volume limit.

Our first step is the study of the generator of Eqs. (2.92) and of the underlying group theoretical structure. Eqs. (2.92) can be recast as [108]:

$$\nu_e^\alpha(x) = G_\theta^{-1}(t)\, \nu_1^\alpha(x)\, G_\theta(t)\,, \tag{2.96a}$$

$$\nu_\mu^\alpha(x) = G_\theta^{-1}(t)\, \nu_2^\alpha(x)\, G_\theta(t)\,, \tag{2.96b}$$

where $G_\theta(t)$ is given by

$$G_\theta(t) = \exp\left[\theta \int d^3x \left(\nu_1^\dagger(x)\nu_2(x) - \nu_2^\dagger(x)\nu_1(x)\right)\right], \qquad (2.97)$$

and is (at finite volume) a unitary operator: $G_\theta^{-1}(t) = G_{-\theta}(t) = G_\theta^\dagger(t)$, preserving the canonical anticommutation relations (2.94). Eq. (2.97) follows from $\frac{d^2}{d\theta^2}\nu_e^\alpha = -\nu_e^\alpha$ and $\frac{d^2}{d\theta^2}\nu_\mu^\alpha = -\nu_\mu^\alpha$ with the initial conditions $\nu_e^\alpha|_{\theta=0} = \nu_1^\alpha$, $\frac{d}{d\theta}\nu_e^\alpha|_{\theta=0} = \nu_2^\alpha$ and $\nu_\mu^\alpha|_{\theta=0} = \nu_2^\alpha$, $\frac{d}{d\theta}\nu_\mu^\alpha|_{\theta=0} = -\nu_1^\alpha$.

Note that G_θ is an element of $SU(2)$ since it can be written as

$$G_\theta(t) = \exp[\theta(S_+(t) - S_-(t))], \qquad (2.98a)$$

$$S_+(t) = S_-^\dagger(t) \equiv \int d^3x \, \nu_1^\dagger(x)\nu_2(x). \qquad (2.98b)$$

By introducing

$$S_3 \equiv \frac{1}{2} \int d^3x \left(\nu_1^\dagger(x)\nu_1(x) - \nu_2^\dagger(x)\nu_2(x)\right), \qquad (2.99)$$

the $su(2)$ algebra is closed (for t fixed):

$$[S_+(t), S_-(t)] = 2S_3 \quad, \quad [S_3, S_\pm(t)] = \pm S_\pm(t). \qquad (2.100)$$

The action of the mixing generator on the vacuum $|0\rangle_{1,2}$ is non-trivial and we have (at finite volume V):

$$|0(t)\rangle_{e,\mu} \equiv G_\theta^{-1}(t) |0\rangle_{1,2}. \qquad (2.101)$$

$|0(t)\rangle_{e,\mu}$ is the *flavor vacuum*, i.e., the vacuum for the flavor fields. Note that the above state is a $SU(2)$ coherent state [519] (see Appendix C). Note also that the flavor vacuum is time-dependent; we will comment more on this aspect in Appendix G and in Section 9.5.

Let us now investigate the infinite volume limit of Eq. (2.101). Using the Gaussian decomposition G_θ^{-1} is written as [519]

$$\exp[\theta(S_- - S_+)] = \exp(-\tan\theta S_+)\exp(-2\ln\cos\theta S_3)\exp(\tan\theta S_-), \qquad (2.102)$$

where $0 \le \theta < \frac{\pi}{2}$. We then compute $_{1,2}\langle 0|0(t)\rangle_{e,\mu}$ and obtain

$$_{1,2}\langle 0|0(t)\rangle_{e,\mu} = \prod_\mathbf{k} \left(1 - \sin^2\theta \, |V_\mathbf{k}|^2\right)^2 \equiv \prod_\mathbf{k} \Gamma(k) = e^{\sum_\mathbf{k} \ln \Gamma(k)}, \qquad (2.103)$$

where the function $|V_\mathbf{k}|^2$ is defined in Eq. (2.108) and plotted in Fig.2.1. By using the continuous limit relation $\sum_\mathbf{k} \to \frac{V}{(2\pi)^3} \int d^3\mathbf{k}$, in the infinite volume limit we obtain (for any t)

$$\lim_{V\to\infty} {}_{1,2}\langle 0|0(t)\rangle_{e,\mu} = \lim_{V\to\infty} e^{\frac{V}{(2\pi)^3} \int d^3k \ln \Gamma(k)} = 0, \qquad (2.104)$$

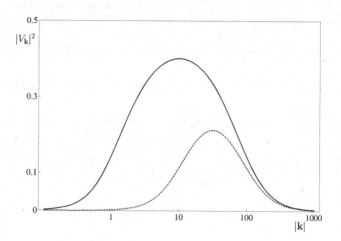

Fig. 2.1   The fermion condensation density $|V_{\mathbf{k}}|^2$ as a function of $|\mathbf{k}|$ for $m_1 = 1$, $m_2 = 100$ (solid line) and $m_1 = 10$, $m_2 = 100$ (dashed line).

since $\Gamma(\mathbf{k}) < 1$ for any value of $\mathbf{k}$ and of m_1 and m_2 (with $m_2 \neq m_1$).

Eq. (2.104) shows that the orthogonality between $|0(t)\rangle_{e,\mu}$ and $|0\rangle_{1,2}$ is due to the infrared contributions which are taken in care by the infinite volume limit, and therefore high momentum contributions do not influence the result (for this reason we do not need to consider the regularization problem of the UV divergence of the integral of $\ln \Gamma(\mathbf{k})$). Of course, this orthogonality disappears when $\theta = 0$ and/or when $m_1 = m_2$ (in this case $V_{\mathbf{k}} = 0$ for any $\mathbf{k}$).

Eq. (2.104) expresses the unitary inequivalence in the infinite volume limit of the flavor and the mass representations, and shows the non-trivial nature of the mixing transformations (2.92) resulting in the condensate structure of the flavor vacuum. We will see that such a vacuum structure leads to phenomenological consequences on neutrino oscillations.

By use of $G_\theta(t)$, the flavor fields can be expanded as:

$$\nu_\sigma(x) = \sum_{r=1,2} \int \frac{d^3k}{(2\pi)^{\frac{3}{2}}} \left[u^r_{\mathbf{k},i}(t)\alpha^r_{\mathbf{k},\sigma}(t) + v^r_{-\mathbf{k},i}(t)\beta^{r\dagger}_{-\mathbf{k},\sigma}(t) \right] e^{i\mathbf{k}\cdot\mathbf{x}}, \quad (2.105)$$

with $(\sigma, i) = (e, 1), (\mu, 2)$. The flavor annihilation operators are defined as $\alpha^r_{\mathbf{k},\sigma}(t) \equiv G_\theta^{-1}(t)\alpha^r_{\mathbf{k},i}G_\theta(t)$ and $\beta^{r\dagger}_{-\mathbf{k},\sigma}(t) \equiv G_\theta^{-1}(t)\beta^{r\dagger}_{-\mathbf{k},i}G_\theta(t)$. In the reference frame such that $\mathbf{k} = (0, 0, |\mathbf{k}|)$, we have the simple expressions:

$$\alpha^r_{\mathbf{k},e}(t) = \cos\theta \, \alpha^r_{\mathbf{k},1} + \sin\theta \left(U^*_{\mathbf{k}}(t) \, \alpha^r_{\mathbf{k},2} + \epsilon^r \, V_{\mathbf{k}}(t) \, \beta^{r\dagger}_{-\mathbf{k},2} \right), \qquad (2.106a)$$

$$\alpha^r_{\mathbf{k},\mu}(t) = \cos\theta \, \alpha^r_{\mathbf{k},2} - \sin\theta \left(U_{\mathbf{k}}(t) \, \alpha^r_{\mathbf{k},1} - \epsilon^r \, V_{\mathbf{k}}(t) \, \beta^{r\dagger}_{-\mathbf{k},1} \right), \qquad (2.106b)$$

$$\beta^r_{-\mathbf{k},e}(t) = \cos\theta \, \beta^r_{-\mathbf{k},1} + \sin\theta \left(U^*_{\mathbf{k}}(t) \, \beta^r_{-\mathbf{k},2} - \epsilon^r \, V_{\mathbf{k}}(t) \, \alpha^{r\dagger}_{\mathbf{k},2} \right), \qquad (2.106c)$$

$$\beta^r_{-\mathbf{k},\mu}(t) = \cos\theta \, \beta^r_{-\mathbf{k},2} - \sin\theta \left(U_{\mathbf{k}}(t) \, \beta^r_{-\mathbf{k},1} + \epsilon^r \, V_{\mathbf{k}}(t) \, \alpha^{r\dagger}_{\mathbf{k},1} \right), \qquad (2.106d)$$

where $\epsilon^r = (-1)^r$ and

$$U_{\mathbf{k}}(t) \equiv u^{r\dagger}_{\mathbf{k},2}(t) u^r_{\mathbf{k},1}(t) = v^{r\dagger}_{-\mathbf{k},1}(t) v^r_{-\mathbf{k},2}(t) = |U_{\mathbf{k}}| \, e^{i(\omega_{k,2} - \omega_{k,1})t}, \qquad (2.107a)$$

$$V_{\mathbf{k}}(t) \equiv \epsilon^r \, u^{r\dagger}_{\mathbf{k},1}(t) v^r_{-\mathbf{k},2}(t) = -\epsilon^r \, u^{r\dagger}_{\mathbf{k},2}(t) v^r_{-\mathbf{k},1}(t) = |V_{\mathbf{k}}| \, e^{i(\omega_{k,2} + \omega_{k,1})t}, \qquad (2.107b)$$

with

$$|U_{\mathbf{k}}| = \frac{|\mathbf{k}|^2 + (\omega_{k,1} + m_1)(\omega_{k,2} + m_2)}{2\sqrt{\omega_{k,1}\omega_{k,2}(\omega_{k,1} + m_1)(\omega_{k,2} + m_2)}}, \qquad (2.108a)$$

$$|V_{\mathbf{k}}| = \frac{(\omega_{k,1} + m_1) - (\omega_{k,2} + m_2)}{2\sqrt{\omega_{k,1}\omega_{k,2}(\omega_{k,1} + m_1)(\omega_{k,2} + m_2)}} \, |\mathbf{k}|, \qquad (2.108b)$$

$$|U_{\mathbf{k}}|^2 + |V_{\mathbf{k}}|^2 = 1. \qquad (2.108c)$$

We thus see that the mixing transformations (2.92) are realized as a combination of a rotation and a Bogoliubov transformation at the level of ladder operators. The Bogoliubov coefficients $U_{\mathbf{k}}$ and $V_{\mathbf{k}}$ are given by combinations of the (spinorial) wave functions with different masses.

The number of particles with definite mass condensed in the flavor vacuum is given by

$$_{e,\mu}\langle 0(t)|\alpha^{r\dagger}_{\mathbf{k},i}\alpha^r_{\mathbf{k},i}|0(t)\rangle_{e,\mu} = \sin^2\theta \, |V_{\mathbf{k}}|^2, \qquad i = 1, 2, \qquad (2.109)$$

with the same result for antiparticles.[1] Note that the $|V_{\mathbf{k}}|^2$ has a maximum at $\sqrt{m_1 m_2}$ and $|V_{\mathbf{k}}|^2 \simeq \frac{(m_2 - m_1)^2}{4|\mathbf{k}|^2}$ for $|\mathbf{k}| \gg \sqrt{m_1 m_2}$.

---

[1] In the case of three flavors [84, 108], the condensation densities are different for different i and for antiparticles (when CP violation is present).

Boson mixing

Let us now consider boson mixing [83] in the case of charged fields. The mixing relations are:

$$\phi_A(x) = \cos\theta\, \phi_1(x) + \sin\theta\, \phi_2(x)\,, \tag{2.110a}$$

$$\phi_B(x) = -\sin\theta\, \phi_1(x) + \cos\theta\, \phi_2(x)\,. \tag{2.110b}$$

The mixed fields have suffixes A and B. The fields $\phi_i(x)$, $i = 1, 2$, are free complex fields with definite masses. Their conjugate momenta are $\pi_i(x) = \partial_0\phi_i^\dagger(x)$ and the commutation relations are the usual ones:

$$[\phi_i(x), \pi_j(y)]_{t=t'} = \left[\phi_i^\dagger(x), \pi_j^\dagger(y)\right]_{t=t'} = i\delta^3(\mathbf{x} - \mathbf{y})\,\delta_{ij}\,, \tag{2.111}$$

with $i, j = 1, 2$ and the other equal-time commutators vanishing. The Fourier expansions of fields and momenta are:

$$\phi_i(x) = \int \frac{d^3k}{(2\pi)^{\frac{3}{2}}} \frac{1}{\sqrt{2\omega_{k,i}}} \left(a_{\mathbf{k},i}\, e^{-i\omega_{k,i}t} + b^\dagger_{-\mathbf{k},i}\, e^{i\omega_{k,i}t}\right) e^{i\mathbf{k}\cdot\mathbf{x}}\,, \tag{2.112}$$

$$\pi_i(x) = i \int \frac{d^3k}{(2\pi)^{\frac{3}{2}}} \sqrt{\frac{\omega_{k,i}}{2}} \left(a^\dagger_{\mathbf{k},i}\, e^{i\omega_{k,i}t} - b_{-\mathbf{k},i}\, e^{-i\omega_{k,i}t}\right) e^{i\mathbf{k}\cdot\mathbf{x}}\,, \tag{2.113}$$

where $\omega_{k,i} = \sqrt{\mathbf{k}^2 + m_i^2}$ and $\left[a_{\mathbf{k},i}, a^\dagger_{\mathbf{p},j}\right] = \left[b_{\mathbf{k},i}, b^\dagger_{\mathbf{p},j}\right] = \delta^3(\mathbf{k} - \mathbf{p})\delta_{ij}$, with $i, j = 1, 2$ and the other commutators vanishing.

We proceed in a similar way as for fermions and write Eqs. (2.110) as[2]

$$\phi_\sigma(x) = \widehat{G}_\theta^{-1}(t)\, \phi_i(x)\, \widehat{G}_\theta(t)\,, \tag{2.114}$$

with $(\sigma, i) = (A, 1), (B, 2)$, and similar expressions for π_A, π_B. We have

$$\widehat{G}_\theta(t) = \exp[\theta(\widehat{S}_+(t) - \widehat{S}_-(t))]\,. \tag{2.115}$$

The operators

$$\widehat{S}_+(t) = \widehat{S}_-^\dagger(t) \equiv -i \int d^3x\, (\pi_1(x)\phi_2(x) - \phi_1^\dagger(x)\pi_2^\dagger(x))\,, \tag{2.116}$$

$$\widehat{S}_3 \equiv \frac{-i}{2} \int d^3x\, \left(\pi_1(x)\phi_1(x) - \phi_1^\dagger(x)\pi_1^\dagger(x) - \pi_2(x)\phi_2(x) + \phi_2^\dagger(x)\pi_2^\dagger(x)\right)\,, \tag{2.117}$$

close the $su(2)$ algebra (at a given t).

As for fermions, the action of the generator of the mixing transformations on the vacuum $|\widehat{0}\rangle_{1,2}$ for the fields $\phi_{1,2}$ is non-trivial and induces on it a $SU(2)$ coherent state structure [519]:

$$|0(t)\rangle_{A,B} \equiv \widehat{G}_\theta^{-1}(t)\, |\widehat{0}\rangle_{1,2}\,. \tag{2.118}$$

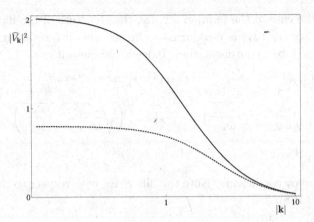

Fig. 2.2   The boson condensation density $|\widehat{V}_{\mathbf{k}}|^2$ as a function of $|\mathbf{k}|$ for $m_1 = 1$, $m_2 = 10$ (solid line) and $m_1 = 2$, $m_2 = 10$ (dashed line).

We will refer to the state $|0(t)\rangle_{A,B}$ as to the flavor vacuum for bosons. The orthogonality between $|0(t)\rangle_{A,B}$ and $|\widehat{0}\rangle_{1,2}$ can be proven in a similar way as in the fermionic case. We have indeed, for any t:

$$\lim_{V \to \infty} {}_{1,2}\langle \widehat{0}|0(t)\rangle_{A,B} = \lim_{V \to \infty} \exp\left\{ \frac{V}{(2\pi)^3} \int d^3k \, \ln\left(\frac{1}{1 + \sin^2 \theta |V_{\mathbf{k}}|^2} \right) \right\} = 0 \,, \tag{2.119}$$

where $|V_{\mathbf{k}}|^2$ is defined in Eq. (2.122) and plotted in Fig. 2.2. The Fourier expansion for the flavor fields are:

$$\phi_\sigma(x) = \int \frac{d^3k}{(2\pi)^{\frac{3}{2}}} \frac{1}{\sqrt{2\omega_{k,i}}} \left(a_{\mathbf{k},\sigma}(t) \, e^{-i\omega_{k,i}t} + b^\dagger_{-\mathbf{k},\sigma}(t) \, e^{i\omega_{k,i}t} \right) e^{i\mathbf{k}\cdot\mathbf{x}} \,, \tag{2.120}$$

with $(\sigma, i) = (A, 1), (B, 2)$, and similar expressions for π_A, π_B.

The annihilation operators for the vacuum $|0(t)\rangle_{A,B}$ are defined as $a_{\mathbf{k},A}(t) \equiv \widehat{G}_\theta^{-1}(t) \, a_{\mathbf{k},1} \, \widehat{G}_\theta(t)$, etc. We have:

$$a_{\mathbf{k},A}(t) = \cos\theta \, a_{\mathbf{k},1} + \sin\theta \left(\widehat{U}_{\mathbf{k}}^*(t) \, a_{\mathbf{k},2} + \widehat{V}_{\mathbf{k}}(t) \, b^\dagger_{-\mathbf{k},2} \right), \tag{2.121a}$$

$$a_{\mathbf{k},B}(t) = \cos\theta \, a_{\mathbf{k},2} - \sin\theta \left(\widehat{U}_{\mathbf{k}}(t) \, a_{\mathbf{k},1} - \widehat{V}_{\mathbf{k}}(t) \, b^\dagger_{-\mathbf{k},1} \right), \tag{2.121b}$$

$$b_{-\mathbf{k},A}(t) = \cos\theta \, b_{-\mathbf{k},1} + \sin\theta \left(\widehat{U}_{\mathbf{k}}^*(t) \, b_{-\mathbf{k},2} + \widehat{V}_{\mathbf{k}}(t) \, a^\dagger_{\mathbf{k},2} \right), \tag{2.121c}$$

$$b_{-\mathbf{k},B}(t) = \cos\theta \, b_{-\mathbf{k},2} - \sin\theta \left(\widehat{U}_{\mathbf{k}}(t) \, b_{-\mathbf{k},1} - \widehat{V}_{\mathbf{k}}(t) \, a^\dagger_{\mathbf{k},1} \right). \tag{2.121d}$$

These operators satisfy the equal-time canonical commutation relations.

[2]Whenever confusion may arise, we use the "hat" to distinguish similar quantities from the fermionic case.

As in the case of the fermion mixing, the structure of the flavor ladder operators Eqs. (2.121) is recognized to be the one of a rotation combined with a Bogoliubov transformation. Indeed, the quantities

$$\widehat{U}_{\mathbf{k}}(t) \equiv |\widehat{U}_{\mathbf{k}}| \, e^{i(\omega_{k,2}-\omega_{k,1})t} \quad , \quad \widehat{V}_{\mathbf{k}}(t) \equiv |\widehat{V}_{\mathbf{k}}| \, e^{i(\omega_{k,1}+\omega_{k,2})t} , \tag{2.122a}$$

$$|\widehat{U}_{\mathbf{k}}| \equiv \frac{1}{2} \left(\sqrt{\frac{\omega_{k,1}}{\omega_{k,2}}} + \sqrt{\frac{\omega_{k,2}}{\omega_{k,1}}} \right) \quad , \quad |\widehat{V}_{\mathbf{k}}| \equiv \frac{1}{2} \left(\sqrt{\frac{\omega_{k,1}}{\omega_{k,2}}} - \sqrt{\frac{\omega_{k,2}}{\omega_{k,1}}} \right), \tag{2.122b}$$

$$|\widehat{U}_{\mathbf{k}}|^2 - |\widehat{V}_{\mathbf{k}}|^2 = 1 , \tag{2.122c}$$

are Bogoliubov coefficients. Note the difference with respect to the fermion case Eq. (2.108).

We have, for any t:

$$_{A,B}\langle 0(t)|a_{\mathbf{k},i}^{\dagger} a_{\mathbf{k},i}|0(t)\rangle_{A,B} = \sin^2\theta \, |\widehat{V}_{\mathbf{k}}|^2, \qquad i = 1, 2, \tag{2.123}$$

and the same result is obtained for antiparticles. The function $|\widehat{V}_{\mathbf{k}}|^2$ is maximal at $|\mathbf{k}| = 0$ ($|\widehat{V}_{max}|^2 = \frac{(m_1-m_2)^2}{4m_1 m_2}$) and $|\widehat{V}_{\mathbf{k}}|^2 \simeq \left(\frac{\Delta m^2}{4|\mathbf{k}|^2} \right)^2$ for $|\mathbf{k}|^2 \gg \frac{m_1^2+m_2^2}{2}$. A plot is given in Fig.2.2 for sample values of the masses.

The QFT field mixing formalism summarized above leads us to consider the energy content of the flavor vacuum condensate. It has been shown [80, 138–140, 465] that such a vacuum energy content turns out to be in agreement with the observed value of the dark energy.

Flavor charges for mixed fields

We now discuss the charges associated to flavor mixing, both for the case of Dirac neutrinos and for charged bosons [83, 101]. We will see that the structure of such flavor charges allows us to select the correct Hilbert space for the calculation of flavor oscillations.

Fermion charges

Let us start by considering the Lagrangian for two free Dirac fields, with masses m_1 and m_2:

$$\mathcal{L}(x) = \bar{\nu}_m(x)\,(i\,\slashed{\partial} - M_d)\,\nu_m(x) , \tag{2.124}$$

where $\nu_m = (\nu_1, \nu_1)^T$ and $M_d = diag(m_1, m_2)$. We introduce a subscript m denoting quantities which are in terms of fields with definite masses.

$\mathcal{L}$ is invariant under global $U(1)$ phase transformations of the type $\nu'_m = e^{i\alpha} \nu_m$. As a result, we have the conservation of the Noether charge $Q = \int d^3\mathbf{x} \, I^0(x)$ (with $I^\mu(x) = \bar{\nu}_m(x) \gamma^\mu \nu_m(x)$) which is indeed the total charge (total lepton number) of the system.

For $m_1 \neq m_2$, $\mathcal{L}$ is not generally invariant under the global $SU(2)$ transformations [101]:

$$\nu'_m(x) = e^{i\alpha_j \tau_j} \nu_m(x), \qquad j = 1, 2, 3, \qquad (2.125)$$

with $\tau_j = \sigma_j/2$ and σ_j being the Pauli matrices. We have indeed:

$$\delta\mathcal{L}(x) = i\alpha_j \, \bar{\nu}_m(x) \, [\tau_j, M_d] \, \nu_m(x) = -\alpha_j \, \partial_\mu J^\mu_{m,j}(x), \quad (2.126a)$$

$$J^\mu_{m,j}(x) = \bar{\nu}_m(x) \gamma^\mu \tau_j \nu_m(x), \qquad j = 1, 2, 3. \qquad (2.126b)$$

The charges $Q_{m,j}(t) \equiv \int d^3\mathbf{x} \, J^0_{m,j}(x)$ satisfy the $su(2)$ algebra (at equal times): $[Q_{m,j}(t), Q_{m,k}(t)] = i \, \epsilon_{jkl} \, Q_{m,l}(t)$. Note that $2Q_{m,2}(t)$ is indeed the generator of mixing transformations introduced above. Also note that the Casimir operator is proportional to the total charge: $C_m = \frac{1}{2}Q$ and that, since $Q_{m,3}$ is conserved in time, we have

$$Q_1 \equiv \frac{1}{2}Q + Q_{m,3} \quad , \quad Q_2 \equiv \frac{1}{2}Q - Q_{m,3}, \qquad (2.127a)$$

$$Q_i = \sum_r \int d^3k \left(\alpha^{r\dagger}_{\mathbf{k},i} \alpha^r_{\mathbf{k},i} - \beta^{r\dagger}_{-\mathbf{k},i} \beta^r_{-\mathbf{k},i} \right), \quad i = 1, 2. \qquad (2.127b)$$

These are nothing but the Noether charges associated with the non-interacting fields ν_1 and ν_2: in the absence of mixing, they are the flavor charges, separately conserved for each generation.

Let us now return to the Lagrangian Eq. (2.124) and write it in the flavor basis (subscript f denotes here flavor), by means of Eqs. (2.92):

$$\mathcal{L}(x) = \bar{\nu}_f(x) \, (i \, \partial\!\!\!/ - M) \, \nu_f(x), \qquad (2.128)$$

where $\nu_f = (\nu_e, \nu_\mu)^T$ and $M = \begin{pmatrix} m_e & m_{e\mu} \\ m_{e\mu} & m_\mu \end{pmatrix}$, with $m_e = m_1 \cos^2 \theta + m_2 \sin^2 \theta$, $m_\mu = m_1 \sin^2 \theta + m_2 \cos^2 \theta$, $m_{e\mu} = (m_2 - m_1) \sin \theta \cos \theta$.

Obviously, $\mathcal{L}$ is still invariant under $U(1)$. We then consider the $SU(2)$ transformation [101]:

$$\nu'_f(x) = e^{i\alpha_j \tau_j} \nu_f(x), \qquad (2.129a)$$

$$\delta\mathcal{L}(x) = i\alpha_j \, \bar{\nu}_f(x) \, [\tau_j, M] \, \nu_f(x) = -\alpha_j \, \partial_\mu J^\mu_{f,j}(x), \qquad (2.129b)$$

$$J^\mu_{f,j}(x) = \bar{\nu}_f(x) \gamma^\mu \tau_j \nu_f(x), \qquad j = 1, 2, 3. \qquad (2.129c)$$

The charges $Q_{f,j}(t) \equiv \int d^3\mathbf{x}\, J^0_{f,j}(x)$ satisfy the $su(2)$ algebra. Note that, because of the off-diagonal (mixing) terms in the mass matrix M, $Q_{f,3}$ is not anymore conserved. This implies an exchange of charge between ν_e and ν_μ, resulting in the phenomenon of flavor oscillations.

Let us indeed define the *flavor charges* for mixed fields as

$$Q_e(t) \equiv \int d^3x\, \nu^\dagger_e(x)\nu_e(x) = \frac{1}{2}Q + Q_{f,3}(t)\,, \qquad (2.130a)$$

$$Q_\mu(t) \equiv \int d^3x\, \nu^\dagger_\mu(x)\nu_\mu(x) = \frac{1}{2}Q - Q_{f,3}(t)\,, \qquad (2.130b)$$

where $Q_e(t) + Q_\mu(t) = Q$. They are related to the Noether charges as

$$Q_\sigma(t) = G^{-1}_\theta(t)\, Q_i\, G_\theta(t)\,, \qquad (2.131)$$

with $(\sigma, i) = (e, 1), (\mu, 2)$. From Eq. (2.131), it follows that the flavor charges are diagonal in the flavor ladder operators:

$$Q_\sigma(t) = \sum_r \int d^3k \left(\alpha^{r\dagger}_{\mathbf{k},\sigma}(t)\alpha^r_{\mathbf{k},\sigma}(t) - \beta^{r\dagger}_{-\mathbf{k},\sigma}(t)\beta^r_{-\mathbf{k},\sigma}(t) \right)\,, \qquad (2.132)$$

with $\sigma = e, \mu$.

The flavor states for mixed neutrinos are defined as eigenstates of the above charges:

$$|\nu^r_{\mathbf{k},\sigma}(t)\rangle = \alpha^{r\dagger}_{\mathbf{k},\sigma}(t)|0(t)\rangle_{e\mu}\,, \qquad \sigma = e, \mu, \qquad (2.133)$$

and similar ones for antiparticles.

Notice that it is not possible to find eigenstates of the flavor charges Eq. (2.132), among the states of the Hilbert space for mass eigenstates, i.e., the one built on $|0\rangle_{1,2}$. The usual QM flavor neutrino states (Pontecorvo states; see Eq. (H.2)), which belong to the Hilbert space for mass eigenstates, lead to a violation of lepton charge conservation in the neutrino production/detection vertices [82, 501].

Boson charges

For the case of bosonic fields, we proceed in a similar way as done above for fermions. We consider the following Lagrangian density:

$$\mathcal{L}(x) = \partial_\mu \Phi^\dagger_f(x)\, \partial^\mu \Phi_f(x) - \Phi^\dagger_f(x)\widehat{M}\Phi_f(x)\,, \qquad (2.134)$$

with $\Phi^T_f = (\phi_A, \phi_B)$, $\widehat{M} = \begin{pmatrix} m^2_A & m^2_{AB} \\ m^2_{AB} & m^2_B \end{pmatrix}$. By means of Eq. (2.110), this is diagonalized as:

$$\mathcal{L}(x) = \partial_\mu \Phi^\dagger_m(x)\, \partial^\mu \Phi_m(x) - \Phi^\dagger_m(x)\widehat{M}_d\Phi_m(x)\,, \qquad (2.135)$$

where $\Phi_m^T = (\phi_1, \phi_2)$, $\widehat{M}_d = diag(m_1^2, m_2^2)$ and $m_A^2 = m_1^2 \cos^2\theta + m_2^2 \sin^2\theta$, $m_B^2 = m_1^2 \sin^2\theta + m_2^2 \cos^2\theta$, $m_{AB}^2 = (m_2^2 - m_1^2)\sin\theta\cos\theta$.

The Lagrangian $\mathcal{L}$ is invariant under the global $U(1)$ phase transformations $\Phi_m' = e^{i\alpha} \Phi_m$ leading to the conservation of the total charge $\widehat{Q} = \int d^3x \, \widehat{I}^0(x)$ (we have $\widehat{I}^\mu(x) = i\,\Phi_m^\dagger(x) \overset{\leftrightarrow}{\partial^\mu} \Phi_m(x)$ with $\overset{\leftrightarrow}{\partial^\mu} \equiv \overset{\rightarrow}{\partial^\mu} - \overset{\leftarrow}{\partial^\mu}$).

Let us now consider the $SU(2)$ transformation

$$\Phi_m'(x) = e^{i\alpha_j \tau_j} \Phi_m(x), \qquad j = 1, 2, 3, \qquad (2.136)$$

with α_j real constants, $\tau_j = \sigma_j/2$ and σ_j being the Pauli matrices. For $m_1 \neq m_2$, the Lagrangian is not generally invariant under (2.136) and we obtain, by use of the equations of motion,

$$\delta\mathcal{L}(x) = -i\,\alpha_j\,\Phi_m^\dagger(x)\,[\widehat{M}_d\,,\,\tau_j]\,\Phi_m(x) = -\alpha_j\,\partial_\mu\,\widehat{J}_{m,j}^\mu(x), \quad (2.137)$$

$$\widehat{J}_{m,j}^\mu(x) = i\,\Phi_m^\dagger(x)\,\tau_j\,\overset{\leftrightarrow}{\partial^\mu}\,\Phi_m(x), \qquad j = 1, 2, 3. \qquad (2.138)$$

The corresponding charges, $\widehat{Q}_{m,j}(t) \equiv \int d^3x \, \widehat{J}_{m,j}^0(x)$, close the $su(2)$ algebra (at each time t). The Casimir operator is $\widehat{C}_m = \frac{1}{2}\widehat{Q}$ and $2\widehat{Q}_{m,2}(t)$ is the generator of the mixing transformations. Observe that the combinations

$$\widehat{Q}_{1,2} \equiv \frac{1}{2}\widehat{Q} \pm \widehat{Q}_{m,3}, \qquad (2.139a)$$

$$\widehat{Q}_i = \int d^3k \left(a_{\mathbf{k},i}^\dagger a_{\mathbf{k},i} - b_{-\mathbf{k},i}^\dagger b_{-\mathbf{k},i} \right), \qquad i = 1, 2, \quad (2.139b)$$

are simply the conserved (Noether) charges for the free fields ϕ_1 and ϕ_2 with $\widehat{Q}_1 + \widehat{Q}_2 = \widehat{Q}$.

The $SU(2)$ transformations on the flavor doublet Φ_f:

$$\Phi_f'(x) = e^{i\alpha_j \tau_j} \Phi_f(x), \qquad j = 1, 2, 3, \qquad (2.140)$$

give

$$\delta\mathcal{L}(x) = -i\,\alpha_j\,\Phi_f^\dagger(x)\,[\widehat{M}, \tau_j]\,\Phi_f(x) = -\alpha_j\,\partial_\mu\widehat{J}_{f,j}^\mu(x), \quad (2.141a)$$

$$\widehat{J}_{f,j}^\mu(x) = i\,\Phi_f^\dagger(x)\,\tau_j\,\overset{\leftrightarrow}{\partial^\mu}\,\Phi_f(x), \qquad j = 1, 2, 3. \qquad (2.141b)$$

The related charges, $\widehat{Q}_{f,j}(t) \equiv \int d^3x \, \widehat{J}_{f,j}^0(x)$, still fulfil the $su(2)$ algebra and $\widehat{C}_f = \widehat{C}_m$. As for fermions, $\widehat{Q}_{f,3}(t)$ is time-dependent and we define the flavor charges for mixed bosons as

$$\widehat{Q}_A(t) \equiv \frac{1}{2}\widehat{Q} + \widehat{Q}_{f,3}(t), \qquad (2.142a)$$

$$\widehat{Q}_B(t) \equiv \frac{1}{2}\widehat{Q} - \widehat{Q}_{f,3}(t), \qquad (2.142b)$$

with $\hat{Q}_A(t) + \hat{Q}_B(t) = \hat{Q}$. We have:

$$\hat{Q}_\sigma(t) = \int d^3k \left(a_{\mathbf{k},\sigma}^\dagger(t) a_{\mathbf{k},\sigma}(t) - b_{-\mathbf{k},\sigma}^\dagger(t) b_{-\mathbf{k},\sigma}(t) \right), \qquad (2.143)$$

with $\sigma = A, B$, which follows from $\hat{Q}_\sigma(t) = \hat{G}_\theta^{-1}(t)\hat{Q}_i\hat{G}_\theta(t)$, where $(\sigma, i) = (A, 1), (B, 2)$.

Finally, we define the flavor states for mixed bosons as eigenstates of the above charges:

$$|a_{\mathbf{k},\sigma}\rangle \equiv a_{\mathbf{k},\sigma}^\dagger(0) |0\rangle_{A,B}, \qquad \sigma = A, B, \qquad (2.144)$$

and similar ones for antiparticles.

Flavor oscillations formulas in QFT

It is now possible to work out the exact QFT oscillation formulas exhibiting corrections with respect to the QM ones [77, 284].

Let us start with the fermion case and consider neutrinos (only two flavors). We work in the Heisenberg picture and take the electron neutrino state at time $t = 0$ (see Eq. (2.133)):

$$|\nu_{\mathbf{k},e}^r\rangle \equiv \alpha_{\mathbf{k},e}^{r\dagger}(0)|0\rangle_{e,\mu}, \qquad (2.145)$$

where $|0\rangle_{e,\mu} \equiv |0(0)\rangle_{e,\mu}$. We have $_{e,\mu}\langle 0|Q_\sigma(t)|0\rangle_{e,\mu} = 0$ and

$$\mathcal{Q}_\sigma^{\mathbf{k}}(t) \equiv \langle \nu_{\mathbf{k},e}^r | Q_\sigma(t) | \nu_{\mathbf{k},e}^r \rangle$$

$$= \left| \left\{ \alpha_{\mathbf{k},\sigma}^r(t), \alpha_{\mathbf{k},e}^{r\dagger}(0) \right\} \right|^2 + \left| \left\{ \beta_{-\mathbf{k},\sigma}^{r\dagger}(t), \alpha_{\mathbf{k},e}^{r\dagger}(0) \right\} \right|^2. \quad (2.146)$$

Charge conservation is ensured at any time: $\mathcal{Q}_e^{\mathbf{k}}(t) + \mathcal{Q}_\mu^{\mathbf{k}}(t) = 1$. The oscillation formulas for the flavor charges are then [93]:

$$\mathcal{Q}_\mu^{\mathbf{k}}(t) = \sin^2(2\theta) \left[|U_{\mathbf{k}}|^2 \sin^2 \left(\frac{\omega_{k,2} - \omega_{k,1}}{2} t \right) + |V_{\mathbf{k}}|^2 \sin^2 \left(\frac{\omega_{k,2} + \omega_{k,1}}{2} t \right) \right],$$

$$\mathcal{Q}_e^{\mathbf{k}}(t) = 1 - \mathcal{Q}_\mu^{\mathbf{k}}(t). \qquad (2.147)$$

This result is exact. There are two differences with respect to the usual formula for neutrino oscillations: the amplitudes are energy-dependent, and there is an additional oscillating term. The usual QM formulas [77, 284] (see Appendices H and Q) are (approximately) recovered in the relativistic limit.

The bosonic counterpart of the above oscillation formulas can be derived in a similar way [83]. By defining the mixed bosonic state for the "A" particle as (see Eq. (2.144)):

$$|a_{\mathbf{k},A}\rangle \equiv a_{\mathbf{k},A}^\dagger(0) |0\rangle_{A,B}, \qquad (2.148)$$

with $|0\rangle_{A,B} \equiv |0(0)\rangle_{A,B}$, we obtain $_{A,B}\langle 0| \widehat{Q}_\sigma(t) |0\rangle_{A,B} = 0$ and

$$\widehat{Q}_\sigma^{\mathbf{k}}(t) \equiv \langle a_{\mathbf{k},A}| \widehat{Q}_\sigma(t) |a_{\mathbf{k},A}\rangle$$

$$= \left| \left[a_{\mathbf{k},\sigma}(t), a_{\mathbf{k},A}^\dagger(0) \right] \right|^2 - \left| \left[b_{-\mathbf{k},\sigma}^\dagger(t), a_{\mathbf{k},A}^\dagger(0) \right] \right|^2. \qquad (2.149)$$

The conservation of the total charge gives $\sum_\sigma \widehat{Q}_\sigma^{\mathbf{k}}(t) = 1$ and the oscillation formulas are:

$$\widehat{Q}_B^{\mathbf{k}}(t) = \sin^2(2\theta) \left[|\widehat{U}_{\mathbf{k}}|^2 \sin^2 \left(\frac{\omega_{k,2} - \omega_{k,1}}{2} t \right) - |\widehat{V}_{\mathbf{k}}|^2 \sin^2 \left(\frac{\omega_{k,2} + \omega_{k,1}}{2} t \right) \right],$$

$$\widehat{Q}_A^{\mathbf{k}}(t) = 1 - \widehat{Q}_B^{\mathbf{k}}(t). \qquad (2.150)$$

Thus also for bosons, the non-trivial structure of the flavor vacuum induces corrections to the usual QM expressions for flavor oscillations. The most obvious difference with respect to fermionic case is in the negative sign which makes a negative value possible for the bosonic flavor charges. As for neutrinos, in the relativistic limit the usual QM formulas are recovered.

The above oscillation formulas express the change of flavor in time. However, for practical purposes, it is more convenient to have formulas for the variation of flavor in function of the distance from a source. Then wave packets must be used in the derivation of oscillation formulas [284]. A treatment of this problem in the QFT framework discussed here can be found in [106]. The wave packet treatment also allows the study of the decoherence of oscillating neutrinos [85, 86]: after a characteristic distance (decoherence length [284]), oscillations are suppressed and survival probabilities become constant.

Appendix E

Computation of $\langle 0|\psi_i(x)|\alpha_n\alpha_m\rangle$

In this Appendix we show how to calculate the matrix element $\langle 0|\psi_i(x)|\alpha_n\alpha_m\rangle$. The outlined calculation completes the discussion of Section 2.1.

We use $p_\mu x^\mu = \mathbf{p}\cdot\mathbf{x} - E_p t$ and the Fourier form of $F_{inm}(x-\xi_1, x-\xi_2)$

$$F_{inm}(x-\xi_1, x-\xi_2)$$
$$= \int \frac{d^4 p_1}{(2\pi)^4} \int \frac{d^4 p_2}{(2\pi)^4} F_{inm}(p_1,p_2) e^{i[p_1^\mu(x-\xi_1)_\mu + p_2^\mu(x-\xi_2)_\mu]} . \qquad (E.1)$$

Use of (2.2), (1.1) and (1.131) gives

$$\langle 0|\psi_i(x)|\alpha_n\alpha_m\rangle = \int \frac{d^4 p_1}{(2\pi)^4} \int \frac{d^4 p_2}{(2\pi)^4} \int \frac{d^3 k_1}{(2\pi)^{\frac{3}{2}}} \int \frac{d^3 k_2}{(2\pi)^{\frac{3}{2}}}$$

$$F_{inm}(p_1,p_2) u(\mathbf{k}_1) u(\mathbf{k}_2) f_m(\mathbf{k}_1) f_n(\mathbf{k}_2) \int_{-\infty}^{t_x} d^4\xi_1 \int_{-\infty}^{t_x} d^4\xi_2$$

$$e^{i\mathbf{k}_1\cdot\xi_1 - i\omega_1 t_{\xi_1}} e^{i\mathbf{k}_2\cdot\xi_2 - i\omega_2 t_{\xi_2}} e^{i[p_1^\mu(x-\xi_1)_\mu + p_2^\mu(x-\xi_2)_\mu]} \pm (n \leftrightarrow m) , \qquad (E.2)$$

where the $(n \leftrightarrow m)$-terms are those obtained by exchanging n with m; $+$ is for fermions, $-$ is for bosons; $\omega_{1(2)} \equiv \omega_{k_{1(2)}}$. Now we note the relation

$$\int dE \int_{-\infty}^{t} dt_1 f(E) e^{iEt_1} = \int dE f(E) \frac{e^{iEt}}{i(E-i\epsilon)} , \qquad (E.3)$$

which holds whenever $f(E)$ is a square-integrable function; indeed, the integration in dt_1 in (E.3) is not defined unless

$$\lim_{t\to-\infty} \int dE f(E) e^{iEt} = 0 , \qquad (E.4)$$

which holds due to the Riemann–Lebesgue theorem when $f(E)$ is a square-integrable function. We also have from (E.3)

$$\lim_{t\to-\infty} \int dE f(E) \frac{e^{iEt}}{i(E-i\epsilon)} = 0 , \qquad (E.5)$$

and

$$\lim_{t\to+\infty} \int dE f(E) \frac{e^{iEt}}{i(E - i\epsilon)} = 2\pi f(0)\,. \tag{E.6}$$

In (E.3-E.6) the limit $\epsilon \to 0$ is understood. We thus obtain

$$\langle 0|\psi_i(x)|\alpha_n\alpha_m\rangle = -\int \frac{dE_1}{2\pi} \int \frac{dE_2}{2\pi} \int \frac{d^3k_1}{(2\pi)^{\frac{3}{2}}} \int \frac{d^3k_2}{(2\pi)^{\frac{3}{2}}} \tag{E.7}$$

$$F_{inm}(\mathbf{k}_1, E_1; \mathbf{k}_2, E_2)u(\mathbf{k}_1)u(\mathbf{k}_2)f_n(\mathbf{k}_1)f_m(\mathbf{k}_2) \tag{E.8}$$

$$e^{i(\mathbf{k}_1+\mathbf{k}_2)\cdot\mathbf{x}} \frac{e^{-i(\omega_1+\omega_2)t_x}}{(E_1 - \omega_1 - i\epsilon)(E_2 - \omega_2 - i\epsilon)} \pm (n \leftrightarrow m)\,, \tag{E.9}$$

and as usual the limit $\epsilon \to 0$ is understood.

In a similar way other matrix elements are obtained.

Appendix F

Computation of $|0(\theta)\rangle$

Let us compute $|0(\theta)\rangle$ given in Eq. (2.52). Eq. (2.63) for fermions is obtained in a similar way. We compute $f_0(\theta) = \langle 0|G^{-1}(\theta)|0\rangle$, where $f_0(\theta)$ is a function θ_k. The functional derivative of $f_0(\theta)$

$$\frac{\delta}{\delta\theta_l} f_0(\theta) \equiv \lim_{\epsilon\to 0} \frac{f_0[\theta_k + \epsilon\delta(\mathbf{k} - \mathbf{l})] - f_0(\theta_k)}{\epsilon}, \qquad (\text{F.1})$$

gives

$$\frac{\delta}{\delta\theta_l} f_0(\theta) = \langle 0|G^{-1}(\theta)b_{-1}^{\dagger}a_1^{\dagger}|0\rangle = -\langle 0|a_1b_{-1}G^{-1}(\theta)|0\rangle. \qquad (\text{F.2})$$

Using Eqs. (2.48) with θ changed in $-\theta$, the last term in (F.2) becomes

$$\langle 0|a_1b_{-1}G^{-1}(\theta)|0\rangle = \sinh^2\theta_l\langle 0|G^{-1}(\theta)b_{-1}^{\dagger}a_1^{\dagger}|0\rangle$$
$$+ \sinh\theta_l\cosh\theta_l\langle 0|G^{-1}(\theta)a_1a_1^{\dagger}|0\rangle. \qquad (\text{F.3})$$

The second equality in (F.2) leads to

$$\frac{\delta}{\delta\theta_l} f_0(\theta) = -\tanh\theta_l\delta(\mathbf{0})f_0(\theta), \qquad (\text{F.4})$$

where we have used $a_1a_1^{\dagger}|0\rangle = [a_1, a_1^{\dagger}]|0\rangle = \delta(\mathbf{0})|0\rangle$, which is not well defined since $\delta(\mathbf{0}) \equiv \delta(\mathbf{k} - \mathbf{l})|_{\mathbf{k}=\mathbf{l}}$ is infinite. For our task, we might ignore such a difficulty and formally proceed to solve the functional differential equations (F.4) subject to the boundary condition $f_0(0) = 1$ (assuming indeed $\delta(\mathbf{0})$ to be a finite quantity). The solution is

$$f_0(\theta) = \exp\left(-\delta(\mathbf{0})\int d^3k\ln\cosh\theta_k\right). \qquad (\text{F.5})$$

Since the generator $G(\theta)$ contains pairs of operators ab and $a^{\dagger}b^{\dagger}$ we need to compute the matrix elements $f_n(\theta) = \langle 0|[a_1b_{-1}]^n G^{-1}(\theta)|0\rangle$. Use of functional derivative leads to the recurrence relation

$$\frac{\delta}{\delta\theta_l} f_n(\theta) = -f_{n+1}(\theta) + n^2[\delta(\mathbf{0})]^2 f_{n-1}(\theta), \qquad (\text{F.6})$$

which is again ill defined. By induction, the formal solution is shown to be

$$f_n(\theta) = n![\delta(\mathbf{0})]^n \tanh^n \theta_l \exp\left(-\delta(\mathbf{0}) \int d^3k \ln \cosh \theta_k\right). \qquad (\text{F.7})$$

Indeed, Eq. (F.7) holds for $n = 1$. Assume it holds for all $n \leq m$, $\forall m$. Then one can show that it holds also for the case $m + 1$. f_0, f_n, for all n, provide the expansion coefficients of $G^{-1}(\theta)|0\rangle$. Thus we arrive at

$$|0(\theta)\rangle = G^{-1}(\theta)|0\rangle \qquad (\text{F.8})$$

$$= \exp\left(-\delta(\mathbf{0}) \int d^3k \ln \cosh \theta_k\right) \exp\left(\int d^3k \tanh \theta_k a^\dagger_{\mathbf{k}} b^\dagger_{-\mathbf{k}}\right)|0\rangle.$$

Eq. (2.52) is then obtained by recalling that

$$\delta(\mathbf{k}) = \frac{1}{(2\pi)^3} \int d^3x e^{i\mathbf{k}\cdot\mathbf{x}}, \qquad (\text{F.9})$$

so that one could formally write

$$\delta(\mathbf{0}) = \frac{1}{(2\pi)^3} \times [\text{volume in which the system is confined}]. \qquad (\text{F.10})$$

In the case of finite volume V, say for a box of sides L_i, $i = 1, 2, 3$, $V = L_1 L_2 L_3$, one deals with discretized momenta $\mathbf{k}$ of components

$$k_{n_i} = \frac{2\pi n_i}{L_i}, \quad i = 1, 2, 3 \quad n_i = 0, \pm 1, \pm 2, \ldots, \qquad (\text{F.11})$$

so that one starts by considering, instead of (2.49), the generator

$$G(\theta) = \exp\left(\sum_{\mathbf{k}} \theta_{\mathbf{k}}(a_{\mathbf{k}} b_{-\mathbf{k}} - b^\dagger_{-\mathbf{k}} a^\dagger_{\mathbf{k}})\right), \qquad (\text{F.12})$$

where the sum is extended to the discrete set of momenta $\mathbf{k}$. Then,

$$|0(\theta)\rangle = G^{-1}(\theta)|0\rangle \qquad (\text{F.13})$$

$$= \exp\left(-\sum_{\mathbf{k}} \ln \cosh \theta_{\mathbf{k}}\right) \exp\left(\sum_{\mathbf{k}} \tanh \theta_k a^\dagger_{\mathbf{k}} b^\dagger_{-\mathbf{k}}\right)|0\rangle,$$

instead of Eq. (2.52). We see that, as far as

$$\sum_{\mathbf{k}} \ln \cosh \theta_{\mathbf{k}} \neq \infty, \qquad (\text{F.14})$$

for any non-vanishing $\theta_{\mathbf{k}}$, $G^{-1}(\theta)$ is a unitary operator mapping vectors of the Fock space into vectors of the same space. However, Eq. (2.52) is recovered in the continuous limit (the limit of infinitely many degrees of freedom) since $d^3k = \frac{(2\pi)^3}{V}$ (cf. Eq. (F.11)). Thus, since in such a limit

$$\sum_{\mathbf{k}} \rightarrow \frac{V}{(2\pi)^3} \int d^3k, \qquad (\text{F.15})$$

unitarily inequivalent representations are obtained in the $V \to \infty$ limit.

Appendix G

Orthogonality of flavor vacua at different times

The flavor vacua discussed in Section 2.7 are time-dependent states. This feature gives rise to orthogonality, in the infinite volume limit, among flavor vacuum states at different times. We show this explicitly in the fermionic case. Let us first report the form of the flavor vacuum making explicit its condensate structure:

$$
|0(t)\rangle_{e,\mu} = \prod_{\mathbf{k},r} \Big[(1 - \sin^2\theta\, |V_\mathbf{k}|^2)
$$

$$
- \epsilon^r \sin\theta \cos\theta\, V_\mathbf{k}(t) \left(\alpha_{\mathbf{k},1}^{r\dagger} \beta_{-\mathbf{k},2}^{r\dagger} + \alpha_{\mathbf{k},2}^{r\dagger} \beta_{-\mathbf{k},1}^{r\dagger} \right)
$$

$$
+ \epsilon^r \sin^2\theta\, V_\mathbf{k}(t) \left(U_\mathbf{k}^*(t)\alpha_{\mathbf{k},1}^{r\dagger} \beta_{-\mathbf{k},1}^{r\dagger} - U_\mathbf{k}(t)\alpha_{\mathbf{k},2}^{r\dagger} \beta_{-\mathbf{k},2}^{r\dagger} \right)
$$

$$
+ \sin^2\theta\, V_\mathbf{k}^2(t)\, \alpha_{\mathbf{k},1}^{r\dagger} \beta_{-\mathbf{k},2}^{r\dagger} \alpha_{\mathbf{k},2}^{r\dagger} \beta_{-\mathbf{k},1}^{r\dagger} \Big] |0\rangle_{1,2}\,. \tag{G.1}
$$

We have

$$
{}_{e,\mu}\langle 0(t')|0(t)\rangle_{e,\mu} = \prod_k C_k^2(t-t') = e^{2\sum_k \ln C_k(t-t')}\,, \tag{G.2}
$$

with

$$
C_k(t-t') \equiv (1 - \sin^2\theta\, |V_k|^2)^2 + 2\,\sin^2\theta\, \cos^2\theta\, |V_k|^2\, e^{i(\omega_{k,2}+\omega_{k,1})(t-t')}
$$

$$
+ \sin^4\theta\, |V_k|^2\, |U_k|^2 \left(e^{2i\omega_{k,1}(t-t')} + e^{2i\omega_{k,2}(t-t')} \right)
$$

$$
+ \sin^4\theta\, |V_k|^4\, e^{2i(\omega_{k,2}+\omega_{k,1})(t-t')}\,. \tag{G.3}
$$

In the infinite volume limit we obtain (note that $|C_k(t)| \leq 1$ for any value of k, t, and of the parameters θ, m_1, m_2):

$$
\lim_{V\to\infty} {}_{e,\mu}\langle 0|0(t)\rangle_{e,\mu}
$$

$$
= \lim_{V\to\infty} \exp\left[\frac{2V}{(2\pi)^3} \int d^3k \,(\ln|C_k(t-t')| + i\alpha_k(t-t')) \right] = 0\,, \tag{G.4}
$$

with $\alpha_k(t) = \tan^{-1}\left(\Im m[C_k(t)]/\Re e[C_k(t)]\right)$.

89

Flavor vacua at different times are thus orthogonal states. From this it follows that the Hilbert spaces built on such vacua are unitarily inequivalent state spaces.

We comment on the different mechanisms underlying the inequivalence among flavor vacua at different times and the one between $|0(t)\rangle_{e,\mu}$ and $|0\rangle_{1,2}$. In the last case, the inequivalence arises due to the difference in the condensate structure of the two states $|0(t)\rangle_{e,\mu}$ and $|0\rangle_{1,2}$. In the former case, the condensate structure is the same at any time; however, the relative phases of the various terms are different. As a consequence, flavor vacua with different times are degenerate in energy while $|0(t)\rangle_{e,\mu}$ and $|0\rangle_{1,2}$ have different energy contents.

Appendix H

Entanglement in neutrino oscillations

It has been shown [86, 87] that flavor states of mixed particles, as those describing flavor neutrinos and quarks (see Section 2.7) can be regarded as entangled states.

Entanglement [215, 552] is a fundamental property of quantum system and is at the basis of important developments in quantum optics and quantum computing [499]. In particular, mode entanglement in single-particle states has been recognized and is well established [605, 624]: it may arise whenever the Hilbert space has a tensor product structure. This is indeed the case with flavor mixing. Here we discuss the simple case of two-flavor neutrino mixing, both in the context of QM and then in QFT. The more realistic case of three-flavor mixing, which involves multipartite entanglement, can be found in [86, 87].

Neutrino entanglement in Quantum Mechanics

Neutrino mixing for the case of three generations is described by the Pontecorvo–Maki–Nakagawa–Sakata (PMNS) matrix [156]. In the case of two flavors, the PMNS matrix reduces to the 2×2 rotation matrix $\mathbf{U}(\theta)$:

$$\mathbf{U}(\theta) = \begin{pmatrix} \cos\theta & \sin\theta \\ -\sin\theta & \cos\theta \end{pmatrix}, \tag{H.1}$$

which connects the neutrino states with definite flavor with those with definite masses [77]:

$$|\underline{\nu}^{(f)}\rangle = \mathbf{U}(\theta) |\underline{\nu}^{(m)}\rangle, \tag{H.2}$$

where $|\underline{\nu}^{(f)}\rangle = (|\nu_e\rangle, |\nu_\mu\rangle)^T$ and $|\underline{\nu}^{(m)}\rangle = (|\nu_1\rangle, |\nu_2\rangle)^T$.

From Eq. (H.2), we see that each flavor state is given by a superposition of mass eigenstates, i.e., $|\nu_\alpha\rangle = U_{\alpha 1}|\nu_1\rangle + U_{\alpha 2}|\nu_2\rangle$, with $\langle \nu_i | \nu_j \rangle = \delta_{i,j}$.

Flavor states can be seen as entangled states, if we establish the following correspondence with two-qubit states:

$$|\nu_1\rangle \equiv |1\rangle_1|0\rangle_2 \equiv |10\rangle, \qquad |\nu_2\rangle \equiv |0\rangle_1|1\rangle_2 \equiv |01\rangle, \tag{H.3}$$

where $|\rangle_i$ denotes states in the Hilbert space for neutrinos with mass m_i. Thus, the occupation number allows us to interpret the flavor states as entangled superpositions of the mass eigenstates [86].

The time evolution of the flavor neutrino states Eq. (H.2) is given by:

$$|\underline{\nu}^{(f)}(t)\rangle = \widetilde{\mathbf{U}}(t)|\underline{\nu}^{(f)}\rangle, \qquad \widetilde{\mathbf{U}}(t) \equiv \mathbf{U}(\theta)\,\mathbf{U}_0(t)\,\mathbf{U}(\theta)^{-1}, \tag{H.4}$$

where $|\underline{\nu}^{(f)}\rangle$ are the flavor states at $t=0$, $\mathbf{U}_0(t) = \mathrm{diag}(e^{-i\omega_1 t}, e^{-i\omega_2 t})$, and $\widetilde{\mathbf{U}}(t=0) = \mathbb{1}$. At time t, the probability associated with the transition $\nu_\alpha \to \nu_\beta$ is

$$P_{\nu_\alpha \to \nu_\beta}(t) = |\langle \nu_\beta|\nu_\alpha(t)\rangle|^2 = |\widetilde{\mathbf{U}}_{\alpha\beta}(t)|^2, \qquad \alpha,\beta = e,\mu. \tag{H.5}$$

The explicit expressions for the above transition probabilities are:

$$P_{\nu_e \to \nu_e}(t) = 1 - \sin^2 2\theta \, \sin^2\left(\frac{\omega_2-\omega_1}{2}t\right), \tag{H.6a}$$

$$P_{\nu_e \to \nu_\mu}(t) = \sin^2 2\theta \, \sin^2\left(\frac{\omega_2-\omega_1}{2}t\right). \tag{H.6b}$$

Let us now establish the following correspondence with two-qubit flavor states:

$$|\nu_e\rangle \equiv |1\rangle_e|0\rangle_\mu, \qquad |\nu_\mu\rangle \equiv |0\rangle_e|1\rangle_\mu. \tag{H.7}$$

States $|0\rangle_\alpha$ and $|1\rangle_\alpha$ correspond, respectively, to the absence and the presence of a neutrino in mode α. Eq. (H.4) can then be recast as

$$|\nu_\alpha(t)\rangle = \widetilde{\mathbf{U}}_{\alpha e}(t)|1\rangle_e|0\rangle_\mu + \widetilde{\mathbf{U}}_{\alpha\mu}(t)|0\rangle_e|1\rangle_\mu, \tag{H.8}$$

with the normalization condition $\sum_\beta |\widetilde{\mathbf{U}}_{\alpha\beta}(t)|^2 = 1$ $(\alpha,\beta = e,\mu)$. The time-evolved states $|\underline{\nu}^{(f)}(t)\rangle$ are entangled superpositions of the two flavor eigenstates with time-dependent coefficients.

In order to quantify the *static entanglement* associated to the flavor neutrino states, we consider explicitly the case of an electron neutrino state at time t, which in terms of the mass eigenstates, reads:

$$|\nu_e(t)\rangle = e^{-i\omega_1 t}\cos\theta\,|\nu_1\rangle + e^{-i\omega_2 t}\sin\theta\,|\nu_2\rangle, \tag{H.9}$$

where $|\nu_i\rangle$ are interpreted as the qubits, see Eq. (H.3).

Following the usual procedure, we construct the density operator $\rho^{(\alpha)} = |\nu_\alpha(t)\rangle\langle\nu_\alpha(t)|$ corresponding to the pure state $|\nu_\alpha(t)\rangle$. Then we consider the

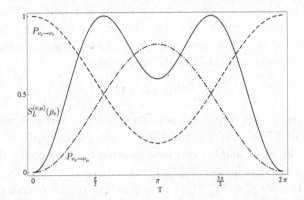

Fig. H.1   Linear entropy $S_L^{(e;\mu)}(\rho_e)$ (full) as a function of the scaled time $T = \frac{2Et}{\Delta m_{12}^2}$. The mixing angle θ is fixed at the experimental value $\sin^2\theta = 0.314$. The transition probabilities $P_{\nu_e \to \nu_e}$ (dashed) and $P_{\nu_e \to \nu_\mu}$ (dot-dashed) are reported as well for comparison.

density matrix $\rho_i^{(\alpha)} = \mathrm{Tr}_j[\rho^{(\alpha)}]$ reduced with respect to the index j. For the specific case of the state Eq. (H.9), we have $\rho^{(e)} = |\nu_e(t)\rangle\langle\nu_e(t)|$ and

$$\rho_1^{(e)} = \mathrm{Tr}_2[\rho^{(e)}] = \cos^2\theta\,|1\rangle_{1\,1}\langle1| + \sin^2\theta\,|0\rangle_{1\,1}\langle0|, \qquad (\text{H.10a})$$

$$\rho_2^{(e)} = \mathrm{Tr}_1[\rho^{(e)}] = \cos^2\theta\,|0\rangle_{2\,2}\langle0| + \sin^2\theta\,|1\rangle_{2\,2}\langle1|. \qquad (\text{H.10b})$$

It is then easy to calculate the corresponding linear entropies, which turn out to be equal:

$$S_L^{(1;2)}(\rho_e) = 2\left(1 - \mathrm{Tr}_1[(\rho_1^{(e)})^2]\right) = \sin^2(2\theta), \qquad (\text{H.11a})$$

$$S_L^{(2;1)}(\rho_e) = 2\left(1 - \mathrm{Tr}_2[(\rho_2^{(e)})^2]\right) = \sin^2(2\theta). \qquad (\text{H.11b})$$

Similar results are found for the muon neutrino state. Note that the above results are particular cases of the more general ones obtained for the three flavor neutrino states in [86], where it was found that such states can be classified as generalized W states [209]. In the present (two-flavor) case, the form of the entangled state is simply that of a Bell state.

Eqs. (H.11) express the fact that flavor neutrino states at any time can be regarded as entangled superpositions of the mass qubits $|\nu_i\rangle$, where the entanglement is a function of the mixing angle only.

Let us now turn to the *dynamic entanglement* arising in connection with flavor oscillations, and rewrite the electron neutrino state $|\nu_e(t)\rangle$ as

$$|\nu_e(t)\rangle = \widetilde{\mathbf{U}}_{ee}(t)\,|\nu_e\rangle + \widetilde{\mathbf{U}}_{e\mu}(t)\,|\nu_\mu\rangle, \qquad (\text{H.12})$$

where $|\nu_e\rangle$, $|\nu_\mu\rangle$ are the flavor neutrino states at time $t = 0$ and are now taken as the relevant qubits (cf. Eq. (H.7)). By proceeding in a similar way as done for the static case, we arrive at the following expression for the linear entropies associated to the above state:

$$S_L^{(\mu;e)}(\rho_e) = S_L^{(e;\mu)}(\rho_e) = 4|\tilde{U}_{ee}(t)|^2\,|\tilde{U}_{e\mu}(t)|^2. \tag{H.13}$$

Eq. (H.13) establishes that the linear entropy of the reduced state is equal to the product of the two-flavor transition probabilities given in Eq. (H.5). It is remarkable that simple expressions similar to those of Eq. (H.13) hold also for the three-flavor case [87].

In Fig. H.1 we plot $S_L^{(e;\mu)}(\rho_e)$ as a function of the scaled, dimensionless time $T = \frac{2Et}{\Delta m_{12}^2}$. In the same figure, we also display the transition probabilities $P_{\nu_e \to \nu_e}$ and $P_{\nu_e \to \nu_\mu}$.

The plots have a clear physical interpretation. At time $T = 0$, the entanglement is zero, the global state of the system is factorized, and the two flavors are not mixed. For $T > 0$, flavors start to oscillate and the entanglement is maximal at largest mixing: $P_{\nu_e \to \nu_e} = P_{\nu_e \to \nu_\mu} = 0.5$, and minimum at $T = \pi$.

Neutrino entanglement in Quantum Field Theory

We consider now the entanglement associated to flavor neutrino states in the context of QFT [88]. In such a case, entanglement can be efficiently quantified by considering the variances of observables [386], such as the charges discussed in Section 2.7.

Let us start with the Noether charges Q_{ν_i}, which are expected to characterize the amount of static entanglement present in the states Eq. (2.133). We obtain indeed:

$$\Delta Q_i(\nu_e)(t) = \langle \nu^r_{\mathbf{k},e}|Q_i^2|\nu^r_{\mathbf{k},e}\rangle - \langle \nu^r_{\mathbf{k},e}|Q_i|\nu^r_{\mathbf{k},e}\rangle^2$$
$$= \frac{1}{4}\sin^2(2\theta)\,, \qquad i = 1, 2. \tag{H.14}$$

in agreement with the quantum mechanical result Eq. (H.11).

Next we consider dynamic entanglement, which is described by the variances of the flavor charges. We have:

$$\Delta Q_e(\nu_e)(t) = \langle \nu^r_{\mathbf{k},e}|Q_e^2(t)|\nu^r_{\mathbf{k},e}\rangle - \langle \nu^r_{\mathbf{k},e}|Q_e(t)|\nu^r_{\mathbf{k},e}\rangle^2$$
$$= \mathcal{Q}^{\mathbf{k}}_{\nu_e \to \nu_e}(t)\,\mathcal{Q}^{\mathbf{k}}_{\nu_e \to \nu_\mu}(t)\,, \tag{H.15}$$

which formally resembles the quantum mechanical result Eq. (H.13). The differences are now due to the presence of the flavor condensate, which affect the oscillation formulas Eqs. (2.147).

The above discussed example of entanglement in QFT, although limited to the very specific situation of particle mixing, offers elements of possible general interest. Indeed, both the static and the dynamical entanglement arise in connection with unitarily inequivalent representations: in the case of the static entanglement, the flavor Hilbert space at time t to which the entangled state $|\nu_\sigma(t)\rangle$ belongs, is unitarily inequivalent to the Hilbert space for the qubit states $|\nu_i\rangle$, see Eq. (2.104); on the other hand, in the case of dynamical entanglement, where the qubits are taken to be the flavor states at time $t = 0$, the inequivalence is among flavor Hilbert spaces at different times, see Eq. (G.4).

Since inequivalent representations are associated with a non-trivial condensate vacuum structure, the above discussion suggests that, in the context of QFT, many interpretational issues connected with entanglement [68] could be revisited in this new light.

Finally, we note that the above results could be of interest for the development of quantum information protocols based on neutrinos [87].

Chapter 3

Spontaneous breakdown of symmetry and the Goldstone theorem

3.1 Introduction

In this Chapter and in the following one we consider the problem of the manifestation of the symmetry properties of the Heisenberg field equations at the observable level. Our interest is focused on continuous symmetries.

The symmetries of the dynamics may appear "broken" or rearranged into different symmetry patterns at the level of the physical asymptotic fields. For example, in the case of the ferromagnets, the dynamics, i.e., the Lagrangian from which the Heisenberg equations are derived, is invariant under the spin rotational $SU(2)$ transformations of the Heisenberg field operators. On the other hand, the observable system, which is described in terms of quasiparticle (or physical) field operators, is characterized by its non-zero magnetization. The appearance of the privileged direction into which the magnetization points and its non-vanishing value signal that the original isotropy, implied by the rotational $SU(2)$ symmetry, has been broken and the observable ferromagnetic ordering has been realized. Order thus appears as a manifestation of the symmetry breaking, or, in other words, as lack of symmetry. Indeed, the equations for the quasiparticle fields are not invariant under the $SU(2)$ group.

We thus have two sets of equations with different invariance properties: the dynamical equations, invariant under the continuous transformation group, say G, and the equations for the quasiparticle field operators, in terms of which observables are expressed, which are invariant under a transformation group G', in general different from G.

The general questions concerning which one is the group G', how the "symmetry rearrangement" $G \to G'$ occurs, and which one is its physical meaning are the objects of study in this Chapter and in the following one.

We note that we can always add to the Lagrangian invariant under G a "symmetry breaking" term. For example, we could add terms describing the coupling of the field operators to some external field (a typical example is switching on a magnetic field in the region where a beam of electrons is passing; as a result, by properly choosing the magnetic field orientation, the beam splits into spin-up and spin-down electron beams, which amounts to the breakdown of the symmetry in the spin orientation in the original beam). In such a case, the symmetry is said to be "explicitly" broken. By adding the symmetry breaking terms to the Lagrangian we actually describe a new system, different from the one described by the original invariant Lagrangian. We are not interested in this type of symmetry breaking in this book. Instead, we will mostly discuss what is called the *spontaneous symmetry breaking* (SSB). This occurs when the ground state or vacuum state $|0\rangle$ is not invariant under the continuous group G under which the Lagrangian is invariant. In other words, it occurs when some of the generators of G do not annihilate the ground state $|0\rangle$.

As explained in the previous Chapters, we must carefully consider which one, among the infinitely many unitarily inequivalent representations of the canonical commutation relations, is the state space where our field operators are realized. It is therefore necessary to consider, along with the symmetry properties of the Lagrangian, the symmetry properties of the space for the physical states we choose to work with. Our choice might be such that the vacuum is itself invariant under the symmetry group G, and then no breakdown of symmetry occurs. We call such a vacuum the "normal" or symmetric vacuum (of course, since all the other states of the system are constructed from the vacuum, we implicitly refer to such states too). However, in the presence of specific boundary conditions and/or specific ranges of values of physically relevant parameters, we could also choose any one of the non-symmetric ground states; each one being associated to one of the infinitely many unitarily inequivalent representations, thus considering the possibility of symmetry breaking.

In principle, under convenient boundary conditions, everyone among the possible non-symmetric vacua can be realized in Nature. In the ferromagnets, for example, the magnetization strength may in principle assume any value up to a saturation limit: the system, driven by its dynamics (Lagrangian), "spontaneously" sets into the state characterized by a specific magnetization under given boundary conditions. This is why the breakdown of the symmetry is said to be "spontaneous". The difference among the magnitude values of the magnetization in magnetized states provides a measure of the "difference" among these states. Since the magnetization fully characterizes the non-symmetric vacuum of a given state space, it acts as a label for the inequivalent representations. Each one of these representations describes a different physical phase of the system. Observables, such as the magnetization, are called order parameters. The same basic dynamics (same Lagrangian invariant under G) may thus manifest itself into a variety of stable symmetry patterns at the level of the observables. Changes occurring in the order parameter describe transitions among the system physical phases (phase transitions).

In this Chapter and in the following one we will not consider temperature effects usually contrasting the emergence of ordered patterns, or inducing symmetry restoration. We will also omit to consider questions related to renormalization problems. Here we are mainly interested in showing that when symmetry is spontaneously broken the dynamics requires the existence of a massless particle, the Nambu–Goldstone (NG) particle, which is the carrier of the long-range correlation among the system elementary constituents [26, 617, 619]. In our discussion we will consider models with $U(1)$ and $SU(2)$ symmetries, with and without gauge fields. We will use the tool of the functional integration which allows to us to obtain general results, independent of the specific form of the considered model Lagrangian. Finite volume effects and spacetime dimensionality are also considered in this Chapter. The rearrangement of the symmetry into the observable symmetry patterns is described in Chapter 4.

3.2 Invariance and symmetry

The terminology "breakdown of symmetry" might suggest that the invariance of the Lagrangian under the continuous symmetry group G is in some way lost when symmetry is broken. However, the invariance of the Lagrangian means that the generators of the group G commute with the Hamiltonian and this determines the constants of motion. Therefore, the internal consistency of the theory requires that the invariance cannot simply disappear. The terminology such as "hidden" or "secret" symmetry has also been used in connection with symmetry breaking. The question arises of which one is the relation between the symmetry group G for the Heisenberg field equations and the one for the physical field equations G': the problem is the one of the mapping between the basic or Heisenberg field language and the physical field language. Such a mapping is displayed through the dynamical map (see Chapter 2). Due to the non-linear character of the dynamical map, which reflects non-linear dynamical effects, one may expect that the symmetry properties at the level of the Heisenberg field operators may manifest themselves through a mechanism of *dynamical rearrangement of symmetry* at the level of physical fields. To be more specific, we remark that one stringent requirement for the dynamical map is that both members of the mapping (cf. Eq. (2.2)) must have the same symmetry properties, although not necessarily term-by-term. The theory is said to be invariant under the transformation $g \in G$ when the basic dynamical equations are form-invariant under g. Thus we need to consider the transformation of the Heisenberg field operator

$$\psi(x) \rightarrow \psi'(x) = g[\psi(x)], \qquad (3.1)$$

where we have dropped the subscripts for brevity. The dynamical map (2.2), which is a weak equality and can be schematically written as

$$\psi(x) = \Psi[\phi(x)], \qquad (3.2)$$

implies that the transformation (3.1) is induced by the transformation, say h, of the physical field ϕ[1]

$$\phi(x) \rightarrow \phi'(x) = h[\phi(x)], \qquad (3.3)$$

with $h \in G'$ and such that

$$g[\psi(x)] = \Psi[h[\phi(x)]]. \qquad (3.4)$$

We say that the dynamical rearrangement of symmetry occurs when $G' \neq G$ [423, 424, 485, 568, 613, 617, 619, 644]. SSB thus appears as a dynamical effect: the Lagrangian (or, equivalently, the field equations for the Heisenberg fields) is and remains fully invariant under G, while, because of the dynamics, the symmetry can appear in a different form at the physical level.

This can be understood by noticing that any observation on a system described by fields (thus endowed with an infinite number of degrees of freedom) is a collection of local observations, so that one may miss an infinitesimal effect of the order of magnitude of $\frac{1}{V}$, with the volume $V \rightarrow \infty$. Such a local infinitesimal effect is called the infrared effect [460, 570]. When it is integrated over the whole system, it produces the difference $G' \neq G$. In the next Chapter we will prove that the original symmetry is restored when the infrared effect is taken into account.

---

[1]As far as no misunderstanding arises, here and in the following we use ϕ instead of ϕ_{in} and/or ϕ_{out}.

We have observed in the previous Section that when symmetry is spontaneously broken, ordered patterns appear in the ground state or vacuum. It is then reasonable to expect that some long-range correlation modes, which create such macroscopic patterns, are present in the system. When the ordered patterns extend over the whole system, the range over which these correlation modes should be able to propagate is of the order of the system size, i.e., infinite (compared to the scale of the system microscopic component size). This implies, as we will see, that they are massless modes. These are in fact the Nambu–Goldstone modes [289,291,489,492,493]. Since they do not carry energy at their lowest (zero) momentum, creation of these modes in the ground state does not modify its (zero) energy value: the ground state where they are created (condensed) is still a stable state, namely a (degenerate) ground state. The ordered pattern becomes observable by exciting these correlation modes out of the ground state. We analyze in detail these points in the following, also considering the effects of gauge fields on symmetry breaking.

3.3   Irreducible representations of the symmetry group

Since invariance cannot be lost, the equations for the physical fields must be invariant under the "rearranged" symmetry. Thus, we are led to study the generators of the symmetry group written in terms of physical fields, i.e., we study the dynamical maps of generators in terms of normal ordered products of physical field operators. Of course, the invariance of the theory under a certain symmetry group G is the invariance of the canonical field equations

$$i[\hat{H}, \psi(x)] = \frac{\partial}{\partial t}\psi(x), \qquad (3.5)$$

where $\hat{H}$ is the Hamiltonian.[2] The invariance of (3.5) under the transformations of $\psi(x)$ belonging to the group G leads to

$$i[\hat{H}, D_i] = \frac{\partial}{\partial t}D_i = 0, \qquad \forall\, i = 1, 2, \ldots, m. \qquad (3.6)$$

where D_i denote the generators of G and m is said to be its dimension. The symmetry generators are thus time-independent and Eq. (3.6) expresses the (symmetry) "charge" conservation.

---

[2]In this and in the following Section, we use a "hat" for the Hamiltonian, in order not to create confusion with the stability subgroup H.

By using the dynamical map, the Heisenberg currents, which here are generically denoted by $j_\mu[\psi(x)]$, can be written [617, 619] in terms of the physical fields: $j_\mu[\psi(x)] = j_\mu[\phi(x), B(x)]$, where we denote massive fields by $\phi(x)$ and massless fields by $B(x)$. This shows that the spacetime dependence of the currents is controlled by the physical fields.

It is interesting to ask which one is the most general set of transformations which leaves invariant the free field equations. We will show in the following Chapter that the group of transformations G' for the physical field equations is, under some quite general constraints, the Inönü–Wigner contraction [321, 336] of the original symmetry group G [182]. In the following Section, we will see that the necessary and sufficient condition for the symmetry to be spontaneously broken is that the symmetry generators expressed in terms of the physical fields must contain terms linear in the massless fields [423, 424, 485, 568, 617].

As a consequence of the time-independence of the symmetry generators, the non-linear part of the generator, if any, can only be a function of fields with the same masses [568]. This means that one cannot mix particles of different masses without supplying energy. Indeed, the time independence of the generator means that it cannot supply energy. We express this fact by saying that there cannot be mass differences among physical fields which belong to the same irreducible representation of the symmetry group [423, 424, 485, 568, 581].

In the case of the mixing of neutrinos and mixing of mesons, considered in Section 2.7, particles with different masses are considered as being members of the same multiplet [77, 156]. Then, the mixing generators turn out to be time-dependent [108] and in the infinite volume limit inequivalent representations are obtained.

3.4   Symmetry and the vacuum manifold

The invariance of the Hamiltonian $\hat{H}$ under the action of the symmetry group G is expressed by Eq. (3.6), or equivalently by

$$D^{-1}(g)\hat{H}D(g) = \hat{H}, \qquad \forall g \in G. \qquad (3.7)$$

Here $D(g)$ is the unitary operator representing the element $g \in G$. SSB occurs, as said above, when the vacuum is invariant *only* under some subgroup H of G:

$$D(g)|0\rangle \equiv |0(g)\rangle \neq |0\rangle, \qquad \forall g \in G/H. \qquad (3.8)$$

The subgroup H is called the vacuum stability group. Of course, we exclude that $|0\rangle$ is the null vector. Eq. (3.8) shows that for any $g \in G/H$ a "different" vacuum or ground state can be obtained. We will see below that these vacua are unitarily inequivalent and that the generators of G/H can be defined only formally. Because of (3.7), the state $|0(g)\rangle$, for any $g \in G/H$, is also an eigenstate of $\hat{H}$ with the same eigenvalue as $|0\rangle$: the ground state is degenerate. Since g belongs to the quotient space G/H (see the Appendix C for the definition of quotient space), we can describe the set of the unitarily inequivalent vacua, i.e., the *vacuum manifold* in terms of coordinates on G/H. So the vacuum manifold can be directly identified with the quotient space G/H.

When SSB occurs, i.e., Eqs. (3.7) and (3.8) hold, some operator $\hat{\Phi}$ exists, whose ground state expectation value, say Φ^0, called the *order parameter* [404] (e.g., magnetization in a ferromagnet) is non-zero: $\langle 0|\hat{\Phi}|0\rangle \equiv \Phi^0 \neq 0$. Whenever $g \in G/H$ changes, then also Φ^0 is changed: thus Φ^0 is not invariant under whole G but only under H. The symmetry H is known as the *broken-phase symmetry* or the *ordered-phase symmetry*.

The order parameter space $\mathcal{M}$ is then defined as a space of all possible values Φ^0. In the following, $\mathcal{M}$ is used both for the vacuum manifold and the order parameter space. See Appendix I for further details on the order parameter space.

From the SSB condition (3.8) we see that the non-invariance of the vacuum $|0\rangle$ means that it is not annihilated by some (or all) of the symmetry generators. For simplicity, we assume that this happens for one of them, say, D_n, with some definite n:

$$D_n|0\rangle \neq 0, \tag{3.9}$$

where, as usual,

$$D_n = \int d^3x \, j_0^{(n)}(x), \tag{3.10}$$

and the integration is extended over all the (infinite) volume. $j_\mu^{(n)}(x)$ is the Heisenberg current for which the conservation law

$$\partial^\mu j_\mu^{(n)}(x) = 0, \tag{3.11}$$

holds as a consequence of the invariance of the field equations. D_n is time-independent due to Eq. (3.11).

Physical (free) fields $\phi(x)$ satisfy linear homogeneous equations, e.g., $\Lambda(\partial)\phi(x) = 0$, where $\Lambda(\partial)$ is a differential operator (cf. Eq. (1.132)). Invariant transformations map linear homogeneous equations into linear homogeneous equations. Therefore, the generators must induce linear transformations of physical fields $\phi(x)$ and this implies that they contain terms at most bilinear in each of the physical fields [617]. Time-independence of the generator is ensured since these bilinear terms do not carry energy dependence [423]. These terms have the (normal ordered) operatorial form $a_i^\dagger a_j \delta_{ij}$ and therefore they annihilate the vacuum (here a_i generically denotes the annihilation operator of the physical fields $\phi_i(x)$). As a consequence, the condition of SSB, Eq. (3.9), cannot be satisfied by generators which are bilinear in the fields. Note that terms of the form $a_i a_j$ and $a_i^\dagger a_j^\dagger$ are excluded for any i, j, since they are not time-independent terms.

On the other hand, terms linear in the field, say $B(x)$, appearing in the generators, are compatible with the invariance of the field equations only in the case $B(x)$ is a massless field. This is because linear terms in the generators induce shift of the field $B(x)$ by a constant c-number c:

$$B(x) \rightarrow B(x) + c\,. \tag{3.12}$$

and only a massless field equation (for which $\Lambda(\partial)$ does not contain the mass term) is invariant under such a shift of the field. Similarly, terms linear in massive fields are excluded since they would violate the invariance of their field equations. Moreover, the hermiticity of the generators requires that both terms, linear in $B(x)$ and linear in $B^\dagger(x)$, appear in the generators. Clearly, generators containing linear terms in the fields can satisfy the SSB condition (3.9). Note that the time-independence of these linear terms in the generator is ensured by the space integration, which picks up the zero momentum contributions.

In conclusion, the necessary condition for the occurrence of SSB, Eq. (3.9), is that D_n contains terms linear in the physical fields. Consistency between invariance and SSB then requires the existence of massless fields among the physical fields, since linear terms in the generators are allowed only for massless fields due to the theory invariance. We remark that the appearance in at least one of the generators of a term linear in the massless fields is also a sufficient condition for the SSB to occur [423, 424, 485, 568].

Eq. (3.9) and the translational invariance of the vacuum imply that

$$\langle 0|D_n D_n|0\rangle = \int d^3x \, \langle 0|j_0^{(n)}(x)D_n|0\rangle\,, \tag{3.13}$$

is divergent. Note that since Eq. (3.9) implies linear terms in the massless fields, a situation similar to the one discussed in Section 2.3 occurs and then we conclude also for the present case that $\langle 0|e^{i\theta D_n}|0\rangle = 0$ and

$$\langle a|e^{i\theta D_n}|b\rangle = 0\,, \tag{3.14}$$

for any state $|a\rangle$ and $|b\rangle$ belonging to the physical space $\mathcal{H}_F$ (see also Appendix J). θ denotes the group parameter involved in the transformation. Eq. (3.14) means that $\exp(i\theta D_n)|b\rangle$, for any $|b\rangle$, does not belong to $\mathcal{H}_F$ [30, 114, 123, 619].

The generator D_n is not mathematically well defined (cf. Eq. (3.13)). It is said to be not unitarily implementable. Thus, much care is required in defining and using symmetry generators in the presence of symmetry breaking. The integration in Eq. (3.10) needs to be regularized and the D_n must be defined as follows:

$$D_n \equiv \lim_{f(x) \to 1} D_{f,n} = \lim_{f(x) \to 1} \int d^3x \, f(x) j_0^{(n)}(x). \tag{3.15}$$

Here $f(x)$ is a square integrable function which is a solution of the equation for the physical massless field. The limit $f(x) \to 1$ must be taken at the end of the computation. The prescription (3.15) is equivalent to perform the finite volume integration in Eq. (3.10) and the limit $V \to \infty$ at the end of the computations. In practice, one needs to consider the limit $f(x) \to 1$ of the commutators of generators.

Eq. (3.15) and the linear structure of D_n imply that the corresponding $B(x)$ field is transformed under D_n as the $f(x) \to 1$ limit of

$$B(x) \to B(x) + cf(x). \tag{3.16}$$

Eqs. (3.12) and (3.16) in turn imply that $B(x)$ must be a boson field since c and $f(x)$ are c-number quantities (not Grassmann quantities, which would imply dealing with anticommuting (fermion) fields).

We thus have reached the conclusion that SSB requires the existence of massless boson fields in the set of the physical fields. This is the content of the Goldstone theorem [289, 291, 489, 492, 493] and the massless bosons are called Nambu–Goldstone (NG) or simply Goldstone modes or particles. In the following Sections we prove the Goldstone theorem in the functional integral formalism.

The transformations (3.12) and (3.16) are canonical transformations. Therefore the fields $(B(x) + c)$ have well defined transformation properties under the original symmetry group, i.e., they form an irreducible representation of such a group. Thus, in the presence of SSB, there must be just enough massless bosons to form an irreducible representation of the symmetry group [423, 424, 485, 568].

The transformations (3.12) and (3.16) can be understood as regulating the condensation of the NG modes in the ground state. Since these transformations play a crucial role in recovering the original invariance of the theory through the dynamical rearrangement of the symmetry, the NG boson condensate adjusts its distribution in such a way as to preserve the conservation of the constants of motion. The dynamical regulation of boson condensation acquires thus the physical meaning of a stability principle (the one expressed by the invariance of the theory).

When a gauge field is present in the theory, the NG boson modes do not appear in the physical particle spectrum. However, as we will see, they are still essential in controlling the structure of the ground state.

We close this Section by observing that NG quanta are not simply mathematical constructs. They are realistic physical boson particles, dynamically generated by SSB. They undergo scattering with other particles of the system or with observational probes, as for example in neutron-phonon scattering in crystals (phonons are NG quanta [617,619]).

3.5   Boson transformation and inequivalent representations

The c-number c in Eq. (3.12) is sometimes called "dynamical spurion" [423, 617, 619]. It carries the symmetry information of the theory. Indeed, it is evident that, if we consider the vacuum expectation of $(B(x) + c)$

$$\langle 0 | (B(x) + c) | 0 \rangle = c, \tag{3.17}$$

c can be considered as the square root of the number of the massless bosons B condensed in the vacuum $|0\rangle$. In other words, the transformation (3.12) by inducing the boson condensation in the ground state of the system, causes the transition from the vacuum $|0\rangle$ with zero B-particle number, $\langle 0 | B^\dagger(x) B(x) | 0 \rangle = 0$, to the "new" vacuum $|0\rangle\rangle$ with non-zero number $|c|^2$ of B-particles condensed in it: $\langle 0 | (B^\dagger(x) + c^*)(B(x) + c) | 0 \rangle = |c|^2 \equiv \langle \langle 0 | B^\dagger(x) B(x) | 0 \rangle \rangle$. Since c is spacetime-independent, (3.12) induces a *homogeneous* boson condensation. The ordered patterns observed in the ground state $|0\rangle\rangle$ as a consequence of symmetry breaking are in this way described as a result of the boson condensation.

In full generality, many c-numbers $c, c', \ldots$ may be allowed. The spurion set $\{c\}$ is the carrier of the original symmetry quantum numbers. The boson condensation acts as a printing process of these symmetry properties on the ground state. This printing does not require any supply of energy since the bosons are massless. It follows that the original conservation laws implied by the invariance, can be recovered only if one takes into account the quantum numbers of these spurions [423, 424, 485, 568].

The fact that the transformation (3.12) is not unitarily implementable (cf. previous Section) means that ground states related to each other by the transformation (3.12) are unitarily inequivalent ground states, each one classified or labeled by c (cf. Eq. (3.17)). This implies that we have a multiplicity of degenerate ground states (zero energy eigenvectors).

Eq. (3.16) is called the *boson transformation*. The transformation (3.16) where the limit $f(x) \to 1$ is not taken is said to induce *non-homogeneous boson condensation*. Of particular interest will be the situation where the function $f(x)$, which is always required to be a solution of the equation for the physical field under consideration, carries some kind of singularity (e.g., a divergent or a topological singularity).

In conclusion, the dynamics of the physical system is described by the basic (Heisenberg) field equations. Different boundary conditions correspond to different solutions of the dynamics. Each solution is described in terms of states belonging to different (i.e., unitarily inequivalent) Fock spaces. Transformations like (3.12) or (3.16) with different choices of (singular) $f(x)$, relate the various dynamical solutions. In such a way the physical meaning of unitary inequivalence is fully manifest. The existence in QFT of infinitely many unitarily inequivalent representations of the canonical commutation relations is thus the pre-requisite for the possibility of symmetry breaking (and thus of physically different phases of the system). In Quantum Mechanics these phenomena cannot occur since all the representations are unitarily equivalent (the von Neumann theorem [648], cf. Chapter 1).

We recall that the generator (1.65) introduced in the discussion of the von Neumann theorem is linear in the field operators. As seen in the present Chapter, such a linearity plays quite a relevant a role in the symmetry breaking phenomenon. Similarly, in the examples of Section 2.3, such a kind of linear generator enters as a part of the model Hamiltonian with the consequences discussed there. Suppose that the shift of the B fields discussed above is not related with changes in the system temperature, but with some parameters appearing in the Hamiltonian (or with some combinations of them such as $g_\mathbf{k}$ and θ_k in the examples of Section 2.3). Then the transitions through the inequivalent representations induced by the B field shift are also called *quantum phase transitions* [545].

Finally, note that the boson condensation consequent to the breakdown of symmetry produces a coherent pattern in the ground state. This follows from the fact that the generators D_n, being linear in the field operator (similar to (1.65)), can be identified with the generator of the coherent states, as remarked at the end of Section 1.6.

3.6 Spontaneous symmetry breaking and functional integrals

The functional integration method is very useful for the study of symmetry properties since field transformations can be dealt with simply as change of integration variables. Moreover, it allows us to obtain results which are model-independent. This is because their derivation depends on the symmetry of the Lagrangian, rather than on its explicit form. The results concerning the symmetry behavior of the theory are thus independent of approximations and of general character.

Here we will not present a general introduction to the path-integral or functional method or formalism [273, 384, 541, 559, 586, 620], which is, however, discussed in some detail in Section 6.3. In this Section, we use it for the study of symmetry breaking.

In the path-integral method one considers the so-called generating functional from which one can derive directly and unambiguously the Green's functions of the theory. However, much care is needed to be able to distinguish between symmetric and asymmetric, or spontaneously broken, solutions [450,451]. One reason for this lies in the fact that the field equations are not sufficient to determine the solution uniquely. One needs to specify the boundary conditions of the problem and also to assign the proper representation among the many unitarily inequivalent ones. In other words, in QFT one should be able to incorporate in the generating functional the information of the chosen representation, e.g., the symmetric or the asymmetric vacuum representation (in QM this problem does not exist and this makes the functional method in QFT slightly different from the path-integral formalism in QM [234, 235, 238, 384]). This is a non-trivial problem to be solved and specific cases exist where ambiguities may emerge. As an example of these cases we consider the model for a real massless scalar field $\varphi(x)$ in the absence of gauge fields.

The Lagrangian density (when no misunderstanding arises we will simply call it the Lagrangian) is:

$$\mathcal{L} = \frac{1}{2}\partial^\mu \varphi(x)\partial_\mu \varphi(x)\,. \tag{3.18}$$

The symmetry of the Lagrangian under the translations in the φ-space

$$\varphi(x) \to \varphi(x) + \lambda\,, \tag{3.19}$$

with λ constant, is not a symmetry for the vacuum. Thus, $\langle 0|\varphi(x)|0\rangle =$ const. $\neq 0$, describes the asymmetric solution, where the constant is related to λ. $\langle 0|\varphi(x)|0\rangle = 0$ corresponds to the symmetric solution.

Our first task is to show how the information of SSB can be incorporated in the functional integral.

The generating functional for the Green's function is

$$Z[J] = \mathcal{N} \int \mathcal{D}\varphi \, \exp\left[i \int d^4x \left(\mathcal{L} + J(x)\varphi(x)\right)\right], \qquad (3.20)$$

where $\int \mathcal{D}\varphi$ denotes functional integration and $\mathcal{N}$, introduced for the normalization $Z[J] = 1$ at $J = 0$, is given by

$$\mathcal{N}^{-1} = \int \mathcal{D}\varphi \, \exp\left[i \int d^4x \mathcal{L}\right]. \qquad (3.21)$$

As usual, the $J(x)$-term has been added to the action. It denotes the source term for the $\varphi(x)$ field. After the computation it should be put equal to zero in order to restore the original dynamical problem specified by the Lagrangian (3.18). We observe that adding the J source term is not simply a mathematical expedient. Actually, the source term does describe the concrete physical operation of preparing our system (e.g., preparing by use of a filter the beam of the wanted particles) [620]. Therefore, the source term has the form of the field φ interacting with an external device described by the field J, the source field, indeed. A different experimental setting, implying a different source, say J', would be described by a different generating functional $Z[J']$. Observables, however, must not depend on the specific experimental arrangement used for the preparation of the system, i.e., they must not depend on the source term used in Z. Thus we put $J = 0$ after the computation. On the other hand, the formal usefulness of the source term resides in the fact that by operating with the functional derivative operator with respect to J on Z we obtain the functional average of the field $\langle\varphi\rangle$ which, in the functional approach, corresponds to the vacuum expectation value of the field operator $\hat{\varphi}$, $\langle 0|\hat{\varphi}|0\rangle$:

$$\left(-i\frac{\delta}{\delta J(x)} Z[J]\right)_{J=0} = \langle\varphi(x)\rangle \equiv \langle 0|\hat{\varphi}(x)|0\rangle. \qquad (3.22)$$

Higher order derivatives naturally give many-point correlation functions.

The points which make the phase stationary are those which mostly contribute to the functional integral $Z[J]$. The condition for the stationary phase $\delta L = 0$, with $L \equiv \int d^3x \left(\mathcal{L} + J\varphi\right)$, computed by neglecting surface terms, is

$$\partial^2\varphi(x) = J(x), \qquad (3.23)$$

where $(\partial^2 \equiv \partial_\mu\partial^\mu)$. Eq. (3.23) expresses the fact that $J(x)$ is the source for the φ field. It is invariant under (3.19). This means that the set of stationary points is an open line in the $\varphi(x)$ space. We will see that the contribution of all the points of this line makes $Z[J]$ undetermined.

Eq. (3.23) is formally solved by

$$\varphi_0(x) = \int d^4y \, G(x,y) J(y) \,, \tag{3.24}$$

with the Green's function $G(x,y) = \langle x \, | 1/\partial^2 | \, y \rangle$, $\partial^2 G(x,y) = \delta^{(4)}(x-y)$. Of course, we need to give a prescription in order to avoid the singularity in the Green's function.

In the numerator of $Z[J]$ given by (3.20) we expand φ around $\varphi_0(x)$

$$\varphi(x) = \varphi'(x) + \varphi_0(x) \,. \tag{3.25}$$

This does not change the generating functional since it amounts to the change of the integration variable (note that $\mathcal{D}\varphi = \mathcal{D}\varphi'$). We then obtain

$$Z[J] = \exp\left[\frac{i}{2} \int d^4x \, d^4y J(x) G(x,y) J(y) \right] \,. \tag{3.26}$$

From this, by use of (3.22), we have

$$\langle 0 | \hat{\varphi}(x) | 0 \rangle = 0 \,. \tag{3.27}$$

We thus obtain the symmetric vacuum expectation value. The broken symmetry condition $\langle 0 | \hat{\varphi}(x) | 0 \rangle \neq 0$ cannot be reached from (3.26).

Actually, we realize that the definition (3.20) of $Z[J]$ is ambiguous since the same expression (3.20) gives us a result different from (3.26) depending on the way we proceed in its computation. Indeed, let us first make the change of integration variable (3.19), $\varphi(x) \to \varphi(x) + \lambda$, in (3.20) and after that the further change of variable (3.25). The final result should not depend on these changes of integration variables. Instead, we now obtain

$$Z[J] = \exp\left[i\lambda \int d^4x \, J(x) \right] \exp\left[\frac{i}{2} \int d^4x \, d^4y J(x) G(x,y) J(y) \right] \,, \tag{3.28}$$

which differs from (3.26) and leads to

$$\langle 0 | \hat{\varphi}(x) | 0 \rangle = \lambda \,. \tag{3.29}$$

In conclusion, the ambiguity arises since the invariance of the Lagrangian under a continuous transformation of the fields implies an extended continuous domain of stationary points for the $Z[J]$ phase, thus making $Z[J]$ undetermined. This suggests that in order to eliminate this ambiguity we need to introduce a term which breaks the symmetry, thus reducing the domain of stationary points to a single point. In this way we introduce in the functional integral the information of the boundary condition selecting the solution we are interested in (in this case the symmetric or the asymmetric solution) [450, 451, 619]. The generating functional is then modified as follows

$$Z[J] = \mathcal{N} \int \mathcal{D}\varphi \exp\left[i\int dx \left(\mathcal{L} + J\varphi + i\frac{\epsilon}{2}(\varphi - v)^2\right)\right], \quad (3.30a)$$

$$\mathcal{N}^{-1} = \int \mathcal{D}\varphi \exp\left[i\int d^4x\mathcal{L} + i\frac{\epsilon}{2}(\varphi - v)^2\right], \quad (3.30b)$$

where v is a constant and the symmetry-breaking term is the ϵ-term. The limit $\epsilon \to 0$ is understood to be taken at the end of the calculations.

The stationary point equation (3.23) now becomes

$$(\partial^2 - i\epsilon)\varphi(x) = J(x) - i\epsilon v, \quad (3.31)$$

which is not invariant under (3.19): the set of stationary points now contains one single point whose position depends on v. Eq. (3.31) is formally solved by

$$\varphi_0(x) = v + \int d^4y \, G(x,y)J(y), \quad (3.32)$$

with the Green's function now given by $G(x,y) = \langle x |1/(\partial^2 - i\epsilon)| y\rangle$, $\partial^2 G(x,y) = \delta^{(4)}(x-y)$. Proceeding as above we now obtain

$$Z[J] = \exp\left[iv\int d^4x J(x)\right] \exp\left[\frac{i}{2}\int d^4xd^4y J(x)\langle x|\frac{1}{\partial^2 - i\epsilon}|y\rangle J(y)\right], (3.33)$$

to be compared with (3.26). On the other hand, performing in (3.30) the transformation (3.19) induces in (3.33) the factor $\exp\left[i\lambda \int d^4x J\right]$ and at the same time v is replaced by $v - \lambda$. Thus $Z[J]$ does not change under (3.19). It is now not ambiguous. From (3.33) we have

$$\langle\varphi(x)\rangle = v, \quad (3.34)$$

which gives us the symmetric solution when $v = 0$; otherwise, when $v \neq 0$, the asymmetric one.

The ϵ-term prescription also implies the correct regularization of the Green's function, which is indeed the Feynman propagator when $\epsilon > 0$.

In conclusion, the role of the ϵ-term is to specify the condition of symmetry breaking under which we want to compute the functional integral [450, 451]. Since it contributes to the source term (cf. (3.31)), it can be interpreted as describing the preparation of the system, namely it may be given the physical meaning of the small external field triggering the breakdown of symmetry. If the dynamics allows the symmetry broken solution, the asymmetric solution must survive to the turning off of such an ϵ-field (the $\epsilon \to 0$ limit).

Thus, we have learned how to cure possible ambiguities in the generating functional by introducing in the functional integral a term carrying the information of the representation where we want to work.

We observe that the ϵ-term may represent the system interaction with a measuring device or observer or the environment, which may turn into pushing the system into a specific inequivalent representation [146].

3.7 The Goldstone theorem

We now prove the Goldstone theorem by use of the functional formalism in theories without gauge fields. We consider the $U(1)$ complex scalar model and, as an example of a non-Abelian theory, the $SU(2)$ ferromagnetic model. In the following Chapter we study the rearrangement of symmetry in these cases. In Section 3.8 we consider the case in which gauge fields are present. In particular, we will study the complex scalar model with $U(1)$ local gauge symmetry.

3.7.1 *U(1) symmetry*

We consider the Heisenberg complex scalar field $\phi(x)$ with Lagrangian $\mathcal{L}[\phi(x)]$ invariant under global phase transformations (the $U(1)$ symmetry):

$$\phi(x) \to e^{i\theta}\phi(x)\,. \tag{3.35}$$

We write $\phi(x)$ as $\phi(x) = \frac{1}{\sqrt{2}}\left[\psi(x) + i\chi(x)\right]$ and assume SSB, i.e.,

$$\tilde{v} \equiv \langle\psi(x)\rangle = \sqrt{2}\,\langle\phi(x)\rangle \neq 0\,. \tag{3.36}$$

The quantity $\tilde{v}$ is called the (renormalized) order parameter. In the following Section we consider the case of local phase invariance where $\theta = \theta(x)$.

Although in the present case the generating functional does not present the ambiguities discussed in the previous example, nevertheless, as we will see below, the asymmetric solution cannot be obtained without the introduction of the ϵ-term [450, 451].

The generating functional is thus defined as [450, 451]:

$$Z[J, J^*] = \mathcal{N} \int \mathcal{D}\phi \mathcal{D}\phi^* \exp\left[i\int d^4x(\mathcal{L} + J^*\phi + J\phi^* + i\epsilon|\phi - v|^2)\right], \tag{3.37a}$$

$$\mathcal{N}^{-1} = \int \mathcal{D}\phi \mathcal{D}\phi^* \exp\left[i\int d^4x(\mathcal{L} + i\epsilon|\phi - v|^2)\right]. \tag{3.37b}$$

In general, v can be a complex number; here for simplicity we assume it is a real constant. The ϵ-term is not invariant under the phase transformation. As usual, the limit $\epsilon \to 0$ must be made at the end of the computations.

We observe that the above is not the only possible choice for the ϵ-term: in general, any ϵ-term breaking the symmetry of the Lagrangian can be used. In our case, the introduction of the ϵ-term is equivalent to the replacement $J \to J - i\epsilon \times$ const. However, the prescription of shifting the source J is not so general as the ϵ-term procedure. For example, the shift cannot be applied when ϕ is a fermion field, since in that case the source is a Grassmann quantity and the shift constant has to be a Grassmann number [450, 451], while the ϵ-term is not a Grassmann quantity.

We use the notation

$$\langle F[\phi]\rangle_{J,\epsilon} \equiv \mathcal{N} \int \mathcal{D}\phi \mathcal{D}\phi^* F[\phi] \exp\left[i \int d^4x \left(\mathcal{L} + J^*\phi + J\phi^* + i\epsilon|\phi - v|^2\right)\right],$$
(3.38)

with $\langle F[\phi]\rangle_\epsilon \equiv \langle F[\phi]\rangle_{J,\epsilon}|_{J=0}$, $\langle F[\phi]\rangle \equiv \lim_{\epsilon\to 0} \langle F[\phi]\rangle_\epsilon$. We also put

$$J(x) = \frac{1}{\sqrt{2}} [J_1(x) + iJ_2(x)],$$
(3.39a)

$$\rho(x) \equiv \psi(x) - \langle \psi(x)\rangle_\epsilon.$$
(3.39b)

Note that $\langle \chi(x)\rangle_\epsilon = 0$ because of the invariance under $\chi \to -\chi$.

Since the functional integral (3.37) is unaltered by the change of variable $\phi(x) \to e^{i\theta}\phi(x)$, then $\partial Z[J]/\partial\theta = 0$. The fundamental identity follows:

$$i \int d^4x \langle J_2(x)\psi(x) - J_1(x)\chi(x)\rangle_{\epsilon,J} = \sqrt{2}\, ev \int d^4x \langle \chi(x)\rangle_{\epsilon,J}.$$
(3.40)

By using $\frac{\delta}{\delta J_2(x)}$ and $\frac{\delta^2}{\delta J_1(x)J_2(y)}$ on Eq. (3.40) we get:

$$\langle \psi(x)\rangle_\epsilon = \sqrt{2}\, ev \int d^4y \, \langle \chi(x)\chi(y)\rangle_\epsilon \equiv \sqrt{2}\, ev\, \Delta_\chi(\epsilon, 0),$$
(3.41)

$$\langle \rho(x)\rho(y)\rangle_\epsilon - \langle \chi(x)\chi(y)\rangle_\epsilon = \sqrt{2}\, ev \int d^4z \langle \chi(z)\chi(x)\rho(y)\rangle_\epsilon,$$
(3.42)

where $\Delta_\chi(\epsilon, 0)$ denotes the two-point function at $p^2 = 0$ and non-zero ϵ. Eqs. (3.41) and (3.42) are Ward–Takahashi identities [598, 619, 656]. Note that without the introduction of the ϵ-term the r.h.s. of Eq. (3.41) would be identically zero and the asymmetric solution (3.36) would never be obtained.

In momentum space the propagator for the field χ has the general form

$$\Delta_\chi(0, p) = \lim_{\epsilon\to 0}\left[\frac{Z_\chi}{p^2 - m_\chi^2 + i\epsilon a_\chi} + (\text{continuum contributions})\right].$$
(3.43)

Z_χ and a_χ are renormalization constants. Due to the presence of $\exp^{-ip(x-y)}$ in $\langle\chi(x)\chi(y)\rangle_\epsilon$, the integration in d^4y in Eq. (3.41) picks up the pole contribution at $p^2 = 0$, and leads to

$$\tilde{v} = \sqrt{2}\,\frac{Z_\chi}{a_\chi}\,v \quad\Leftrightarrow\quad m_\chi = 0\,, \tag{3.44a}$$

$$\tilde{v} = 0 \quad\Leftrightarrow\quad m_\chi \neq 0\,. \tag{3.44b}$$

This is nothing but the Goldstone theorem [289, 291, 492, 493]: if the symmetry is spontaneously broken ($\tilde{v} \neq 0$), a massless mode exists, whose interpolating Heisenberg field is $\chi(x)$ (see Section 2.2). This is the NG boson mode: since it is massless it manifests as a long-range correlation mode. Notice that in the present case of a complex scalar field model the NG mode is an elementary field. In other models it may appear as a bound state, e.g., the magnon in ferromagnets [570].

Eqs. (3.44) show that real v implies real $\tilde{v}$. One can show [450, 451] that

$$\frac{\partial}{\partial v}\langle\psi(x)\rangle_\epsilon = \sqrt{2}\,\epsilon \int d^4y\,\langle\rho(x)\rho(y)\rangle_\epsilon\,, \tag{3.45}$$

by using Eqs. (3.41) and (3.42). Because $m_\rho \neq 0$, the r.h.s. of Eq. (3.45) vanishes in the limit $\epsilon \to 0$; therefore $\tilde{v}$ is independent of the magnitude of v: as in ferromagnets, once an external magnetic field is switched on, the system is magnetized independently of the strength of the external field.

When v is assumed to be complex, then one can show that the phase of v determines the one of $\tilde{v}$. We also observe that Eq. (3.42) gives

$$\Delta_\chi^{-1}(p) - \Delta_\rho^{-1}(p) = \tilde{v}\Gamma_{\chi\chi\rho}(0, p, -p)\,, \tag{3.46}$$

which is the usual WT-identity relating propagators and vertex function. Note that, when ρ is massive, it is unstable due to the $\chi\chi\rho$ coupling.

We remark that by assuming that an expansion of asymmetric Green's functions in terms of symmetric ones exists, each term of the expansion should carry a power of asymmetric parameters. Then the expansion would be a power expansion in ϵv. Thus, each term would approach to zero in the $\epsilon \to 0$ limit, meaning that the expansion would not make any sense. This confirms the impossibility of expanding asymmetric states in terms of symmetric ones (cf. Sections 1.7 and 2.3).

3.7.2 *SU(2) symmetry*

As a physically interesting example we consider the itinerant electron model of ferromagnetism [570, 571]. Another interesting example is the one of the isospin vector fields reported in [460].

Let $\psi(x)$ denote the electron field operator[3]:

$$\psi(x) = \begin{pmatrix} \psi_\uparrow(x) \\ \psi_\downarrow(x) \end{pmatrix}, \tag{3.47}$$

with $\uparrow$ and $\downarrow$ denoting the field spin up or down, respectively. Under $SU(2)$ $\psi(x)$ transforms as

$$\psi(x) \rightarrow \psi'(x) = \exp(i\theta_i\tau_i)\psi(x), \qquad i = 1,2,3, \tag{3.48}$$

with $\tau_i = \frac{\sigma_i}{2}$, σ_i the Pauli matrices, and θ_i a triplet of real continuous group parameters (the rotation angles in the spin-space).

We do not need to specify the explicit form of the Lagrangian. We only require that it is invariant under the $SU(2)$ group of rotations (3.48) in the spin space. We denote by $S^{(i)}(x)$, $i = 1,2,3$, the $SU(2)$ generators:

$$[S^{(i)}(x), S^{(j)}(x)] = i\epsilon_{ijk}S^{(k)}(x). \tag{3.49}$$

The explicit form of the generators $S^{(i)}(x)$ in terms of the anticommuting fields $\psi(x)$ can be given for example by $S_\psi^{(i)}(x) = \psi^\dagger(x)\frac{\sigma_i}{2}\psi(x)$. Most of our conclusions will be, however, independent of the specific form of $S^{(i)}(x)$. In the case of localized spins, we may introduce $S^{(i)}(x_l)$, where x_l denotes the lth lattice site, and the (total) $SU(2)$ generators

$$S^{(i)} = \sum_l S^{(i)}(x_l), \qquad i = 1,2,3, \tag{3.50a}$$

$$[S^{(i)}, S^{(j)}] = i\epsilon_{ijk}S^{(k)}. \tag{3.50b}$$

The invariance of the Lagrangian under $SU(2)$ implies: $\mathcal{L}[\psi(x)] = \mathcal{L}[\psi'(x)]$; the ground state, which we denote by $|0\rangle$, is, however, assumed to be not invariant under the full $SU(2)$ group but only under the subgroup $U(1)$ of the rotations around the 3rd axis in the spin-space.

The Green's function generating functional is

$$Z[J,j,n] = \mathcal{N} \int \mathcal{D}\psi\mathcal{D}\psi^\dagger \, \exp\left[i \int dt \, \left(\mathcal{L}[\psi(x)] + J^\dagger(x)\psi(x) + \psi^\dagger(x)J(x)\right.\right.$$
$$\left.\left. + j^\dagger(x)S_\psi^{(-)}(x) + S_\psi^{(+)}(x)j(x) + S_\psi^{(3)}(x)n(x) - i\epsilon S_\psi^{(3)}(x)\right)\right], \tag{3.51a}$$

$$\mathcal{N}^{-1} = \int \mathcal{D}\psi\mathcal{D}\psi^\dagger \, \exp\left[i \int dt \, \left(\mathcal{L}[\psi(x)] - i\epsilon S_\psi^{(3)}(x)\right)\right]. \tag{3.51b}$$

---

[3]Throughout this and the following Chapter, the same symbols will often be used to denote both Heisenberg field operators and c-number fields. Which one of the two cases applies will be clear from the context.

Here the spin densities, made of $\psi(x)$, are $S_\psi^{(\alpha)}(x)$, $\alpha = +, -, 3$, with $S_\psi^{(\pm)}(x) \equiv S_\psi^{(1)}(x) \pm i S_\psi^{(2)}(x)$. The electron fields ψ, $\psi^\dagger$ and their sources J, $J^\dagger$ anticommute; the sources $j, j^\dagger$, n are commuting c-numbers. The ϵ-term has been discussed above and the limit $\epsilon \to 0$ has to be taken at the end of the computation. We then introduce a notation similar to the one introduced in (3.38) and recall that in the functional integral formalism the functional average $\langle F[\psi] \rangle$ agrees with the ground state expectation value of $T(F[\psi])$ where T denotes the time-ordered products of the Heisenberg fields ψ and $\psi^\dagger$:

$$\langle F[\psi] \rangle = \langle 0|T(F[\psi])|0 \rangle . \tag{3.52}$$

The ground state expectation values of time-ordered products of ψ and $\psi^\dagger$, i.e., the Green's functions, are now obtained by repeated functional derivatives of $Z[J, j, n]$ with respect to the respective sources $\frac{\partial}{\partial J^\dagger}$ and $\frac{\partial}{\partial J}$ followed by the limits of J, j and n going to zero. We observe that the presence of the source terms with j and n allows the study of the behavior of the spin densities without specifying the dependence of $S_\psi^{(i)}$ on ψ.

For θ_i infinitesimal, $S_\psi^{(i)}$ transforms as

$$S_\psi^{(i)}(x) \quad \rightarrow \quad S_\psi^{(i)'}(x) = S_\psi^{(i)}(x) - \epsilon_{ijk}\, \theta_j\, S_\psi^{(k)}(x). \tag{3.53}$$

Let us now put $J = 0 = n$ and perform the change of variables (3.48) in the numerator of (3.51). Since a change of variables does not change the integration, we have

$$\frac{\partial Z}{\partial \theta_l} = 0 . \tag{3.54}$$

By operating with $\frac{\delta}{\delta j(y)}$ on this and putting $j = 0$ we then obtain

$$(\epsilon_{1lk} + i\epsilon_{2lk})\langle S_\psi^{(k)}(y) \rangle_\epsilon = -\epsilon\, \epsilon_{3lk} \int d^4x \langle S_\psi^{(k)}(x) S_\psi^{(+)}(y) \rangle_\epsilon . \tag{3.55}$$

Similarly, operating with $\frac{\delta}{\delta j^\dagger(y)}$ and putting $j = 0$ leads to

$$(\epsilon_{1lk} - i\epsilon_{2lk})\langle S_\psi^{(k)}(y) \rangle_\epsilon = -\epsilon\, \epsilon_{3lk} \int d^4x \langle S_\psi^{(k)}(x) S_\psi^{(-)}(y) \rangle_\epsilon . \tag{3.56}$$

These last two equations lead to

$$\epsilon_{1lk}\langle S_\psi^{(k)}(y) \rangle_\epsilon = -\epsilon\, \epsilon_{3lk} \int d^4x \langle S_\psi^{(k)}(x) S_\psi^{(1)}(y) \rangle_\epsilon . \tag{3.57}$$

$$\epsilon_{2lk}\langle S_\psi^{(k)}(y) \rangle_\epsilon = -\epsilon\, \epsilon_{3lk} \int d^4x \langle S_\psi^{(k)}(x) S_\psi^{(2)}(y) \rangle_\epsilon . \tag{3.58}$$

From these equations we have for $l = 1$ and $l = 2$:

$$0 = \epsilon \int d^4x \langle S_\psi^{(2)}(x) S_\psi^{(1)}(y) \rangle_\epsilon, \qquad (3.59)$$

$$\langle S_\psi^{(3)}(y) \rangle_\epsilon = \epsilon \int d^4x \langle S_\psi^{(1)}(x) S_\psi^{(1)}(y) \rangle_\epsilon, \qquad (3.60)$$

$$\langle S_\psi^{(3)}(y) \rangle_\epsilon = \epsilon \int d^4x \langle S_\psi^{(2)}(x) S_\psi^{(2)}(y) \rangle_\epsilon, \qquad (3.61)$$

and for and $l = 3$:

$$\langle S_\psi^{(1)}(y) \rangle_\epsilon = \langle S_\psi^{(2)}(y) \rangle_\epsilon = 0. \qquad (3.62)$$

We now write

$$\langle S_\psi^{(i)}(x) S_\psi^{(i)}(y) \rangle_\epsilon = i \int \frac{d^4p}{(2\pi)^4} e^{-ip(x-y)} \rho_i(p)$$

$$\times \left(\frac{1}{p_0 - \omega_p + i\epsilon a_i} - \frac{1}{p_0 + \omega_p - i\epsilon a_i} \right) + \text{c.c.}, \quad i = 1, 2, 3. \qquad (3.63)$$

Here our notation is $p(x-y) = p_0(t_x - t_y) - \mathbf{p} \cdot (\mathbf{x} - \mathbf{y})$ and ω_p is the energy of a quasiparticle which is a bound state of electrons. a_i is a renormalization constant. We will prove that the spectral density $\rho_i(p)$ is not zero, which proves the existence of such a bound state. The corresponding field will be denoted by $B(x)$ in Section 4.2.1. The explicit dynamical calculation can be done provided the specific form of the Lagrangian is assigned (see, e.g., [571] and [619]). It is, however, remarkable that our general treatment based on symmetry considerations proves the existence of such a bound state in a model-independent way, since we have not specified the Lagrangian form except for its invariance properties. The continuum contribution ("c.c." in (3.63)) comes from states which contain more than one quasiparticle. The singularities in the Feynman Green's functions are defined as usual by $\omega_p - i\eta$ with infinitesimal η. In (3.63) we have introduced $a_i = \frac{\eta}{\epsilon}$. We note that since $S_\psi^{(i)}$ are hermitian, $\rho_i(p)$ cannot be negative.

By operating with $[\frac{\delta}{\delta j^\dagger(z)}][\frac{\delta}{\delta j(y)}]$ and $[\frac{\delta}{\delta j(z)}][\frac{\delta}{\delta j^\dagger(y)}]$ on (3.54), putting then $j = 0$ and subtracting we have

$$\langle S_\psi^{(1)}(x) S_\psi^{(1)}(y) \rangle_\epsilon = \langle S_\psi^{(2)}(x) S_\psi^{(2)}(y) \rangle_\epsilon, \qquad (3.64)$$

which gives $\rho_1(p) = \rho_2(p)$ and $a_1 = a_2$. The magnetization is given by $g\mu_B \langle S_\psi^{(3)}(x) \rangle_\epsilon$ with μ_B the Bohr magneton. We will use the notation

$$M(\epsilon) = \langle S_\psi^{(3)}(x) \rangle_\epsilon \qquad (3.65)$$

and

$$M = \lim_{\epsilon \to 0} M(\epsilon). \qquad (3.66)$$

Eqs. (3.60) and (3.61) give

$$M(\epsilon) = i\epsilon\Delta_i(\epsilon, 0), \qquad i = 1, 2, \tag{3.67a}$$

$$\Delta_i(\epsilon, p) = \rho_i(p) \left(\frac{1}{p_0 - \omega_p + i\epsilon a_i} - \frac{1}{p_0 + \omega_p - i\epsilon a_i} \right), \tag{3.67b}$$

which shows that non-vanishing M in the $\epsilon \to 0$ limit is allowed only provided $\omega_p = 0$ at $p = 0$, i.e., only provided a gapless boson exists. We further have

$$M = \frac{2\rho}{a}, \tag{3.68}$$

where $\rho = \rho_1 = \rho_2$ and $a = a_1 = a_2$.

In the case of localized spins, the integration of $\mathbf{p}$ is confined to the domain

$$-\frac{\pi}{d} < p_i < \frac{\pi}{d}, \tag{3.69}$$

where d is the lattice length and use of the formula

$$\frac{v}{(2\pi)^3} \sum_l e^{-i\mathbf{p}\cdot\mathbf{x}_1} = \delta^{(3)}(p), \tag{3.70}$$

with v the volume of unit lattice, is required in the derivation of Eq. (3.67).

In conclusion, we have shown that Eq. (3.65), along with non-zero M, requires the existence of gapless bosons, i.e., the magnons, which are the NG bosons of the breakdown of the spin $SU(2)$ symmetry. In practical computations the magnon is a bound state of electrons and is treated by the Bethe–Salpeter equation [619]. As is well known, the magnons are the long-range correlations responsible for ferromagnetic ordering [322, 462, 619, 652]. They are the spin wave quanta. Thus, ordering is originated from the spontaneous breakdown of the $SU(2)$ symmetry, through the dynamical generation of the NG gapless bound states (the magnons).

We now calculate ρ. M is the local spin density in the third direction. The total spin in this direction is then NM, where N is the number of lattice points. Thus, the ground state expectation value of $\mathbf{S}^2$ is given by

$$\langle 0|\mathbf{S}^2|0\rangle = NM(NM + 1). \tag{3.71}$$

By assuming $t_k < t_l$ in Eq. (3.63), with $i = 1, 2$, and performing the limit $t_k \to t_l$ (same result is obtained by assuming $t_l < t_k$), we find, using Eq. (3.70), that

$$\langle 0|S^i S^i|0\rangle = \rho N \quad \text{for} \quad i = 1, 2. \tag{3.72}$$

Therefore, $\langle 0|\mathbf{S}^2|0\rangle = 2\rho N + (NM)^2$ and comparing with (3.71) we obtain

$$\rho = \frac{1}{2}M\,, \tag{3.73}$$

which gives

$$\frac{M}{2\rho^{1/2}} = \left(\frac{M}{2}\right)^{1/2}\,, \tag{3.74}$$

and $a = 1$.

3.8   Spontaneous symmetry breaking in local gauge theories

In this Section we study SSB in models where a gauge field is present. In particular, we study the complex scalar field model with $U(1)$ local gauge symmetry and the chiral gauge model. We show that the Goldstone theorem holds and that the gauge field acquires a mass, a feature which, together with the disappearance of the NG modes from the physical spectrum, which is discussed in Chapter 4, is referred to as the Englert–Brout–Higgs–Guralnik–Hagen–Kibble mechanism [24, 25, 220, 305, 323–325, 370, 490], or sometimes [449, 617, 619] as the Anderson–Higgs–Kibble mechanism. In the following, however, for brevity we adopt the short naming "Higgs mechanism", which is usually adopted in particle physics literature.

3.8.1   *The $U(1)$ local gauge model*

We consider as an example the model with $U(1)$ local gauge symmetry, where the complex scalar field $\phi(x)$ is interacting with a gauge field $A_\mu(x)$ (the so-called Goldstone–Higgs-type model) [24, 325, 370, 449]. The Lagrangian density $\mathcal{L}[\phi(x), \phi^*(x), A_\mu(x)]$ is invariant under the global and the local gauge transformations:

$$\phi(x) \rightarrow e^{i\theta}\phi(x)\,, \qquad A_\mu(x) \rightarrow A_\mu(x)\,, \tag{3.75}$$

$$\phi(x) \rightarrow e^{ie_0\lambda(x)}\phi(x)\,, \qquad A_\mu(x) \rightarrow A_\mu(x) + \partial_\mu\lambda(x)\,, \tag{3.76}$$

respectively, where $\lambda(x) \rightarrow 0$ for $|x_0| \rightarrow \infty$ and/or $|\mathbf{x}| \rightarrow \infty$.

We also put

$$\phi(x) = \frac{1}{\sqrt{2}}\left[\psi(x) + i\chi(x)\right]\,, \qquad \rho(x) = \psi(x) - \langle\psi(x)\rangle_\epsilon\,. \tag{3.77}$$

We will use the Lorentz gauge:

$$\partial_\mu A_H^\mu(x) = 0\,, \tag{3.78}$$

where $A_H^\mu(x)$ is the Heisenberg operator gauge field.[4] SSB is introduced through the condition

$$\langle 0|\phi_H(x)|0\rangle \equiv \tilde{v} \neq 0\,, \tag{3.79}$$

with $\tilde{v}$ constant and $\phi_H(x)$ the Heisenberg field operator for $\phi(x)$. Notice that here and in the following, whenever no confusion may arise, we use the same symbols ϕ, A_μ for the Heisenberg field operators as well as for the corresponding c-number fields appearing in the functional integral.

The generating functional is [449]

$$Z[J,K] = \mathcal{N} \int \mathcal{D}A_\mu \mathcal{D}\phi \mathcal{D}\phi^* \mathcal{D}B \, \exp\left[i \int d^4x \Big(\mathcal{L}(x) + B(x)\partial^\mu A_\mu(x)\right.$$

$$\left. + K^*(x)\phi(x) + K(x)\phi^*(x) + J^\mu(x)A_\mu(x) + i\epsilon|\phi(x) - v|^2 \Big) \right], \tag{3.80a}$$

$$\mathcal{N}^{-1} = \int \mathcal{D}A_\mu \mathcal{D}\phi \mathcal{D}\phi^* \mathcal{D}B \, \exp\left[i \int d^4x \Big(\mathcal{L}(x) + i\epsilon|\phi(x) - v|^2 \Big) \right]. \tag{3.80b}$$

$B(x)$ is an auxiliary field which does not appear in the Lagrangian and which is introduced[5] in order to guarantee the gauge condition (3.78). The gauge constraint is implemented by means of the identity:

$$\int \mathcal{D}B \, \exp\left[i \int d^4x B(x)\partial^\mu A_\mu(x) \right] \propto \delta\left[\partial^\mu A_\mu(x) \right]. \tag{3.81}$$

Below we obtain the Ward–Takahashi identities by exploiting the properties of the functional integration as already done in the previous examples.

Ward–Takahashi identities

Since the functional integral is unaltered by the change of variable (3.75), then $\partial Z[J,K]/\partial\theta = 0$. This leads to

$$i \int d^4x \langle K_2(x)\psi(x) - K_1(x)\chi(x)\rangle_{\epsilon,J,K} = \sqrt{2}\,\epsilon v \int d^4x \langle \chi(x)\rangle_{\epsilon,J,K}\,, \tag{3.82}$$

where we have put $K(x) = \frac{1}{\sqrt{2}}[K_1(x) + iK_2(x)]$.

---

[4]Here and in the following Chapter, symbols carrying subscripts "*in*" and "*H*" denote always field operators.

[5]The $B(x)\partial^\mu A_\mu$ term was first introduced by Nakanishi [486, 487] in the operator formalism of gauge theories.

By acting on the above relation with $\frac{\delta}{\delta K_2(x)}$ and $\frac{\delta^2}{\delta K_1(x) K_2(y)}$, we obtain:

$$\langle\psi(x)\rangle_\epsilon = \sqrt{2}\,\epsilon\,v\int d^4y\langle\chi(x)\chi(y)\rangle_\epsilon = \sqrt{2}\,\epsilon\,v\,\Delta_\chi(\epsilon,0)\,, \qquad (3.83)$$

$$\langle\rho(x)\rho(y)\rangle_\epsilon - \langle\chi(x)\chi(y)\rangle_\epsilon = \sqrt{2}\,\epsilon\,v\int d^4z\langle\chi(z)\chi(x)\rho(y)\rangle_\epsilon\,. \qquad (3.84)$$

Similarly, the invariance under the transformations (3.76) gives

$$i\int d^4x\langle\partial^2 B(x) - \partial^\mu J_\mu(x) + e_0(K_2(x)\psi(x) - K_1(x)\chi(x))\rangle_{\epsilon,J,K}\,\lambda(x)$$

$$= \sqrt{2}\,\epsilon\,v\,e_0\int d^4x\langle\chi(x)\rangle_{\epsilon,J,K}\,\lambda(x)\,. \qquad (3.85)$$

The invariance under the transformation $B(x) \to B(x) + \lambda(x)$ gives

$$\langle\partial^\mu A_\mu(x)\rangle_{\epsilon,J,K} = 0\,. \qquad (3.86)$$

Another identity is obtained when the global phase transformations are performed in the Green's function $\langle\partial^2_{x_1}B(x_1)\ldots\partial^2_{x_n}B(x_n)\rangle_{\epsilon,J,K}$:

$$i\int d^4x\langle\partial^2_{x_1}B(x_1)\ldots\partial^2_{x_n}B(x_n)\Big[\partial^2 B(x) - \partial^\mu J_\mu(x)$$

$$+ e_0(K_2(x)\psi(x) - K_1(x)\chi(x))\Big]\rangle_{\epsilon,J,K}\,\lambda(x)$$

$$= \sqrt{2}\,\epsilon\,v\,e_0\int d^4x\langle\partial^2_{x_1}B(x_1)\ldots\partial^2_{x_n}B(x_n)\chi(x)\rangle_{\epsilon,J,K}\,\lambda(x)\,. \qquad (3.87)$$

Operating with $\delta/i\delta K_2(y)$ and $\delta/i\delta J_\mu(y)$ on Eq. (3.85), we get

$$i\partial^2_x\langle B(x)\chi(y)\rangle_\epsilon + e_0\langle\psi(x)\rangle_\epsilon\delta(x-y) = \sqrt{2}\,\epsilon\,v\,e_0\langle\chi(x)\chi(y)\rangle_\epsilon\,, \qquad (3.88\text{a})$$

$$i\partial^2_x\langle B(x)A^\mu(y)\rangle_\epsilon - \partial^\mu_x\delta(x-y) = \sqrt{2}\,\epsilon\,v\,e_0\langle\chi(x)A^\mu(y)\rangle_\epsilon\,. \qquad (3.88\text{b})$$

Eq. (3.87) with $n = 1$ and $J = K = 0$ gives

$$i\partial^2_x\langle B(x)\partial^2_y B(y)\rangle_\epsilon = \sqrt{2}\,\epsilon\,v\,e_0\partial^2_x\langle B(x)\chi(y)\rangle_\epsilon \qquad (3.89)$$

and, from Eq. (3.86)

$$\langle\partial_\mu A^\mu(x)\,\chi(y)\rangle_\epsilon = 0\,, \qquad (3.90)$$

$$\langle\partial_\mu A^\mu(x)\,A_\nu(y)\rangle_\epsilon = 0\,. \qquad (3.91)$$

Because of the invariance under $(\chi, A_\mu, B) \to (-\chi, -A_\mu, -B)$, all the other two-point functions vanish:

$$\langle\rho(x)B(y)\rangle_\epsilon = \langle\rho(x)\chi(y)\rangle_\epsilon = \langle\rho(x)A^\mu(y)\rangle_\epsilon = 0\,. \qquad (3.92)$$

Pole structure of two-point functions

Eq. (3.83) leads us again to the Goldstone theorem, as discussed in Section 3.7.1. Moreover, the above relations allow us to extract information on the pole structure of the remaining two-point functions. We obtain [449]:

$$\langle\chi(x)\chi(y)\rangle = i \lim_{\epsilon\to 0} \int \frac{d^4p}{(2\pi)^4}\, e^{-ip(x-y)} \frac{Z_\chi}{p^2 + i\epsilon a_\chi} + \text{cont. contr.}, \qquad (3.93)$$

$$\langle B(x)\chi(y)\rangle = -i \lim_{\epsilon\to 0} \int \frac{d^4p}{(2\pi)^4}\, e^{-ip(x-y)} \frac{e_0\tilde v}{p^2 + i\epsilon a_\chi}, \qquad (3.94)$$

$$\langle B(x)A^\mu(y)\rangle = i\,\partial_x^\mu \int \frac{d^4p}{(2\pi)^4}\, e^{-ip(x-y)} \frac{1}{p^2}, \qquad (3.95)$$

$$\langle B(x)B(y)\rangle = -i \lim_{\epsilon\to 0} \int \frac{d^4p}{(2\pi)^4}\, e^{-ip(x-y)} \frac{(e_0\tilde v)^2}{Z_\chi}\left[\frac{1}{p^2 + i\epsilon a_\chi} - \frac{1}{p^2}\right]. \qquad (3.96)$$

Use of Eq. (3.93) in Eq. (3.83) shows that the field χ is the NG massless mode. In the present case, Eq. (3.96) shows that the model also contains a massless negative norm state (ghost) (in this connection see also [488]). The absence of cut singularities in the last three of these propagators suggests that $B(x)$ obeys a free field equation.

Summarizing, the above discussion leads us to the following results:

Eqs. (3.93)–(3.96) imply the existence of two asymptotic fields: the NG field χ_{in} and the negative norm (ghost) field b_{in}. They satisfy the free field equations:

$$\partial^2 \chi_{in}(x) = 0, \qquad (3.97a)$$

$$\partial^2 b_{in}(x) = 0. \qquad (3.97b)$$

The field operator $B_H(x)$ is then given by

$$B_H(x) = e_0\tilde v\, Z_\chi^{-\frac{1}{2}} \left[b_{in}(x) - \chi_{in}(x)\right], \qquad (3.98)$$

and the field equations for B_H and $A_{\mu H}$ are

$$\partial^2 B_H(x) = 0, \qquad (3.99)$$

$$-\partial^2 A_{\mu H}(x) = j_{\mu H}(x) - \partial_\mu B_H(x), \qquad (3.100)$$

where the equation for $A_{\mu H}(x)$ is the one compatible with Eq. (3.80) and $j_{\mu H}(x) = \delta\mathcal{L}(x)/\delta A_H^\mu(x)$ has been used (here $\mathcal{L}(x)$ is the Lagrangian written for the field operators).

We also assume that a (massive) stable asymptotic field ρ_{in} corresponding to $\rho(x)$ does exist (see comment in Section 3.7.1). We therefore include it in the list of the in-fields. The corresponding free field equation is

$$(\partial^2 + m_\rho^2)\rho_{in}(x) = 0. \qquad (3.101)$$

We are now ready to consider the gauge field.

The massive vector field

Eqs. (3.90), (3.92) and (3.95) show that in the dynamical map for $A_{\mu H}$ there are no terms linear in χ_{in} or ρ_{in}; instead, the term $Z_\chi^{\frac{1}{2}}/(e_0\tilde{v})\partial_\mu b_{in}(x)$ appears. In full generality, we may also assume that an asymptotic massive vector field $U_{in}^\mu(x)$ exists. As we will see, this is indeed the case. The dynamical map then takes the form:

$$A_H^\mu(x) = \frac{Z_\chi^{\frac{1}{2}}}{e_0\tilde{v}}\partial^\mu b_{in}(x) + Z_3^{\frac{1}{2}}U_{in}^\mu(x) + \dots . \tag{3.102}$$

We examine whether the renormalization factor Z_3 vanishes or not. The (Proca) equations for U_{in}^μ are

$$(\partial^2 + m_V^2)U_{in}^\mu(x) = 0, \tag{3.103a}$$

$$\partial_\mu U_{in}^\mu(x) = 0. \tag{3.103b}$$

Let us define

$$\langle A_\mu(x)A_\nu(y)\rangle = i\int \frac{d^4p}{(2\pi)^4}\, e^{-ip(x-y)}\Delta_{\mu\nu}(p). \tag{3.104}$$

Eq. (3.102) implies the following structure for the propagator $\Delta_{\mu\nu}(p)$:

$$\Delta_{\mu\nu}(p) = -\frac{Z_\chi}{(e_0\tilde{v})^2}\frac{p_\mu p_\nu}{p^2} + Z_3\frac{-g_{\mu\nu} + m_V^{-2}p_\mu p_\nu}{p^2 - m_V^2} + \dots . \tag{3.105}$$

Here dots stand for cut contributions. From the identity (3.91), written as

$$p^\mu \Delta_{\mu\nu}(p) = 0, \tag{3.106}$$

it follows that $Z_3 \neq 0$ and

$$m_V^2 = \frac{Z_3}{Z_\chi}(e_0\tilde{v})^2. \tag{3.107}$$

Eq. (3.107) shows that the non-vanishing value of the mass of the vector boson field is due to the SSB condition, Eq. (3.79) [449]. This, together with the fact that NG and ghost modes do not appear in the physical particle spectrum, which is discussed in Chapter 4, is called *the Higgs mechanism* [24, 25, 220, 305, 323–325, 370, 490].

The dynamical maps

The discussion presented above, allows us to anticipate the expressions of dynamical maps in their partial form [449]:

$$\chi_H(x) = Z_\chi^{\frac{1}{2}} \chi_{in}(x) + \dots , \tag{3.108a}$$

$$\rho_H(x) \equiv \psi_H(x) - \tilde{v} = Z_\rho^{\frac{1}{2}} \rho_{in}(x) + \dots , \tag{3.108b}$$

$$B_H(x) = e_0 \tilde{v} \, Z_\chi^{-\frac{1}{2}} \left[b_{in}(x) - \chi_{in}(x) \right] + \dots , \tag{3.108c}$$

$$A_H^\mu(x) = \frac{Z_\chi^{\frac{1}{2}}}{e_0 \tilde{v}} \partial^\mu b_{in}(x) + Z_3^{\frac{1}{2}} U_{in}^\mu(x) + \dots , \tag{3.108d}$$

where dots denote higher order normal product terms. We will present the complete form of the dynamical maps in Chapter 4, where we also discuss the macroscopic effects due to the condensation of the NG particles.

3.8.2   The chiral gauge model

We turn to the chiral gauge model, which is a generalization of the Nambu–Jona–Lasinio model [449, 492, 493]. We have now the spin-$\frac{1}{2}$ field $\varphi(x)$ and a chiral gauge field $A_\mu(x)$. The Lagrangian is invariant under the chiral transformations :

$$\varphi(x) \to e^{i\theta\gamma_5}\varphi(x), \qquad A_\mu(x) \to A_\mu(x), \tag{3.109}$$

$$\varphi(x) \to e^{ie_0\lambda(x)\gamma_5}\varphi(x), \quad A_\mu(x) \to A_\mu(x) + \partial_\mu\lambda(x), \tag{3.110}$$

where $\lambda(x) \to 0$ for $|x_0| \to \infty$ and/or $|\mathbf{x}| \to \infty$.

We put

$$\psi(x) = \overline{\varphi}(x)\varphi(x), \tag{3.111a}$$

$$\chi(x) = i\,\overline{\varphi}(x)\gamma_5\varphi(x), \tag{3.111b}$$

$$\phi(x) = \frac{1}{\sqrt{2}}\left[\psi(x) + i\chi(x)\right], \tag{3.111c}$$

where $\overline{\varphi} = \varphi^\dagger\gamma_0$. When $\varphi(x)$ undergoes the transformations (3.109) and (3.110), ϕ changes as

$$\phi(x) \to e^{2i\theta}\phi(x), \tag{3.112a}$$

$$\phi(x) \to e^{2ie_0\lambda(x)}\phi(x), \tag{3.112b}$$

respectively.

We use the Lorentz gauge:

$$\partial_\mu A_H^\mu(x) = 0, \tag{3.113}$$

and SSB is introduced through the condition

$$\langle 0|\overline{\varphi}_H(x)\varphi_H(x)|0\rangle \equiv \sqrt{2}\tilde{v} \neq 0, \tag{3.114}$$

with real $\tilde{v}$, i.e.,

$$\langle 0|\phi_H(x)|0\rangle \equiv \tilde{v}. \tag{3.115}$$

The similarity with the $U(1)$ gauge model previously considered suggests to us that we can formulate the chiral model in a similar way, provided the following changes are made [449]:

- the constant v in the generating functional is now $\frac{1}{\sqrt{2}}v$.
- e_0 should be replaced by $2e_0$ in the relations involving A_μ, ϕ or $B(x)$.
- there exists a fermion in-field $\varphi_{in}(x)$.
- the existence of an asymptotic $\rho_{in}(x)$ field, corresponding to a scalar bound state, is not required in general; we shall assume that it does not exist in our model.
- The NG particle $\chi_{in}(x)$ appears as a fermion-antifermion bound state.

In addition there is the massless ghost field $b_{in}(x)$ and the massive vector field $U_{in}^\mu(x)$ whose mass is $m_V^2 = \frac{4Z_3}{Z_\chi}(e_0\tilde{v})^2$.

Here we also present the dynamical maps in their general form, to be compared with the ones of the Goldstone model presented in Section 4.2. The explicit form of the functionals F and F^μ in the following relations is defined once the explicit form of the Lagrangian is assigned.

We have [449, 452]:

$$\phi_H(x) =: \exp\left\{i\frac{Z_\chi^{\frac{1}{2}}}{\tilde{v}}\chi_{in}(x)\right\}[\tilde{v} + F[\varphi_{in}, U_{in}^\mu, \partial(\chi_{in} - b_{in})]]:, \tag{3.116}$$

$$A_H^\mu(x) = Z_3^{\frac{1}{2}}U_{in}^\mu(x) + \frac{Z_\chi^{\frac{1}{2}}}{2e_0\tilde{v}}\partial^\mu b_{in}(x) + : F^\mu[\varphi_{in}, U_{in}^\mu, \partial(\chi_{in} - b_{in})] :, \tag{3.117}$$

$$B_H(x) = \frac{2e_0\tilde{v}}{Z_\chi^{\frac{1}{2}}}[b_{in}(x) - \chi_{in}(x)] + \text{const.} \tag{3.118}$$

The in-field expansion for the fermion φ_H field and the S-matrix are:

$$\varphi_H(x) =: \exp\left\{i\frac{Z_\chi^{\frac{1}{2}}}{2\tilde{v}}\chi_{in}(x)\gamma_5\right\}[Z_\phi^{\frac{1}{2}}\varphi_{in}(x) + F[\varphi_{in}, U_{in}^\mu, \partial(\chi_{in} - b_{in})]]:, \tag{3.119}$$

$$S =: S[\varphi_{in}, U_{in}^\mu, \partial(\chi_{in} - b_{in})] :. \tag{3.120}$$

Eq. (3.120) shows that the NG field and the ghost field enter the S-matrix only through the term $\partial(\chi_{in} - b_{in})$ where ∂ denotes at least first order derivative. The S-matrix, due to the presence of the derivative in such a term, is not affected by their homogeneous condensate (described by a constant c-number). Moreover, also due to such a derivative, the S-matrix is independent of soft (small momentum) NG fields and ghost fields. This is called the *low energy theorem* and its meaning is that the S-matrix is stable against NG field and ghost field fluctuations. Notice, however, that although the NG field (and the ghost field) disappears from the observable spectrum and does not participate to observable reactions, its role is crucial for the internal consistency between the invariance of the theory and the SSB, as the dynamical maps derived above show. See the following Chapter for further comments on the role of the NG boson field in local gauge theories.

3.9   Finite volume effects

In the proof of the Goldstone theorem presented in Section 3.7, the system is considered without boundaries, i.e., its volume is considered to be infinite. This is a reasonable working assumption since observations are always local and therefore the volume V of the system may be taken to be infinite. This is also the case with the so-called thermodynamic limit, where the limit to the infinite number of degrees of freedom and to the infinite volume is taken in a way that the density remains finite. The spatial integration domain thus extends to infinity and this is crucial in picking up the zero-momentum contribution in the two-point Green's function (cf. Eqs. (3.41)–(3.44)). In Chapter 4 we will see that the dynamical rearrangement of the symmetry occurs since terms of the order of $\frac{1}{V}$ are missing in local observations.

Sometimes it is, however, interesting to consider the effects of boundaries on the dynamics. For example, in some cases it is necessary to consider how the ordering induced by the NG condensation gets distorted in the vicinity of the system boundaries and how "defects" (non-homogeneous condensation) appear. In this Section we study the effect of finite volume and show that the NG particle acquires an effective non-zero mass due to finite volume effects. Such a non-zero mass reflects on the correlation length and thus it is directly related to the size of the ordered domain. We also relate volume effects with temperature effects.

We consider the $U(1)$ complex field model. However, our conclusions apply to many systems in a wide range of energy scales.

For large but finite volume, the order parameter is expected to be constant inside the bulk *far* from the boundaries. However, as already mentioned, it might present distortions *near* the boundaries: near the boundaries non-homogeneities in the order parameter may thus appear, $\tilde{v} = \tilde{v}(x)$ (or even $\tilde{v} \to 0$), which are "smoothed out" in the $V \to \infty$ limit.

Consider Eq. (3.41). Restrict now the space integration over the finite (but large) volume $V \equiv \eta^{-3}$ and use for each component of $\mathbf{p}$:

$$\delta_\eta(p) \equiv \frac{1}{2\pi} \int_{-\frac{1}{\eta}}^{\frac{1}{\eta}} dx\, e^{ipx} = \frac{1}{\pi p} \sin \frac{p}{\eta}, \tag{3.121}$$

which approaches $\delta(p)$ as $\eta \to 0$: $\lim_{\eta \to 0} \delta_\eta(p) = \delta(p)$. Consider that

$$\lim_{\eta \to 0} \int dp\, \delta_\eta(p)\, f(p) = f(0) = \lim_{\eta \to 0} \int dp\, \delta(p - \eta)\, f(p). \tag{3.122}$$

Then, using $\delta_\eta(p) \simeq \delta(p - \eta)$ for small η, one obtains

$$\tilde{v}(y, \epsilon, \eta) = i\, \epsilon\, v\, e^{-i\eta \cdot y}\, \Delta_\chi(\epsilon, \eta, p_0 = 0), \tag{3.123}$$

where

$$\Delta_\chi(\epsilon, \eta, p_0 = 0) = \frac{Z_\chi}{-\omega^2_{\mathbf{p}=\eta} + i\epsilon a_\chi} + \text{cont. contr.}, \tag{3.124}$$

and $\omega^2_{\mathbf{p}=\eta} = \eta^2 + m_\chi^2$. Thus, $\lim_{\epsilon \to 0} \lim_{\eta \to 0} \tilde{v}(y, \epsilon, \eta) \neq 0$ only if $m_\chi = 0$, otherwise $\tilde{v} = 0$. Note that the Goldstone theorem is recovered in the infinite volume limit ($\eta \to 0$).

On the other hand, suppose $m_\chi = 0$, but η is given a finite non-zero value (i.e., volume is finite, namely we are in the presence of boundaries), then $\omega_{\mathbf{p}=\eta} \neq 0$ and it acts as an "effective mass" for the χ bosons. Therefore, in such a case, the order parameter $\tilde{v}$ is different from zero, provided ϵ is kept as non-zero.

As in the case of localized spins (cf. Eqs. (3.69) and (3.70)), each component of $\mathbf{p}$ is confined to the domain $-\frac{\pi}{d} < p_i < \frac{\pi}{d}$ also when the volume is finite, and thus the formula (3.70) needs to be used. We do not analyze further the consequences of the discretized momentum in the finite volume systems. See [221, 456], where such a a problem is considered in connection with systems of trapped atoms in the Bose–Einstein condensates (BEC). We only remark that one should first perform the infinite volume limit ($\eta \to 0$) and then the $\mathbf{p} \to 0$ limit. In Chapter 4 we will come back to the non-commutativity of the two limits $\eta \to 0$ and $\epsilon \to 0$. The above discussion also sheds some light on the fact that the Goldstone theorem holds in QFT (i.e., for infinite volume), but not in QM (finite volume systems).

In conclusion, near the boundaries ($\eta \neq 0$) the NG bosons acquire an effective mass $m_{eff} \equiv \omega_{\mathbf{p}=\eta}$. They will then propagate over a range of the order of $\xi \equiv \frac{1}{\eta}$, which is the linear size of the condensation domain.

We stress that if $\eta \neq 0$, then the order parameter is different from zero provided ϵ is kept non-zero: $\tilde{v} \neq 0$ (at least locally). In such a case, the breakdown of symmetry is sustained because $\epsilon \neq 0$: ϵ acts as the coupling with an external field (the pump) providing energy. Energy supply is required in order to condensate modes of non-zero lowest energy $\omega_{\mathbf{p}=\eta}$. In summary, boundary effects are in competition with the breakdown of symmetry [15, 17, 637]. They may preclude its occurrence or, if symmetry is already broken, they may reduce to zero the order parameter (symmetry restoration).

In Chapter 5 we will see that temperature may have similar effects on the order parameter, namely a critical temperature T_C may exist, such that, for $T > T_C$, the order parameter goes to zero, i.e., symmetry is restored. Since the order parameter goes to zero when NG modes acquire non-zero effective mass (unless, as seen, external energy is supplied), we may then represent the effect of thermalization in terms of finite volume effects and put, e.g., $\eta \propto \sqrt{\frac{|T-T_C|}{T_C}}$. In this way temperature fluctuations around T_C may produce fluctuations in the size ξ of the condensed domain (the size of ordered domain and/or of the domain where non-homogeneous condensation occurs, namely of the defect). The converse is also true: fluctuations in the finite system size may manifest as thermal fluctuations.

3.10　Space-time dimensionality

In the next Chapter we will see that NG particles of very low energy are responsible for the rearrangement of the symmetry. In this Section we show that when their influence becomes too strong the ordered state is destroyed, i.e., the order parameter becomes zero. This effect occurs in low dimensions (one or two spatial dimensions, see below).

It is known that there is a critical dimension D_c such that no SSB is possible for any short-range Hamiltonian at $T \neq 0$. The dimension is $D_c = 1$ for discrete symmetries while $D_c = 2$ for continuous symmetries. In fact, the absence of ferromagnetism in one- and two-dimensional isotropic Heisenberg model was originally observed by Mermin and Wagner [472]. Hohenberg [330] and Mermin [470] generalized this result into a statement about the non-existence of a long-range order in the one- and two-dimensional systems with continuous symmetry. Coleman [164] reached similar conclusion in the framework of QFT for $(1+1)$-spacetime dimensional systems in the absence of gauge fields. Infrared NG bosons are also known to destroy one-dimensional superconductivity [422]. Here we will assume that the order parameter is non-zero and we will show that the effects of the infrared NG bosons cause the order parameter to vanish under specified conditions of temperature and dimensionality. We follow closely the discussion presented in [453].

We decompose the NG field, say χ_{in}, in a "hard" part $\chi_{in,t}(x)$ and a "soft" or infrared part $\chi_{in,\eta}(x)$ with η infinitesimal, $\chi_{in} = \chi_{in,t}(x) + \chi_{in,\eta}(x)$. $\chi_{in,t}(x)$ contains only momenta larger than η, while momenta in $\chi_{in,\eta}(x)$ are smaller than, or equal to η and $\chi_{in,\eta}(x)$ can be written as [453]

$$
\chi_{in,\eta}(x) = \frac{1}{2}\eta \int_{-\infty}^{+\infty} dt\, e^{-\eta|t|} \chi_{in}(\mathbf{x}, t),
$$

$$
= \frac{\eta}{(2\pi)^{3/2}} \int \frac{d^3k}{\sqrt{2k}} \frac{\eta}{\mathbf{k}^2 + \eta^2} (a_{\mathbf{k}} e^{i\mathbf{k}\cdot\mathbf{x}} + a_{\mathbf{k}}^\dagger e^{-i\mathbf{k}\cdot\mathbf{x}}), \quad (3.125)
$$

which for $\eta \to 0$ gives

$$
\chi_{in,\eta}(x) \to \frac{\eta}{2\sqrt{2\pi}} \int \frac{d^3k}{\sqrt{2k}} \delta(\mathbf{k})(a_{\mathbf{k}} e^{i\mathbf{k}\cdot\mathbf{x}} + a_{\mathbf{k}}^\dagger e^{-i\mathbf{k}\cdot\mathbf{x}}). \quad (3.126)
$$

Therefore $\chi_{in,\eta}$ is of order of η and independent of x in the limit $\eta \to 0$. Derivatives of $\chi_{in,\eta}$ can thus be neglected.

Let us consider as an example the chiral gauge model considered in Section 3.8.2. We have (cf. Eq. (3.114))

$$
\sqrt{2}\tilde{v} = \langle 0| : \overline{\tilde{F}}_t[\varphi_{in}, U_{in}^\mu, \chi_{in,t}, b_{in,t}] :: \tilde{F}_t[\varphi_{in}, U_{in}^\mu, \chi_{in,t}, b_{in,t}] : |0\rangle
$$

$$
\times \langle 0| : \exp\left\{ i\frac{Z_\chi^{\frac{1}{2}}}{2\tilde{v}} \chi_{in,\eta}(x)\gamma_5 \right\} :: \exp\left\{ i\frac{Z_\chi^{\frac{1}{2}}}{2\tilde{v}} \chi_{in,\eta}(x)\gamma_5 \right\} : |0\rangle
$$

$$
= \tilde{v}_t \exp\left[-\frac{1}{2}\left(\frac{Z_\chi^{\frac{1}{2}}}{2\tilde{v}} \right)^2 D_\eta(0) \right], \quad (3.127)
$$

where $\tilde{F}_t$ denotes what remains of the dynamical mapping without the infrared contribution coming from $\chi_{in,\eta}$ (cf. Eq. (3.119)). In Eq. (3.127)

$$
D_\eta(0) = \langle 0|\chi_{in,\eta}^2(x)|0\rangle \simeq \frac{1}{2(2\pi)^n} \int_{|\mathbf{p}|<\eta} d^n p \, \frac{1}{|\mathbf{p}|}, \quad (3.128)
$$

for spacetime dimensions $n + 1$. If the system is at a non-vanishing temperature T, instead of Eq. (3.128) we have

$$D_\eta(0) = \frac{1}{2(2\pi)^n} \int_{|\mathbf{p}|<\eta} d^n p \, \frac{1}{|\mathbf{p}|(1 - e^{-\beta|\mathbf{p}|})},$$

$$\simeq \frac{1}{2(2\pi)^n \beta} \int_{|\mathbf{p}|<\eta} d^n p \, \frac{1}{|\mathbf{p}|^2}, \qquad (3.129)$$

where $\beta = (k_B T)^{-1}$. These expressions show that $D_\eta(0) \to \infty$ for $n = 1$ for any T and for $n = 2$ for $T \neq 0$. For $n \geq 3$, $D_\eta(0) \to 0$. We note that $\tilde{v}_t$ is the order parameter without the infrared contribution and is a finite quantity after renormalization. Therefore, owing to the infrared effects,

$$\tilde{v} = 0 \qquad \text{for } n = 1, \quad \forall \, T, \qquad (3.130a)$$

$$\tilde{v} = 0 \qquad \text{for } n = 2, \quad \text{at } T \neq 0, \qquad (3.130b)$$

i.e., SSB is impossible in $1 + 1$ spacetime dimensions always, and in $2 + 1$ dimensions for $T \neq 0$. The same result also holds when no gauge field is present in the chiral model. Moreover, it also holds in the $U(1)$ self-interacting complex field model (the Goldstone model) studied in Section 3.8.1. In this case the order parameter is given by $\tilde{v} = \langle 0|\phi_H(x)|0\rangle$ and the in-field expansion by

$$\phi_{Ht}(x) = : \exp\left\{ i \frac{Z_\chi^{\frac{1}{2}}}{\tilde{v}} \chi_{in,\eta}(x) \right\} \phi_t(x) :, \qquad (3.131)$$

$$\phi_t(x) = \tilde{v}_t + \ldots.. \qquad (3.132)$$

Therefore, $\tilde{v} = \tilde{v}_t$. We must then consider the self-consistent condition which determines $\tilde{v}_t$. The Heisenberg equation of motion for $\phi_H(x)$ is

$$(-\partial^2)\phi_H(x) = -\mu_0^2 \phi_H(x) + \lambda \phi_H(x)\phi_H^\dagger(x)\phi_H(x), \qquad (3.133)$$

which gives

$$-\mu_0^2 \tilde{v}_t = \lambda \langle 0|\phi_H(x)\phi_H^\dagger(x)\phi_H(x)|0\rangle. \qquad (3.134)$$

Use of Eq. (3.131) in (3.134) leads to

$$-\mu_0^2 \tilde{v}_t = \lambda \langle 0|\phi_t(x)\phi_t^\dagger(x)\phi_t(x)|0\rangle \exp\left\{ -\frac{1}{2}\frac{Z_\chi}{\tilde{v}^2} D_\eta(0) \right\}. \qquad (3.135)$$

The quantities $\mu_0^2 \tilde{v}_t$ and $\lambda \langle 0|\phi_t(x)\phi_t^\dagger(x)\phi_t(x)|0\rangle$ will become finite after renormalization. Instead $D_\eta(0)$ is infinite for a one-dimensional space and two-dimensional space dimension if $T \neq 0$. Thus, again it is $\tilde{v} = 0$.

Consider the case where the NG boson satisfies the equation [453]

$$\left[i\frac{\partial}{\partial t} - \omega(-i\nabla)\right]\chi_{in}(x) = 0\,.$$ (3.136)

We then have

$$D_\eta(0) = \langle 0|\chi_{in,\eta}^2(x)|0\rangle = \int_{|\mathbf{p}|<\eta} \frac{d^n p}{(2\pi)^n} \frac{1}{(1 - e^{-\beta\omega(\mathbf{p})})}$$

$$\simeq \frac{1}{\beta} \int_{|\mathbf{p}|<\eta} \frac{d^n p}{(2\pi)^n} \frac{1}{\omega(\mathbf{p})}\,.$$ (3.137)

In the case of ferromagnets, $\omega(\mathbf{p}) \sim c\mathbf{p}^2$ for $|\mathbf{p}| \to 0$ and the result obtained above is again valid for $T \neq 0$.

In superconductors, the equation of motion is [453]

$$\left[i\frac{\partial}{\partial t} - \omega(-i\nabla^2)\right]\chi_{in}(x) = 0\,,$$ (3.138)

with $\omega(\mathbf{p}^2) \sim c\mathbf{p}$ for $|\mathbf{p}| \to 0$, and the above result also holds.

In finite volume systems we must consider the effects of surfaces as explained in the previous Section.

See Appendix J for further discussion on dimensionality and SSB.

Appendix I

The order parameter space

In Section 3.4 we hae seen that key concept in the discussion of SSB is the so-called *order parameter space*. For the reader's convenience we provide here a mathematical introduction into order parameter spaces.

For definiteness, consider a multiplet of operators $\hat{\boldsymbol{\Phi}} = \{\hat{\Phi}_i; i = 1, \ldots, n\}$ transforming under some n-dimensional representation S of G

$$D^{-1}(g)\hat{\Phi}_i D(g) = \sum_{j=1}^{n} S_{ij}(g)\hat{\Phi}_j. \tag{I.1}$$

Due to Eq. (3.8), for $g \in G/H$

$$\langle 0|D^{-1}(g)\hat{\Phi}_i D(g)|0\rangle = \sum_{j=1}^{n} S_{ij}(g)\Phi_j^0 \neq \Phi_i^0, \tag{I.2}$$

which means that whenever $g \in G/H$ changes, then also $\boldsymbol{\Phi}^0$ is changed: $\boldsymbol{\Phi}^0$ is not invariant under whole G but only under H.

Eq. (3.8) implies that for $g_1 \neq g_2$, with $g_1, g_2 \in G/H$, $|0(g_1)\rangle \neq |0(g_2)\rangle$, which in turn implies $\langle 0(g_1)|\hat{\boldsymbol{\Phi}}|0(g_1)\rangle \neq \langle 0(g_2)|\hat{\boldsymbol{\Phi}}|0(g_2)\rangle$, i.e., $\boldsymbol{\Phi}_1^0 \neq \boldsymbol{\Phi}_2^0$. This shows that the order parameter space $\mathcal{M}$ is isomorphic to G/H: $\mathcal{M} \cong G/H$. Thus the vacuum manifold can be formally identified with the order parameter space.

We now assume that G acts transitively on $\mathcal{M}$, i.e., the action of G on $\mathcal{M}$ generates the whole of $\mathcal{M}$ from any given point of it. So for any $\mathbf{x}_1, \mathbf{x}_2 \in \mathcal{M}$, there exits $g \in G$, such that

$$g\mathbf{x}_1 = \mathbf{x}_2. \tag{I.3}$$

Thus the order parameter space is the set of points $\mathcal{M} = \{g\mathbf{x}; g \in G\}$ for any fixed $\mathbf{x} \in \mathcal{M}$. We introduce the *stability* (or little) *group* $H_\mathbf{x}$ consisting of transformations leaving $\mathbf{x}$ invariant, i.e.,

$$H_\mathbf{x} = \{h \in G : h\mathbf{x} = \mathbf{x}\}. \tag{I.4}$$

It is clear that $H_{\mathbf{x}}$ is a subgroup of G, for if h_1 and h_2 leave $\mathbf{x}$ unchanged so does also $h_1 h_2^{-1}$. In general $H_{\mathbf{x}}$ changes with $\mathbf{x} \in \mathcal{M}$. Indeed, if $\mathbf{x}_2 = g\mathbf{x}_1$, since $h\mathbf{x}_1 = \mathbf{x}_1$ and $gh\mathbf{x}_1 = \mathbf{x}_2$, we have from (I.3) that $ghg^{-1}\mathbf{x}_2 = \mathbf{x}_2$, which shows that $gH_{\mathbf{x}_1}g^{-1} \subset H_{\mathbf{x}_2}$. The converse inclusion would follow if we were started with $\mathbf{x}_2$ instead. Thus,

$$H_{\mathbf{x}_2} \;=\; gH_{\mathbf{x}_1}g^{-1}, \tag{I.5}$$

and $H_{\mathbf{x}_1}$ and $H_{\mathbf{x}_2}$ are *conjugate* subgroups of G. Since the conjugate subgroups $H_{\mathbf{x}}$ are *isomorphic* for different $\mathbf{x} \in \mathcal{M}$, it is customary to use simply H to denote such groups.

The structure of the order parameter space $\mathcal{M}$ can be related to G and H. Indeed, given a reference point $\mathbf{x} \in \mathcal{M}$, $g\mathbf{x}$ spans the whole of $\mathcal{M}$ as g ranges over G. However, in general each point of $\mathcal{M}$ will be traversed many times during this procedure because the map $g \to g\mathbf{x}$ is typically not a bijection from $G \to \mathcal{M}$. Two different elements $g_1, g_2 \in G$ may yield the same point in $\mathcal{M}$ when $g_1\mathbf{x} = g_2\mathbf{x}$. This happens precisely when $g_1^{-1}g_2 \in H_{\mathbf{x}}$, i.e., when both g_1 and g_2 belong to the same coset of $H_{\mathbf{x}}$ in G (see Appendix C for the coset definition). As a result, the order parameter space can be considered to be the coset space $G/H_{\mathbf{x}}$ with $\mathbf{x}$ being some reference order parameter. Since the isomorphism class of $H_{\mathbf{x}}$ does not depend on $\mathbf{x} \in \mathcal{M}$, one can simply write $\mathcal{M} \cong G/H$. The dimension of this space is $\dim G - \dim H$ and so this is also the dimension of the order parameter space.

Appendix J

The Mermin–Wagner–Coleman theorem

We have seen in Section 3.4 that the generators $D_n \in G/H$, $n = 1, 2, \ldots, \dim G/H$, are linear operators in the asymptotic NG fields $\phi_{\mathrm{NG}}^{(n)}$:

$$D_n = \int d^3x \, \partial_0 \phi_{\mathrm{NG}}^{(n)}(x), \qquad n = 1, 2, \ldots, \dim G - \dim H. \qquad (\mathrm{J}.1)$$

Here $\partial_0 \phi_{\mathrm{NG}}^{(n)}$ corresponds to $j_0^{(n)}$ in Eq. (3.10). Let us consider the state

$$|0(\boldsymbol{\theta})\rangle = \exp\left(-i \sum_n \theta_n \int d^3x \, \partial_0 \phi_{\mathrm{NG}}^{(n)}(x) \right) |0\rangle, \qquad (\mathrm{J}.2)$$

where $\langle 0|\phi_{\mathrm{NG}}^{(n)}|0\rangle = 0$, and the exponential represents the generator of the shifts $\phi_{\mathrm{NG}}^{(n)} \to \phi_{\mathrm{NG}}^{(n)} + \theta_n$.

For simplicity, we consider shift only of the n-th component $\phi_{\mathrm{NG}}^{(n)}$ of the NG multiplet. The overlap of $|0\rangle$ with $|0(\theta_n)\rangle$ is given by (no summation over n)

$$
\begin{aligned}
\langle 0|0(\theta_n)\rangle &= \langle 0| \exp\left(-i\theta_n \int d^3x \, \partial_0 \phi_{\mathrm{NG}}^{(n)}(x) \right) |0\rangle \\
&= \exp\left(-\theta_n^2 \int d^3x \, d^3y \, \langle 0|\partial_0 \phi_{\mathrm{NG}}^{(n)}(x) \partial_0 \phi_{\mathrm{NG}}^{(n)}(y)|0\rangle \right) \\
&= \exp\left(-\theta_n^2 \langle 0|D_n D_n|0\rangle \right), \qquad (\mathrm{J}.3)
\end{aligned}
$$

which is equal to zero due to the translational invariance of the vacuum (cf. Eq. (3.11)).

According to the discussion of Section 3.4, the generators D_n must be regularized (see Eq. (3.15)). As a regulating function $f(x)$, we can take $e^{-\mathbf{x}^2/L^2}$ where we set the volume $V = L^3$. Note that $f(x)$ satisfies the equation for the massless field (d'Alambert equation), in the large volume limit. In this case we have

$$D_n = \lim_{L \to \infty} \int d^3x \, e^{-\mathbf{x}^2/L^2} \partial_0 \phi_{\mathrm{NG}}^{(n)}. \qquad (\mathrm{J}.4)$$

Using

$$\phi_{\text{NG}}^{(n)} = \int \frac{d^3k}{(2\pi)^{\frac{3}{2}}} \frac{1}{\sqrt{2\omega_{\mathbf{k}}}} \left(a_{\mathbf{k}} e^{i\mathbf{k}\cdot\mathbf{x} - i\omega_{\mathbf{k}} x_0} + a_{\mathbf{k}}^{\dagger} e^{-i\mathbf{k}\cdot\mathbf{x} + i\omega_{\mathbf{k}} x_0} \right), \qquad (\text{J.5})$$

with $\omega_{\mathbf{k}=0} = 0$, the overlap (J.3) reads

$$
\begin{aligned}
\langle 0|0(\theta_n)\rangle &= \lim_{L\to\infty} \exp\left(-\frac{\theta_n^2}{2} \int \frac{d^3k}{(2\pi)^3} \, d^3x \, d^3y \, e^{-\mathbf{x}^2/L^2 - \mathbf{y}^2/L^2} e^{i\mathbf{k}\cdot(\mathbf{x}-\mathbf{y})} \omega_{\mathbf{k}} \right) \\
&= \lim_{L\to\infty} \exp\left(-\frac{\pi}{2} \theta_n^2 L^2 \right) = 0.
\end{aligned}
\qquad (\text{J.6})
$$

The previous result applies to any two vacuum states $|0(\theta_1)\rangle$ and $|0(\theta_2)\rangle$ with $\theta_1 \neq \theta_2$. In addition, it is straightforward to show that the generating function $G(\boldsymbol{J}) = \langle 0(\boldsymbol{\theta}_1)|e^{\phi_{\text{NG}}\cdot\boldsymbol{J}}|0(\boldsymbol{\theta}_2)\rangle$ is 0 for $\boldsymbol{\theta}_1 \neq \boldsymbol{\theta}_2$. This implies that $\langle 0(\boldsymbol{\theta}_1)|\hat{O}|0(\boldsymbol{\theta}_2)\rangle$ with an arbitrary *local* operator $\hat{O}$ must be zero. Ground states are thus orthogonal in the limit of the large volumes and there are no transitions between them induced by local operators.

In n-spatial dimensions, the overlap $\langle 0|0(\boldsymbol{\theta})\rangle$ goes as $e^{-cL^{n-1}}$, with c a positive constant. We thus see that only for $n = 1$ the vacuum states $|0(\boldsymbol{\theta})\rangle$ are not unitarily inequivalent and thus one cannot have physically distinct phases. Consequently one cannot have SSB of *continuous* symmetries in one spatial dimension (at $T = 0$). This result is known as the Mermin–Wagner–Coleman theorem.

Chapter 4

Dynamical rearrangement of symmetry and macroscopic manifestations of QFT

4.1 Introduction

In this Chapter we study how the invariance of theory manifests itself at the level of the physical (quasiparticle) fields when the symmetry is spontaneously broken. Our aim is to find the symmetry group G' under which the free field equations are invariant. As said in the previous Chapter, when symmetry is spontaneously broken, the group G' is different from the original symmetry group G under which the equations for the Heisenberg fields are invariant. As examples we consider $U(1)$ and $SU(2)$ global and local symmetry models where we explicitly show the phenomenon of the dynamical rearrangement of symmetry. In these examples G' turns out to be the group contraction of G (i.e., of $U(1)$ and $SU(2)$). This means that G' contains a subgroup of transformations which induce translations of the Nambu–Goldstone (NG) boson fields and thus it describes the condensation of the NG modes in the ground state. We provide explicit representations of the contractions of $U(1)$ and $SU(2)$ and discuss their physical meaning with respect to the boson condensation and their relation with low energy theorems. Ordered patterns are generated through the transformations of the contracted groups. These ordered patterns constitute the macroscopic manifestation of symmetry breaking.

Infrared effects of the order of $\frac{1}{V}$, with V the volume, are shown to be responsible for the rearrangement of symmetry and for the origin of the ordered patterns.

The occurrence of group contraction in spontaneously broken symmetry theories is thus a central issue in our discussion in this Chapter.

Here we also discuss local gauge theories and remark how, also in such a context, the NG field plays a crucial role in determining the structure of the

137

vacuum condensate and making the invariance of the theory under gauge transformations mathematically consistent with the condition of SSB. The occurrence of the Higgs mechanism introduced in the previous Chapter, predicting that the gauge field acquires a mass and the NG field disappears from the physical particle spectrum, does not imply that the NG field have no physical relevance (on this point see also the remarks at the end of Section 3.8.2). Boson condensation of NG fields controls many physically relevant quantities, such as the boson current which enters as a source term in the field equation for the massive vector gauge field. We will obtain indeed such a Maxwell-like equation for field operators and the associated c-number (vacuum expectation value) fields. These equations are actually the *classical Maxwell equations* for the massive vector field (Proca equations). They are macroscopic manifestations of the microscopic dynamics.

Following the path of our discussion, we consider then *non-homogeneous* boson condensation. This is generated by translating an asymptotic boson by a spacetime-dependent function $f(x)$. A space- and/or time-dependent vacuum is obtained as a result of the non-homogeneous condensation. We will prove the boson transformation theorem, which states that when the boson transformation is performed in the dynamical map of the Heisenberg fields, these resulting Heisenberg fields satisfy the same field equations they satisfied before the transformation was performed. The same Heisenberg field equations may thus describe homogeneous and non-homogeneous ground states. When $f(x)$ carries some singularity (divergence or topological singularities) the boson condensation may give rise to the formation of topologically non-trivial extended objects (also called defects), such as vortices, monopoles, etc. These aspects will be treated in more detail in Chapters 7 and 8.

Finally, we discuss, by means of an explicit model, how the stability of the macroscopic structures originated from the group contraction, which ultimately describes the boson condensation mechanism, is protected against quantum fluctuations.

In conclusion, we discuss the occurrence of macroscopic (classically behaving) quantities and how the microscopic dynamics manifests itself into *macroscopic quantum systems*, namely systems whose behavior cannot be understood without recourse to the underlying quantum dynamics.

4.2 Dynamical rearrangement of symmetry

The dynamical rearrangement of symmetry has been studied in several cases of physical interest [182, 619]. In the following we consider the examples of $SU(2)$ symmetry in ferromagnets and the global and local $U(1)$ symmetry, whose breakdown has been discussed in Chapter 3.

4.2.1 *SU(2) symmetry*

The dynamical rearrangement consists in the change of the continuous symmetry group $G \equiv SU(2)$, under which the equations for the Heisenberg electron field operator $\psi(x)$ are invariant, into G', the symmetry group for the quasiparticle field operator equations:

$$\psi(x) \to \psi'(x) = \Psi\left[g[\phi(x), B(x)]\right], \qquad g \in G'. \qquad (4.1)$$

Here $\psi'(x)$ is the transformed of $\psi(x)$ under $SU(2)$ (cf. Eq. (3.48))

$$\psi(x) \to \psi'(x) = \exp(i\theta_i \lambda_i)\psi(x), \qquad i = 1, 2, 3, \qquad (4.2)$$

with $\lambda_i = \frac{\sigma_i}{2}$, σ_i the Pauli matrices, and θ_i a triplet of real continuous group parameters (the rotation angles in the spin-space). In Eq. (4.1) $\phi(x)$ is the quasielectron field operator and $B(x)$ is the magnon field operator (the NG field operator), whose existence has been shown in Section 3.7.2.[1] In the present discussion we omit considering other fields such as the electromagnetic field. We want to know which one is the group G'. The boson field operator for the magnons is introduced as

$$B(x) = \int \frac{d^3k}{(2\pi)^{3/2}} B_{\mathbf{k}}\, e^{i\mathbf{k}\cdot\mathbf{x} - i\omega_k t}, \qquad (4.3a)$$

$$B^\dagger(x) = \int \frac{d^3k}{(2\pi)^{3/2}} B_{\mathbf{k}}^\dagger\, e^{-i\mathbf{k}\cdot\mathbf{x} + i\omega_k t}, \qquad (4.3b)$$

with commutation relations

$$[B(x), B^\dagger(y)]_{t_x = t_y} = \delta(\mathbf{x} - \mathbf{y}), \qquad (4.4a)$$

$$[B(x), B(y)] = [B^\dagger(x), B^\dagger(y)] = 0. \qquad (4.4b)$$

The fields (4.3) satisfy the equations

$$K(\overrightarrow{\partial})B^\dagger(x) = 0, \qquad B(x)K(\overleftarrow{\partial}) = 0, \qquad (4.5)$$

---

[1] As already done in the previous Chapter, here we use ϕ and $B(x)$ instead of ϕ_{in} and $B_{in}(x)$ and/or ϕ_{out} and $B_{out}(x)$, as far as no misunderstanding arises. Note that symbols carrying subscripts "*in*" and "*H*" denote always field operators.

with

$$K(\overrightarrow{\partial}) = -\left(i\frac{\overrightarrow{\partial}}{\partial t} + \omega\right). \tag{4.6}$$

In explicit model computations $B(x)$ is found to be a gapless bound state of the electron field [570,619]. The free field equations for the quasielectron are

$$\Lambda(\overrightarrow{\partial})\phi(x) = 0, \qquad \phi^\dagger(x)\Lambda(\overleftarrow{\partial}) = 0, \tag{4.7}$$

where $\Lambda(\overrightarrow{\partial})$ denotes the partial derivative operator, including the quasi-electron mass term, which is appropriate for the model specified by the Lagrangian one considers. In the present discussion we do not need to specify the Lagrangian, except for its symmetry properties under the $SU(2)$ group. For the explicit form of $\Lambda(\overrightarrow{\partial})$ in the model of the itinerant-electron ferromagnet see [571,619].

In the following it is convenient to work with the Heisenberg spin operator densities $S_\psi^{(i)}(x)$, $i = 1, 2, 3$, which, for infinitesimal θ_i, transform under $SU(2)$ as (cf. Eq. (3.53))

$$S_\psi^{(i)}(x) \rightarrow S_\psi'^{(i)}(x) = S_\psi^{(i)}(x) - \epsilon_{ijk}\,\theta_j S_\psi^{(k)}(x). \tag{4.8}$$

We look for the expressions of the S-matrix and spin densities in terms of ϕ and B fields: $\mathcal{S}(\phi, \phi^\dagger, B, B^\dagger)$ and $S^{(i)}(x, \phi, \phi^\dagger, B, B^\dagger)$, respectively. For that we resort to functional formalism together with the Lehmann–Symanzik–Zimmermann (LSZ) formula. We have[2] [570]

$$\mathcal{S}(\phi, \phi^\dagger, B, B^\dagger) = \langle : \exp[-iA(\phi, \phi^\dagger, B, B^\dagger)] : \rangle, \tag{4.9}$$

$$S^{(i)}(\phi, \phi^\dagger, B, B^\dagger) = \langle S_\psi^{(i)}(x) : \exp[-iA(\phi, \phi^\dagger, B, B^\dagger)] : \rangle, \tag{4.10}$$

where $i = 1, 2, 3$ and

$$\begin{aligned}
A(\phi, \phi^\dagger, B, B^\dagger) = \int d^4x \Big[&\rho^{-1/2}B(x)K(\overrightarrow{\partial})S_\psi^{(-)}(x) \\
&+ \rho^{-1/2}S_\psi^{(+)}(x)K(-\overleftarrow{\partial})B^\dagger(x) + Z^{-1/2}\phi^\dagger(x)\Lambda(-\overrightarrow{\partial})\psi(x) \\
&+ Z^{-1/2}\psi^\dagger(x)\Lambda(-\overleftarrow{\partial})\phi(x) \Big].
\end{aligned} \tag{4.11}$$

Here Z is the wave function renormalization of the electron and ρ is given by Eq. (3.73): $\rho = \frac{1}{2}M$. The symbol $: \cdots :$ denotes normal product ordering and $\langle \ldots \rangle$ denotes functional average as defined in Section 3.7.2 with $\psi(x)$ and $\psi^\dagger(x)$ the functional integration variables (cf. Eqs. (3.51) and

---

[2]The same symbols are used to denote both operators and c-number fields; one can see from the context which one of the two cases applies.

(3.52)). Our task is to find the transformations for $\phi, \phi^\dagger, B, B^\dagger$ in (4.9) and (4.10) which leave invariant their field equations and such that the transformation (4.8) of $S^{(i)}(\phi, \phi^\dagger, B, B^\dagger)$ is induced. Let $\phi_\theta, \phi_\theta^\dagger, B_\theta, B_\theta^\dagger$ denote the transformed fields. We require they satisfy the free equations:

$$K(\overrightarrow{\partial})B_\theta^\dagger(x) = 0, \quad B_\theta(x)K(\overleftarrow{\partial}) = 0, \tag{4.12}$$

$$\Lambda(\overrightarrow{\partial})\phi_\theta(x) = 0, \quad \phi_\theta^\dagger(x)\Lambda(\overleftarrow{\partial}) = 0, \tag{4.13}$$

and that

$$\frac{\partial}{\partial\theta_l}S(\phi_\theta, \phi_\theta^\dagger, B_\theta, B_\theta^\dagger) = 0, \tag{4.14}$$

$$\frac{\partial}{\partial\theta_l}S^i(x, \phi_\theta, \phi_\theta^\dagger, B_\theta, B_\theta^\dagger) = -\epsilon_{ilk}S^k(x, \phi_\theta, \phi_\theta^\dagger, B_\theta, B_\theta^\dagger). \tag{4.15}$$

We now use the transformed fields in Eq. (4.11) and obtain equations for these fields implied by (4.14) and (4.15) by following steps which here we omit, but that can be found in great detail in [570]. In such a derivation, the reduction formulas of the LSZ formalism are also used. The equations eventually obtained are:

$$\frac{\partial}{\partial\theta_1}B_\theta(x) = i\sqrt{M/2}, \qquad \frac{\partial}{\partial\theta_1}B_\theta^\dagger(x) = -i\sqrt{M/2},$$

$$\frac{\partial}{\partial\theta_1}\phi_\theta(x) = 0, \qquad \frac{\partial}{\partial\theta_1}\phi_\theta^\dagger(x) = 0, \tag{4.16a}$$

$$\frac{\partial}{\partial\theta_2}B_\theta(x) = -\sqrt{M/2}, \qquad \frac{\partial}{\partial\theta_2}B_\theta^\dagger(x) = -\sqrt{M/2},$$

$$\frac{\partial}{\partial\theta_2}\phi_\theta(x) = 0, \qquad \frac{\partial}{\partial\theta_2}\phi_\theta^\dagger(x) = 0, \tag{4.16b}$$

$$\frac{\partial}{\partial\theta_3}B_\theta(x) = -iB_\theta(x), \qquad \frac{\partial}{\partial\theta_3}B_\theta^\dagger(x) = iB_\theta^\dagger(x),$$

$$\frac{\partial}{\partial\theta_3}\phi_\theta(x) = i\lambda_3\phi_\theta(x), \qquad \frac{\partial}{\partial\theta_2}\phi_\theta^\dagger(x) = -i\phi_\theta^\dagger(x)\lambda_3, \tag{4.16c}$$

and using the conditions

$$\phi_\theta(x) = \phi(x), \quad B_\theta(x) = B(x), \text{ etc., at } \theta = 0, \tag{4.17}$$

we arrive at

$$\phi(x) \to \phi_\theta(x) = \phi(x), \quad \phi^\dagger(x) \to \phi_\theta^\dagger(x) = \phi^\dagger(x), \tag{4.18a}$$

$$B(x) \to B_\theta(x) = B(x) + i\theta_1\sqrt{M/2}, \tag{4.18b}$$

$$B^\dagger(x) \to B_\theta^\dagger(x) = B^\dagger(x) - i\theta_1\sqrt{M/2}, \tag{4.18c}$$

for $\theta_2 = \theta_3 = 0$, and

$$\phi(x) \to \phi_\theta(x) = \phi(x), \quad \phi^\dagger(x) \to \phi^\dagger_\theta(x) = \phi^\dagger(x), \qquad (4.19\text{a})$$

$$B(x) \to B_\theta(x) = B(x) - \theta_2\sqrt{M/2}, \qquad (4.19\text{b})$$

$$B^\dagger(x) \to B^\dagger_\theta(x) = B^\dagger(x) - \theta_2\sqrt{M/2}, \qquad (4.19\text{c})$$

for $\theta_1 = \theta_3 = 0$, and

$$\phi(x) \to \phi_\theta(x) = e^{i\theta_3\lambda_3}\phi(x), \quad \phi^\dagger(x) \to \phi^\dagger_\theta(x) = \phi^\dagger(x)e^{-i\theta_3\lambda_3}, \quad (4.20\text{a})$$

$$B(x) \to B_\theta(x) = e^{-i\theta_3}B(x), \quad B^\dagger(x) \to B^\dagger_\theta(x) = e^{i\theta_3}B^\dagger(x), \qquad (4.20\text{b})$$

for $\theta_1 = \theta_2 = 0$.

The transformations (4.18)–(4.20) belong to the $E(2)$ group, which is the Inönü–Wigner group contraction of $SU(2)$ [182, 336, 638]. Eqs. (4.18)–(4.20) express the dynamical rearrangement of symmetry: when the quasiparticle field operators $\phi, \phi^\dagger, B, B^\dagger$ undergo the $E(2)$ transformations (4.18)–(4.20), the $SU(2)$ transformations (4.2) and (4.8) of the Heisenberg field operators $\psi, \psi^\dagger, S^i$ are induced, and vice-versa. Note that (4.20) represents the unbroken rotation around the third axis. In the following we will see that the Inönü–Wigner group contraction [336] is the mathematical mechanism determining the rearranged symmetry group [182, 638].

Consistently with our observation in the previous Chapter, we remark that the c-number translations of the field operator $B(x)$ (and $B^\dagger(x)$) in Eqs. (4.18) and (4.19) must be understood as the limit for $f(x) \to 1$ of the transformations

$$B(x) \to B_\theta(x) = \lim_{f(x) \to 1}\left[B(x) + if(x)\theta_1\sqrt{M/2}\right], \qquad (4.21\text{a})$$

$$B(x) \to B_\theta(x) = \lim_{f(x) \to 1}\left[B(x) - f(x)\theta_2\sqrt{M/2}\right], \qquad (4.21\text{b})$$

(and h.c.), respectively. Here the function $f(x)$ is any square-integrable function which satisfies the magnon equation (4.5). As already mentioned in Chapter 3, the role of $f(x)$ is to regularize infrared (infinite volume) divergences acting as an infrared cut-off (much like usual infrared regularization in QFT [112, 558]). Without such a function, terms like $\theta_l(\frac{M}{2\rho})K(\partial)S^{(-)}_\psi(x)$ would be contained in the quantity $A(\phi_\theta, \phi^\dagger_\theta, B_\theta, B^\dagger_\theta)$ (cf. Eq. (4.11)), contributing, in $S(\phi_\theta, \phi^\dagger_\theta, B_\theta, B^\dagger_\theta)$ and $S^{(i)}(\phi_\theta, \phi^\dagger_\theta, B_\theta, B^\dagger_\theta)$, to Feynman diagrams by energyless and momentumless external lines. Such diagrams can thus contain powers of zero energy singularities. In order to avoid such an infrared catastrophe we substitute θ_i by $f(x)\theta_i$, $i = 1, 2$, and, since

$B_\theta(x)$ must satisfy the magnon equation, it is necessary that $f(x)$ satisfy the magnon equation. The limit $f(x) \to 1$ must be taken at the end of the computations. Note that the magnon equations are invariant under (4.21) even before the limit $f(x) \to 1$, thus exhibiting the $E(2)$ invariance.

The generators of the transformations (4.18)–(4.20) (with θ_i, $i = 1, 2$, replaced by $f(x)\theta_i$) are

$$s_f^{(1)} = \sqrt{\frac{M}{2}} \int d^3x \left[B(x)f(x) + B^\dagger(x)f^*(x) \right] , \qquad (4.22a)$$

$$s_f^{(2)} = -i\sqrt{\frac{M}{2}} \int d^3x \left[B(x)f(x) - B^\dagger(x)f^*(x) \right] , \qquad (4.22b)$$

$$s_f^{(3)} = \int d^3x \left[\phi^\dagger(x)\lambda_3\phi(x) - B^\dagger(x)B(x) \right] . \qquad (4.22c)$$

The introduction of the square-integrable function $f(x)$ is essential in order for the generators (4.22) to be well defined. Moreover, we see that these generators are time-independent since $f(x)$ satisfies the magnon equation. The generators (4.22) have commutation relations:

$$[s_f^{(1)}, s_f^{(2)}] = iM \int d^3x |f(x)|^2 = (\text{const.})\mathbb{1} , \qquad (4.23a)$$

$$[s_f^{(3)}, s_f^{(1)}] = is_f^{(2)} , \qquad [s_f^{(3)}, s_f^{(2)}] = -is_f^{(1)} , \qquad (4.23b)$$

which, in terms of $s_f^{(\pm)} = s_f^{(1)} \pm is_f^{(2)}$, read as

$$[s_f^{(+)}, s_f^{(-)}] = 2M \int d^3x |f(x)|^2 = (\text{const.})\mathbb{1} , \qquad (4.24a)$$

$$[s_f^{(3)}, s_f^{(\pm)}] = \pm s_f^{(\pm)} . \qquad (4.24b)$$

The algebra is thus the $e(2)$ algebra and we see that the generators $s_f^{(i)}$, $i = 1, 2$, (or $s_f^{(\pm)}$) exhibit their boson character (compare the above algebra (4.24) with the Weyl–Heisenberg algebra) when expressed in terms of the quasiparticle fields ϕ and $B(x)$.

4.2.2   Global $U(1)$ symmetry

We now study the dynamical rearrangement of symmetry in the model discussed in Section 3.7.1 for the complex scalar field $\phi(x) = \frac{1}{\sqrt{2}}[\psi(x) + i\chi(x)]$, when the global phase symmetry $\phi(x) \to e^{i\theta}\phi(x)$ is assumed to be spontaneously broken (cf. Eq. (3.36)): $\langle\psi(x)\rangle = \sqrt{2}\langle\phi(x)\rangle = \tilde{v} \neq 0$. Notice that now the notation is different from the one used above. Here $\phi(x)$ does not denote the physical (or quasiparticle) field, but the Heisenberg field, consistently with the notation of Section 3.7.1. In the following, in order to avoid misunderstanding we therefore restore the suffix "*in*" for physical fields.

We observe that when the field $\rho(x) = \psi(x) - \langle \psi(x) \rangle_\epsilon$ (cf. Eqs. (3.39)) is assumed to be massive it can decay into two χ's. In this case we do not need to include it in the dynamical map. The results presented below are in any case independent of such an assumption.

In the functional formalism, the S-matrix and the dynamical map for the Heisenberg field ϕ are introduced as:

$$ \mathcal{S} = \langle : \exp[-i\,A(\chi_{in})] : \rangle , \qquad \mathcal{S}\phi(x) = \langle \phi(x) : \exp[-i\,A(\chi_{in})] : \rangle , \qquad (4.25) $$

with

$$ A(\chi_{in}) = -Z_\chi^{\frac{1}{2}} \int d^4x \, \chi_{in}(x) \overset{\leftrightarrow}{\partial^2} \chi(x) . \qquad (4.26) $$

Here χ_{in} denotes the physical (or quasiparticle) field operator for the Goldstone boson. The field χ denotes the Heisenberg interpolating field (cf. Section 2.2) and it acts as the c-number integration variable in the functional integral. Z is the wave function renormalization constant. The LSZ reduction formulas [115, 343, 558] are used in Eqs. (4.25). χ_{in} satisfies the free field equation

$$ \partial^2 \chi_{in}(x) = 0 . \qquad (4.27) $$

We want to determine the specific form of the physical field transformation $\chi_{in}(x) \to \chi_{in}(x, \theta)$ implied by the Heisenberg field global phase transformation (3.35): $\phi(x) \to e^{i\theta}\phi(x)$. Our procedure is fully analogous to the one followed in the $SU(2)$ case considered above. The following conditions have to be satisfied by $\chi_{in}(x, \theta)$:

$$ \partial^2 \chi_{in}(x, \theta) = 0 , \qquad \mathcal{S}[\chi_{in}(x, \theta)] = \mathcal{S}[\chi_{in}(x)] ; \qquad (4.28) $$

$$ \mathcal{S}\phi[\chi_{in}(x, \theta)] = e^{i\theta} \mathcal{S}\phi[\chi_{in}(x)] . \qquad (4.29) $$

Eqs. (4.28) are consequences of the original invariance of the Heisenberg field dynamics. This implies the invariance also of the physical field equations and of the S-matrix. Indeed, as already remarked above, even at the level of the physical fields the original invariance cannot be lost, although at that level it can manifest itself in a different symmetry group structure. Eq. (4.29) is a consequence of the dynamical map.

We now use the transformed field $\phi[\chi_{in}(x,\theta)]$ in Eq. (4.25) and by imposing the invariance constraints (4.28) and (4.29) we obtain the equation for $\phi[\chi_{in}(x,\theta)]$. By omitting details fully given in [450], we obtain:

$$\chi_{in}(x,\theta) = \chi_{in}(x) + \theta\tilde{v}Z_\chi^{-\frac{1}{2}}, \tag{4.30}$$

i.e., the physical field χ_{in} translates by a constant when the Heisenberg field ϕ undergoes the global phase transformation, and vice-versa (cf. Eq. (4.29)). The global $U(1)$ invariance group is thus dynamically rearranged into the one-parameter constant translation group, which is the $U(1)$ group contraction.

Again, the translation of a field by a constant, such as in (4.30), must be actually understood as the limit for $f(x) \to 1$ of the transformation

$$\chi_{in}(x,\theta) = \chi_{in}(x) + \theta\tilde{v}Z_\chi^{-\frac{1}{2}}f(x), \tag{4.31}$$

with $f(x)$ a square-integrable function satisfying the χ_{in} field equation: $\partial^2 f(x) = 0$. Eq. (4.26) shows that $\mathcal{S}[\chi_{in}(x,\theta)]$ and $\mathcal{S}\phi[\chi_{in}(x,\theta)]$ are not well defined due to infrared singularities contained in Feynman diagrams with many momentumless and energyless lines when $\chi_{in}(x)$ undergoes the transformation (4.30). These infrared divergencies are cured by the introduction of $f(x)$ [450, 460, 570].

4.2.3 *Local $U(1)$ symmetry and the emergence of classical Maxwell equations*

On the basis of the relations obtained in Section 3.8.1, we may introduce the fields $b_H(x)$ and $U_{\mu H}(x)$ and rewrite the dynamical maps, Eqs. (3.108), as

$$\chi_H(x) = Z_\chi^{\frac{1}{2}}\chi_{in}(x) + \cdots, \tag{4.32a}$$

$$\rho_H(x) \equiv \psi_H(x) - \tilde{v} = Z_\rho^{\frac{1}{2}}\rho_{in}(x) + \cdots, \tag{4.32b}$$

$$b_H(x) \equiv B_H(x) + \frac{e_0\tilde{v}}{Z_\chi}\chi_H(x) = \frac{e_0\tilde{v}}{Z_\chi^{\frac{1}{2}}}b_{in}(x) + \cdots, \tag{4.32c}$$

$$U_H^\mu(x) \equiv A_H^\mu(x) - \frac{Z_\chi}{(e_0\tilde{v})^2}\partial^\mu B_H(x) - \frac{1}{e_0\tilde{v}}\partial^\mu\chi_H(x)$$

$$= Z_3^{\frac{1}{2}}U_{in}^\mu(x) + \cdots. \tag{4.32d}$$

As in the examples considered in the previous Sections, compact expressions of the dynamical maps are given by use of the functional formalism and the LSZ reduction formula [115, 343, 558]. Thus we define

$$A(\chi_{in}, b_{in}, \rho_{in}, U_{in}^\mu) = Z_\chi^{-1/2}\chi_{in}(x)(-\partial^2)\chi(x)$$
$$-\frac{Z_\chi^{-1/2}}{e_0\tilde{v}}b_{in}(x)(-\partial^2)b(x) + Z_\rho^{-1/2}\rho_{in}(x)(-\partial^2 - m_\rho^2)\rho(x)$$
$$-Z_3^{-1/2}U_{in}^\mu(x)(-\partial^2 - m_V^2)U_\mu(x), \tag{4.33}$$

where Z_χ and Z_3 are the wave function renormalization constants for the χ and ρ fields, respectively, and we have introduced the fields b and U_μ in analogy with Eqs. (4.32):

$$b(x) = B(x) + \frac{e_0\tilde{v}}{Z_\chi}\chi(x), \tag{4.34}$$

$$U_\mu(x) = A_\mu(x) - \frac{Z_\chi}{e_0\tilde{v}}\partial_\mu B(x) - \frac{1}{e_0\tilde{v}}\partial_\mu\chi(x). \tag{4.35}$$

Then, the dynamical maps are given by

$$\mathcal{S} = \langle : \exp[-i\int d^4x\, A(\chi_{in}, b_{in}, \rho_{in}, U_{in}^\mu)] : \rangle, \tag{4.36a}$$

$$\mathcal{S}A_{\mu H}(x) = \langle A_\mu(x) : \exp[-i\int d^4y\, A(\chi_{in}, b_{in}, \rho_{in}, U_{in}^\mu)] : \rangle, \tag{4.36b}$$

$$\mathcal{S}\phi_H(x) = \langle \phi(x) : \exp[-i\int d^4y\, A(\chi_{in}, b_{in}, \rho_{in}, U_{in}^\mu)] : \rangle :, \tag{4.36c}$$

$$\mathcal{S}B_H(x) = \langle B(x) : \exp[-i\int d^4y\, A(\chi_{in}, b_{in}, \rho_{in}, U_{in}^\mu)] : \rangle : . \tag{4.36d}$$

We now observe that

$$\int d^4x\, A(\chi_{in}, b_{in}, \rho_{in}, U_{in}^\mu) = \int d^4x\, A^0(\chi_{in}, b_{in}, \rho_{in}, U_{in}^\mu)$$
$$- \int d^4x\, \frac{Z_\chi^{1/2}}{e_0\tilde{v}}b_{in}(x)(-\partial^2)B(x), \tag{4.37}$$

where

$$A^0(\chi_{in}, b_{in}, \rho_{in}, U_{in}^\mu) = Z_\chi^{-1/2}(\chi_{in}(x) - b_{in}(x))(-\partial^2)\chi(x)$$
$$+ Z_\rho^{-1/2}\rho_{in}(x)(-\partial^2 - m_\rho^2)\rho(x)$$
$$- Z_3^{-1/2}U_{in}^\mu(x)(-\partial^2 - m_V^2)U_\mu(x). \tag{4.38}$$

We then introduce the sources

$$J_\mu(x) = Z_3^{-1/2}U_{in}^\mu(x)(-\partial^2 - m_V^2), \tag{4.39a}$$
$$K_1(x) = -Z_\rho^{-1/2}\rho_{in}(x)(-\partial^2 - m_\rho^2), \tag{4.39b}$$
$$K_2(x) = -Z_\chi^{-1/2}(\chi_{in}(x) - b_{in}(x))(-\partial^2), \tag{4.39c}$$

and rewrite Eqs. (4.36) as

$$S = \langle : \exp\left\{ i \int d^4x \frac{Z_\chi^{1/2}}{e_0\tilde{v}} b_{in}(x)(-\partial^2)B(x) \right\} : \rangle_{J,K}, \qquad (4.40a)$$

$$SG(x) = \langle G(x) : \exp\left\{ i \int d^4y \frac{Z_\chi^{1/2}}{e_0\tilde{v}} b_{in}(y)(-\partial^2)B(y) \right\} : \rangle_{J,K}, \qquad (4.40b)$$

where G is any of the Heisenberg fields A_H^μ, ϕ_H, B_H or any functional of them. Notice that the fields $\chi_{in}(x)$ and $b_{in}(x)$ appear in the sources (4.39) always in the combination $\chi_{in}(x) - b_{in}(x)$. We may proceed now by introducing a parameter, say σ, so to write

$$A^\sigma(\chi_{in}, b_{in}, \rho_{in}, U_{in}^\mu) = A^0(\chi_{in}, b_{in}, \rho_{in}, U_{in}^\mu)$$
$$- \sigma \frac{Z_\chi^{1/2}}{e_0\tilde{v}} b_{in}(x)(-\partial^2)B(x). \qquad (4.41)$$

We then use A^σ in Eqs. (4.40), differentiate them with respect to σ and put $\sigma = 1$ at the end. By use of convenient further derivatives with respect to the sources and other identities obtained in the functional formalism (see [449]) we eventually obtain

$$S = \langle : \exp\left\{ i \int d^4x A^0 \right\} : \rangle, \qquad (4.42a)$$

$$SA_{\mu H}(x) = SA_{\mu H}^0(x) + \frac{Z_\chi^{\frac{1}{2}}}{e_0\tilde{v}} : S\partial_\mu b_{in}(x) :, \qquad (4.42b)$$

$$S\phi_H(x) = : \exp\left\{ i \frac{Z_\chi^{\frac{1}{2}}}{\tilde{v}} b_{in}(x) \right\} S\phi_H^0(x) :, \qquad (4.42c)$$

$$B_H(x) = \frac{e_0\tilde{v}}{Z_\chi^{\frac{1}{2}}}(b_{in}(x) - \chi_{in}(x)) + c, \qquad (4.42d)$$

where c is a c-number constant, which is irrelevant since only derivatives of $B_H(x)$ appear in the Heisenberg equation for $A_{\mu H}(x)$ (cf. Eq. (3.100)). In Eqs. (4.42)

$$SA_{\mu H}^0(x) = : \langle A_\mu(x) \rangle_{J,K} :, \qquad (4.43a)$$

$$S\phi_H^0(x) = : \langle \phi(x) \rangle_{J,K} : . \qquad (4.43b)$$

By inspection of the dynamical maps (4.42) we see that the local gauge transformations of the Heisenberg fields

$$\phi_H(x) \rightarrow e^{ie_0\lambda(x)}\phi_H(x), \qquad (4.44a)$$

$$A_H^\mu(x) \rightarrow A_H^\mu(x) + \partial^\mu\lambda(x), \qquad (4.44b)$$

$$B_H(x) \rightarrow B_H(x), \qquad (4.44c)$$

are induced by the in-field transformations

$$\chi_{in}(x) \to \chi_{in}(x) + e_0 \tilde{v} Z_\chi^{-\frac{1}{2}} \lambda(x), \tag{4.45a}$$

$$b_{in}(x) \to b_{in}(x) + e_0 \tilde{v} Z_\chi^{-\frac{1}{2}} \lambda(x), \tag{4.45b}$$

$$\rho_{in}(x) \to \rho_{in}(x), \qquad U_{in}^\mu(x) \to U_{in}^\mu(x). \tag{4.45c}$$

Note that under the above in-field transformations the S-matrix, $SA_{\mu H}^0(x)$ and $S\phi_H^0(x)$ are invariant.

The global transformation $\phi_H(x) \to e^{i\theta} \phi_H(x)$ is induced by

$$\chi_{in}(x) \to \chi_{in}(x) + \theta \tilde{v} Z_\chi^{-\frac{1}{2}} f(x), \tag{4.46a}$$

$$b_{in}(x) \to b_{in}(x), \quad \rho_{in}(x) \to \rho_{in}(x), \quad U_{in}^\mu(x) \to U_{in}^\mu(x), \tag{4.46b}$$

with $\partial^2 f(x) = 0$ and the limit $f(x) \to 1$ to be performed at the end of the computation. We use indeed the transformations (4.46) in the sources (4.39) and obtain

$$J_\mu \to J_\mu, \qquad K_1 \to K_1, \qquad K_2 \to K_2 - \theta \tilde{v} Z_\chi^{-\frac{1}{2}} f(x)(-\partial^2). \tag{4.47}$$

We then compute the derivatives with respect to θ of the S-matrix, $SA_{\mu H}^0(x)$ and $S\phi_H^0(x)$ and get [449]

$$\frac{\partial}{\partial \theta} S = 0, \tag{4.48a}$$

$$\frac{\partial}{\partial \theta} \phi_H^0(x) = i\phi_H^0(x), \tag{4.48b}$$

$$\frac{\partial}{\partial \theta} A_{\mu H}^0(x) = 0. \tag{4.48c}$$

Moreover, from the fourth of Eqs. (4.42) we see that B_H is changed by an irrelevant c-number under Eqs. (4.46). From Eqs. (4.48) we see that the global transformation of $\phi_H(x)$ is indeed induced by the in-field transformation (4.46). Under these the S-matrix and $A_{\mu H}(x)$ are instead left invariant. Eqs. (4.48) also imply that

$$S = : S[\rho_{in}, U_{\mu in}, \partial(\chi_{in}(x) - b_{in}(x))] :, \tag{4.49a}$$

$$\phi_H^0(x) = : \exp\left\{ i\frac{Z_\chi^{\frac{1}{2}}}{\tilde{v}} (\chi_{in}(x) - b_{in}(x)) \right\} \left\{ \tilde{v} + Z_\chi^{\frac{1}{2}} \rho_{in} \right.$$

$$\left. + F^0[\rho_{in}, U_{\mu in}, \partial(\chi_{in}(x) - b_{in}(x))] \right\} :, \tag{4.49b}$$

$$A_{\mu H}^0(x) = Z_3^{\frac{1}{2}} U_{\mu in} + : F_\mu^0[\rho_{in}, U_{\mu in}, \partial(\chi_{in}(x) - b_{in}(x))] :, \tag{4.49c}$$

where $\partial(\chi_{in}(x) - b_{in}(x))$ stands for at least first order derivative of the combination $(\chi_{in}(x) - b_{in}(x))$. These relations used in Eqs. (4.42) then give

$$S = \, : S[\rho_{in}, U_{in}^{\mu}, \partial(\chi_{in} - b_{in})] \, :, \qquad (4.50a)$$

$$\phi_H(x) = \, : \exp\left\{ i \frac{Z_{\chi}^{\frac{1}{2}}}{\tilde{v}} \chi_{in}(x) \right\}$$

$$\times \left(\tilde{v} + Z_{\rho}^{\frac{1}{2}} \rho_{in}(x) + F[\rho_{in}, U_{in}^{\mu}, \partial(\chi_{in} - b_{in})] \right) \, :, \qquad (4.50b)$$

$$A_H^{\mu}(x) = Z_3^{\frac{1}{2}} U_{in}^{\mu}(x) + \frac{Z_{\chi}^{\frac{1}{2}}}{e_0 \tilde{v}} \partial^{\mu} b_{in}(x) + \, : F^{\mu}[\rho_{in}, U_{in}^{\mu}, \partial(\chi_{in} - b_{in})] \, : . \quad (4.50c)$$

The various functionals F appearing in Eqs. (4.49) and (4.50) are to be determined within a particular model and are at least bilinear in their arguments. From Eqs. (4.50) we see that the global transformations of the Heisenberg fields are also induced by

$$\chi_{in}(x) \to \chi_{in}(x) + \theta \, \tilde{v} \, Z_{\chi}^{-\frac{1}{2}} f(x), \qquad (4.51a)$$

$$b_{in}(x) \to b_{in}(x) + \theta \, \tilde{v} \, Z_{\chi}^{-\frac{1}{2}} \beta f(x), \qquad (4.51b)$$

with β any real constant and $f(x) \to 1$.

Since the S-matrix involves the NG field $\chi_{in}(x)$ and the ghost field $b_{in}(x)$ only through the combination $\chi_{in}(x) - b_{in}(x)$, these fields do not participate to any observable reaction. However, the discussion in the previous and in the present Section shows they are crucial in maintaining the invariance of the theory when SSB occurs. Moreover, their transformations control the boson condensation in the ground state.

The disappearance of the NG and ghost fields from the observable spectrum and the appearance of the massive vector field $U_{in}(x)$ is called the Higgs mechanism.

We remark that under the transformations (4.45), Eqs. (3.78) and (3.79) change into

$$\langle 0 | \partial_{\mu} A_H^{\mu}(x) | 0 \rangle = \partial^2 \lambda(x), \qquad (4.52)$$

$$\langle 0 | \phi_H(x) | 0 \rangle = e^{i e_0 \lambda(x)} \tilde{v}, \qquad (4.53)$$

respectively. Therefore the condition (3.79) is not sufficient to determine the physical content of the theory: the gauge should also be specified.

Finally, we recall that the dynamical rearrangement of symmetry for the chiral gauge model has the same form as in the $U(1)$ gauge model (cf. Section 3.8.2).

The classical Maxwell equations and the physical state condition

The field equation for $A_{\mu H}(x)$ is given by Eq. (3.100):

$$-\partial^2 A_{\mu H}(x) = j_{\mu H}(x) - \partial_\mu B_H(x),\qquad(4.54)$$

with $j_{\mu H}(x) = \delta \mathcal{L}(x)/\delta A_H^\mu(x)$. From the second of Eqs. (4.42) we see that

$$\partial^2 A_{\mu \dot{H}}(x) = \partial^2 A_{\mu H}^0(x),\qquad(4.55\text{a})$$

$$\partial^\mu A_{\mu H}^0(x) = 0,\qquad(4.55\text{b})$$

and we define

$$F_{\mu\nu H} \equiv \partial_\mu A_{\nu H}(x) - \partial_\nu A_{\mu H}(x) = \partial_\mu A_{\nu H}^0(x) - \partial_\nu A_{\mu H}^0(x).\qquad(4.56)$$

Then Eq. (4.54) can be written as

$$-\partial^2 A_{\mu H}^0(x) = j_{\mu H}(x) - \partial_\mu B_H(x),\qquad(4.57)$$

and we can use $A_{\nu H}^0(x)$ as the vector potential in the Lorentz gauge. Use of the fourth of Eqs. (4.42) and the third of Eqs. (4.49) in Eq. (4.57) gives the dynamical map of $j_{\mu H}(x)$:

$$j_{\mu H}(x) = Z_3^{\frac{1}{2}} m_V^2 U_{in}^\mu(x) - : \partial^2 F_\mu^0 : + \frac{e_0 \tilde{v}}{Z_\chi^{1/2}} \partial_\mu(b_{in} - \chi_{in}),\qquad(4.58)$$

which shows that the current $j_{\mu H}(x)$ depends on b_{in} and χ_{in} only through the difference $b_{in} - \chi_{in}$. The correspondence with the *classical Maxwell equations* requires that, for any pairs of physical states $|a\rangle$ and $|b\rangle$, we must have (see also [488])

$$-\partial^2 \langle a|A_{\mu H}^0(x)|b\rangle = \langle a|j_{\mu H}(x)|b\rangle.\qquad(4.59)$$

Eq. (4.57) shows that this is equivalent to impose the condition:

$$\langle b|\partial_\mu B_H(x)|a\rangle = 0,\qquad(4.60)$$

namely the Gupta–Bleuler-like condition on the physical states

$$\left[\chi_{in}^{(-)}(x) - b_{in}^{(-)}(x)\right]|a\rangle = 0.\qquad(4.61)$$

Here $\chi_{in}^{(-)}$ and $b_{in}^{(-)}$ are the positive-frequency parts (the annihilator operators) of the corresponding fields. This means that the physical states contain an equal number of χ_{in} and b_{in} modes.

Under the transformation $\chi_{in}(x) \to \chi_{in}(x) + \tilde{v} Z_\chi^{-\frac{1}{2}} f(x)$ (cf. Eqs. (4.46)), $A_{\mu H}^0(x)$ in Eqs. (4.49) is invariant and $B_H(x)$ changes as (cf. the fourth of Eqs. (4.42))

$$B_H(x) = e_0 \tilde{v}\, Z_\chi^{-\frac{1}{2}} [b_{in}(x) - \chi_{in}(x)] \to B_H(x) - \frac{e_0 \tilde{v}^2}{Z_\chi} f(x),\qquad(4.62)$$

where $\partial^2 f(x) = 0$. The physical state condition (4.61) and the field equation for $A^0_{\mu H}(x)$, Eq. (4.59), are then violated. In order to restore them, we must compensate the change in $B_H(x)$ by another transformation. We observe that the S-matrix and the various Heisenberg fields other than $B_H(x)$, Eqs. (4.50), contain $U^\mu_{in}(x)$ and $B(x)$ in the combination

$$\int d^4x \left[Z_3^{-\frac{1}{2}} U^\mu_{in}(x)(-\partial^2 - m_V^2)A_\mu(x) + B(x)\partial^\mu A_\mu(x) \right].$$ (4.63)

Thus, transforming the field $U^\mu_{in}(x)$ as

$$U^\mu_{in}(x) \to U^\mu_{in}(x) + u^\mu(x), \quad \partial^\mu u_\mu(x) = 0,$$ (4.64)

with $u_\mu(x)$ a c-number Fourier transformable function, is equivalent to transform $B(x)$ in the generating functional (neglecting irrelevant surface terms which do not affect the motion equations), as

$$B(x) \to B(x) + \lambda(x),$$ (4.65)

with $\lambda(x)$ satisfying the equation

$$Z_3^{-1/2}(-\partial^2 - m_V^2)u_\mu(x) = -\partial_\mu \lambda(x),$$ (4.66)

and $\partial^2 \lambda(x) = 0$, since $\partial^\mu u_\mu(x) = 0$. Note that, if we assume $u_\mu(x)$ solution of the Proca equation, then $\partial_\mu \lambda(x) = 0$; in such a case, (4.64) is a transformation which leaves invariant the $U^\mu_{in}(x)$ field equation. Here we consider $\partial_\mu \lambda(x) \neq 0$. We see that the combined transformations of $U^\mu_{in}(x)$, Eq. (4.64), and $B_H(x)$, Eq. (4.62), compensate each other provided we set

$$\lambda(x) = \frac{e_0 \tilde{v}^2}{Z_\chi} f(x),$$ (4.67)

(compare Eqs. (4.62) and (4.65)). Thus we obtained that under the combined transformations (4.62) and (4.64) the physical state condition (4.61) and the classical Maxwell equation for $A^0_{\mu H}(x)$, Eq. (4.59), are not violated.

The classical equation for the massive vector potential and the classical ground state current

By using Eq. (4.67), Eq. (3.107), i.e., $m_V^2 = \frac{Z_3}{Z_\chi}(e_0 \tilde{v})^2$, and setting

$$a_\mu(x) \equiv Z_3^{\frac{1}{2}} u_\mu(x),$$ (4.68)

Eq. (4.66) becomes

$$(-\partial^2 - m_V^2)a_\mu(x) = -\frac{m_V^2}{e_0} \partial_\mu f(x),$$ (4.69)

which is *the classical Maxwell (Proca) equation for the massive vector potential* a_μ [449, 452]. The classical ground state current $j_{\mu,cl}$ turns out to be

$$j_{\mu,cl}(x) \equiv \langle 0|j_{\mu H}(x)|0\rangle = m_V^2 \left[a_\mu(x) - \frac{1}{e_0} \partial_\mu f(x) \right]. \qquad (4.70)$$

The term $m_V^2 a_\mu(x)$ is the *Meissner current*, $\frac{m_V^2}{e_0} \partial_\mu f(x)$ is the *boson current*.

The macroscopic vector field and current are thus given in terms of the variations of the boson transformation function $\partial_\mu f(x)$. Eq. (4.69) shows that $\partial_\mu f$ is a regular function since a_μ is a regular function (u_μ has been assumed to be a Fourier transformable function).

Under the combined transformations Eq. (4.64) and Eq. (4.46) (or Eq. (4.62)), $A_{\mu H}^0(x)$ changes into

$$A_{\mu H}^0(x) = a_\mu(x) + Z_3^{\frac{1}{2}} U_{\mu in} \qquad (4.71)$$

$$+ : F_\mu^0 \left[\rho_{in}, U_{\mu in} + u_\mu, \partial(\chi_{in}(x) - b_{in}(x)) + \frac{\tilde{v}}{Z_\chi^{\frac{1}{2}}} \partial f(x) \right] : .$$

$A_{\mu H}^0(x)$ and $a_\mu(x)$ remain unchanged under the gauge transformations (4.45). Remarkably, the classical equation (4.69) and the classical ground state current (4.70) are invariant under the gauge transformations

$$a_\mu(x) = a_\mu(x) + \partial_\mu \theta(x)), \qquad (4.72a)$$

$$f(x) = f(x) + e_0 \partial_\mu \theta(x), \qquad (4.72b)$$

where $\theta(x)$ satisfies $\partial^2 \theta(x) = 0$.

In conclusion, the boson condensation induced in the ground state by the combined transformations Eq. (4.46) and Eq. (4.64) generates the classical ground state current $j_{\mu,cl}(x) \equiv \langle 0|j_{\mu H}(x)|0\rangle$. This is the source of the classical vector potential $a_\mu(x) = \langle 0|A_{\mu H}^0(x)|0\rangle$ which fulfills $-\partial^2 a_\mu(x) = j_{\mu,cl}(x)$. The condensation of the NG field thus determines the structure of the vacuum condensate and makes the condition of SSB compatible with the invariance of the theory under gauge transformations. The classical field equations appear to be the macroscopic manifestations of the microscopic dynamics.

In Section 4.3.1 we find that $j_{\mu,cl}(x)$ does actually exist only provided $f(x)$ is a singular function.

The formalism here developed can be extended also to other gauges, such as the radiation gauge. The reader can find details in [449, 452].

4.3 The boson transformation theorem and the non-homogeneous boson condensation

In the previous Sections, the introduction of the regularizing function $f(x)$ has been required in order to avoid the infrared catastrophe and obtain a mathematically well defined expression for the generator of NG translations by a constant c-number. The limit $f(x) \to 1$ is then to be taken at the end of the computations. In this Section we show that translations of fields are, however, very interesting even if such a limit is not performed.

Let $\chi_{i,in}(x)$, $i = 1, 2, \cdots$ denote an irreducible set of m (not necessarily massless) boson field operators satisfying the free field equations

$$K_i(\partial)\chi_{i,in}(x) = 0 . \tag{4.73}$$

Consider then the transformation

$$\chi_{i,in}(x) \to \chi_{i,in}(x) + \alpha_i(x), \tag{4.74}$$

where the functions $\alpha_i(x)$ satisfy the field equations for $\chi_{i,in}(x)$:

$$K_i(\partial)\alpha_i(x) = 0 . \tag{4.75}$$

Eq. (4.74) is called the *boson transformation* [450, 617, 619]. Eqs. (3.16) and (4.46) are thus examples of boson transformations (with boson transformation function $\alpha(x) \equiv \theta\tilde{v}/Z_\chi^{\frac{1}{2}} f(x)$) (see also (4.21)).

Consider the Heisenberg field operator $\phi_H(x)$, which, for simplicity, in the following we will denote by $\phi(x)$. Let its field equation be denoted by

$$\Lambda(\partial)\phi(x) = J[\phi(x)], \tag{4.76}$$

where J denotes terms non-linear in ϕ. The dynamical map of $\phi(x)$ in terms of normal products of $\chi_{i,in}(x)$ field operators is written as

$$\phi(x) = \Phi[x; \chi_{i,in}(x)] . \tag{4.77}$$

Notice that in computing products of ϕ fields one first computes the normal products of the $\chi_{i,in}(x)$ fields in the dynamical map and then computes the integrations and other operations on the products of ϕ fields.

Let ϕ' denote the transformed field operator obtained through the dynamical mapping when the physical field $\chi_{i,in}(x)$ undergoes the boson transformation:

$$\phi'(x) = \Phi[x; \chi_{i,in}(x) + \alpha_i(x)] . \tag{4.78}$$

The *boson transformation theorem* then holds, which states that ϕ' is also solution of the Heisenberg field equation for ϕ:

$$\Lambda(\partial)\phi'(x) = J[\phi'(x)] . \tag{4.79}$$

Let us follow closely [619] in presenting the general proof of the theorem, see also [450, 617]. First, we note that

$$J[\phi'(x)] = J'[\phi(x)] \,, \qquad (4.80)$$

where $F'[\phi(x)]$ denotes that the boson transformation has been applied to the dynamical map of $F[\phi(x)]$.[3] Next, we observe that any derivative or integration operation, generically denoted by the symbol $\mathcal{O}$, performed on $\chi'_{i,in}(x)$ fields, commutes with the boson transformation, i.e.,

$$\mathcal{O}(\chi_{i,in}(x) + \alpha_i(x)) = [\mathcal{O}\chi_{i,in}(x)]' \,, \qquad (4.81)$$

where $[\mathcal{O}\chi_{i,in}(x)]'$ denotes that the boson transformation has been applied after $\mathcal{O}$ has operated on $\chi_{i,in}(x)$.

Since in solving the Heisenberg equations one only needs to perform derivatives and/or integrations on normal products of $\chi_{i,in}(x)$ and/or on product of normal products of $\chi_{i,in}(x)$, we see that

$$\Lambda(\partial)\phi'(x) = [\Lambda(\partial)\phi(x)]' = J'[\phi(x)] = J[\phi'(x)] \,, \qquad (4.82)$$

which proves the theorem.

We remark that the boson transformation functions α_i are not required to be regular (Fourier transformable) functions. They are required to solve the free field equations (4.73) for the $\chi_{i,in}(x)$ field operators.

Since translation of a boson field describes boson condensation, the boson transformation describes *non-homogeneous*, i.e., spacetime-dependent, boson condensation. The boson transformation theorem then shows that the same set of Heisenberg field equations may describe homogeneous and non-homogeneous phenomena: as we will see in the following Chapters, this directly leads us to the mechanism of formation of extended objects (defect formation) through non-homogeneous boson condensation.

As an application of the boson transformation theorem, we can mention that the boson-transformed S-matrix $\mathcal{S}[\chi_{in}(x) + \alpha(x)]$ and field $\mathcal{S}\phi[x; \chi_{in}(x) + \alpha(x)]$ differ from $\mathcal{S}[\chi_{in}(x)]$ and $\mathcal{S}\phi[x; \chi_{in}(x)]$ only by an ϵ-dependent factor [452]. For example, in the complex scalar field model (Goldstone-type model) presented in Section 3.7.1, this factor is proportional to (see [619] for details):

$$C \equiv \lim_{\epsilon \to 0} \exp\left\{ \sqrt{2}\epsilon v \frac{Z_\chi^{\frac{1}{2}}}{\tilde{v}} \int d^4x\, \alpha(x)\chi(x) \right\} . \qquad (4.83)$$

---

[3]For example, if $F[a] =: aa^\dagger a :$ and $a' = a + \alpha$, then $F'[a] = (: aa^\dagger a :)' = a'^\dagger a' a'$ and $F[a'] =: a'a'^\dagger a' := a'^\dagger a' a'$.

Thus, since the two fields $\mathcal{S}\phi[x; \chi_{in}(x)]$ and $\mathcal{S}\phi[x; \chi_{in}(x) + \alpha(x)]$ differ only by the ϵ-term, they are solutions of the same field equation. Under the boson transformation (4.74), the order parameter $\tilde{v} \equiv \langle 0|\phi[x; \chi_{in}(x)]|0\rangle$ gets spacetime dependence: $\tilde{v}(x) \equiv \langle 0|\phi[x; \chi_{in}(x) + \alpha(x)]|0\rangle$.

When no gauge fields are present, $\tilde{v}(x)$ can be shown to be given by [450]

$$\tilde{v}(x) \equiv \langle 0|\phi[x; \chi_{in}(x) + \alpha(x)]|0\rangle$$
$$= \exp\left\{i\frac{Z_\chi^{\frac{1}{2}}}{\tilde{v}}\alpha(x)\right\}\left[\tilde{v} + V\left(i\frac{Z_\chi^{\frac{1}{2}}}{\tilde{v}}\partial_\mu\alpha(x)\right)\right], \qquad (4.84)$$

where the expansion $\tilde{v}(x) = \tilde{v} + V(i\frac{Z_\chi^{\frac{1}{2}}}{\tilde{v}}\partial_\mu\alpha(x))$, with $V(i\frac{Z_\chi^{\frac{1}{2}}}{\tilde{v}}\partial_\mu\alpha(x)) \to 0$ when $\partial_\mu\alpha(x) \to 0$, is used. Note that, in the limit $\alpha \to const.$, only the phase is changed.

In the case of a local gauge theory and for regular $\alpha(x)$, the only effect of the boson transformation is the appearance of a phase factor in the order parameter: $\tilde{v}(x) = e^{ic\alpha(x)}\tilde{v}$, with c a constant [449, 452]. Therefore, any spacetime dependence of the ϵ-term can be eliminated by a gauge transformation when $\alpha(x)$ is a regular function. If in local gauge theory the function $\alpha(x)$ carries some divergence and/or topological singularities, then one has to carefully exclude the singularity regions when integrating on space and/or time. For example, if $\alpha(x)$ is singular on the axis of a cylinder (at $r = 0$) one must exclude the singular line $r = 0$ by means of a cylindrical surface of infinitesimal radius. The phase of the order parameter will be singular on that line. This means that SSB does not occur in that region: there we have the "normal" state rather than the ordered one. Provided one uses such a care, the boson transformation can be safely used also in the case of singular $\alpha(x)$.

In Chapter 3, we have seen that changes in the order parameter describe transitions among unitarily inequivalent representations and therefore among physically distinct phases of the system (phase transitions). The conclusions reached above show that, when a gauge field is present in the theory, phase transitions are induced only by boson transformations with singular condensation function $\alpha(x)$.

Summarizing, when a theory allows SSB, there always exist solutions of the field equations with space and/or time-dependent vacuum. They are obtained from the translationally invariant ones by means of boson transformations. In a local gauge theory, regular boson transformation functions can be gauged away unless they carry divergence or topological singularities. Thus changes in the order parameter due to boson condensation may result in a phase transformation only when the condensation function is singular. On the other hand, in the case we have only global gauge invariance, regular boson transformations can produce non-trivial physical effects (like linear flow in superfluidity) [619].

4.3.1 *Topological singularities, gapless modes and macroscopic observables*

We now show that boson transformation functions carrying topological singularities are only allowed for massless bosons [457, 619, 650, 651]. Thus we may expect topologically non-trivial non-homogeneous boson condensations when massless bosons exist, as it happens in theories with SSB where the existence of gapless NG modes is required (although, in the presence of a gauge field, they do not appear in the physical spectrum).

Let the boson transformation function $f(x)$ for the field χ_{in} carry a topological singularity. It is then not single-valued and thus path-dependent:

$$G^{\dagger}_{\mu\nu}(x) \equiv [\partial_{\mu}, \partial_{\nu}] f(x) \neq 0, \qquad \text{for certain } \mu, \nu, x. \tag{4.85}$$

On the other hand, $\partial_{\mu} f$, which is related with observables since these may be influenced by gradients in the Bose condensate, is single-valued, i.e., $[\partial_{\rho}, \partial_{\nu}] \partial_{\mu} f(x) = 0$ (see also the comment after Eq. (4.70)). Let χ_{in} have non-zero mass m and recall that $f(x)$ is solution of the χ_{in} equation:

$$(\partial^2 + m^2) f(x) = 0. \tag{4.86}$$

From the definition of $G^{\dagger}_{\mu\nu}$ and the regularity of $\partial_{\mu} f(x)$ it follows, by computing $\partial^{\mu} G^{\dagger}_{\mu\nu}$, that

$$\partial_{\mu} f(x) = \frac{1}{\partial^2 + m^2} \partial^{\lambda} G^{\dagger}_{\lambda\mu}(x). \tag{4.87}$$

This leads to $\partial^2 f(x) = 0$, which in turn implies $m = 0$. Thus we conclude that (4.85) is only compatible with massless χ_{in}.

The quantity $\partial_{\mu} f(x)$ is given by (4.87) with $m = 0$. $f(x)$ can be determined from this equation. The topological charge is defined as

$$N_T = \int_C dl^{\mu} \partial_{\mu} f = \int_S dS_{\mu} \epsilon^{\mu\nu\sigma} \partial_{\nu} \partial_{\sigma} f = \frac{1}{2} \int_S dS^{\mu\nu} G^{\dagger}_{\mu\nu}. \tag{4.88}$$

Here C is a contour enclosing the singularity and S a surface with C as boundary. N_T does not depend on the path C provided this does not cross the singularity.

The dual tensor $G^{\mu\nu}$ is defined as

$$G^{\mu\nu}(x) \equiv -\frac{1}{2} \epsilon^{\mu\nu\lambda\rho} G^{\dagger}_{\lambda\rho}(x) \tag{4.89}$$

and satisfies the continuity equation

$$\partial_\mu G^{\mu\nu}(x) = 0 \qquad \Leftrightarrow \qquad \partial_\mu G^\dagger_{\lambda\rho} + \partial_\rho G^\dagger_{\mu\lambda} + \partial_\lambda G^\dagger_{\rho\mu} = 0 . \qquad (4.90)$$

This equation completely characterizes the topological singularity [619,650, 651].

We are now ready to prove, by considering the local $U(1)$ model in Section 4.2.3, that in the context of gauge theories, all the macroscopic ground state effects do not occur for regular $f(x)$. In fact, from (4.69) we obtain $a_\mu(x) = \frac{1}{e_0}\partial_\mu f(x)$ for regular f since $\partial^2 f(x) = 0$ and $G^\dagger_{\mu\nu} = 0$. This implies zero classical current $(j_{\mu,cl} = 0)$ since the Meissner and the boson current cancel each other, and zero classical field $(F_{\mu\nu} = \partial_\mu a_\nu - \partial_\nu a_\mu)$.

In conclusion, the vacuum current and the classical gauge field appear only when $f(x)$ has topological singularities and these can be created only by condensation of massless bosons, e.g., when SSB occurs. On the other hand, the appearance of spacetime order parameter in a gauge theory is no guarantee that persistent ground state currents (and fields) will exist: if f is a regular function, the spacetime dependence of $\tilde{v}$ can be gauged away by an appropriate gauge transformation.

Since the boson transformation with regular f does not affect observable quantities (because it is equivalent to a gauge transformation), the S-matrix (4.50) must depend on U^μ_{in} and χ_{in} through the combination $U^\mu_{in} - \frac{1}{m_V}\partial(\chi_{in} - b_{in})$:

$$\mathcal{S} =: \mathcal{S}\left[\rho_{in}, U^\mu_{in} - \frac{1}{m_V}\partial(\chi_{in} - b_{in})\right] : . \qquad (4.91)$$

This is in fact independent of the boson transformation with regular f:

$$\mathcal{S} \to \mathcal{S}' =: \mathcal{S}\left[\rho_{in}, U^\mu_{in} - \frac{1}{m_V}\partial(\chi_{in} - b_{in}) + Z_3^{-\frac{1}{2}}(a^\mu - \frac{1}{e_0}\partial^\mu f)\right] :, \qquad (4.92)$$

since $a_\mu(x) = \frac{1}{e_0}\partial_\mu f(x)$ for regular f. However, $\mathcal{S}' \neq \mathcal{S}$ for singular f. Eq. (4.92) shows that $\mathcal{S}'$ includes the interaction of the quanta U^μ_{in} and ρ_{in} with the classical field and current appearing for singular $f(x)$ and associated to non-homogeneous condensation (extended objects).

In the following Section we comment on the appearance of macroscopically behaving extended objects (defects), for example vortices, as an observable effect of NG boson condensation.

4.3.2   *Defect formation in the process of symmetry breaking phase transitions*

The conclusions reached in the previous Sections shed some light on the macroscopic behavior of extended objects, the fact that they are observed in systems exhibiting ordered patterns and their formation occurs during the process of non-equilibrium phase transitions (limited to the case of the existence of the order parameter and in the presence of a gauge field).

Extended objects are described by solutions of classical field equations. These may be seen as the macroscopic dynamical manifestation of the microscopic boson condensation dynamics, in a way similar to the one by which classical field equations and currents have been derived in the previous Section. The extended object appears then as the macroscopic envelope of the non-homogeneous condensate localized over a finite domain and described by the boson transformation function $f(x)$. This plays the role of "form factor" and carries the topological charge singularity. In Chapters 7 and 8 we will present explicit examples of topological defects described in terms of non-homogeneous boson condensation [449, 452, 619, 637].

The questions concerning the connection between topological defect formation and symmetry breaking can then be addressed considering that the boson transformation function $f(x)$ is allowed to carry topological singularities only for the case of condensation of massless bosons (see also [617, 619]), such as the NG bosons, whose existence is required by SSB. This explains why topological defects are observed in systems exhibiting ordered patterns, namely in the presence of condensation of NG bosons sustaining the long-range ordering correlation.

We have seen that, in a gauge theory, symmetry breaking phase transitions, characterized by macroscopic ground state current and field (e.g., in superconductors), can occur only when there are non-zero gradients of topologically non-trivial condensation function $f(x)$, which constitutes, on the other hand, the condition to be met for the formation of topological defects [15]. Thus, we see why topological defects are observed during the process of symmetry breaking phase transitions (*"where the defects come from"*) (see Section 10.5). Since the picture so obtained is model-independent, it may account for the fact that some features of topological defects are shared by quite different systems, from condensed matter to cosmology. It might help to depict in a unified theoretical scheme the macroscopic behavior of topological defects and their interaction with quanta. For a detailed account see Chapters 7 and 8.

We close this Section with a comment on homotopy groups, which turn out to be useful in the study of topological defects (see Appendix T for essential notions on homotopy). In topologically non-trivial condensation at finite temperature the order parameter $v(x, \beta)$, with $\beta \equiv \frac{1}{k_B T}$, provides a mapping between the domain of variation of (x, β) and the *space of the unitarily inequivalent representations* of the canonical commutation relations, i.e., the set of Hilbert spaces where the field operators are realized. We thus have non-trivial homotopy mappings between the (x, β) variability domain and the group manifold. In the vortex case, for example, one has the mapping π of S^1, surrounding the $r = 0$ singularity, to the group manifold of $U(1)$ which is topologically characterized by the winding number $n \in \mathbb{Z} \in \pi_1(S^1)$. It is such a singularity which is carried by the boson condensation function. In the monopole case [440], the mapping π is the one of the sphere S^2, surrounding the singularity $r = 0$, to $SO(3)/SO(2)$ group manifold, with homotopy classes of $\pi_2(S^2) = \mathbb{Z}$. The same situation occurs in the sphaleron case [440] (see also Section 8.3.4), provided one replaces $SO(3)$ and $SO(2)$ with $SU(2)$ and $U(1)$, respectively.

4.4 Group contraction and spontaneous symmetry breaking

We have mentioned in previous Sections that the symmetry group G' under which physical field equations are invariant is the group contraction [336] of the invariance group G of the Heisenberg field equations. We now present the explicit proof, in terms of functional integral formalism, of the group contraction occurring in the case of $G = SU(2)$ and then we list other physical examples in which group contraction occurs. The mathematical mechanism of group contraction [336] thus turns out to account for the dynamical rearrangement of symmetry characterizing the phenomenon of SSB.

4.4.1 The infrared effect

We want to study the origin of the change of the group in the process of the rearrangement of symmetry. We consider the case of $SU(2)$ with reference to the example of the ferromagnet. Our conclusions will be, however, general. They hold for any continuous compact symmetry group of the Lagrangian. We will show that infrared contributions to the commutations relations among the generators of the symmetry group are missed in the process of going from the Heisenberg fields to the asymptotic fields. This results in the rearrangement of the symmetry.

Let us decompose the magnon field (cf. Eqs. (4.3)) into the sum of an "hard" part $B_t(x)$ and a "soft" or infrared part $B_\eta(x)$ with η infinitesimal (see also Section 3.10)

$$B(x) = B_t(x) + B_\eta(x). \tag{4.93}$$

$B_t(x)$ contains only momenta larger than η, while momenta in $B_\eta(x)$ are smaller than, or equal to η. One possible representation of $B_\eta(x)$ is [621]

$$B_\eta(x) = \frac{1}{2}\eta \int_{-\infty}^{+\infty} dt\, e^{-\eta|t|} B(x)$$

$$= \frac{1}{2(2\pi)^{1/2}} \eta \int d^3k\, \delta_\eta(k)\, B_{\mathbf{k}}\, e^{i\mathbf{k}\cdot\mathbf{x}}, \tag{4.94}$$

where we have used Eqs. (4.3), $\omega_k = k$, and

$$\delta_\eta(\omega_k) = \frac{1}{2\pi} \int_{-\infty}^{+\infty} dt\, e^{-\eta|t|-i\omega_k t}. \tag{4.95}$$

The function $\delta_\eta(k)$ approaches to $\delta(k)$ in the limit $\eta \to 0$. Therefore $B_\eta(x)$ is of order of η and independent of x in the limit $\eta \to 0$.

Now we use the field $B(x)$ written as in Eq. (4.93) in the expression (4.10) and follow the contribution of $B_\eta(x)$ to $S^{(i)}(\phi, \phi^\dagger, B, B^\dagger)$. We obtain

$$S^{(i)}(y, \phi, \phi^\dagger, B, B^\dagger) = \langle S_\psi^{(i)}(y) : \exp\left[-iA(\phi, \phi^\dagger, B_t + B_\eta, B_t^\dagger + B_\eta^\dagger)\right] :\rangle$$

$$= s_t^{(i)}(y) - i\rho^{-1/2} B_\eta \int d^4x\, \langle S_\psi^{(i)}(y) K(\overrightarrow{\partial}) S_\psi^{(-)}(x) : e^{-iA_t} :\rangle$$

$$- i\rho^{-1/2} B_\eta^\dagger \int d^4x\, \langle S_\psi^{(i)}(y) S_\psi^{(+)}(x) K(-\overleftarrow{\partial}) : e^{-iA_t} :\rangle. \tag{4.96}$$

$s_t^{(i)}(y)$ is obtained from $S^{(i)}(y)$ by ignoring the infrared fields B_η and $B_\eta^\dagger$:

$$s_t^{(i)}(y) = \langle S_\psi^{(i)}(y) : e^{-iA_t} :\rangle. \tag{4.97}$$

A_t denotes A with the infrared fields disregarded. By a straightforward computation whose details are given in [570], one then obtains:

$$S^{(1)}(y) = s_t^{(1)}(y) + \frac{1}{\sqrt{2M}}(B_\eta + B_\eta^\dagger)\, s_t^{(3)}(y), \tag{4.98a}$$

$$S^{(2)}(y) = s_t^{(2)}(y) - i\frac{1}{\sqrt{2M}}(B_\eta - B_\eta^\dagger)\, s_t^{(3)}(y), \tag{4.98b}$$

$$S^{(3)}(y) = s_t^{(3)}(y)$$

$$+ \frac{1}{\sqrt{2M}}\left[i(B_\eta - B_\eta^\dagger)\, s_t^{(2)}(y) - (B_\eta + B_\eta^\dagger)\, s_t^{(1)}(y)\right]. \tag{4.98c}$$

These are the spin density operators expressed in terms of the quasiparticle fields and we see that when we ignore the infrared contributions coming from the operators B_η and $B_\eta^\dagger$ we obtain $s_t^{(i)}$. We remark that the matrix elements of $S^{(i)}(y)$ are equal to those of $s_t^{(i)}$:

$$\langle i|S^{(i)}(y)|j\rangle = \langle i|s_t^{(i)}(y)|j\rangle\,, \tag{4.99}$$

which shows that taking the matrix elements between physical states, i.e., smeared out (localized) states, causes the missing of the infrared contributions. In other words, the limit $\eta \to 0$ is automatically implied in the computation of the matrix elements between physical states due to the fact that physical states carry in their definition smearing out functions which act as a cutoff at infinite volume, namely, as a cutoff for infrared momenta. For $i = 3$ Eq. (4.99) gives

$$\langle i|S^{(3)}(y)|j\rangle = \langle i|s_t^{(3)}(y)|j\rangle = M\,, \tag{4.100}$$

and therefore we can write

$$s_t^{(3)}(y) = M + \,:\, s_t^{(3)}(y)\,:\,. \tag{4.101}$$

We conclude that the space integration of $s_t^{(i)}(y)$ must give the generators (4.22) with B and $B^\dagger$ substituted by B_t and $B_t^\dagger$, respectively, since the cutoff $f(x)$ excludes contributions at infinite volume (the infrared contributions). We eventually get:

$$S_f^{(1)} = s_f^{(1)} + \frac{1}{\sqrt{2M}}(B_\eta + B_\eta^\dagger)\,:\, s_t^{(3)}\,:\,, \tag{4.102a}$$

$$S_f^{(2)} = s_f^{(2)} - i\frac{1}{\sqrt{2M}}(B_\eta - B_\eta^\dagger)\,:\, s_t^{(3)}\,:\,, \tag{4.102b}$$

$$S_f^{(3)} = s_t^{(3)} + \frac{1}{\sqrt{2M}}\left[i(B_\eta - B_\eta^\dagger)\,s_t^{(2)} - (B_\eta + B_\eta^\dagger)\,s_t^{(1)}\right]\,. \tag{4.102c}$$

We now show that the spin operators $S_f^{(i)}$ satisfy the algebra for the $SU(2)$ group when the limit $f \to 1$ is taken: $[S^{(i)}, S^{(j)}] = i\epsilon_{ijk}S^{(k)}$. We have

$$[s_f^{(1)}, B_\eta^\dagger(x)] = \sqrt{\frac{M}{2}}f_\eta(x)\,, \tag{4.103}$$

and similar commutators for $s_f^{(i)}$ with $i = 2, 3$. Here we have used

$$f(x) = f_t(x) + f_\eta(x)\,, \tag{4.104}$$

with the same meaning for the notation as in Eq. (4.93). $f_\eta(x)$ contains only momenta smaller than, or equal to η and thus it has a spatial domain of range $\frac{1}{\eta}$ and vanishes as $\eta \to 0$ since $f(x)$ is square-integrable. To take into account the locally infinitesimal effect, the space integration must extend to infinity. Therefore, the limit $f \to 1$ must be performed before the limit $\eta \to 0$ in order to recognize the differences between $S_f^{(i)}$ and $s_f^{(i)}$. We find

$$[S_f^{(1)}, S_f^{(2)}] = iM \int d^3x |f(x)|^2$$

$$+i \left(\frac{1}{2}\right) [f_\eta^*(x) + f_\eta(x) + f_{\bar\eta}^*(x) + f_{\bar\eta}(x)] : s_t^{(3)} :$$

$$-\left(\frac{1}{2M}\right)^{1/2} \left[i(B_{\bar\eta} - B_{\bar\eta}^\dagger) s_t^{(2)} - i(B_\eta + B_\eta^\dagger) s_t^{(1)}\right] . \quad (4.105)$$

We have two limits of the infrared cutoffs η and $\bar\eta$ to be performed successively (no matter in which order) since two generators (i.e., two successive rotations) are involved in the commutator. The $f \to 1$ limit then gives

$$[S^{(1)}, S^{(2)}] = iS^{(3)} , \qquad (4.106a)$$

$$[S^{(3)}, S^{(1)}] = iS^{(2)} , \quad [S^{(3)}, S^{(2)}] = -iS^{(1)} . \qquad (4.106b)$$

Therefore we find that

$$\lim_{\eta,\bar\eta \to 0} \lim_{f \to 1} [S_f^{(i)}, S_f^{(j)}] = i\epsilon_{ijk} S^{(k)} . \qquad (4.107)$$

On the other hand,

$$\lim_{f \to 1} \lim_{\eta,\bar\eta \to 0} [S_f^{(1)}, S_f^{(2)}] = iM \lim_{f \to 1} \int d^3x |f(x)|^2 = (\text{const})\mathbb{1}, \quad (4.108a)$$

$$\lim_{f \to 1} \lim_{\eta,\bar\eta \to 0} [S_f^{(3)}, S_f^{(1)}] = \lim_{f \to 1} \lim_{\eta,\bar\eta \to 0} iS^{(2)} , \qquad (4.108b)$$

$$\lim_{f \to 1} \lim_{\eta,\bar\eta \to 0} [S_f^{(3)}, S_f^{(2)}] = i \lim_{f \to 1} \lim_{\eta,\bar\eta \to 0} iS^{(1)} . \qquad (4.108c)$$

Thus, if the limit $f \to 1$ is performed before the limits $\eta \to 0$ and $\bar\eta \to 0$, then the $SU(2)$ symmetry group is obtained, while the (projective) $E(2)$ group, the group contraction of $SU(2)$, is obtained by inverting the order in which the limits are performed: limit $f \to 1$ and limit $\bar\eta \to 0$ are not commutable. The infrared term, although locally infinitesimal, gives, however, a finite global contribution to the commutators of the generators $S^{(i)}$. Its locally infinitesimal nature makes it, instead, commutable with any local operator and therefore it does not contribute to the commutators of the generators for the quasiparticle fields, which are directly related to (local) observations (the quasiparticle fields are related to the observable energy levels). The algebra which is related to the experimental observable results is the group contraction algebra: quasiparticles form an irreducible representation of the contraction group.

The result obtained above is an *exact* result, obtained without any approximation in a model-independent derivation, and therefore it must not be confused with the linear approximations used sometimes in the literature, as for example when using the Holstein–Primakoff representation [331] of the $SU(2)$ group. For the reader's convenience we will also consider such linear approximations in Section 4.4.3.

Low energy theorems

The so-called low-energy theorems [9, 10, 210, 657], according to which low momentum NG modes do not affect the S-matrix, as indeed observed in solid state physics and in high energy physics, are observable manifestations of the group contraction mechanism. This can be seen in the following way. Translations of the NG modes by a constant quantity implied by the group contraction transformations are invariant transformations for the quasiparticle field equations. Therefore, also the S-matrix has to be invariant under such transformations. This implies that the NG modes $B(x)$ always appear with their derivatives in the S-matrix, i.e., in the form $\partial_\mu B(x)$, and thus the NG mode interaction disappears in the zero-momentum limit. The Dyson low-energy theorem for magnons in ferromagnets, the Adler theorem in high-energy physics, the soft boson limit of current algebra theory find in this way their root in the invariant properties of S-matrix under the transformations belonging to the group contraction.

Other remarkable consequences of the group contraction process are some of the relations in the current algebra formalism, e.g., the partially conserved axial vector currents (PCAC) and the Goldberger–Treiman relations [274, 288, 432, 619].

In conclusion, one reason why the observable symmetry group G' can be different from the Lagrangian invariance group G is based on the fact that macroscopic observations are always a collection of local observations and therefore there always exists a possibility that in each local observation one misses an infinitesimal contribution of the order of magnitude of $\frac{1}{V}$, with the volume V going to infinity. Such missing contributions can be accumulated as a finite amount when integrated over the whole system. This locally infinitesimal effect, called the infrared effect, is responsible for the origin of the difference between the algebra of the generators written in terms of the Heisenberg fields and the one of the generators written in terms of physical fields. These carry, indeed, square integrable functions (with finite spatial support) (see the discussion in Chapter 1).

The occurrence of group contraction in SSB has also been proven by means of projective geometry arguments [183] and the contraction of group representation has been shown to provide non-linear realizations [6, 124, 125, 133, 149, 154, 168, 474, 506, 546] of the $SU(2)$ doublet and $SO(n)$ vector gauge theory models. Non-linear realizations provide a powerful tool of investigation in phenomenological theories where effective Lagrangians are used and SSB occurs. By singling out the NG modes they make explicit the low-energy behavior of the theory [6, 474, 506]. They have been used in the determination of the extrema of the most general renormalizable Higgs potentials in elementary particle physics [124, 125].

Further examples of group contraction

In all the cases where a subgroup is preserved in the process of SSB (the stability, or little group, see Section 3.4), it has been found that the group G', contraction of the Lagrangian invariance group G, is the one relevant to the observations. The infrared effect is the origin of the rearrangement of G into the Inönü–Wigner group contraction G'. The contraction parameter can thus be taken to be $\frac{1}{V}$ with the volume $V \to \infty$.

Group contraction has been shown to occur in models invariant under $SU(n)$, $SO(n)$, chiral $SU(2) \times SU(2)$, $SU(3) \times SU(3)$, etc. (see [182] and [619] and references quoted therein). The different cases may be classified into three categories R_i, $i = 1, 2, 3$ [182]. We always refer to continuous invariance groups and we assume that the stability group is a maximal subgroup [619].

In the case R_1, the dynamical groups are Abelian compact groups; the phenomenological groups also have an Abelian algebraic structure, but are non-compact groups. Heisenberg and asymptotic fields provide different realizations of the Abelian algebraic structure. Examples include spontaneous breakdown of phase, chiral phase and scale invariance [450, 614].

In the case R_2, the rearrangement of the basic non-Abelian symmetry group leads to its contraction, by which an Abelian subgroup is introduced. Examples are given in [362, 454, 460, 570, 571, 614, 639, 640] and include, e.g., the case of a scalar isotriplet, the ferromagnet discussed above, the chiral $SU(2) \times SU(2)$ symmetry [362] realized by non-linear transformation of the pion field, the $SU(3)$ group in a linear approximation of solid state systems as $T - t$ Jahn-Teller systems [640, 653].

In the case R_3, the phenomenological symmetry generators do not form a closed algebra. By enlarging the set of generators, their algebra can be completed, then resulting in the contraction of the basic symmetry algebra. In [455] such a rearrangement is studied for an $SU(2)$ invariant model.

In Appendix K we briefly discuss the relation of the group contractions $SU(2) \to E(2)$ and $SO(4) \to E(3)$ to the loop-antiloop symmetry breaking and to the Virasoro algebra.

The rearrangement of the group of continuous space translations into the group of discrete space translations, leading to lattice structures, has also been studied with particular reference to crystal formation [617, 619], where the NG bosons are the phonons, i.e., the quanta of the elastic waves.

4.4.2 Group contraction, boson condensation and macroscopic quantum systems

In SSB, the contraction mechanism offers a powerful tool to compute the number of gapless modes whose existence is required by the Goldstone theorem. For example, it allows us to compute the number of the degrees of freedom of the instanton solutions in a non-Abelian gauge theory (see Chapter 8). Actually, SSB implies that the NG fields must form an irreducible representation of the invariance group of the theory [162, 182, 230, 423, 449, 613].

Notice that while G is in general a compact group, the contraction group G' is not a compact group, since it contains an (Abelian) subgroup of generators of field translations. These are $2(N-1)$ and $N-1$ for the contraction of $G = SU(N)$ and $G = SO(N)$, respectively [182]. The number of gapless modes is given by the number of the generators closing the Abelian subalgebra. They are, indeed, linear in the fields of the NG modes and thus induce their translations by constant quantities. Notice also that the equation for the field χ_{in}, e.g., in the model with global $U(1)$ group (cf. Section 4.2.2), is invariant under the translation of the field by a constant quantity, Eq. (4.30), if and only if χ_{in} is a massless field, $\partial^2 \chi_{in}(x, \theta) = 0$. Therefore, since the mechanism of group contraction implies translations of fields by a constant and G' is the symmetry group for the physical field equations, it actually implies the existence of massless fields, i.e., the mechanism of group contraction has the same content as the Goldstone theorem [182, 638]. In Chapter 1 the translation of field operators by a constant has been recognized to generate coherent states. Group contraction thus implies that coherent condensation of NG bosons occurs in the system ground state. In Section 4.2.3, such a condensation has been related to the appearance of macroscopic vacuum currents which are controlled by classical equations.

Group contraction thus plays an important role in the passage to the macroscopic phenomena: the basic symmetry is rearranged to a contraction at observable level; in this way Abelian (boson) transformations are introduced, which regulate classical macroscopic phenomena through boson condensation. When a large number of bosons is condensed, observable symmetry patterns appear in ordered states, the quantum fluctuations become very small ($\frac{\Delta n}{n} \ll 1$) and the system behaves as a classical one [182, 619]. *Macroscopic quantum systems* thus emerge from the quantum dynamics. As already observed, these are quantum systems not in the trivial sense that they are made, as any other physical system, by quantum components, but in the sense that their macroscopic features, such as ordering and stability, cannot be explained without recourse to the undergoing quantum dynamics. Similarly, the order parameter, which characterizes the macroscopic states of such systems, is a classical field in the sense that its measurable value (far from the critical region of phase transition) is not affected by quantum fluctuations Δn in the condensate. These results seem to support the conjecture [182, 547, 565, 638, 658] that the passage from quantum to classical physics involves group contraction phenomena (for reviews on this subject see [240]).

4.4.3   *The collective behavior of quantum components and group contraction*

In order to better clarify the mechanism of group contraction, we now consider the Holstein–Primakoff boson realization of the $SU(2)$ group, as an example of non-linear realization of the invariance group. It was originally introduced in order to diagonalize the Hamiltonian in the exchange interaction model of a ferromagnet [331]. The approximation used by Holstein and Primakoff was to neglect all terms which are not bilinear in the boson operators. In this way they were able to construct a linear formalism suitable for computations. In Section 4.2.1 we have proven that it is an *exact* result that the spin operator densities $S^i, i = 1, 2$, are linear in the boson field $B(x)$. Therefore, such a result is not to be confused with the linear approximation used by Holstein and Primakoff. It is nevertheless useful to see how the Holstein–Primakoff $SU(2)$ non-linear boson representation is related to the $E(2)$ contraction. Instead of the ferromagnetic system, we consider a system of N two-level atoms under the action of an external driving field able to excite them. Such a system is physically interesting in a number of problems and in quantum optics. The atom system may be described as a (fermion-like) system of non-interacting N electrical dipole doublets under the influence of an electric field. Below we will follow closely [64, 197].

The ground state and the excited state of each of the N two-level atoms are denoted by $|0\rangle_i$ and $|1\rangle_i$, $i = 1, \ldots, N$, respectively, associated to the eigenvalues $\mp\frac{1}{2}$ of the operator $\sigma_{3i} = \frac{1}{2}(|1\rangle_{ii}\langle 1| - |0\rangle_{ii}\langle 0|)$, no-summation on i. The operators $\sigma_i^+ = |1\rangle_{ii}\langle 0|$ and $\sigma_i^- = (\sigma_i^+)^\dagger$ generate the transitions between the two levels induced by the action of the electric field. The N-atom system is thus described by $\sigma^\pm = \sum_{i=1}^N \sigma_i^\pm$, $\sigma_3 = \sum_{i=1}^N \sigma_{3i}$ with the fermion-like su(2) (rotational) algebra

$$[\sigma_3, \sigma^\pm] = \pm\sigma^\pm, \qquad [\sigma^-, \sigma^+] = -2\sigma_3. \tag{4.109}$$

The interaction $H = -\mathbf{d} \cdot \mathbf{E}$ of the atoms with the electrical field $\mathbf{E}$, where $\mathbf{d}$ is the atomic electric dipole moment, can be written [279] as

$$H = \hbar\gamma(b^\dagger\sigma^- + b\sigma^+), \tag{4.110}$$

which is a Jaynes–Cummings-like Hamiltonian [279, 353]. The coupling constant γ is proportional to the matrix element of the atomic dipole moment and to the inverse of the volume square root $V^{-1/2}$, b is the electric field quantum operator, $\sigma^\pm$ are the atomic polarization operators.

Suppose that the electric field induces the transition $|0\rangle_i \to |1\rangle_i$ for a certain number l of atoms, originally assumed for simplicity in the ground state (in a realistic system, of course, not all the atoms are initially in their respective ground state; our conclusions, however, do not depend on these more realistic initial conditions). The system state may then be represented as the normalized superposition $|l\rangle$:

$$|l\rangle \equiv \binom{N}{l}^{-\frac{1}{2}} (|0\rangle_1 \ldots |0\rangle_{N-l}|1\rangle_{N-l+1} \ldots |1\rangle_N + \ldots + |1\rangle_1 \ldots |1\rangle_l|0\rangle_{l+2} \ldots |0\rangle_N).$$
$$\tag{4.111}$$

The difference between the number of atoms in the excited state and the ones in the ground state is measured by σ_3:

$$\langle l|\sigma_3|l\rangle = l - \frac{1}{2}N \tag{4.112}$$

and its non-zero value (proportional to the system polarization in the case of dipoles) signals that the rotational $SU(2)$ symmetry is broken. Operating with $\sigma^\pm$ on $|l\rangle$ gives:

$$\sigma^+|l\rangle = \sqrt{l+1}\sqrt{N-l}\,|l+1\rangle, \qquad \sigma^-|l\rangle = \sqrt{N-(l-1)}\sqrt{l}\,|l-1\rangle. \tag{4.113}$$

Eqs. (4.112) and (4.113) show that σ_3 and $\sigma^\pm$ are represented on $|l\rangle$ by

$$\sigma_3 = S^+S^- - \tfrac{1}{2}N, \qquad (4.114a)$$

$$\frac{\sigma^+}{\sqrt{N}} = S^+\sqrt{1 - \frac{S^+S^-}{N}}, \qquad \frac{\sigma^-}{\sqrt{N}} = \sqrt{1 - \frac{S^+S^-}{N}}\,S^-, \qquad (4.114b)$$

where $S^- = (S^+)^\dagger$, $[S^-,S^+] = 1$, $S^+|l\rangle = \sqrt{l+1}|l+1\rangle$ and $S^-|l\rangle = \sqrt{l}\,|l-1\rangle$, for any l. Eqs. (4.114) are the Holstein–Primakoff non-linear boson realization of $SU(2)$ [331,570]. The σ's in Eqs. (4.114) still satisfy the su(2) algebra (4.109). However, Eqs. (4.113) give for $N \gg l$

$$\frac{\sigma^\pm}{\sqrt{N}}|l\rangle = S^\pm|l\rangle. \qquad (4.115)$$

By defining $S_3 \equiv \sigma_3$, the su(2) algebra (4.109) therefore contracts to the (projective) e(2) (or Weyl–Heisenberg) algebra in the large N limit [64,182,336]

$$[S_3, S^\pm] = \pm S^\pm, \qquad [S^-, S^+] = 1, \qquad (4.116)$$

the contraction parameter being $\frac{1}{\sqrt{N}}$. Thus, for large N, $S^\pm$ act as *boson* operators. They behave as the NG boson modes B and $B^\dagger$ (or $s_f^\pm$, cf. Eq. (4.116) with Eqs. (4.24)) associated to the collective electric dipole wave quanta. The interaction (4.110), written in terms of $S^\pm$, is

$$\mathcal{H} = \hbar\sqrt{N}\gamma\left(b^\dagger S^- + b S^+\right). \qquad (4.117)$$

In summary, in the large N limit the collection of single two-level (fermion-like) atoms manifests itself as a collective bosonic system (by assuming finite density, the large N limit implies the large volume limit (the thermo-dynamic limit)). Remarkably, the original coupling of the individual atoms to the field gets enhanced by the factor $\sqrt{N}$ and manifests itself as the coupling of the collective modes $S^\pm$ to the field. For large N the system of atoms thus behaves as a collective whole. The breakdown of symmetry, with the consequent phenomenon of group contraction, thus provides a *change of scale* [182,638,641,645], from the microscopic quantum dynamics to the macroscopic quantum system behavior. For large N, the coupling enhancement by the factor $\sqrt{N}$ implies that for the collective interaction the energy gap is quite large and the time-scale is much shorter (indeed by the factor $\frac{1}{\sqrt{N}}$) than for the short range interactions among the atoms. Hence the macroscopic stability of the system vs quantum fluctuations in the short range interactions among of the microscopic components [197].

We remark that, when the system has a finite size, a persistent polarization cannot survive in the limit of a vanishing electric field [15,17]. In such a limit, the dipole rotational symmetry is restored. More details on the relation between group contraction and the Holstein and Primakoff [331] $su(2)$ non-linear realization can be found in [182,570].

As a closing remark, we observe that boson condensation and quantum fluctuation of NG modes in the ground state of SSB theories have been described [188] in terms of jump (Poisson) Markovian stochastic processes. The associated *classical* probability satisfies the Fokker–Planck equation [160,188]. The generators of the contracted group (e.g., $E(2)$ in the case of $SU(2)$ invariance) may be interpreted in terms of probability theory, in the sense that the emission and the absorption of quanta in the coherent ground state are Poisson processes with constant probability and probability rate proportional to the condensed number of quanta, respectively. This sheds further light on the above-mentioned change of scale from the quantum dynamics to the macroscopic (classic) behavior.

4.5 Quantum fluctuations and macroscopic stability

We have seen that in gauge theories with SSB the Higgs mechanism plays a crucial role: the gauge field is expelled out of the ordered domains and confined into "normal" regions having a vanishing order parameter, i.e., where long-range correlation modes (the NG modes) responsible for the ordering are damped away. Now we focus our attention on the dynamics governing the radiative gauge field and its role in the onset of phase locking among the e.m. modes and the matter components. In the Higgs mechanism the gauge field removes the order in the regions where it penetrates. Here our purpose is to study the role of the radiative gauge field in sustaining the phase locking in the coherent regions.

In gauge theories, the multitude of quantum configurations of the system is described by the local gauge freedom through the local gauge transformations. The independence of the system observables from the gauge transformations guarantees the stability of the observable properties of the system against the many accessible microscopic configurations: although quantum fluctuations are the dominant feature at the microscopic scale of the quantum components, physical systems, however, are usually remarkably stable at macroscopic spacetime scales. In QFT, one takes into account such a double feature by prescribing that the Lagrangian of the system should be invariant under the local phase transformation of the quantum component field $\psi(\mathbf{x}, t) \rightarrow \psi'(\mathbf{x}, t) = \exp(ig\theta(\mathbf{x}, t))\psi(\mathbf{x}, t)$. Local phase invariance is the QFT solution to the problem of building a stable system out of fluctuating components. In order that the Lagrangian be invariant under local gauge transformation it is necessary, as well known, to introduce the gauge fields, e.g., the electromagnetic (e.m.) field $A_\mu(\mathbf{x}, t)$, such that the Lagrangian be also invariant under the gauge field transformation $A_\mu(\mathbf{x}, t) \rightarrow A'_\mu(\mathbf{x}, t) - \partial_\mu \theta(\mathbf{x}, t)$. Such a transformation is devised to compensate terms proportional to $\partial_\mu \theta(\mathbf{x}, t)$ arising in the Lagrangian from the kinetic term for the matter field $\psi(\mathbf{x}, t)$. In Section 5.6 we will see that the gauge field may be described, indeed, as a compensating "reservoir" against variations in the many accessible microscopic configurations.

Our model system is an ensemble of two-level atoms, say N atoms per unit volume, which may represent rigid rotators endowed with an electric dipole. We consider the interaction of these atoms with the e.m. quantum radiative modes and disregard the static dipole-dipole interaction. Transitions between the atomic levels are radiative dipole transitions. The system is assumed to be spatially homogeneous and to be in a thermal bath kept at a non-vanishing temperature T.

The N atoms are collectively described by the complex dipole wave field $\psi(\mathbf{x}, t)$. In the previous Section such a system of two-level atoms has been described as a system of $\frac{1}{2}$ spins, according to the known formal equivalence among the two descriptions [311]. In the following, we will use natural units $\hbar = 1 = c$. By denoting with $d\Omega = \sin\theta d\theta d\phi$ the element of solid angle and with (r, θ, ϕ) the polar coordinates of $\mathbf{r}$, the dipole wave field $\psi(\mathbf{x}, t)$ integrated over the sphere of unit radius $\mathbf{r}$ gives:

$$\int d\Omega |\psi(\mathbf{x}, t)|^2 = N, \tag{4.118}$$

which, in terms of the rescaled field $\chi(\mathbf{x}, t) = \frac{1}{\sqrt{N}}\psi(\mathbf{x}, t)$, also reads as

$$\int d\Omega |\chi(\mathbf{x}, t)|^2 = 1. \tag{4.119}$$

The system is invariant under dipole rotations and since the atom density is assumed to be spatially uniform, the only relevant variables are the angular ones. In full generality, the field $\chi(\mathbf{x}, t)$ may thus be expanded in the unit sphere in terms of spherical harmonics and we set the amplitudes $\alpha_{l,m}(t) = 0$ for $l \neq 0$, 1, reducing the expansion to the four levels $(l, m) = (0, 0)$ and $(1, m), m = 0, \pm 1$. Thermal equilibrium and the dipole rotational invariance imply that there is no preferred direction in the dipole orientation. Thus, the amplitudes $\alpha_{1,m}(t)$ for any m have in the average the same value, independent of m. We write

$$\alpha_{0,0}(t) \equiv a_0(t) \equiv A_0(t) \, e^{i\delta_0(t)} , \tag{4.120a}$$

$$\alpha_{1,m}(t) \equiv A_1(t) \, e^{i\delta_{1,m}(t)} \, e^{-i\omega_0 t} \equiv a_{1,m}(t) \, e^{-i\omega_0 t} . \tag{4.120b}$$

In Eqs. (4.120) we have used $a_{1,m}(t) \equiv A_1(t) \, e^{i\delta_{1,m}(t)}$; $A_0(t)$, $A_1(t)$, $\delta_0(t)$ and $\delta_{1,m}(t)$ are real quantities; $\omega_0 \equiv I^{-1}$, where I denotes the moment of inertia of the atom, which gives a relevant scale for the system: $\omega_0 \equiv k = \frac{2\pi}{\lambda}$ (the eigenvalue of $\frac{\mathbf{L}^2}{2I}$ on the state $(1, m)$ is $\frac{l(l+1)}{2I} = I^{-1} = \omega_0$, $\mathbf{L}^2$ is the squared angular momentum operator). We also put $\omega t \equiv \delta_{1,0}(t) - \delta_0(t)$. In the assumed conditions, no permanent polarization may develop for our system. This confirms that the three levels $(1, m)$, $m = 0, \pm 1$ are in the average equally populated under normal conditions and that we can safely write $\sum_m |\alpha_{1,m}(t)|^2 = 3 \, |a_1(t)|^2$. In full generality, consistently with Eq. (L.2) of Appendix L, we can set the initial conditions at $t = 0$ as

$$|a_0(0)|^2 = \cos^2 \theta_0 , \qquad |a_1(0)|^2 = \frac{1}{3} \sin^2 \theta_0 , \qquad 0 < \theta_0 < \frac{\pi}{2} . \tag{4.121}$$

The values zero and $\frac{\pi}{2}$ are excluded since it is physically unrealistic for the state $(0,0)$ to be completely filled or completely empty, respectively. We will find that the lower bound for θ_0 is imposed by the dynamics in a self-consistent way. The field equations for our system are [279, 317]:

$$i\frac{\partial \chi(\mathbf{x}, t)}{\partial t} = \frac{\mathbf{L}^2}{2I} \, \chi(\mathbf{x}, t) \tag{4.122a}$$

$$-i\sum_{\mathbf{k}, r} d \, \sqrt{\rho} \, \sqrt{\frac{k}{2}} \, (\epsilon_r \cdot \mathbf{x})[u_r(\mathbf{k}, t) \, e^{-ikt} - u_r^{\dagger}(\mathbf{k}, t) \, e^{ikt}] \, \chi(\mathbf{x}, t) ,$$

$$i\frac{\partial u_r(\mathbf{k}, t)}{\partial t} = i \, d\sqrt{\rho} \, \sqrt{\frac{k}{2}} \, e^{ikt} \int d\Omega (\epsilon_r \cdot \mathbf{x}) |\chi(\mathbf{x}, t)|^2 , \tag{4.122b}$$

where $u_r(\mathbf{k}, t) = \frac{1}{\sqrt{N}} c_r(\mathbf{k}, t)$, and $c_r(\mathbf{k}, t)$ denotes the radiative e.m. field operator with polarization r; d is the magnitude of the electric dipole moment, $\rho \equiv \frac{N}{V}$ and ϵ_r is the polarization vector of the e.m. mode (for which we assume the transversality condition $\mathbf{k} \cdot \epsilon_r = 0$). Notice the enhancement by the factor $\sqrt{N}$ appearing in the coupling $d \sqrt{\rho}$ in Eqs. (4.122) due to the the rescaling of the fields. In obtaining Eqs. (4.122) we have restricted ourselves to the resonant radiative e.m. modes, i.e., those for which $k = \frac{2\pi}{\lambda} = \omega_0$, and we have used the dipole approximation, i.e., $\exp(i\mathbf{k} \cdot \mathbf{x}) \approx 1$, since we are interested in the macroscopic behavior of the system. This means that the wavelengths of the e.m. modes we consider, of the order of $\frac{2\pi}{\omega_0}$, are larger than (or comparable to) the system linear size.

From Eqs. (4.122), by using $a_{1,m}(t) \equiv \alpha_{1,m}(t)\, e^{i\omega_0 t}$, we obtain

$$\dot{a}_0(t) = \Omega \sum_m u_m^*(t)\, a_{1,m}(t)\,, \tag{4.123a}$$

$$\dot{a}_{1,m}(t) = -\Omega\, u_m(t)\, a_0(t)\,, \tag{4.123b}$$

$$\dot{u}_m(t) = 2\,\Omega\, a_0^*(t)\, a_{1,m}(t)\,, \tag{4.123c}$$

where $\Omega \equiv \frac{2d}{\sqrt{3}}\sqrt{\frac{\rho}{2\omega_0}}\,\omega_0 \equiv G\,\omega_0$ and u_m is the amplitude of the e.m. mode coupled to the transition $(1,m) \leftrightarrow (0,0)$. We introduce the real quantities $U(t)$ and $\varphi_m(t)$ and put

$$u_m(t) = U(t)e^{i\varphi_m(t)}\,. \tag{4.124}$$

Notice that Eqs. (4.123) and (4.122), are not invariant under time-dependent phase transformations of the field amplitudes.

Our task is to investigate how the gauge symmetry can be recovered.

In Appendix L, the phases of the amplitudes in Eqs. (4.120) are shown to be independent of m. We therefore put $\varphi \equiv \varphi_m$, $\delta_1(t) \equiv \delta_{1,m}(t)$, $\alpha \equiv \alpha_m$ (α_m is given in Eq. (L.8)) and also $u(t) \equiv u_m(t)$, $a_1(t) \equiv a_{1,m}(t)$.

The study of the system ground states for each of the modes $a_0(t)$, $a_1(t)$ and $u(t)$ shows that spontaneous breakdown of the $SO(2)$ symmetry (the phase symmetry) in the plane $(a_{0,R}(t), a_{0,I}(t))$ occurs [197] (the indexes R and I denote the real and the imaginary component, respectively, of the field). In the semiclassical approximation [343], one finds [197] that for the mode $a_0(t)$ there is the quasi-periodic mode with pulsation $m_0 = 2\Omega\sqrt{(1 + \cos^2\theta_0)}$ (the "massive" mode with real mass $2\Omega\sqrt{(1 + \cos^2\theta_0)}$) and a zero-frequency mode $\delta_0(t)$ corresponding to a massless mode playing the role of the NG field. It is remarkable that the value $a_0 = 0$ consistently appears to be the relative maximum for the potential, and therefore an instability point out of which the system (spontaneously) runs away.

One also finds [197] that $a_1(t)$ is a massive field with (real) mass (pulsation) $\sigma^2 = 2\,\Omega^2(1 + \sin^2\theta_0)$.

For the $u(t)$ field, we derive [197] that the $SO(2)$ cylindrical symmetry around an axis orthogonal to the plane $(u_R(t), u_I(t))$ can be spontaneously broken or not, according to the negative or positive value of the field squared mass $\mu^2 = 2\Omega^2 \cos 2\theta_0$, respectively. For $\mu^2 < 0$, i.e., $\theta_0 > \frac{\pi}{4}$, the potential has a relative maximum at $u_0 = 0$ and a (continuum) set of minima for

$$|u(t)|^2 = -\frac{1}{3}\cos 2\theta_0 = -\frac{\mu^2}{6\Omega^2} \equiv v^2(\theta_0)\,, \quad \theta_0 > \frac{\pi}{4}\,. \tag{4.125}$$

They represent (infinitely many) possible vacua for the system and they transform into each other under shifts of the field φ: $\varphi \to \varphi + \alpha$. The phase symmetry is broken, the order parameter is given by $v(\theta_0) \neq 0$ and one specific ground state is singled out by fixing the value of the φ field. We have a "massive" mode, as indeed expected in the Higgs mechanism, with real mass $\sqrt{2|\mu^2|} = 2\Omega\sqrt{|\cos 2\theta_0|}$ (a quasi-periodic mode) and the zero-frequency mode $\varphi(t)$ (the massless NG collective field, also called the "phason" field [424]). The fact that in such a case $u_0 = 0$ is a maximum for the potential means that the system dynamically evolves away from it, consistently with the similar situation noticed for the a_0 mode. We find that dynamical consistency requires $\theta_0 > \frac{\pi}{4}$.

Provided $\theta_0 > \frac{\pi}{4}$, a time-independent amplitude $U(t) \equiv \overline{U}$ is compatible with the system dynamics (e.g., the ground state value of $A_0 \neq 0$ implies $\overline{U} = const.$). Eqs. (L.7) and (L.7) show that a time-independent amplitude $\overline{U} = const.$ indeed exists, $\dot{U}(t) = 0$, if and only if the phase locking relation

$$\alpha = \delta_1(t) - \delta_0(t) - \varphi(t) = \frac{\pi}{2} \tag{4.126}$$

holds. Then we have

$$\dot{\varphi}(t) = \dot{\delta}_1(t) - \dot{\delta}_0(t) = \omega, \tag{4.127}$$

and this shows that any change in time of the difference between the phases of the amplitudes $a_1(t)$ and $a_0(t)$ is compensated by the change of the phase of the e.m. field. When Eq. (4.126) holds we also have $\dot{A}_0 = 0 = \dot{A}_1$ (cf. Eqs. (L.7), (L.7)). Eq. (4.126) shows that, provided $\theta_0 > \frac{\pi}{4}$, $\dot{\alpha} = 0$. It expresses nothing but the gauge invariance of the theory. Since δ_0 and φ are the NG modes, Eqs. (4.126) and (4.127) exhibit the coherent feature of the collective dynamical regime. The system of N dipoles and of the e.m. field is characterized by the "in phase" dynamics expressed by Eq. (4.126) (phase locking): the gauge invariance is preserved by the dynamical emergence of the coherence between the matter field and the e.m. field.

The link between the phase of the matter field and the gauge of A_μ is stated by the equation $A_\mu = \partial_\mu \varphi$ (A_μ is a pure gauge field). When $\varphi(x,t)$ is a regular (continuous differentiable) function then it can be easily shown that $\mathbf{E} = 0 = \mathbf{B}$, namely the potentials and not the fields are present in the coherent region. The existence of non-vanishing fields $\mathbf{E} \neq 0$ and $\mathbf{B} \neq 0$ is connected to the topological singularities of the gauge function $\varphi(x,t)$ [15].

The collective dynamical features presented above protect the macroscopic stability of the system vs quantum fluctuations in the short range dynamics of the microscopic components. This has already been observed in the previous Section in connection with the enhancement of the coupling by the factor $\sqrt{N}$. For the same reason, for sufficiently large N, the collective interaction is protected against thermal fluctuations. Of course, the larger the energy gap, the more robust the protection against thermal fluctuations.

An analysis of energy losses when the system is enclosed in a cavity is presented in [64] in connection with the problem of efficient cooling of an ensemble of N atoms. A problem which we have not considered above is the one related to how much time the system demands to set up the collective regime. This problem is a central one in the domain formation in the Kibble–Zurek mechanism [371, 372, 672] (see also Section 10.5). We finally remark that in the discussion presented above, since the correlation among the elementary constituents is kept by a pure gauge field, the communication among them travels at the phase velocity of the gauge field [197].

4.5.1   *Quantum mechanical decoherence and stability of macroscopic quantum systems*

In recent years it has been recognized that the phenomenon of quantum mechanical decoherence [283, 361, 667, 673, 674] (see also [28, 427–430, 540]) signals the *appearance of a classical world in quantum theory* [283]. In general, decoherence in Quantum Mechanics (QM) is triggered by the interaction of the system with the environment and it formally consists in suppressing the off-diagonal elements of the reduced density matrix. Quantum superposition in the system wave function is then destroyed and thus, provided the time-scale τ_{dyn} characteristic of the dynamics is much greater than the decoherence time-scale τ_{dec}, $\tau_{dyn} \gg \tau_{dec}$, classical behavior may be approached. The reduced density matrix is the one obtained by tracing over the degrees of freedom of the environment (the rest of the Universe).

On the other hand, it is also a fact that macroscopic quantum systems, such as superconductors, ferromagnets, crystals, and in general systems presenting ordered patterns, exhibit coherence over macroscopically large distances which is extraordinarily robust against environment perturbations (see, e.g., the comment on the robustness of coherence in Section 4.4.2). Moreover, as we will see in the following Chapters, there are a number of topologically non-trivial "objects", such as vortices, domain walls, etc., created by the quantum dynamics, which also behave as macroscopic classical objects [161, 617, 619]. The question of the consistency between the decoherence phenomenon in QM and the existence of macroscopic quantum systems and their robust stability needs then to be addressed, which we do in the present Section. Our conclusion is that decoherence characterizes QM by designing its borderlines with QFT, besides those with classical mechanics. In other words, our result suggests the use of decoherence as a useful *criterion* [16] to scan the border between QM and QFT.

As an example of a macroscopic quantum system, we consider the crystal. The Lagrangian for a gas of atoms (or molecules), which below a critical temperature may be arranged in a crystal phase, is invariant under the continuous space translations. However, the crystalline phase of such a system of atoms is not invariant under continuous space translations, but under discrete translations of the length of the lattice spacing or of its multiples. The continuous space translations are spontaneously broken. The Goldstone theorem then predicts the existence of massless NG boson particles, which in the crystal case are the phonons. These quanta are of dynamical origin, namely they are not found in the symmetric or normal phase. The phonons are the quanta of the long-range correlation among the system components (atoms, or molecules), sitting in the crystal lattice sites, which is responsible for the crystal ordering. They are collective modes. It is in this way that the crystal order is generated by the quantum dynamics and appears as a macroscopic property of the system. In the ordered phase the system components get, so to speak, "trapped" by the long-range correlation: they cannot behave as individual particles. Some of their degrees of freedom get frozen by the NG long-range correlation and this manifests itself as the system macroscopically observable ordered pattern: in this specific sense the crystal is a macroscopic quantum system. Since the phonons are massless their coherent condensation in the lowest energy mode does not add energy to the ground state and the observed high stability of the crystal phase is thus explained. The formalism of the many-body theory of crystals is given in [619] and therefore we do not report it here. We only recall that in systems such as the crystal, degenerate ground states, each corresponding to a physically different phase, are described by unitarily inequivalent representations of the canonical commutation relations. Thus one cannot express the ground state of a specific phase in terms of the ground state of another different phase: the crystal ground state cannot be expressed in terms of the amorphous ground state (and vice-versa).

We now consider decoherence in QM. One finds [16] that the result implied by QM decoherence prevents the formation of the crystal, in clear contradiction with the familiar commonly observed process of the formation of a crystal out of an ionic solution. The reason is that the computed decoherence time for the ions in the solution is much too short with respect to the time one has to wait before the crystal gets formed in usual observations. This does not mean that the QM decoherence formulas are wrong, neither, of course, does it mean that our system (the crystal) is a classical one. In fact, once one starts with the ionic solution phase, decoherence tells us that no other phase (the crystal phase) is reachable, which indeed is in perfect agreement with what the von Neumann theorem states in QM. What is wrong is to blindly apply QM decoherence formulas to the ions in solution, in order to study the process of *phase transition,* to the crystal phase (the crystal formation). What one needs is not QM but QFT, since the crystal state is generated by the SSB which cannot occur in QM.

To be specific, we focus on the formation of the binary crystals such as $NaCl, KCl, AgBr$, etc. [16]. We stress that the system we study is *not* the *already formed* crystal: we study the solution of ions out of which one observes that the crystal is formed. Thus, according to the standard chemical recipe, we consider a solution (typically, water is the solvent) of the constituent elements (e.g., a solution of Na^+ ions and Cl^- ions) and wait, in specific conditions of temperature, density, etc., till the crystallization occurs. This happens when the saturation of the solution is reached. At the crystallization point, the saturation concentrations can be quite different in different cases, depending on the crystal one wants to obtain, ranging, for example, from 1 ion of K^+ for 4 molecules of water for KF, or 1 ion of Na^+ for 10 molecules of water for $NaCl$, to 1 ion of Ag^+ for 10^8 molecules of water for $AgBr$, to 1 ion of Pb^+ for 10^{15} molecules of water for PbS [401]. The ions in the solution are normally bound to water molecules by Coulomb forces. For example the Na^+ ion is surrounded by four water molecules. The shielding of the ionic charge by the surrounding water molecules lowers the intensity of the Coulomb interaction among ions. Sometimes one adds a "germ", namely a small crystal of the same kind of the one to be formed. Such a germ acts as a catalytic agent making more favorable the aggregation, in the wanted crystal structure, of the ions in the solution. Sometimes the nucleation is simply produced by some "defect" or "impurity", e.g., on the walls of the bowl or container of the solution. One observes the crystal formation in the vicinity of these defects. At the crystallization point, lowering the temperature of the solution normally helps the crystal formation, which occurs within a short lapse of time (from fractions of a second to several seconds) or in a longer one (from minutes to hours).

The Coulomb type interaction among the ions in the solution controls the ion-ion collisions and the interactions with distant ions. These are two possible sources of decoherence. Other sources of decoherence are, e.g., the interaction with the environment (the water in our case), with the crystal germ or with the defects or impurities, or else with dipole and higher moments of charge distribution. However, one can show [16] that the decoherence effects from ion-ion collisions and distant ion interactions are so strong that we can neglect any other decoherence source. Moreover, one finds [16] that the decoherence time does not strongly depend on the different concentrations of the different ionic solutions mentioned above.

The final result [16] is that the computed decoherence time is so short that there would be no possibility for the formation of the crystals to occur, in contradiction with common experience. These contradictions find their origin in the fact that the QM is not adequate for the description of the crystal formation and its observed stability: these must be studied by the use of QFT. The binding of the atoms in the crystalline lattice is due to the long-range correlation mediated by the phonons (the NG bosons) [619]. The decoherence mechanism thus points to the borderline between QM and QFT.

Appendix K

Group contraction and Virasoro algebra

Loop algebras and Virasoro algebra play a central role in conformal field theories. It is therefore interesting to point out a relation between Euclidean algebras, obtained in the contraction procedure, and loop algebras.

In connection with the locality of observations, implying the missing of infrared effects, we recall the known picture of the group contraction $SU(2) \to E(2)$ in terms of the projection of the $SU(2)$ sphere on the plane tangent to one of the poles. The radius ρ of the sphere acts as a "scale": the $E(2)$ translations in the tangent plane are good approximations to rotations for distances much smaller than ρ (namely, in the limit $\rho \to \infty$). The physical meaning of this is that the $SU(2)$ contraction to $E(2)$ manifests itself in local, with respect to the ρ scale, observations [182]. In the local observation process on the tangent plane, the *orientation* of the x_3 axis is "locked", which amounts to the loss of symmetry under the $x_3 \to -x_3$ (breakdown of the loop-antiloop symmetry). As a matter of fact, specifying the direction of the x_3 axis produces topologically inequivalent configurations [468].

Also in the case of $SO(4) \to E(3)$, such a contraction manifests itself in local observations and the x_4 axis orientation gets "locked". Again, the breakdown of the $x_4 \to -x_4$ symmetry is built in the geometrical structure of the $E(3)$ group (breakdown of the loop-antiloop symmetry).

The breakdown of the loop-antiloop symmetry then leads us to investigate the relation between the Euclidean algebras and the loop algebras. For the case of the Virasoro algebra this goes as follows [18].

Let us first recall that the $e(3)$ algebra has six generators P_i and M_i, $i = 1, 2, 3$, corresponding to the translation and to the rotation generators, respectively. The commutation relations are:

$$[P_i, P_j] = 0, \quad [M_i, M_j] = \epsilon_{ijk}M_k, \quad [P_i, M_j] = \epsilon_{ijk}P_k; \qquad \text{(K.1)}$$

In the contraction process, the $SO(3)$ subgroup generated by the M_is is left unchanged. The algebra $e(3)$ has two invariants, $P^2 = \Sigma P_i^2$ and $\Sigma P_i \cdot M_i$.

By means of the so-called graded contraction method [190, 481], it is possible to map the Virasoro algebra into a sort of generalization of the Euclidean algebra $e(3)$. The commutation relations of the Virasoro algebra $\mathcal{L}$ of central charge c (c commuting with all the T's) are

$$[T_n, T_m] = (n - m)T_{n+m} + \frac{c}{12}(n^3 - n)\delta_{n+m,0}, \quad m, n \in \mathbb{Z}. \quad (K.2)$$

The $\mathbb{Z}_2$-grading of the algebra consists in dividing the set of the T_n generators into an even set $L_0 \equiv \{A_n, c\}$ and an odd set $L_1 \equiv \{B_n\}$, with

$$A_n = \frac{1}{2}\left(T_{2n} + \frac{c}{8}\delta_{n,0}\right), \quad B_n = \frac{1}{2}T_{2n+1}, \quad (K.3)$$

so that $\mathcal{L} = L_0 \oplus L_1$ and

$$[L_0, L_0] \subseteq L_0, \quad [L_0, L_1] \subseteq L_1, \quad [L_1, L_1] \subseteq L_0. \quad (K.4)$$

The commutation relations of the graded generators are given by [390]

$$[A_n, A_m] = (n - m)A_{n+m} + \frac{2c}{12}(n^3 - n)\delta_{n+m,0}, \quad (K.5a)$$

$$[B_n, B_m] = (n - m)A_{n+m+1} + \frac{2c}{12}(n - \frac{1}{2})(n + \frac{1}{2})(n + \frac{3}{2})\delta_{n+m+1,0}, \quad (K.5b)$$

$$[A_n, B_m] = (n - m - \frac{1}{2})B_{n+m}. \quad (K.5c)$$

Eqs. (K.5) show that $\{A_n, c\}$ is again a Virasoro algebra but with central charge $2c$. We can then consider the $\mathbb{Z}_2$-graded contraction of the algebra (K.5) [390] (see also [190, 481]):

$$[A_n, A_m] = (n - m)A_{n+m} + \frac{2c}{12}(n^3 - n)\delta_{n+m,0}, \quad (K.6a)$$

$$[B_n, B_m] = 0, \quad (K.6b)$$

$$[A_n, B_m] = (n - m - \frac{1}{2})B_{n+m}. \quad (K.6c)$$

Note now that, in the centerless case ($c = 0$), the A_0 and $A_{\pm 1}$ generators close the algebra isomorphic to $so(3) \sim su(2)$ and the set of these three generators and $B_{-\frac{1}{2}}$, $B_{\frac{1}{2}}$, $B_{-\frac{3}{2}}$ close the $e(3)$ isomorphic algebra [18]. The commutation relations (K.1) are then obtained by setting:

$$M_+ \equiv A_1, \quad M_- \equiv A_{-1}, \quad M_3 \equiv iA_0,$$
$$P_+ \equiv B_{\frac{1}{2}}, \quad P_- \equiv B_{-\frac{3}{2}}, \quad P_3 \equiv iB_{-\frac{1}{2}}. \quad (K.7)$$

This result has a general extension, i.e., the algebra

$$\mathcal{E}_n \equiv \{A_0, A_{\pm n}\} \oplus \{B_{-\frac{1}{2}}, B_{\pm n - \frac{1}{2}}\}, \quad (K.8)$$

reproduces the $e(3)$ algebra for each integer value of n, provided the following positions are assumed:

$$M_+ \equiv \frac{1}{n} A_n, \quad M_- \equiv \frac{1}{n} A_{-n}, \quad M_3 \equiv \frac{i}{n} A_0,$$
$$P_+ \equiv B_{n-\frac{1}{2}}, \quad P_- \equiv B_{-n-\frac{1}{2}}, \quad P_3 \equiv i B_{-\frac{1}{2}}. \quad \text{(K.9)}$$

As a final remark we notice that the $e(2)$ algebra can be obtained [18] as a subalgebra of (K.9) by choosing $A_{\pm n} = 0$, for non-zero values of n.

We conclude that the extension of the Virasoro algebra by means of its $\mathbb{Z}_2$-grading with the subsequent step of the $\mathbb{Z}_2$-graded contraction appears as a n-graded hierarchy of Euclidean algebras. This establishes the relation between the contraction to the Euclidean groups and the Virasoro algebras.

Appendix L

Phase locking in the N atom system

Under the conditions mentioned in Section 4.5, the field $\chi(\mathbf{x}, t)$ may be expanded in the unit sphere in terms of spherical harmonics:

$$\chi(\mathbf{x}, t) = \sum_{l,m} \alpha_{l,m}(t) Y_l^m(\theta, \phi), \tag{L.1}$$

which, by setting $\alpha_{l,m}(t) = 0$ for $l \neq 0$, 1, reduces to the expansion in the four levels $(l, m) = (0, 0)$ and $(1, m), m = 0, \pm 1$. Their populations are given by $N|\alpha_{l,m}(t)|^2$ and at thermal equilibrium, in the absence of interaction, they follow the Boltzmann distribution. The normalization condition (4.119) gives

$$Q \equiv |\alpha_{0,0}(t)|^2 + \sum_m |\alpha_{1,m}(t)|^2 = |a_0(t)|^2 + 3\,|a_1(t)|^2 = 1, \quad \forall\, t \tag{L.2}$$

and therefore $\dot{Q} = 0$ (the dot as usual denotes time derivative), i.e.,

$$\frac{\partial}{\partial t}|a_1(t)|^2 = -\frac{1}{3}\,\frac{\partial}{\partial t}|a_0(t)|^2. \tag{L.3}$$

As explained in Section 4.5, the amplitudes do not depend on m. The conservation law $\dot{Q} = 0$ (and Eq. (L.3)) expresses nothing but the conservation of the total number N of atoms; it also means that, due to the rotational invariance, the rate of change of the population in each of the levels $(1, m)$, $m = 0, \pm 1$, equally contributes, in the average, to the rate of change in the population of the level $(0, 0)$, at each time t.

When the initial conditions (4.121) are used, one can properly tune the parameter θ_0 in its range of definition. For example, $\theta_0 = \frac{\pi}{3}$ describes the equipartition of the field modes of energy $E(k)$ among the four levels $(0, 0)$ and $(1, m)$, $|a_0(0)|^2 \simeq |a_{1,m}(0)|^2$, $m = 0, \pm 1$, as given by the Boltzmann distribution when the temperature T is high enough, $k_B T \gg E(k)$.

Eqs. (4.123) are of course consistent with the conservation law $\dot{Q} = 0$ and they also show that

$$\frac{\partial}{\partial t}|u_m(t)|^2 = -2\,\frac{\partial}{\partial t}|a_{1,m}(t)|^2\,, \qquad\qquad \text{(L.4)}$$

from which we see that $|u_m(t)|$ does not depend on m since $|\alpha_{1,m}(t)| = |a_{1,m}(t)|$ does not depend on m. Also, another conservation law holds, i.e.,

$$|u(t)|^2 + 2\,|a_1(t)|^2 = \frac{2}{3}\sin^2\theta_0\,, \qquad \forall\, t\,, \qquad\qquad \text{(L.5)}$$

where $|u(t)| \equiv |u_m(t)|$, $|a_1(t)| \equiv |a_{1,m}(t)|$, the initial condition (4.121) has been used and we have set

$$|u(0)|^2 = 0\,. \qquad\qquad \text{(L.6)}$$

By using Eqs. (4.120) and (4.124), Eqs. (4.123) give

$$\dot{A}_0(t) = \Omega U(t) A_1(t) \cos\alpha_m(t)\,, \qquad\qquad \text{(L.7a)}$$

$$\dot{A}_1(t) = -\Omega U(t) A_0(t) \cos\alpha_m(t)\,, \qquad\qquad \text{(L.7b)}$$

$$\dot{U}(t) = 2\Omega A_0(t) A_1(t) \cos\alpha_m(t)\,, \qquad\qquad \text{(L.7c)}$$

$$\dot{\varphi}_m(t) = 2\Omega \frac{A_0(t) A_1(t)}{U(t)} \sin\alpha_m(t)\,, \qquad\qquad \text{(L.7d)}$$

where

$$\alpha_m \equiv \delta_{1,m}(t) - \delta_0(t) - \varphi_m(t)\,. \qquad\qquad \text{(L.8)}$$

Equations for $\dot{\delta}_{1,m}$ and $\dot{\delta}_0$ can be derived in a similar way. Eqs. (L.7) show that phases turn out to be independent of m. Indeed, the r.h.s. of these equations have to be independent of m since their l.h.s. are independent of m, so either $\cos\alpha_m(t) = 0$ for any m at any t, or α_m is independent of m at any t. In both cases, Eq. (L.7) shows that φ_m is then independent of m, which in turn implies, together with Eq. (L.8), that $\delta_{1,m}(t)$ is independent of m. We therefore put $\varphi \equiv \varphi_m$, $\delta_1(t) \equiv \delta_{1,m}(t)$, $\alpha \equiv \alpha_m$, $u(t) \equiv u_m(t)$ and $a_1(t) \equiv a_{1,m}(t)$. Use of Eqs. (L.7) leads to the phase locking Eq. (4.126).

In general, one can always change the phases by arbitrary constants. However, if they are equal in one frame they are unequal in a rotated frame and gauge invariance is lost. The independence of m of the phases is here of dynamical origin and the phase locking among $\delta_0(t)$, $\delta_1(t)$ and $\varphi(t)$ has indeed the meaning of recovering the gauge symmetry.

Chapter 5

Thermal field theory and trajectories in the space of the representations

5.1 Introduction

In this and in the next ·Chapter we present some essential aspects of finite temperature QFT formalism. We will see that the characteristic feature of QFT, the existence of infinitely many unitarily inequivalent representations of the canonical (anti-) commutation relations, perfectly fits the non-zero temperature formalism, each representation being labeled by a definite value of the temperature. Variations of the temperature, and the consequent process of phase transition, if any, are thus described by "trajectories" or paths through the set of unitarily inequivalent representations.

Our discussion also shows that in full generality the algebraic structure underlying the space of the inequivalent representations is actually the one of the deformed Hopf algebra. The representations are labeled by the deformation parameter, which can be temperature-dependent, so that, in this sense, QFT appears to be an "intrinsically" thermal field theory. This clarifies and reinforces the known connection between the QFT formalism and the one of statistical mechanics [300, 508, 509, 511].

Such an intrinsic thermal nature of QFT contrasts with the still prevailing feeling that QFT is nothing but zero temperature formalism developed in order to describe particle physics, despite the fact that QFT has revealed itself to be the theoretical framework necessary to study also condensed matter physics. Even in high energy physics there are many situations where the zero temperature approximation is inadequate. Examples include cosmology, astrophysics and quark-gluon plasma. More generally, in all cases where one studies systems near the critical temperature (and so neither low or high temperature expansions properly describe the basic features) the finite temperature treatment is necessary.

The introduction of temperature into QFT was originally pioneered by Matsubara in 1955 [446]. Soon after that, Ezawa, Tomozawa and Umezawa [228] applied the Matsubara approach, also called *the imaginary-time formalism*, to a statistical model of multiple pion production. They observed that, in the frame of the Matsubara approach, the Fourier representation involved a summation over the discrete frequency spectrum which is usually referred to as summation of the Matsubara frequencies (see Section 6.5). The same observations were independently made by Abrikosov, Gor'kov and Dzyaloshinski [5] in their application of the Matsubara method to the theory of superconductivity.

In the Matsubara method, the evaluation of thermal averages is performed by replacing the time t by the imaginary time $i\beta$, according to the formal analogy between imaginary time and inverse temperature first noticed by F. Bloch. Time integration is restricted to a finite domain along the imaginary axis, from 0 to $i\beta$. Therefore, time does not appear explicitly as a parameter in the Matsubara Green's functions. In order to consider time-dependent phenomena one must perform analytic continuation on the complex time plane. The extension of the domain of time integration to the complex plane, thus generalizing the Matsubara method to the so-called closed time-path (CTP) formalism, has been considered in [171, 367, 475, 560]. In this generalization, the choice of the integration path is constrained by the requirement that the imaginary part of the complex time variable must be monotonically decreasing. The integration path must contain a section which lies along the real axis in order to compute the real-time Green's functions. We will discuss the CTP formalism in the following Chapter.

The Matsubara method and its generalization is by its nature a theory of Green's functions. On the other hand, the axiomatic field theory, in its C^*−algebra formulation extended to the interacting field with the introduction of the Kubo–Martin–Schwinger conditions, shows that the effect of temperature may be taken into account by doubling the field degrees of freedom [31, 32, 309]. Inspired by this last specific feature of C^*−algebra, in 1975 Takahashi and Umezawa formulated the so-called Thermo Field Dynamics (TFD) [600, 617, 619], where the doubling of the field degrees of freedom plays a central role and there is no need of extension to the complex time plane. Subsequently, Ojima [505] clarified the relation between the C^*−algebra formalism and TFD. An early formulation of real-time formalism, although not yet in a systematic form, was applied to superconductivity [422]. TFD has been successfully applied since then to a number of physical problems at non-zero temperature [617, 619] (see also [14, 151, 153, 444, 615, 641] and references quoted therein), in condensed matter physics and in particle physics, nuclear physics and cosmology (see, e.g., [369]). The advantages of TFD over previous methods rely on the fact that in TFD one can study algebraic features of the theory as well as Green's function method and Feynman diagrams. Also, the concept of finite temperature vacuum is available. The relation of TFD with the CTP formalism has been analyzed in detail and, at least for the case of thermal equilibrium, a full equivalence seems to be established between the two formalisms [369, 628]. The formalism for non-equilibrium TFD (NETFD) has also been developed in detail [34–38, 49, 118, 227, 314, 333, 410, 523] and the relation of TFD with the renormalization group theory has been investigated, leading to a generalization of the renormalization group techniques [447]. However, in this book we will not enter into the presentation of these topics.

The study of topologically non-trivial soliton solutions and extended objects which are formed in the processes of phase transition has also been developed in the frame of TFD [617, 619]. This will be discussed in the following Chapters.

In this Chapter, before presenting some essential aspects of TFD, we will discuss the occurrence of the doubling of the system degrees of freedom in the Wigner function formalism and in the density matrix formalism in Quantum Mechanics. We will show that QFT is structurally related with the doubling of the field degrees of freedom, and this will lead us to recognize that the quantum deformed Hopf algebra is the basic algebraic structure of QFT. We will also discuss the Kählerian nature of the space of the inequivalent representations and show that the trajectories in such a space are classical trajectories which may have chaotic character. We also discuss temperature effects on boson condensation and thermal features of gauge theories.

In this book we do not give a complete account of the formalism of thermal field theory and TFD. Their detailed presentation may be found in several textbooks. We introduce only essential notions in order to better depict the general structural aspects of QFT. Also, we do not discuss renormalization at finite temperature. A procedure has been developed [448] which shows that ultraviolet divergences at finite temperature may be eliminated by counter-terms prepared at zero temperature, provided the theory is renormalizable at zero temperature, and that finite renormalization relates theories at different temperatures.

5.2   Doubling the degrees of freedom

In the present Section we consider the doubling of the degrees of freedom of the system under study. This is not simply a mathematical tool useful to describe our system. It appears to be an essential feature of QFT [633]. In the following Sections we will see that the formalism of TFD is built on the doubling of the field degrees of freedom.

As a preliminary discussion, we show that such a doubling is also present in the standard QM formalism of the density matrix [239, 560] and the associated Wigner function [237]. To see this, by following closely [576], see also [633], we observe that the Wigner function formalism suggests that the $x(t)$ coordinate of a quantum particle may be split into two coordinates $x_+(t)$ (going forward in time) and $x_-(t)$ (going backward in time). Indeed, the standard expression for the Wigner function is [237],

$$W(p, x, t) = \frac{1}{2\pi\hbar} \int \psi^* \left(x - \frac{1}{2}y, t \right) \psi \left(x + \frac{1}{2}y, t \right) e^{-i\frac{py}{\hbar}} dy, \qquad (5.1)$$

where

$$x_\pm = x \pm \frac{1}{2}y. \qquad (5.2)$$

The density matrix associated to the Wigner function (5.1), which appears in the mean value of a quantum operator A

$$\bar{A}(t) = \langle \psi(t)|A|\psi(t)\rangle = \int\int \psi^*(x_-, t) \langle x_-|A|x_+\rangle \psi(x_+, t)dx_+dx_-$$

$$= \int\int \langle x_+|\rho(t)|x_-\rangle\langle x_-|A|x_+\rangle dx_+dx_-, \qquad (5.3)$$

is [237]

$$W(x_+, x_-, t) \equiv \langle x_+|\rho(t)|x_-\rangle = \psi^*(x_-, t)\psi(x_+, t). \qquad (5.4)$$

The evolution of the density matrix $W(x_+, x_-, t)$ follows two copies of the Schrödinger equation, the forward in time motion and the backward in time motion, controlled by the two Hamiltonian operators $H_\pm$, respectively:

$$i\hbar\frac{\partial\psi(x_\pm, t)}{\partial t} = H_\pm\psi(x_\pm, t), \qquad (5.5)$$

i.e.,

$$i\hbar\frac{\partial\langle x_+|\rho(t)|x_-\rangle}{\partial t} = \mathcal{H}\,\langle x_+|\rho(t)|x_-\rangle, \qquad (5.6)$$

where

$$\mathcal{H} = H_+ - H_-. \qquad (5.7)$$

Using two copies of the Hamiltonian (i.e., $H_\pm$) operating on the outer product of two Hilbert spaces $\mathcal{F}_+ \otimes \mathcal{F}_-$ has been implicitly required in QM since the very beginning of the theory. For example, from Eqs. (5.6), (5.7) one finds immediately that the eigenvalues of $\mathcal{H}$ are directly the Bohr transition frequencies $\hbar\omega_{nm} = E_n - E_m$, which was the first clue to the explanation of spectroscopic structure.

The notion that a quantum particle has two coordinates $x_\pm(t)$ moving at the same time is therefore central [576]. The density matrix and the Wigner function *require* the introduction of a "doubled" set of coordinates, $(x_\pm, p_\pm)$ (or (x, p_x) and (y, p_y)).

The mathematical role and the physical meaning of the doubling of the degrees of freedom fully appears in dealing with phase transitions, with equilibrium and non-equilibrium thermal field theories and with dissipative, open systems [632,633,636]. In these cases the doubling of the degrees of freedom appears to be a structural feature of QFT since it strictly relates with the existence of the unitarily inequivalent representations of the canonical commutation relations (CCR) in QFT. In Chapter 9 we will show that the doubling of the degrees of freedom is related with quantum noise (the vacuum quantum fluctuations). We thus conclude that any microscopic system is in fact an intrinsically open system permanently interacting with the vacuum quantum fluctuations.

It is instructive to see how the doubling of the coordinates works in the remarkable QM example of the two-slit diffraction experiment. We will shortly summarize the discussion reported in [107] in the following subsection.

5.2.1 *The two-slit experiment*

We want to derive the diffraction pattern observed in the two-slit experiment. The density matrix in terms of the wave function $\psi_0(x)$ of the particle when it "passes through the slits" at time zero is [107]

$$\langle x_+ | \rho_0 | x_- \rangle = \psi_0^*(x_-)\psi_0(x_+). \tag{5.8}$$

The probability density for the electron to be found at position x at the detector screen at a later time t is written as

$$P(x,t) = \langle x | \rho(t) | x \rangle = \psi^*(x,t)\psi(x,t). \tag{5.9}$$

Here $\psi(x,t)$ is the solution to the free particle Schrödinger equation,

$$\psi(x,t) = \left(\frac{M}{2\pi\hbar it}\right)^{1/2} \int_{-\infty}^{\infty} e^{\frac{i}{\hbar}A(x-x',t)}\psi_0(x')dx', \tag{5.10}$$

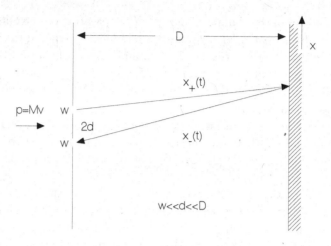

Fig. 5.1   Two slit experiment.

where

$$A(x - x', t) = \frac{M(x - x')^2}{2t} \qquad (5.11)$$

is the Hamilton-Jacobi action for a classical free particle to move from x' to x in a time t. Eqs. (5.8)–(5.11) imply that

$$P(x, t) = \frac{M}{2\pi\hbar t} \int_{-\infty}^{\infty} \int_{-\infty}^{\infty} e^{i\frac{M}{2\hbar t}[(x - x_+)^2 - (x - x_-)^2]} \langle x_+ | \rho_0 | x_- \rangle dx_+ dx_-. $$

$$(5.12)$$

From Eq. (5.12) we see that if $x_+ = x_-$, $P(x, t)$ would not oscillate in x, i.e., there would not be the usual quantum diffraction. In Eq. (5.12), in order to have quantum interference the forward in time action $A(x - x_+, t)$ must be different from the backward in time action $A(x - x_-, t)$: the non-trivial dependence of the density matrix on the difference $(x_+ - x_-)$ crucially determines the quantum nature of the phenomenon.

In the quantum diffraction experiment the experimental apparatus is prepared so that $w \ll d \ll D$, with w the opening of the slits which are separated by a distance $2d$. D is the distance between the slits and the screen (Fig. 5.1). By defining $K = \frac{Mvd}{\hbar D}$, $\beta = \frac{w}{d}$, with $v = D/t$ the velocity of the incident electron, Eq. (5.12) with $|x| \gg |x_\pm|$ leads [107] to the usual diffraction pattern result

$$P(x, D) \approx \frac{4}{\pi\beta K x^2} \cos^2(Kx) \sin^2(\beta K x), \qquad (5.13)$$

where the initial wave function

$$\psi_0(x) = \frac{1}{\sqrt{2}}\Big[\phi(x-d) + \phi(x+d)\Big], \tag{5.14}$$

with $\phi(x) = \frac{1}{\sqrt{w}}$ if $|x| \leq \frac{w}{2}$ and zero otherwise, has been used. From Eqs. (5.8) and (5.14) we have

$$\langle x_+|\rho_0|x_-\rangle = \frac{1}{2}\Big\{\phi(x_+ - d)\phi(x_- - d) + \phi(x_+ + d)\phi(x_- + d)$$

$$+\phi(x_+ - d)\phi(x_- + d) + \phi(x_+ + d)\phi(x_- - d)\Big\}. \tag{5.15}$$

In the r.h.s. of Eq. (5.15) the first and the second terms describe the classical processes of the particle going forward and backward in time through slit 1 and going forward and backward in time through slit 2, respectively. In these processes it is $x_+(t) = x_-(t)$ and in such cases no diffraction is observed on the screen. The third term and the fourth term describe the particle going forward in time through slit 1 and backward in time through slit 2, or forward in time through slit 2 and backward in time through slit 1, respectively. These are the terms generating quantum interference since $|x_+(t) - x_-(t)| > 0$.

In conclusion, when forward in time and backward in time motions are (at the same time) unequal $x_+(t) \neq x_-(t)$, then the system is behaving in a quantum mechanical fashion and interference patterns are observed. If, on the contrary, $x(t) \equiv x_+(t) \equiv x_-(t)$, then the system behavior appears to be a classical one. This confirms that the system quantum behavior is strictly dependent on the doubling of the degrees of freedom.

5.3 Thermo Field Dynamics: A brief introduction

In this Section we present essential notions of TFD [369, 600, 617, 619] and in this connection we discuss the doubling of the degrees of freedom. In Section 5.4 we will show that the q-deformed Hopf algebras for bosons and for fermions naturally provide the algebraic structure on which the TFD formalism is based.

In the following we follow closely [600] where TFD was originally presented. The central point in the TFD formalism is the possibility to express the statistical average $\langle \mathcal{A} \rangle$ of an observable $\mathcal{A}$ as the expectation value in the temperature-dependent vacuum $|0(\beta)\rangle$:

$$\langle \mathcal{A} \rangle \equiv \frac{Tr[\mathcal{A}\,e^{-\beta H}]}{Tr[e^{-\beta H}]} = \langle 0(\beta)|\mathcal{A}|0(\beta)\rangle, \tag{5.16}$$

where, in general, H includes also the term μN, with μ the chemical potential and N the number operator.

The thermal vacuum

The first problem is therefore to construct a suitable temperature-dependent state $|0(\beta)\rangle$ which satisfies Eq. (5.16), namely

$$\langle 0(\beta)|\mathcal{A}|0(\beta)\rangle = \frac{1}{\text{Tr}[e^{-\beta H}]} \sum_n \langle n|\mathcal{A}|n\rangle e^{-\beta E_n} , \qquad (5.17)$$

with

$$H|n\rangle = E_n|n\rangle , \quad \langle n|m\rangle = \delta_{nm} . \qquad (5.18)$$

Here $E_n \equiv \omega - \mu$. However, in the following we neglect to consider the chemical potential μ for simplicity. Such a state cannot be constructed as long as one remains in the original Fock space $\{|n\rangle\}$. By following [600], one can write $|0(\beta)\rangle$ in terms of the states $|n\rangle$ as

$$|0(\beta)\rangle = \sum_n f_n(\beta)|n\rangle . \qquad (5.19)$$

Then, use of this equation into (5.17) gives

$$f_n^*(\beta)f_m(\beta) = \frac{1}{\text{Tr}[e^{-\beta H}]} e^{-\beta E_n} \delta_{nm} , \qquad (5.20)$$

which is impossible to be satisfied by c-number functions $f_n(\beta)$. However, Eq. (5.20) can be regarded as the orthogonality condition in a Hilbert space in which the expansion coefficient $f_n(\beta)$ is a vector. In order to realize such a representation it is convenient to introduce a system identical to the one under study, namely to double the given system. The quantities associated with the doubled system are denoted by the tilde in the usual notation of TFD [600]. Thus the tilde-system is characterized by the Hamiltonian $\tilde{H}$ and the states are denoted by $|\tilde{n}\rangle$, with

$$\tilde{H}|\tilde{n}\rangle = E_n|\tilde{n}\rangle , \quad \langle \tilde{n}|\tilde{m}\rangle = \delta_{nm} . \qquad (5.21)$$

where E_n is by definition the same as the one appearing in Eq. (5.18). It is also assumed that non-tilde and tilde operators are commuting (anti-commuting) boson (fermion) operators. One then considers the space spanned by the direct product $|n\rangle \otimes |\tilde{m}\rangle \equiv |n, \tilde{m}\rangle$. The matrix element of a Bose-like operator $\mathcal{A}$ is thus

$$\langle \tilde{m}, n|\mathcal{A}|n', \tilde{m}'\rangle = \langle n|\mathcal{A}|n'\rangle \delta_{mm'} , \qquad (5.22)$$

and the one of the corresponding $\tilde{\mathcal{A}}$ is

$$\langle \tilde{m}, n | \tilde{\mathcal{A}} | n', \tilde{m}' \rangle = \langle \tilde{m} | \tilde{\mathcal{A}} | \tilde{m}' \rangle \delta_{nn'} . \qquad (5.23)$$

It turns out to be convenient to identify

$$\langle m | \mathcal{A} | n \rangle = \langle \tilde{n} | \tilde{\mathcal{A}}^{\dagger} | \tilde{m} \rangle . \qquad (5.24)$$

Eq. (5.20) is satisfied if one defines

$$f_n(\beta) = \frac{1}{\sqrt{\text{Tr}[e^{-\beta H}]}} e^{\frac{-\beta E_n}{2}} |\tilde{n}\rangle , \qquad (5.25)$$

and Eq. (5.17) is obtained by using the definition (5.25) in (5.19):

$$|0(\beta)\rangle = \frac{1}{\sqrt{\text{Tr}[e^{-\beta H}]}} \sum_n e^{\frac{-\beta E_n}{2}} |n, \tilde{n}\rangle . \qquad (5.26)$$

The vectors $|n\rangle$ and $|\tilde{n}\rangle$ thus appear as a pair in $|0(\beta)\rangle$. We remark that the formal role of the "doubled" states $|\tilde{n}\rangle$ is merely to pick up the diagonal matrix elements of $\mathcal{A}$. In this connection, thinking of the role of the environment, which is able to reduce the system density matrix to its diagonal form in the QM decoherence processes [673, 674] (cf. Section 4.5.1), it is remarkable that the doubled degrees of freedom in TFD are indeed susceptible to being interpreted as the environment degrees of freedom, as better specified in the following.

It is useful to consider, as an example, the case in which the operator $\mathcal{A}$ is the number operator $N = a^{\dagger}a$. Here and in the following, for notational simplicity, provided no misunderstanding arises, we are omitting the momentum index $\mathbf{k}$ and related summations over $\mathbf{k}$ and denote by $a^{\dagger}$ and a the operators $a_{\mathbf{k}}^{\dagger}$ and $a_{\mathbf{k}}$, respectively. For definiteness we consider the boson case. The statistical average of N is the Bose–Einstein distribution $f_B(\omega)$, where ω denotes the energy of the quantum a,

$$\langle N \rangle \equiv \frac{\text{Tr}[N\, e^{-\beta H}]}{\text{Tr}[e^{-\beta H}]} = \langle 0(\beta) | N | 0(\beta) \rangle = \frac{1}{e^{\beta \omega} - 1} = f_B(\omega) , \qquad (5.27)$$

where

$$H = \omega a^{\dagger} a . \qquad (5.28)$$

We now introduce the bosonic ladder operators $\tilde{a}$, $\tilde{a}^{\dagger}$ for the tilde states. The commutation relations are:

$$[a, a^{\dagger}] = 1 = [\tilde{a}, \tilde{a}^{\dagger}] , \qquad (5.29)$$

with all other commutators vanishing and a and $\tilde{a}$ commuting among themselves. One then can show [600] that, by setting

$$u(\beta) \equiv \sqrt{1 + f_B(\omega)}, \quad v(\beta) \equiv \sqrt{f_B(\omega)} , \qquad (5.30)$$

$$u^2(\beta) - v^2(\beta) = 1 \,, \tag{5.31}$$

so that

$$u(\beta) = \cosh \theta(\beta) \,, \quad v(\beta) = \sinh \theta(\beta) \,, \tag{5.32}$$

and defining

$$\mathcal{G} = -i(a^\dagger \tilde{a}^\dagger - a \tilde{a}) \,, \tag{5.33}$$

the state $|0(\beta)\rangle$ is formally given (at finite volume) by

$$|0(\beta)\rangle = e^{i\theta(\beta)\mathcal{G}}|0\rangle = \frac{1}{u(\beta)} \exp\left(\frac{v(\beta)}{u(\beta)}\right) a^\dagger \tilde{a}^\dagger |0\rangle \,. \tag{5.34}$$

Here and in Eq. (5.32), $\theta(\beta)$ is the Bogoliubov parameter and we will simply write θ instead of $\theta(\beta)$ whenever no misunderstanding arises. We also put $|0(\theta(\beta))\rangle \equiv |0(\beta)\rangle$ and we will write in the following $|0(\theta)\rangle$ instead of $|0(\theta(\beta))\rangle$. In TFD it is customary to use the notation $f = e^{-\beta\omega}$, so that $\langle N \rangle = \frac{f}{1-f}$ (cf. Eq. (5.27)) and the density matrix is $\rho = f^{a^\dagger a}$.

It is clear that the state $|0(\theta)\rangle$ is not annihilated by a and $\tilde{a}$. However, it is annihilated by the "new" set of operators $a(\theta)$ and $\tilde{a}(\theta)$,

$$a(\theta)|0(\theta)\rangle = 0 = \tilde{a}(\theta)|0(\theta)\rangle \,, \tag{5.35}$$

with

$$a(\theta) = \exp(i\theta\mathcal{G}) \, a \, \exp(-i\theta\mathcal{G}) = a \cosh \theta - \tilde{a}^\dagger \sinh \theta \,, \tag{5.36a}$$

$$\tilde{a}(\theta) = \exp(i\theta\mathcal{G}) \, \tilde{a} \, \exp(-i\theta\mathcal{G}) = \tilde{a} \cosh \theta - a^\dagger \sinh \theta \,, \tag{5.36b}$$

$$[a(\theta), a^\dagger(\theta)] = 1 = [\tilde{a}(\theta), \tilde{a}^\dagger(\theta)] \,. \tag{5.37}$$

All other commutators are equal to zero and $a(\theta)$ and $\tilde{a}(\theta)$ commute among themselves. Eqs. (5.36) are nothing but the Bogoliubov transformations of the $(a, \tilde{a})$ pair into a new set of creation, annihilation operators. We will show in Section 5.4 that the Bogoliubov-transformed operators $a(\theta)$ and $\tilde{a}(\theta)$ are linear combinations of the deformed coproduct operators.

The state $|0(\theta)\rangle$ is not the vacuum (zero energy eigenstate) of H and of $\tilde{H}$. It is, however, the zero energy eigenstate for the "Hamiltonian" $\hat{H}$, $\hat{H}|0(\theta)\rangle = 0$, with

$$\hat{H} \equiv H - \tilde{H} = \omega(a^\dagger a - \tilde{a}^\dagger \tilde{a}) \,. \tag{5.38}$$

The state $|0(\theta)\rangle$ is called the thermal vacuum.

When the proper field notation is restored, a and $\tilde{a}$ carry dependence on the momentum $\mathbf{k}$. Then the Bogoliubov transformation should be thought of as inner automorphism of the algebra $su(1,1)_\mathbf{k}$ (or $su(2)_\mathbf{k}$ for fermions, see below). This shows that one is globally dealing with $\oplus_{\mathbf{k}} su(1,1)_{\mathbf{k}}$ (or $\oplus_{\mathbf{k}} su(2)_{\mathbf{k}}$). Therefore one is led to consider $\mathbf{k}$-dependence also for the θ parameter. For many degrees of freedom the generator $\mathcal{G}$ is,

$$\mathcal{G} = -i \sum_{\mathbf{k}} \theta_k \left(a_{\mathbf{k}}^{\dagger} \tilde{a}_{\mathbf{k}}^{\dagger} - a_{\mathbf{k}} \tilde{a}_{\mathbf{k}} \right). \tag{5.39}$$

The (bosonic) thermal operators and the thermal vacuum are

$$a_{\mathbf{k}}(\theta) = e^{i\mathcal{G}} \, a_{\mathbf{k}} \, e^{-i\mathcal{G}} = a_{\mathbf{k}} \cosh \theta_k - \tilde{a}_{\mathbf{k}}^{\dagger} \sinh \theta_k, \tag{5.40a}$$

$$\tilde{a}_{\mathbf{k}}(\theta) = e^{i\mathcal{G}} \, \tilde{a}_{\mathbf{k}} \, e^{-i\mathcal{G}} = \tilde{a}_{\mathbf{k}} \cosh \theta_k - a_{\mathbf{k}}^{\dagger} \sinh \theta_k, \tag{5.40b}$$

$$|0(\theta)\rangle = \prod_{k} \frac{1}{\cosh \theta_k} \exp \left(\tanh \theta_k \, a_{\mathbf{k}}^{\dagger} \tilde{a}_{\mathbf{k}}^{\dagger} \right) |0\rangle, \tag{5.41}$$

with $\langle 0(\theta)|0(\theta)\rangle = 1$, $\forall \theta$. Note that the thermal vacuum $|0(\theta)\rangle$ is a $SU(1,1)$ generalized coherent state [519].

Using the continuous limit relation $\sum_{\mathbf{k}} \to \frac{V}{(2\pi)^3} \int d^3 k$, in the infinite volume limit we have (for $\int d^3 k \, \ln \cosh \theta_k$ finite and positive)

$$\langle 0(\theta)|0\rangle \to 0 \quad \text{as } V \to \infty \quad \forall \, \theta \equiv \{\theta_k\} \neq 0, \tag{5.42}$$

and in general, $\langle 0(\theta)|0(\theta')\rangle \to 0$ as $V \to \infty$ $\forall \, \theta$ and θ', $\theta' \neq \theta$. For each $\theta \equiv \{\theta_k\}$ a representation $\{|0(\theta)\rangle\}$ of the CCR is defined and it turns out to be unitarily inequivalent to any other representation $\{|0(\theta')\rangle, \forall \theta' \neq \theta\}$ in the infinite volume limit. Also note that

$$\mathcal{N}_{a_{\mathbf{k}}}(\theta) = \langle 0(\theta)|a_{\mathbf{k}}^{\dagger} a_{\mathbf{k}}|0(\theta)\rangle = \sinh^2 \theta_k, \tag{5.43}$$

and similar expression for $\mathcal{N}_{\tilde{a}_{\mathbf{k}}}(\theta)$. Thus, according to our discussion (cf. Eq. (5.27)), we have

$$\mathcal{N}_{a_{\mathbf{k}}}(\theta) = \sinh^2 \theta_k = \frac{1}{e^{\beta \omega_k} - 1}. \tag{5.44}$$

For each θ one has a copy $\{a_{\mathbf{k}}(\theta), a_{\mathbf{k}}^{\dagger}(\theta), \tilde{a}_{\mathbf{k}}(\theta), \tilde{a}_{\mathbf{k}}^{\dagger}(\theta) \, ; \, |0(\theta)\rangle \, | \, \forall \, \mathbf{k}\}$ of the original algebra induced by the Bogoliubov generator which can thus be thought of as a generator of the group of automorphisms of $\oplus_{\mathbf{k}} su(1,1)_{\mathbf{k}}$ parametrized by θ_k (we have a realization of the operator algebra at each θ, which can be implemented by Gel'fand–Naimark–Segal construction in the C*-algebra formalism [123,505]). Notice that the various copies become unitarily inequivalent in the infinite volume limit, as shown by Eqs. (5.42): the space of the states splits into unitarily inequivalent representations of the CCR, each one labeled by the parameter θ. As usual, one works at finite volume and only at the end of the computations the limit $V \to \infty$ is performed.

The state $|0(\theta)\rangle$ in Eq. (5.41) is a condensate of pairs of particles $a_{\mathbf{k}}$ and $\tilde{a}_{\mathbf{k}}$, so that the so-called [617, 619] thermal vacuum condition holds

$$(a_{\mathbf{k}}^{\dagger} a_{\mathbf{k}} - \tilde{a}_{\mathbf{k}}^{\dagger} \tilde{a}_{\mathbf{k}})|0(\theta)\rangle = 0\,. \tag{5.45}$$

Since the Hamiltonian (5.38) is invariant under the $SU(1,1)$ transformations, but the vacuum $|0,0\rangle$ is not invariant, as Eq. (5.43) shows, we have spontaneous breakdown of the $SU(1,1)$ symmetry. In particular, the vacuum is not invariant under the transformation induced by $\exp(i\theta\mathcal{G})$. The role of the NG particle is played in this case by the couple of a and $\tilde{a}$ particles condensed in the vacuum $|0(\theta)\rangle$.

We also observe that since $a_{\mathbf{k}}(\theta)|0(\theta)\rangle = 0 = \tilde{a}_{\mathbf{k}}(\theta)|0(\theta)\rangle$, creation or annihilation of an $a_{\mathbf{k}}$ particle in the state $|0(\theta)\rangle$ is equivalent, up to a c-number factor, to the annihilation or creation, respectively, of a tilde $\tilde{a}_{\mathbf{k}}$ particle in the same state (cf., e.g., Eq. (5.36)). The $\tilde{a}_{\mathbf{k}}$ particle can thus be thought of as the "hole", or the anti-particle, of the $a_{\mathbf{k}}$ particle. When considering non-equilibrium systems or dissipative systems, the energy flowing out of the a-system can be shown to be flowing into the $\tilde{a}$-system [153, 617, 619]. This last one may therefore represent the thermal bath or the environment into which the a-system is embedded.

In conclusion, in the infinite volume limit $|0(\theta)\rangle$ becomes orthogonal to $|0,0\rangle$ and we have that the whole Hilbert space $\{|0(\theta)\rangle\}$, constructed by operating on $|0(\theta)\rangle$ with $a_{\mathbf{k}}^{\dagger}(\theta)$ and $\tilde{a}_{\mathbf{k}}^{\dagger}(\theta)$, is asymptotically (i.e., in the infinite volume limit) orthogonal to the space generated over $\{|0,0\rangle\}$. In general, for each value of $\theta(\beta)$, i.e., for each value of the temperature $T = \frac{1}{k_B\beta}$, one obtains in the infinite volume limit a representation of the CCR unitarily inequivalent to the other ones, associated with different values of T. In other words, the parameter $\theta(\beta)$ (or the temperature T) acts as a label for the inequivalent representations [153].

A similar construction [600] can be carried out for fermion operators, here denoted as α, $\tilde{\alpha}$. Their anticommutators are

$$\{\alpha_{\mathbf{k}}, \alpha_{\mathbf{p}}^{\dagger}\} = \delta_{\mathbf{k},\mathbf{p}} = \{\tilde{\alpha}_{\mathbf{k}}, \tilde{\alpha}_{\mathbf{p}}^{\dagger}\}\,, \tag{5.46}$$

with all the other anticommutators vanishing. The thermal Bogoliubov transformations are now:

$$\alpha_{\mathbf{k}}(\theta) = e^{i\mathcal{G}_F}\alpha_{\mathbf{k}}e^{-i\mathcal{G}_F} = \alpha_{\mathbf{k}}\cos\theta_{\mathbf{k}} - \tilde{\alpha}_{\mathbf{k}}^{\dagger}\sin\theta_{\mathbf{k}}\,, \tag{5.47a}$$

$$\tilde{\alpha}_{\mathbf{k}}(\theta) = e^{i\mathcal{G}_F}\tilde{\alpha}_{\mathbf{k}}e^{-i\mathcal{G}_F} = \tilde{\alpha}_{\mathbf{k}}\cos\theta_{\mathbf{k}} + \alpha_{\mathbf{k}}^{\dagger}\sin\theta_{\mathbf{k}}\,, \tag{5.47b}$$

with generator $\mathcal{G}_F$ given by

$$\mathcal{G}_F = -i\sum_{\mathbf{k}}\theta_k\left(\alpha_{\mathbf{k}}^{\dagger}\tilde{\alpha}_{\mathbf{k}}^{\dagger} - \tilde{\alpha}_{\mathbf{k}}\alpha_{\mathbf{k}}\right)\,. \tag{5.48}$$

The fermionic thermal vacuum has the form of a $SU(2)$ generalized coherent state [519]:

$$|0(\theta)\rangle = e^{i\mathcal{G}_F}|0\rangle = \prod_{\mathbf{k}}\left(\cos\theta_k + \sin\theta_k \alpha_{\mathbf{k}}^\dagger \tilde{\alpha}_{\mathbf{k}}^\dagger\right)|0\rangle, \qquad (5.49)$$

with

$$\alpha_{\mathbf{k}}(\theta)|0(\theta)\rangle = \tilde{\alpha}_{\mathbf{k}}(\theta)|0(\theta)\rangle = 0. \qquad (5.50)$$

We also have

$$\mathcal{N}_{\alpha_k}(\theta) = \langle 0(\theta)|\alpha_{\mathbf{k}}^\dagger \alpha_{\mathbf{k}}|0(\theta)\rangle = \sin^2\theta_k = \frac{1}{e^{\beta\omega_k}+1}, \qquad (5.51)$$

with a similar result for the tilde particles.

Let us now omit again the momentum suffices for simplicity. It is interesting to observe that, for bosons, the operator $N - \tilde{N} \equiv a^\dagger a - \tilde{a}^\dagger \tilde{a}$ is invariant under the Bogoliubov transformations (5.36), i.e.:

$$a^\dagger a - \tilde{a}^\dagger \tilde{a} = a^\dagger(\theta)a(\theta) - \tilde{a}(\theta)^\dagger \tilde{a}(\theta), \quad \forall\, \theta. \qquad (5.52)$$

Thus, $\frac{\delta}{\delta\theta}(N(\theta) - \tilde{N}(\theta)) = 0$, with $N(\theta) - \tilde{N}(\theta) \equiv a^\dagger(\theta)a(\theta) - \tilde{a}^\dagger(\theta)\tilde{a}(\theta)$, consistently with the fact that $\frac{1}{4}(N - \tilde{N})^2$ is the Casimir operator of the $su(1,1)$ algebra closed by the generators

$$J_1 \equiv \frac{1}{2}(a^\dagger \tilde{a}^\dagger + a\tilde{a}), \quad J_2 \equiv \frac{1}{2}\mathcal{G}, \quad J_3 \equiv \frac{1}{2}(N + \tilde{N} + 1). \qquad (5.53)$$

In the fermion case,

$$J_{F1} \equiv \frac{1}{2}\mathcal{G}_F, \quad J_{F2} \equiv \frac{1}{2}(a^\dagger \tilde{a}^\dagger + a\tilde{a}), \quad J_{F3} \equiv \frac{1}{2}(N_F + \tilde{N}_F - 1), \quad (5.54)$$

close the algebra $su(2)$. Also in this case $\frac{\delta}{\delta\theta}(N_F(\theta) - \tilde{N}_F(\theta)) = 0$, with $N_F(\theta) - \tilde{N}_F(\theta) \equiv \alpha^\dagger(\theta)\alpha(\theta) - \tilde{\alpha}^\dagger(\theta)\tilde{\alpha}(\theta) = N_F - \tilde{N}_F$, again consistently with the fact that $\frac{1}{4}(N_F - \tilde{N}_F)^2$ is related to the $su(2)$ Casimir operator.

The normal ordering

We have seen that the thermal vacuum $|0(\theta(\beta))\rangle$ is not annihilated by $a_{\mathbf{k}}$ and $\tilde{a}_{\mathbf{k}}$. Therefore, the normal ordered product of, say, the operators $a_{\mathbf{k}}$'s does not have zero expectation value in the state $|0(\theta(\beta))\rangle$, as Eq. (5.43) shows. Indeed, use of Eq. (5.36) (where the subscript $\mathbf{k}$ is introduced) gives

$$: a_{\mathbf{k}} a_{\mathbf{q}}^\dagger : = a_{\mathbf{q}}^\dagger a_{\mathbf{k}} = a_{\mathbf{q}}(\theta)^\dagger a_{\mathbf{k}}(\theta)\cosh\theta_q \cosh\theta_k$$
$$+ a_{\mathbf{q}}^\dagger(\theta)\tilde{a}_{\mathbf{k}}^\dagger(\theta)\cosh\theta_q \sinh\theta_k + \tilde{a}_{\mathbf{q}}(\theta)a_{\mathbf{k}}(\theta)\sinh\theta_q \cosh\theta_k$$
$$+ \tilde{a}_{\mathbf{q}}(\theta)\tilde{a}_{\mathbf{k}}^\dagger(\theta)\sinh\theta_q \sinh\theta_k. \qquad (5.55)$$

As a matter of fact, normal ordering is defined with respect to a specific representation [90, 438]: thus, normal ordering with respect to the representation built on the vacuum $|0\rangle$ is not normal ordering with respect to the one built on $|0(\theta(\beta))\rangle$:

$$\langle 0| : a_{\mathbf{k}} a_{\mathbf{q}}^\dagger : |0\rangle = 0 \,, \tag{5.56a}$$

$$\langle 0(\theta)| : a_{\mathbf{k}} a_{\mathbf{q}}^\dagger : |0(\theta)\rangle = \sinh^2 \theta_k \cdot \delta(\mathbf{k} - \mathbf{q}) \,, \tag{5.56b}$$

We will denote by $: A :$ and by $: A :_{\theta(\beta)}$ normal ordering of any product A of a (or $\tilde{a}$) operators with respect to $|0\rangle$ and $|0(\theta(\beta))\rangle$, respectively:

$$: a_{\mathbf{k}}(\theta) a_{\mathbf{k}}^\dagger(\theta) :_{\theta(\beta)} = a_{\mathbf{k}}^\dagger(\theta) a_{\mathbf{k}}(\theta) \,. \tag{5.57}$$

Thus, we obtain

$$: a_{\mathbf{k}} a_{\mathbf{k}}^\dagger : = : a_{\mathbf{k}} a_{\mathbf{k}}^\dagger :_{\theta(\beta)} + \sinh^2 \theta_k \cdot \delta(\mathbf{k} - \mathbf{q}) \,. \tag{5.58}$$

We finally observe that due to Eq. (5.52) the two operations $: :$ and $: :_{\theta(\beta)}$ give the same result for combinations of the form $a_{\mathbf{k}} a_{\mathbf{k}}^\dagger - \tilde{a}_{\mathbf{k}} \tilde{a}_{\mathbf{k}}^\dagger$:

$$: a_{\mathbf{k}} a_{\mathbf{k}}^\dagger - \tilde{a}_{\mathbf{k}} \tilde{a}_{\mathbf{k}}^\dagger : = a_{\mathbf{k}}^\dagger(\theta) a_{\mathbf{k}}(\theta) - \tilde{a}_{\mathbf{k}}(\theta)^\dagger \tilde{a}_{\mathbf{k}}(\theta) = : a_{\mathbf{k}} a_{\mathbf{k}}^\dagger - \tilde{a}_{\mathbf{k}} \tilde{a}_{\mathbf{k}}^\dagger :_{\theta(\beta)} \,. \tag{5.59}$$

for any θ. Normal ordering of mixed product of $a_{\mathbf{k}}$ and $\tilde{a}_{\mathbf{k}}$ is also to be defined according to the reference vacuum, since results are in general different for different $\theta(\beta)$'s (see also Appendix M).

The tilde-conjugation rule

We now recall the so-called "tilde-conjugation rules" which are defined in TFD. In the following, again we omit for simplicity the subscript $\mathbf{k}$.

For any two bosonic (respectively, fermionic) operators $\mathcal{O}$ and $\mathcal{O}'$ and any two c-numbers α and β the tilde-conjugation rules of TFD are postulated to be the following [369, 600, 617, 619]:

$$(\mathcal{O}\mathcal{O}')\tilde{} = \tilde{\mathcal{O}}\tilde{\mathcal{O}}' \,, \tag{5.60a}$$

$$(\alpha\mathcal{O} + \beta\mathcal{O}')\tilde{} = \alpha^* \tilde{\mathcal{O}} + \beta^* \tilde{\mathcal{O}}' \,, \tag{5.60b}$$

$$(\mathcal{O}^\dagger)\tilde{} = \tilde{\mathcal{O}}^\dagger \,, \tag{5.60c}$$

$$(\tilde{\mathcal{O}})\tilde{} = \mathcal{O} \,. \tag{5.60d}$$

According to (5.60) the tilde-conjugation does not change the order among operators. Furthermore, it is required that tilde and non-tilde operators are mutually commuting (or anti-commuting) operators and that the thermal vacuum $|0(\theta)\rangle$ is invariant under tilde-conjugation:

$$[\mathcal{O}, \tilde{\mathcal{O}}']_{\mp} = 0 = [\mathcal{O}, \tilde{\mathcal{O}}'^\dagger]_{\mp} \,, \tag{5.61}$$

$$|0(\theta))^{\sim} = |0(\theta))\,. \tag{5.62}$$

It is useful to introduce the label σ defined by $\sqrt{\sigma} \equiv +1$ for bosons and $\sqrt{\sigma} \equiv +i$ for fermions. We shall therefore simply write commutators as $[\mathcal{O}, \mathcal{O}']_{-\sigma} \doteq \mathcal{O}\mathcal{O}' - \sigma\mathcal{O}'\mathcal{O}$, and $(\mathbb{1} \otimes \mathcal{O})(\mathcal{O}' \otimes \mathbb{1}) \equiv \sigma(\mathcal{O}' \otimes \mathbb{1})(\mathbb{1} \otimes \mathcal{O})$, without further specification of whether $\mathcal{O}$ and $\mathcal{O}'$ (which are equal products of a, $a^\dagger$ in all possible ways) are fermions or bosons.

Upon identifying from now on $a_1 \equiv a$, $a_1^\dagger \equiv a^\dagger$, one easily checks that the TFD tilde-operators (consistent with (5.60) – (5.62)) are straightforwardly recovered by setting $a_2 \equiv \tilde{a}$, $a_2^\dagger \equiv \tilde{a}^\dagger$. According to such identification, it is the action of the $1 \leftrightarrow 2$ permutation π: $\pi a_i = a_j$, $i \neq j$, $i, j = 1, 2$, that realizes the operation of "tilde-conjugation" defined in (5.60):

$$\pi a_1 = \pi(a \otimes \mathbb{1}) = \mathbb{1} \otimes a = a_2 \equiv \tilde{a} \equiv (a)^{\sim}\,, \tag{5.63a}$$

$$\pi a_2 = \pi(\mathbb{1} \otimes a) = a \otimes \mathbb{1} = a_1 \equiv a \equiv (\tilde{a})^{\sim}\,. \tag{5.63b}$$

In particular, since the permutation π is involutive, i.e., $\pi^2 = \mathbb{1}$, tilde-conjugation also turns out to be involutive, as in fact required by the rule (5.60). Notice that, as $(\pi a_i)^\dagger = \pi(a_i^\dagger)$, it is also $((a_i)^{\sim})^\dagger = ((a_i)^\dagger)^{\sim}$, i.e., tilde-conjugation commutes with hermitian conjugation. Furthermore, from (5.63), one has

$$(ab)^{\sim} = [(a \otimes \mathbb{1})(b \otimes \mathbb{1})]^{\sim} = (ab \otimes \mathbb{1})^{\sim} = \mathbb{1} \otimes ab = (\mathbb{1} \otimes a)(\mathbb{1} \otimes b) = \tilde{a}\tilde{b}\,. \tag{5.64}$$

Rules (5.60) are thus obtained. (5.61) is insured by the σ-commutativity of a_1 and a_2. The vacuum of TFD, $|0(\theta)\rangle$, is a condensed state of an equal number of tilde and non-tilde particles [369, 600, 617, 619], thus (5.62) requires no further conditions: Eqs. (5.63) are sufficient to show that the rule (5.62) is satisfied. Finally, the tilde-rule (5.60) can also be recovered [141] as we will mention in the following Section. There we discuss the q-deformed algebraic structure underlying TFD.

5.3.1 *The propagator structure in TFD*

Let us now consider the simplest situation of a (real) scalar boson field in thermal equilibrium. This is sufficient to illustrate the propagator structure of a thermal theory. For fermion fields see [617, 619]. We have:

$$\phi(x) = \int \frac{d^3k}{(2\pi)^{\frac{3}{2}}(2\omega_k)^{\frac{1}{2}}} \left[a_\mathbf{k}(t)e^{i\mathbf{kx}} + a_\mathbf{k}^\dagger(t)e^{-i\mathbf{kx}} \right]\,, \tag{5.65a}$$

$$\tilde{\phi}(x) = \int \frac{d^3k}{(2\pi)^{\frac{3}{2}}(2\omega_k)^{\frac{1}{2}}} \left[\tilde{a}_\mathbf{k}(t)e^{-i\mathbf{kx}} + \tilde{a}_\mathbf{k}^\dagger(t)e^{i\mathbf{kx}} \right]\,, \tag{5.65b}$$

where $a_{\mathbf{k}}(t) = e^{-i\omega_k t} a_{\mathbf{k}}$. $\phi(x)$ and $\tilde{\phi}(x)$ commute and

$$[\phi(t, \mathbf{x}), \partial_t \phi(t, \mathbf{x}')] = i\delta(\mathbf{x} - \mathbf{x}'), \qquad (5.66a)$$

$$\left[\tilde{\phi}(t, \mathbf{x}), \partial_t \tilde{\phi}(t, \mathbf{x}')\right] = -i\delta(\mathbf{x} - \mathbf{x}'). \qquad (5.66b)$$

Here we consider on-shell energy for our fields

$$\omega_k = \sqrt{\mathbf{k}^2 + m^2}, \qquad (5.67)$$

where m is the particle mass. As we will see in the next Section, at finite temperature a perturbative approach based on (on-shell) free fields does not hold: nevertheless, we assume that the temperature is low enough so that we can consider excitations approximatively described by free fields.

In TFD, and more generally in a thermal field theory (TFT) [408], the two-point functions (propagators) have a matrix structure, arising from the various possible combinations of physical and tilde fields in the (thermal) vacuum expectation value. Notice that in TFD, although the physical and tilde particles are not coupled in the Hamiltonian $\hat{H}$, nevertheless they do couple in the vacuum state $|0(\theta)\rangle$. The finite temperature causal propagator for a free boson field $\phi(x)$ is[1]

$$D_0^{\alpha\beta}(x, x') = -i\langle 0(\theta)| T \left[\phi^\alpha(x) \phi^{\beta\dagger}(x') \right] |0(\theta)\rangle, \qquad (5.68)$$

where D_0 is the free field propagator, T denotes time ordering and the a, b indices refer to the thermal doublet $\phi^1 = \phi$, $\phi^2 = \tilde{\phi}^\dagger$. In the present case of a real field, we will use the above definition with $\phi^\dagger = \phi$.

A remarkable feature of the above propagator is that it can be cast (in momentum representation) in the following form [619]:

$$D_0^{\alpha\beta}(k_0, \mathbf{k}) = \left(\mathcal{B}_k^{-1} \left[\frac{1}{k_0^2 - (\omega_k - i\epsilon\tau_3)^2} \right] \mathcal{B}_k \tau_3 \right)^{\alpha\beta}, \qquad (5.69)$$

where τ_3 is the Pauli matrix $\mathrm{diag}(1, -1)$. The internal, or "core" matrix, is diagonal (cf. next Section) and coincides with the zero temperature Feynman propagator. Thus the thermal propagator is obtained from the zero temperature one by the action of the Bogoliubov matrix:

$$\mathcal{B}_k = \frac{1}{\sqrt{1 - f_k}} \begin{pmatrix} 1 & -\sqrt{f_k} \\ -\sqrt{f_k} & 1 \end{pmatrix} = \begin{pmatrix} \cosh\theta_k & -\sinh\theta_k \\ -\sinh\theta_k & \cosh\theta_k \end{pmatrix}, \qquad (5.70)$$

where, at the equilibrium, $f_k = e^{-\beta\omega_k}$ (the Boltzmann factor). The Bogoliubov matrix does affect only the imaginary part of $D_0(k)$; for example the

---

[1]Thermal propagators are here defined with a "-i" factor, in contrast with the definition given in Chapter 6.

$(1,1)$ component is $D_0^{11}(k) = \left[k_0^2 - (\omega_k - i\epsilon)^2\right]^{-1} - 2\pi i n_k \delta(k_0^2 - \omega_k^2)$, where n_k is the number of the a_k excitations.

Another fundamental property of the matrix propagator (5.68) is that only three elements are independent, since the linear relation holds:

$$D^{11} + D^{22} - D^{12} - D^{21} = 0. \tag{5.71}$$

This relation can be verified easily by using the annihilation of the thermal operators on $|0(\beta)\rangle$. Note also that the above relation has a more general validity, being true also for a different (gauge) parametrization of the thermal Bogoliubov matrix (see next Section).

5.3.2   *Non-hermitian representation of TFD*

The statistical average of a generic observable A can be written by means of the density matrix ρ, as [319]

$$\langle A \rangle = \frac{\mathrm{Tr}[A\rho]}{\mathrm{Tr}[\rho]} = \frac{\mathrm{Tr}[\rho^{(1-\alpha)}A\rho^{\alpha}]}{\mathrm{Tr}[\rho^{(1-\alpha)}\rho^{\alpha}]}, \tag{5.72}$$

where the cyclic property of the trace and the positiveness of ρ have been used. The parameter α is in the range $\alpha = [0,1]$. The operator ρ can be seen as a vector of a (doubled) Hilbert space, called Liouville space (see [319, 548, 628] for a detailed description of the Liouville space). The thermal average is then written as expectation value in the Liouville space as

$$\langle A \rangle = \frac{(\!(\rho^L \| A \| \rho^R)\!)}{(\!(\rho^L \| \rho^R)\!)}, \tag{5.73}$$

where, for a single oscillator, the left and right statistical states are coherent states given by

$$\| \rho^R)\!) = \exp\left(f^{\alpha} a^{\dagger} \tilde{a}^{\dagger}\right) \| 0,0)\!), \tag{5.74a}$$

$$(\!(\rho^L \| = (\!(0,0\| \exp\left(f^{(1-\alpha)} a \tilde{a}\right). \tag{5.74b}$$

Here f is a statistical weight factor and $\| 0,0)\!)$ is the vacuum state of Liouville space, annihilated by the $a, \tilde{a}$ operators. Note that the time evolution in Liouville space is controlled by $\hat{H} = \omega\left(a^{\dagger}a - \tilde{a}^{\dagger}\tilde{a}\right)$. In thermal equilibrium $f = e^{-\beta(\omega - \mu)}$ and for $\alpha = 1/2$, we recover the picture of Section 5.3.1: in particular the states $\| \rho^R)\!)$ and $(\!(\rho^L \|$ become hermitian conjugates. The choice $\alpha = 1/2$ is called in TFD the *symmetric gauge*.

Let us now introduce the thermal transformation as

$$\begin{pmatrix} \xi \\ \tilde{\xi}^\sharp \end{pmatrix} = \mathcal{B} \begin{pmatrix} a \\ \tilde{a}^\dagger \end{pmatrix}, \qquad \begin{pmatrix} \xi^\sharp \\ -\sigma\tilde{\xi} \end{pmatrix}^T = \begin{pmatrix} a^\dagger \\ -\sigma\tilde{a} \end{pmatrix}^T \mathcal{B}^{-1}, \qquad (5.75)$$

where $\sigma = 1$ for bosons and $\sigma = -1$ for fermions. The ξ operators satisfy canonical commutation relations

$$\left[\xi, \xi^\sharp\right]_\sigma = \xi\xi^\sharp - \sigma\xi^\sharp\xi = 1, \qquad \left[\tilde{\xi}, \tilde{\xi}^\sharp\right]_\sigma = 1, \qquad (5.76a)$$

$$\left[\tilde{\xi}, \xi^\sharp\right]_\sigma = 0, \qquad \left[\xi, \tilde{\xi}^\sharp\right]_\sigma = 0, \qquad (5.76b)$$

and the *thermal state condition*:

$$\xi\|\rho^R)) = 0, \qquad \tilde{\xi}\|\rho^R)) = 0, \qquad ((\rho^L\|\xi^\sharp = 0, \qquad ((\rho^L\|\tilde{\xi}^\sharp = 0. \qquad (5.77)$$

Again, in case of thermal equilibrium and in the symmetric gauge $\alpha = 1/2$, the above Bogoliubov matrix $\mathcal{B}$ coincides with that of Eq. (5.70), the ξ operators with the thermal operators $a(\theta)$ of Eq. (5.36), and the $\sharp$ conjugation reduces to the usual hermitian conjugation.

An important parametrization of the Bogoliubov matrix is that arising in the so-called *linear gauge*, corresponding to $\alpha = 1$ (actually, the Bogoliubov matrix also has a phase factor, which here is fixed conveniently). This is an essential choice in case of non-equilibrium systems [319], and in the following Sections we will therefore adopt $\alpha = 1$. In this gauge the Bogoliubov matrix reads

$$\mathcal{B} = \begin{pmatrix} 1 + \sigma n & -n \\ & \\ -\sigma & 1 \end{pmatrix}, \qquad (5.78)$$

where $n = \frac{f}{1-f}$ is the number density. With this choice of the Bogoliubov matrix the propagator defined in (5.68) still satisfies Eq. (5.71).

5.3.3   *TFD for fields with continuous mass spectrum*

In the previous Sections we have seen how to incorporate in an (operatorial) QFT framework, the statistical properties of a thermal system, and we have considered the case of a free neutral boson field in thermal equilibrium.

For interacting fields, however, a field theory at finite temperature requires additional considerations. It has indeed been stated that a perturbation theory in terms of asymptotically stable quasiparticle fails at finite temperature (and/or density) [495].

The point is that the spacetime symmetry group for $T > 0$ is not anymore the Poincaré group, as for $T = 0$: rather, the spacetime symmetry restricts to $SO(3) \otimes T_4$, i.e., the semidirect product of the rotation group and the four-dimensional translation group. Thus, while in the vacuum the representations of the Poincaré group are (stable) particles characterized by *one* continuous parameter ($0 \leq p < \infty$), at finite temperature the symmetry group admit representations labeled by *two* continuous parameters ($0 \leq p < \infty$ and $0 \leq E < \infty$). This means that for $T \neq 0$ the excitations do not live on the mass-shell $\omega_k = \sqrt{\mathbf{k}^2 + m^2}$, rather they have a continuous distribution of their energy around this value, i.e., they have a *spectral width*.

A formalism able to deal with such a situation has been developed [407] in terms of generalized free fields, containing an appropriate spectral function. Here we present a very short summary of such a formalism. The reader may find the details and the mathematical derivations in [319, 407].

For scalar uncharged bosons, generalized free fields have the following expansion [319]

$$\phi(x) = \int \frac{d^3k}{(2\pi)^{\frac{3}{2}}} \left(a_{\mathbf{k}}^\dagger(t) e^{-i\mathbf{k}\mathbf{x}} + a_{\mathbf{k}}(t) e^{i\mathbf{k}\mathbf{x}} \right), \qquad (5.79)$$

with a similar expression for $\tilde{\phi}$ and

$$\begin{pmatrix} a_{\mathbf{k}}(t) \\ \tilde{a}_{\mathbf{k}}^\dagger(t) \end{pmatrix} = \int_0^\infty dE \left[\mathcal{A}_B^+(E, \mathbf{k}) \right]^{\frac{1}{2}} \mathcal{B}^{-1}(n_B(E)) \begin{pmatrix} \xi_{E\mathbf{k}} \\ \tilde{\xi}_{E\mathbf{k}}^\sharp \end{pmatrix} e^{-iEt}, \qquad (5.80a)$$

$$\begin{pmatrix} a_{\mathbf{k}}^\dagger(t) \\ -\tilde{a}_{\mathbf{k}}(t) \end{pmatrix}^T = \int_0^\infty dE \left[\mathcal{A}_B^+(E, \mathbf{k}) \right]^{\frac{1}{2}} \begin{pmatrix} \xi_{E\mathbf{k}}^\sharp \\ -\tilde{\xi}_{E\mathbf{k}} \end{pmatrix}^T \mathcal{B}(n_B(E)) e^{iEt}, \qquad (5.80b)$$

where the Bogoliubov matrix is given, in linear gauge, by

$$\mathcal{B}(n_B(E)) = \begin{pmatrix} 1 + n_B(E) & -n_B(E) \\ -1 & 1 \end{pmatrix}, \qquad (5.81)$$

with $n_B(E)$ denoting the boson occupation number.

From Eqs. (5.79) and (5.80) we see that the generalized field ϕ contains an energy integration trough the above generalized Bogoliubov transformation. The weight function $\mathcal{A}_B^+(E, \mathbf{k})$ is the spectral function, containing the full information about the single particle spectrum. It is normalized as

$$\int_0^\infty dE E \mathcal{A}_B^+(E, \mathbf{k}) = \frac{1}{2}. \qquad (5.82)$$

The thermal quasiparticle operators ξ satisfy the thermal state conditions (5.77) for any E and $\mathbf{k}$ and obey the following commutation relations

$$\left[\xi_{E\mathbf{k}}, \xi_{E'\mathbf{k}'}^\sharp \right] = \delta(E - E') \delta(\mathbf{k} - \mathbf{k}') = \left[\tilde{\xi}_{E\mathbf{k}}, \tilde{\xi}_{E'\mathbf{k}'}^\sharp \right]. \qquad (5.83)$$

The equal-time commutation relations for the a operators are the usual ones

$$\left[a_{\mathbf{k}}(t), a^{\dagger}_{\mathbf{k}'}(t)\right] = \delta(\mathbf{k} - \mathbf{k}') = \left[\tilde{a}_{\mathbf{k}}(t), \tilde{a}^{\dagger}_{\mathbf{k}'}(t)\right] . \tag{5.84}$$

It is important to observe that the generalized field operators $\phi(x)$ are non-trivial objects: their commutation relations are in general not known [319]. However, since the quantities of interest are generally given in terms of Green's functions, it is sufficient to know these commutators at the level of expectation value [319].

The above spectral function $\mathcal{A}^{+}_{B}(E, \mathbf{k})$ can be continued to negative energies as

$$\mathcal{A}_{B}(E, \mathbf{k}) = \theta(E)\mathcal{A}^{+}_{B}(E, \mathbf{k}) - \theta(-E)\mathcal{A}^{+}_{B}(-E, -\mathbf{k}) . \tag{5.85}$$

In thermal equilibrium and for the real boson field under consideration, it holds $n_B(-E) = -(1 + n_B(E))$. Using this relation and the generalized field (5.79), the matrix boson propagator defined in (5.68) becomes

$$D^{ab}(k_0, \mathbf{k}) = \int_{-\infty}^{\infty} dE\, \mathcal{A}_{B}(E, \mathbf{k})\mathcal{B}^{-1}(n_B(E)) \left[\frac{1}{k_0 - E + i\epsilon\tau_3}\right] \mathcal{B}(n_B(E))\tau_3 , \tag{5.86}$$

which is again a diagonal matrix transformed under Bogoliubov matrices. It is important to stress that the above propagator is the *full* interacting propagator: all the information regarding interaction is contained in the spectral function. Observe also that the free particle propagator (5.69) is recovered in the limit

$$\mathcal{A}_{B}(E, \mathbf{k}) \to \text{sign}(E)\delta(E^2 - \omega_k^2) , \tag{5.87}$$

i.e., when the spectral function is peaked around the free particle energy.

Retarded and advanced propagators

The retarded and advanced propagators are respectively $D^R = D^{11} - D^{12}$ and $D^A = D^{11} - D^{21}$; they are given by dispersion integral in the spectral function

$$D^{R,A}(k_0, \mathbf{k}) = \Re e\, (D(k_0, \mathbf{k})) \mp i\pi \mathcal{A}_{B}(k_0, \mathbf{k}) = \int_{-\infty}^{\infty} dE\, \frac{\mathcal{A}_{B}(E, \mathbf{k})}{k_0 - E \pm i\epsilon} . \tag{5.88}$$

Thus the spectral function is proportional to the imaginary part of the retarded and advanced Green's functions:

$$\mathcal{A}_{B}(E, \mathbf{k}) = \mp \frac{1}{\pi}\Im m\, \left(D^{R,A}(E, \mathbf{k})\right) . \tag{5.89}$$

In thermal equilibrium the off-diagonal components of the full propagator D^{ab} satisfy the Kubo–Martin–Schwinger (KMS) boundary condition

$$(1 + n_B(k_0))\, D^{12}(k_0, \mathbf{k}) - n_B(k_0) D^{21}(k_0, \mathbf{k}) = 0\,. \qquad (5.90)$$

The above relation allows us to diagonalize the matrix propagator in the following form

$$D^{ab}(k_0, \mathbf{k}) = \mathcal{B}^{-1}(n_B(k_0)) \begin{pmatrix} D^R(k_0, \mathbf{k}) & 0 \\ 0 & D^A(k_0, \mathbf{k}) \end{pmatrix} \mathcal{B}(n_B(k_0))\tau_3\,. \quad (5.91)$$

It is important to stress that, in contrast to what happens in the symmetric gauge, in the linear gauge the propagator is not automatically diagonal; in general the propagator matrix is triangular. However, as we see from the above relations, it can be diagonalized in the case of thermal equilibrium, holding the KMS condition (5.90).

The above theoretical scheme now has to be related to the quantity specific of a given physical situation, i.e., to the self-energy. The knowledge of the self-energy indeed allows us to calculate the full propagator starting from the free one. This is done by using the Schwinger–Dyson equation, which is an integro-differential equation in coordinate space (see Section 6.4), reducing, however, for homogeneous systems to an algebraic equation in momentum space:

$$D(k) = D_0(k) + D_0(k)\Pi(k)D(k) \qquad (5.92)$$
$$= D_0(k) + D_0(k)\Pi(k)D_0(k) + D_0(k)\Pi(k)D_0(k)\Pi(k)D_0(k) + \dots$$

where $\Pi(k)$ is the boson self-energy (polarization tensor) in momentum representation. Since Eq. (5.71) is satisfied both for the free and the full propagator in the linear gauge, the relation for the matrix self-energy follows:

$$\Pi^{11} + \Pi^{22} + \Pi^{12} + \Pi^{21} = 0\,. \qquad (5.93)$$

This implies that Eq. (5.92) has the structure of a triangular matrix.

Let us now introduce the retarded and advanced self-energies $\Pi^R = \Pi^{11} + \Pi^{12}$ and $\Pi^A = \Pi^{22} + \Pi^{21}$, and their expression in terms of the self-energy spectral function σ

$$\Pi^{R,A}(k) = \Re e(\Pi(k)) \mp i\pi\sigma(k) = \int dE\, \frac{\sigma(E, \mathbf{k})}{k_0 - E \pm i\epsilon}\,. \qquad (5.94)$$

The Schwinger–Dyson equation (5.92) can then be solved [319, 407, 408]. The diagonal components give the result

$$A_B(E, \mathbf{k}) = \frac{\sigma(E, \mathbf{k})}{\left[E^2 - \omega_k^2 - \Re e\left(\Pi(E, \mathbf{k})\right)\right]^2 + \pi^2\sigma^2(E, \mathbf{k})}\,. \qquad (5.95)$$

The off-diagonal component instead relates the Bogoliubov parameter $n_B(E)$ to the environment through the relation

$$(n_B(E) + 1)\,\Pi^{12}(E, \mathbf{k}) - n_B(E)\Pi^{21}(E, \mathbf{k}) = 0\,, \qquad (5.96)$$

which is again the KMS condition (see Eq. (5.90)).

Thus the spectral function and the number density are determined from the above two equations. Of course, this is not sufficient to solve a particular problem, since one needs to calculate explicitly the self-energy. This can be done by perturbation theory at finite temperature [615, 617].

Up to now, the form of the spectral function is completely unspecified: we know, however, that the free particle limit of $\mathcal{A}_B(E, \mathbf{k})$ is a delta function (see Eq. (5.87)). Thus a reasonable approximated form for the bosonic spectral function $\mathcal{A}_B$ is the following expansion in simple poles in the complex E-plane around the quasiparticle energy $\omega_B = \sqrt{\mathbf{k}^2 + m_B^2}$:

$$
\begin{aligned}
\mathcal{A}_B(E, \mathbf{k}) &= \frac{1}{\pi} \frac{2E\gamma_B}{(E^2 - \gamma_B^2 - \omega_B^2)^2 + 4\gamma_B^2 E^2} \\
&= \frac{1}{4\pi i \omega_B} \left[\frac{1}{E - \omega_B - i\gamma_B} - \frac{1}{E - \omega_B + i\gamma_B} \right. \\
&\quad \left. + \frac{1}{E + \omega_B + i\gamma_B} - \frac{1}{E + \omega_B - i\gamma_B} \right].
\end{aligned}
\qquad (5.97)
$$

In such an approximation, the spectral function is parametrized by two quantities: the quasiparticle energy ω_B and the spectral width γ_B. Both these quantities can be in general momentum- and energy-dependent (in the case of non-homogeneous systems, they are also coordinate-dependent). Note also that the above expression (a Lorentzian curve) reduces to the delta function of the free case when $\gamma_B \to 0$, i.e., when the (quasi-)particle has an infinite lifetime.

5.4 The q-deformed Hopf algebra and the doubling of the field degrees of freedom

In this Section we show that the doubling of the degrees of freedom is intimately related to the structure of the space of the states in QFT [141, 189]. This brings us to consider the q-deformed Hopf algebra [150, 155, 207, 359, 393, 437]. In our presentation we follow closely [633]. We will show that the doubling of the degrees of freedom on which the TFD formalism is based finds its natural realization in the coproduct map.

One key ingredient of Hopf algebra [48] is the coproduct operation, i.e., the operator doubling implied by the coalgebra. The coproduct operation is indeed a map $\Delta : \mathcal{A} \to \mathcal{A} \otimes \mathcal{A}$, which duplicates the algebra $\mathcal{A}$. Coproducts are commonly used in the familiar addition of energy, momentum, angular momentum and of other so-called primitive operators. The coproduct of a generic operator $\mathcal{O}$ is a homomorphism defined as $\Delta\mathcal{O} = \mathcal{O} \otimes \mathbb{1} + \mathbb{1} \otimes \mathcal{O} \equiv \mathcal{O}_1 + \mathcal{O}_2$, with $\mathcal{O} \in \mathcal{A}$. Since additivity of observables such as energy, momentum, angular momentum, etc. is an essential requirement, the coproduct, and therefore the Lie–Hopf algebra structure, appears to provide an essential algebraic tool in QM and in QFT.

Thermal systems, and other systems where the duplication of the degrees of freedom reveals to be central in their treatment, are natural candidates to be described by the Lie–Hopf algebra. The remarkable result holds [141] according to which the infinitely many unitarily inequivalent representations of the CCR are classified by use of the *q-deformed* Hopf algebra. Quantum deformations of Hopf algebra thus have a deeply nontrivial physical meaning in QFT.

In the following we consider boson operators. The discussion and the conclusions can be easily extended to the case of fermion operators [141]. For notational simplicity we will omit the momentum suffix **k**.

Consider the bosonic algebra, called $h(1)$, generated by the set of operators $\{a, a^\dagger, H, N\}$ with commutation relations:
$$[a, a^\dagger] = 2H, \quad [N, a] = -a, \quad [N, a^\dagger] = a^\dagger, \quad [H, \bullet] = 0. \quad (5.98)$$
H is a central operator, constant in each representation and "$\bullet$" denotes any of the a, $a^\dagger$, H, N. The Casimir operator is given by $\mathcal{C} = 2NH - a^\dagger a$. The algebra $h(1)$ is a Hopf algebra, equipped with the coproduct operation, defined by

$$\Delta a = a \otimes \mathbb{1} + \mathbb{1} \otimes a \equiv a_1 + a_2, \qquad (5.99a)$$

$$\Delta a^\dagger = a^\dagger \otimes \mathbb{1} + \mathbb{1} \otimes a^\dagger \equiv a_1^\dagger + a_2^\dagger, \qquad (5.99b)$$

$$\Delta H = H \otimes \mathbb{1} + \mathbb{1} \otimes H \equiv H_1 + H_2, \qquad (5.99c)$$

$$\Delta N = N \otimes \mathbb{1} + \mathbb{1} \otimes N \equiv N_1 + N_2. \qquad (5.99d)$$

Note that $[a_i, a_j] = [a_i, a_j^\dagger] = 0$, $i, j = 1, 2$, $i \neq j$. The coproduct provides the prescription for operating on two modes. As mentioned, one familiar example of coproduct is the addition of the angular momentum J^α, $\alpha = 1, 2, 3$, of two particles: $\Delta J^\alpha = J^\alpha \otimes \mathbb{1} + \mathbb{1} \otimes J^\alpha \equiv J_1^\alpha + J_2^\alpha$, $J^\alpha \in SU(2)$.

The q-deformation of $h(1)$ is the Hopf algebra $h_q(1)$:
$$[a_q, a_q^\dagger] = [2H]_q, \quad [N, a_q] = -a_q, \quad [N, a_q^\dagger] = a_q^\dagger, \quad [H, \bullet] = 0,$$
$$(5.100)$$

where $N_q \equiv N$, $H_q \equiv H$ and $[x]_q = \dfrac{q^x - q^{-x}}{q - q^{-1}}$. The Casimir operator C_q is given by $C_q = N[2H]_q - a_q^\dagger a_q$. The deformed coproduct is defined by

$$\Delta a_q = a_q \otimes q^H + q^{-H} \otimes a_q \,, \tag{5.101a}$$

$$\Delta a_q^\dagger = a_q^\dagger \otimes q^H + q^{-H} \otimes a_q^\dagger \,, \tag{5.101b}$$

$$\Delta H = H \otimes \mathbb{1} + \mathbb{1} \otimes H \,, \tag{5.101c}$$

$$\Delta N = N \otimes \mathbb{1} + \mathbb{1} \otimes N \,, \tag{5.101d}$$

whose algebra is isomorphic with (5.100): $[\Delta a_q, \Delta a_q^\dagger] = [2\Delta H]_q$, etc. Note that $h_q(1)$ is a structure different from the commonly considered q-deformation of the harmonic oscillator [75, 433] (cf. Section 1.8).

We denote by $\mathcal{F}_1$ the single mode Fock space, i.e., the fundamental representation $H = 1/2$, $C = 0$. In such a representation $h(1)$ and $h_q(1)$ coincide as it happens for $su(2)$ and $su_q(2)$ for the spin-$\frac{1}{2}$ representation. The differences appear in the coproduct and in the higher spin representations.

As is customary, we require that a and $a^\dagger$, and a_q and $a_q{}^\dagger$, are adjoint operators. This implies that q can only be real (or of modulus one in the fermionic case; in the two mode Fock space $\mathcal{F}_2 = \mathcal{F}_1 \otimes \mathcal{F}_1$, for $|q| = 1$, the hermitian conjugation of the coproduct must be supplemented by the inversion of the two spaces for consistency with the coproduct isomorphism).

Summarizing, one can write for both bosons (and fermions) on $\mathcal{F}_2 = \mathcal{F}_1 \otimes \mathcal{F}_1$:

$$\Delta a = a_1 + a_2 \,, \qquad \Delta a^\dagger = a_1^\dagger + a_2^\dagger \,, \tag{5.102a}$$

$$\Delta a_q = a_1 q^{1/2} + q^{-1/2} a_2 \,, \qquad \Delta a_q^\dagger = a_1^\dagger q^{1/2} + q^{-1/2} a_2^\dagger \,, \tag{5.102b}$$

$$\Delta H = 1 \,, \qquad \Delta N = N_1 + N_2 \,. \tag{5.102c}$$

Now, the key point is [141] that the full set of infinitely many unitarily inequivalent representations of the CCR in QFT are classified by use of the q-deformed Hopf algebra. Since Bogoliubov transformations relate different (i.e., unitarily inequivalent) representations, it is sufficient to show that the Bogoliubov transformations are directly obtained by use of the deformed coproduct operation. We consider therefore the following operators (cf. (5.101) with $q(\theta) \equiv e^{2\theta}$ and $H = 1/2$):

$$\alpha_{q(\theta)} \equiv \frac{\Delta a_q}{\sqrt{[2]_q}} = \frac{1}{\sqrt{[2]_q}} (e^\theta a_1 + e^{-\theta} a_2) \,, \tag{5.103a}$$

$$\beta_{q(\theta)} \equiv \frac{1}{\sqrt{[2]_q}} \frac{\delta}{\delta \theta} \Delta a_q = \frac{2q}{\sqrt{[2]_q}} \frac{\delta}{\delta q} \Delta a_q = \frac{1}{\sqrt{[2]_q}} (e^\theta a_1 - e^{-\theta} a_2) \,, \tag{5.103b}$$

and h.c. A set of commuting operators with CCR is given by

$$\alpha(\theta) \equiv \frac{\sqrt{[2]_q}}{2\sqrt{2}}[\alpha_{q(\theta)} + \alpha_{q(-\theta)} - \beta^\dagger_{q(\theta)} + \beta^\dagger_{q(-\theta)}], \qquad (5.104a)$$

$$\beta(\theta) \equiv \frac{\sqrt{[2]_q}}{2\sqrt{2}}[\beta_{q(\theta)} + \beta_{q(-\theta)} - \alpha^\dagger_{q(\theta)} + \alpha^\dagger_{q(-\theta)}], \qquad (5.104b)$$

and h.c. One then introduces [141]

$$A(\theta) \equiv \frac{1}{\sqrt{2}}(\alpha(\theta) + \beta(\theta)) = A\cosh\theta - B^\dagger\sinh\theta , \qquad (5.105a)$$

$$B(\theta) \equiv \frac{1}{\sqrt{2}}(\alpha(\theta) - \beta(\theta)) = B\cosh\theta - A^\dagger\sinh\theta , \qquad (5.105b)$$

with

$$[A(\theta), A^\dagger(\theta)] = 1 , \qquad [B(\theta), B^\dagger(\theta)] = 1 . \qquad (5.106)$$

All other commutators are equal to zero and $A(\theta)$ and $B(\theta)$ commute among themselves. Eqs. (5.105) are nothing but the Bogoliubov transformations for the (A, B) pair. In other words, Eqs. (5.105) show that the Bogoliubov-transformed operators $A(\theta)$ and $B(\theta)$ are linear combinations of the co-product operators defined in terms of the deformation parameter $q(\theta)$ and of their θ-derivatives.

From this point on, by setting $\theta = \theta(\beta)$, one can reconstruct the TFD formalism presented in the previous Section.

The generator of (5.105) is $\mathcal{G} \equiv -i(A^\dagger B^\dagger - AB)$:

$$-i\frac{\delta}{\delta\theta}A(\theta) = [\mathcal{G}, A(\theta)] , \qquad -i\frac{\delta}{\delta\theta}B(\theta) = [\mathcal{G}, B(\theta)] , \qquad (5.107)$$

and h.c. Compare this generator with $\mathcal{G}$ in Eq. (5.33). By taking derivatives with respect to θ of the Bogoliubov transformed operators, one can show that the tilde-rule (5.60) can also be recovered [141].

Let $|0\rangle \equiv |0\rangle \otimes |0\rangle$ denote the vacuum annihilated by A and B, $A|0\rangle = 0 = B|0\rangle$. By introducing the suffix $\mathbf{k}$ (till now omitted for simplicity), at finite volume V one obtains

$$|0(\theta)\rangle = e^{i\sum_{\mathbf{k}}\theta_k \mathcal{G}_{\mathbf{k}}}|0\rangle = \prod_{\mathbf{k}}\frac{1}{\cosh\theta_k}\exp(\tanh\theta_k A^\dagger_{\mathbf{k}}B^\dagger_{\mathbf{k}})|0\rangle , \qquad (5.108)$$

to be compared with Eq. (5.34). θ denotes the set $\{\theta_k = \frac{1}{2}\ln q_{\mathbf{k}}, \forall\mathbf{k}\}$ and $\langle 0(\theta)|0(\theta)\rangle = 1$. The underlying group structure is $\bigotimes_{\mathbf{k}} SU(1,1)_{\mathbf{k}}$ and the vacuum $|0(\theta)\rangle$ is an $SU(1,1)$ generalized coherent state [519] (see Appendix C). The q-deformed Hopf algebra is thus intrinsically related to coherence and to the vacuum structure in QFT.

In the infinite volume limit, the number of degrees of freedom becomes uncountable infinite, and thus one obtains [153,617,619] $\langle 0(\theta)|0(\theta')\rangle \to 0$ as $V \to \infty$, $\forall\, \theta, \theta'$, $\theta \neq \theta'$. By denoting with $\mathcal{H}_\theta$ the Hilbert space with vacuum $|0(\theta)\rangle$, $\mathcal{H}_\theta \equiv \{|0(\theta)\rangle\}$, this means that $\mathcal{H}_\theta$ and $\mathcal{H}_{\theta'}$ become unitarily inequivalent. In this limit, the "points" of the space $\mathcal{H} \equiv \{\mathcal{H}_\theta, \forall\, \theta\}$ of the infinitely many unitarily inequivalent representations of the CCR are labeled by the deformation parameter θ [141,153]. The space $\mathcal{H} \equiv \{\mathcal{H}_\theta, \forall\, \theta\}$ is called the space of the representations.

We observe that $p_\theta \equiv -i\dfrac{\delta}{\delta\theta}$ can be regarded [141] as the momentum operator "conjugate" to the "degree of freedom" θ. For an assigned fixed value $\bar{\theta}$, it is

$$e^{i\bar{\theta}p_\theta}A(\theta) = e^{i\bar{\theta}\mathcal{G}}A(\theta)e^{-i\bar{\theta}\mathcal{G}} = A(\theta + \bar{\theta}), \qquad (5.109)$$

and similarly for $B(\theta)$. The "conjugate thermal momentum" p_θ generates transitions among inequivalent (in the infinite volume limit) representations: $\exp(i\bar{\theta}p_\theta)\,|0(\theta)\rangle = |0(\theta + \bar{\theta})\rangle$. The notion of thermal degree of freedom [615] thus acquires formal definiteness in the sense of the canonical formalism. Notice that the derivative with respect to the θ parameter is actually a derivative with respect to the system temperature T. This sheds some light on the role of θ in thermal field theories for non-equilibrium systems and phase transitions. We shall comment more on this point in the following Sections.

It is interesting to consider the case of time-dependent deformation parameter. This immediately relates to the dissipative systems [153]. The Heisenberg equation for $A(t, \theta(t))$ is

$$-i\dot{A}(t, \theta(t)) = -i\frac{\delta}{\delta t}A(t, \theta(t)) - i\frac{\delta\theta}{\delta t}\frac{\delta}{\delta\theta}A(t, \theta(t))$$

$$= [H, A(t, \theta(t))] + \frac{\delta\theta}{\delta t}[\mathcal{G}, A(t, \theta(t))] = [H + Q, A(t, \theta(t))], \quad (5.110)$$

and $Q \equiv \dfrac{\delta\theta}{\delta t}\mathcal{G}$ plays the role of the heat-term in dissipative systems. H is the Hamiltonian responsible for the time variation in the explicit time dependence of $A(t, \theta(t))$. $H + Q$ can be therefore identified with the free energy [153]: variations in time of the deformation parameter involve dissipation. In thermal theories and in dissipative systems the doubled modes B play the role of the thermal bath or environment [153,633] (see Chapter 9).

Summarizing, we know that QFT is characterized by the existence of unitarily inequivalent representations of the CCR, as seen in the previous

Chapters, which are related among themselves by the Bogoliubov transformations. These, as seen above, are obtained as linear combinations of the deformed coproduct maps which express the doubling of the degrees of freedom. Therefore one may conclude that the intrinsic algebraic structure of QFT (independent of the specificity of the system under study) is the one of the q-deformed Hopf algebra. The unitarily inequivalent representations existing in QFT are related and labeled by means of such an algebraic structure.

It should be stressed that the coproduct map is also essential in QM in order to deal with a many mode system (typically, with identical particles). However, in QM all the representations of the CCR are unitarily equivalent and therefore the Bogoliubov transformations induce unitary transformations among the representations, thus preserving their physical content. The q-deformed Hopf algebra therefore does not have that physical relevance in QM, which it has, on the contrary, in QFT. Here, the representations of the CCR, related through Bogoliubov representations, are unitarily *inequivalent* and therefore physically inequivalent: they represent different physical phases of the system corresponding to different boundary conditions, such as, for example, the system temperature. Typical examples are the superconducting and the normal phase, the ferromagnetic and the non-magnetic (i.e., zero magnetization) phase, the crystal and the gaseous phase, etc. The physical meaning of the deformation parameter q in terms of which unitarily inequivalent representations of the CCR are labeled is thus recognized.

When the above discussion is applied to non-equilibrium (e.g., thermal and/or dissipative) field theories it appears that the couple of conjugate variables θ and $p_\theta \equiv -i\frac{\partial}{\partial\theta}$, with $\theta = \theta(\beta(t))$ ($\beta(t) = \frac{1}{k_B T(t)}$), related to the q-deformation parameter, describe trajectories in the space $\mathcal{H}$ of the representations. In [632] it has been shown that there is a symplectic structure associated to the "degrees of freedom" θ and that the trajectories in the $\mathcal{H}$ space may exhibit properties typical of chaotic trajectories in classical non-linear dynamics. We will discuss this in Section 5.8. In the next Section we present further characterizations of the vacuum structure in TFD.

5.5   Free energy, entropy and the arrow of time. Intrinsic thermal nature of QFT

The state $|0(\theta)\rangle$ in Eq. (5.41) may be written as:

$$|0(\theta)\rangle = \exp\left(-\frac{1}{2}S_a\right)|\mathcal{I}\rangle = \exp\left(-\frac{1}{2}S_{\tilde{a}}\right)|\mathcal{I}\rangle, \qquad (5.111)$$

$$S_a \equiv -\sum_{\mathbf{k}}\left(a_{\mathbf{k}}^{\dagger}a_{\mathbf{k}}\log\sinh^2\theta_k - a_{\mathbf{k}}a_{\mathbf{k}}^{\dagger}\log\cosh^2\theta_k\right). \qquad (5.112)$$

Here $|\mathcal{I}\rangle \equiv \exp\left(\sum_{\mathbf{k}} a_{\mathbf{k}}^{\dagger}\tilde{a}_{\mathbf{k}}^{\dagger}\right)|0\rangle$ and $S_{\tilde{a}}$ is given by an expression similar to S_a, with $\tilde{a}_{\mathbf{k}}$ and $\tilde{a}_{\mathbf{k}}^{\dagger}$ replacing $a_{\mathbf{k}}$ and $a_{\mathbf{k}}^{\dagger}$, respectively. Eq. (5.111) has been obtained by using the relation

$$e^{-\frac{1}{2}S_a}\, a_{\mathbf{k}}^{\dagger}\, e^{\frac{1}{2}S_a} = a_{\mathbf{k}}^{\dagger}\tanh\theta_k. \qquad (5.113)$$

We can also write [153, 617, 619]:

$$|0(\theta)\rangle = \sum_{n=0}^{+\infty}\sqrt{W_n}\,(|n\rangle\otimes|n\rangle), \qquad (5.114a)$$

$$W_n = \prod_{\mathbf{k}}\frac{\sinh^{2n_{\mathbf{k}}}\theta_k}{\cosh^{2(n_{\mathbf{k}}+1)}\theta_k}, \qquad (5.114b)$$

where n denotes the set $\{n_{\mathbf{k}}\}$, $0 < W_n < 1$ and $\sum_{n=0}^{+\infty} W_n = 1$. Then

$$\mathcal{S}_a \equiv \langle 0(\theta)|S_a|0(\theta)\rangle = \sum_{n=0}^{+\infty} W_n\,\log W_n. \qquad (5.115)$$

It is useful [153, 617, 619] to introduce the functional $\mathcal{F}_a$ for the a-modes

$$\mathcal{F}_a \equiv \langle 0(\theta)|\left(H_a - \frac{1}{\beta}S_a\right)|0(\theta)\rangle, \qquad (5.116)$$

where β is the inverse temperature ($\beta = \frac{1}{k_B T}$) and $H_a = \sum_{\mathbf{k}} \hbar\omega_k a_{\mathbf{k}}^{\dagger}a_{\mathbf{k}}$.

The parameter θ may depend on time: $\theta_k = \theta_k(\beta(t))$. One then considers the extremal condition $\frac{\partial \mathcal{F}_a}{\partial \theta_k} = 0, \forall\, \mathbf{k}$, to be satisfied in each representation, and using $E_k \equiv \hbar\omega_k$, one finds

$$\mathcal{N}_{a_{\mathbf{k}}}(t) = \sinh^2\theta_k = \frac{1}{e^{\beta(t)E_k}-1}, \qquad (5.117)$$

which is the Bose distribution for $a_{\mathbf{k}}$ at time t. Inspection of Eqs. (5.112)–(5.116) then suggests that $\mathcal{F}_a$ and $\mathcal{S}_a$ can be interpreted as the *free energy* and the *entropy*, respectively.[2]

---

[2]The TFD entropy operator for fermions [600, 617, 619] is defined in a similar way to S_a, namely as: $S_\alpha \equiv -\sum_{\mathbf{k}}\left(\alpha_{\mathbf{k}}^{\dagger}\alpha_{\mathbf{k}}\log\sin^2\theta_k + \alpha_{\mathbf{k}}\alpha_{\mathbf{k}}^{\dagger}\log\cos^2\theta_k\right).$

We remark that in full generality one can proceed to the construction of $|0(\theta)\rangle$ for a generic θ parameter, without specifying any dependence of θ on temperature, or on any other quantity. One could have introduced Eq. (5.116) in a strict formal sense, assuming that there β is just a non-zero c-number parameter. Eq. (5.117) gives then the Bose–Einstein distribution only *provided* β is the inverse temperature. In this way, β could be recognized to be the inverse temperature and $|0(\theta)\rangle = |0(\theta(\beta))\rangle$ to be the thermal vacuum. Since the construction of $|0(\theta)\rangle$ is fully general and it amounts to labeling by the θ parameter the QFT unitarily inequivalent representations, one could reach in this way the conclusion that QFT is an intrinsically thermal field theory.

Assuming that θ is time-dependent, use of Eq. (5.111) shows that

$$\frac{\partial}{\partial t}|0(t)\rangle = -\left(\frac{1}{2}\frac{\partial S_a}{\partial t}\right)|0(t)\rangle. \qquad (5.118)$$

One thus sees that $i\left(\frac{1}{2}\hbar\frac{\partial S_a}{\partial t}\right)$ is the generator of time translations, namely time evolution is controlled by the entropy variations [153, 185]. It is remarkable that the same dynamical variable S_a whose expectation value is formally the entropy also controls time evolution: damping (or, more generally, dissipation) implies indeed the choice of a privileged direction in time evolution (*arrow of time*) with a consequent breakdown of time-reversal invariance.

One may also show that $d\mathcal{F}_a = dE_a - \frac{1}{\beta}d\mathcal{S}_a = 0$, which expresses the first principle of thermodynamics for a system coupled with environment at constant temperature and in absence of mechanical work. As usual, one may define heat as $dQ_a = \frac{1}{\beta}d\mathcal{S}_a$ and see that the change in time $d\mathcal{N}_a$ of particles condensed in the vacuum turns out into heat dissipation dQ_a:

$$dE_a = \sum_k \hbar\omega_k \dot{\mathcal{N}}_{a_k}(t)dt = \frac{1}{\beta}d\mathcal{S}_a = dQ_a. \qquad (5.119)$$

Here $\dot{\mathcal{N}}_{a_k}$ denotes the time derivative of $\mathcal{N}_{a_k}$.

We observe that the thermodynamic arrow of time, whose direction is defined by the increasing entropy direction, points in the same direction as the cosmological arrow of time, namely the inflating time direction for the expanding Universe [14, 19] (see also [444]). The concordance between the two arrows of time (and also with the psychological arrow of time, see [641]) is not at all granted and is a subject of an ongoing debate (see, e.g., [316]).

In Chapter 9 we show that quantum dissipation induces a dissipative phase interference [107], analogous to the Bohm–Aharonov phase [11, 23], and non-commutative geometry in the plane (x_+, x_-) [574].

Moreover, the generator of the Bogoliubov transformations discussed above is strictly related with the squeezed coherent states in quantum optics and with the quantum Brownian motion [107].

As a final comment we observe that the generator $\mathcal{G} = -i \sum_{\mathbf{k}} (a_{\mathbf{k}}^{\dagger} \tilde{a}_{\mathbf{k}}^{\dagger} - a_{\mathbf{k}} \tilde{a}_{\mathbf{k}})$, considered above and related to the entropy operator S_a, is nothing but the counter-rotating term in quantum optics Hamiltonian models. It is usually neglected in the rotating wave approximation when resonance conditions are met. From the above discussion, it appears, however, that it plays a crucial role in off-resonant regimes, where it is indeed related to the system entropy and phase transitions. For a detailed discussion on this point see [397, 398].

5.5.1   *Entropy and system-environment entanglement*

In the previous Section we have shown that the time evolution of the state $|0(\theta)\rangle$ is actually controlled by the entropy variations (cf. Eq. (5.118)). We will shortly comment on the entropy in this Section from a more general point of view, also in connection with the entanglement of the $a - \tilde{a}$ modes, since it appears as a structural aspect of QFT related with the existence of the unitarily inequivalent representations of the CCR.

The state $|0(\theta)\rangle$ in Eq. (5.41) can also be written as

$$|0(\theta)\rangle = \left(\prod_{\mathbf{k}} \frac{1}{\cosh \theta_k} \right) \left[|0\rangle \otimes |0\rangle + \sum_{\mathbf{k}} \tanh \theta_k |a_{\mathbf{k}}\rangle \otimes |\tilde{a}_{\mathbf{k}}\rangle + \dots \right], \quad (5.120)$$

which clearly cannot be factorized into the product of two single-mode states. There is thus entanglement between the modes a and $\tilde{a}$: $|0(\theta)\rangle$ is an entangled state. Eqs. (5.114) and (5.115) then show that S provides a measure of the degree of entanglement.

We remark that the entanglement is truly realized in the infinite volume limit where

$$\langle 0(\theta)|0\rangle = e^{-\frac{V}{(2\pi)^3} \int d^3 k \ln \cosh \theta_k} \to 0 \quad \text{for } V \to \infty, \quad (5.121)$$

provided $\int d^3 k \ln \cosh \theta_k$ is not identically zero. The probability of having the component state $|n\rangle \otimes |n\rangle$ in the state $|0(\theta)\rangle$ is W_n (see Eq. (5.114)). Since W_n is a decreasing monotonic function of n, the contribution of the states $|n\rangle \otimes |n\rangle$ would be suppressed for large n at finite volume. In that case, the transformation induced by the unitary operator $G^{-1}(\theta) \equiv \exp(-i \sum_{\mathbf{k}} \theta_k \mathcal{G}_{\mathbf{k}})$ could disentangle the a and $\tilde{a}$ sectors. However, this is not the case in the infinite volume limit, where the summation extends to an infinite number of components and Eq. (5.121) holds (in such a limit Eq. (5.41) is only a formal relation since $G^{-1}(\theta)$ does not exist as a unitary operator) [339]. The robustness of the entanglement is rooted in the fact that, once the infinite volume limit is reached, there is no unitary generator able to disentangle the $a - \tilde{a}$ coupling.

We also remark that usually one thinks that the entanglement phenomenon concerns two or more particles. On the other hand, the entanglement as discussed above is among different field modes; thus one can also speak of single-particle entanglement [624], as for example, the one found in the context of particle mixing and oscillations [86,87] (see Appendix H).

5.6   Thermal field theory and the gauge field

In this Section we focus our attention on the structure of TFD from the point of view of gauge theories. We find [147] that in TFD the matrix elements of the Lagrangian are invariant under local gauge (i.e., local phase) transformations provided the space of the physical thermal states is restricted to the subspace H_{th} of states constrained by the thermal state condition $(a^\dagger a - \tilde{a}^\dagger \tilde{a})|phys\rangle = 0$. It is the presence of the tilde-kinematical term in the TFD Lagrangian which is responsible for the local gauge invariance of the theory in H_{th}. We can indeed show [147] that in H_{th} the tilde-kinematical term in the Lagrangian may be replaced by the minimal coupling between the system field and the vector field $A_\mu(x)$ which may be identified with the conventional vector field associated with a compact Lie gauge group ($U(1)$ in the Abelian case or $SU(n)$ in the non-Abelian one).

As a result, in H_{th} TFD appears to be a gauge theory, the gauge field A_μ being related with the thermal bath represented by the tilde-degrees of freedom. In this Section, for simplicity we only consider equilibrium systems and therefore Bogoliubov transformations which are independent of space and time. The result, however, also holds [148] for spacetime-dependent Bogoliubov transformations: the gauge structure of TFD thus emerges for equilibrium and non-equilibrium systems, with or without space inhomogeneities.

For simplicity, we consider the Lagrangian of the massless free Dirac field in standard non-thermal QFT:

$$L = -\overline{\psi}\gamma^\mu \partial_\mu \psi. \tag{5.122}$$

The $U(1)$ local gauge transformation is

$$\psi(x) \to \exp\left[ig\alpha(x)\right]\psi(x). \tag{5.123}$$

Under (5.123) L transforms as

$$L \to L' = L - ig\partial^\mu \alpha(x)\overline{\psi}(x)\gamma_\mu\psi(x).\tag{5.124}$$

The well-known story is that in order to make L invariant under (5.123), we have to introduce in L the coupling of the current $j_\mu = i\overline{\psi}\gamma_\mu\psi$ with the gauge vector A_μ such that, when $\psi(x)$ tansforms as in Eq. (5.123), $j^\mu(x)A_\mu(x)$ transforms as

$$gj^\mu(x)A_\mu(x) \to gj^\mu(x)A_\mu(x) + gj^\mu(x)\partial_\mu\alpha(x)\,,\tag{5.125}$$

i.e.,

$$A_\mu(x) \to A_\mu(x) + \partial_\mu\alpha(x).\tag{5.126}$$

The Lagrangian L_g so modified is invariant under the $U(1)$ local gauge transformations (5.123) and (5.126):

$$L_g = -\overline{\psi}\gamma^\mu\partial_\mu\psi + ig\overline{\psi}\gamma^\mu\psi A_\mu\,,\tag{5.127a}$$

$$L_g \to L'_g = L_g.\tag{5.127b}$$

The kinematical term $-\frac{1}{4}F^{\mu\nu}F_{\mu\nu}$ has to be added to L_g if one wants A_μ to be a dynamical field.

We now consider TFD for equilibrium systems and follow closely [147]. Introduction of the doubled degree of freedom, as explained in previous Sections, is done by introducing the tilde-field $\tilde{\psi}(x)$ and the tilde-system is a copy (with the same spectrum and couplings) of the physical system (cf. Eqs. (5.21) and (5.38) where the tilde-Hamiltonian $\tilde{H}$ has been introduced). The TFD Lagrangian for the massless free Dirac theory at finite temperature is then given by

$$\hat{L} = L - \tilde{L} = -\overline{\psi}\gamma^\mu\partial_\mu\psi + \overline{\tilde{\psi}}\gamma^\mu\partial_\mu\tilde{\psi}.\tag{5.128}$$

Coupling of physical field $\psi(x)$ with the tilde-field $\tilde{\psi}(x)$ is not allowed in $\hat{L}$. The thermal vacuum $|0(\theta)\rangle$ appears as a condensate of couples of physical and tilde-field quanta; in the present case, in a simplified notation it is written as (cf. Eq. (5.49)):

$$|0(\theta)\rangle = \prod_k \left[\cos\theta_k + \sin\theta_k a_{\mathbf{k}}^\dagger \tilde{a}_{\mathbf{k}}^\dagger\right] |0\rangle.\tag{5.129}$$

Here $a_{\mathbf{k}}^\dagger$ and $\tilde{a}_{\mathbf{k}}^\dagger$ denote the creation operators associated to ψ and $\tilde{\psi}$, respectively; all quantum number indices are suppressed except momentum. $|0(\theta)\rangle$ is the vacuum with respect to the fields $\psi(\theta;x)$ and $\tilde{\psi}(\theta;x)$ obtained by the Bogoliubov transformation from $\psi(x)$ and $\tilde{\psi}(x)$, respectively:

$$\psi(\theta;x) = B^{-1}(\theta)\psi(x)B(\theta),\tag{5.130a}$$

$$\tilde{\psi}(\theta;x) = B^{-1}(\theta)\tilde{\psi}(x)B(\theta)\,.\tag{5.130b}$$

Note that $|0(\theta)\rangle$ in Eq. (5.129) is a BCS-like state, similar to the one in the theory of superconductivity [536]. As said, here we only consider Bogoliubov transformations whose parameter θ is independent of spacetime. The space of states constructed out of $|0(\theta)\rangle$ by repeated applications of creations operators of $\psi(\theta;x)$ and $\tilde{\psi}(\theta;x)$ is called the finite temperature representation $\{|0(\theta)\rangle\}$. The thermal vacuum condition (5.45) holds: $[a_{\mathbf{k}}^{\dagger}a_{\mathbf{k}} - \tilde{a}_{\mathbf{k}}^{\dagger}\tilde{a}_{\mathbf{k}}]|0(\theta)\rangle = 0$ for any $\mathbf{k}$. In the following we consider the subspace $H_{th} \subset \{|0(\theta)\rangle\}$ made of all the states $|a\rangle_{th}$, including $|0(\theta)\rangle$, such that the thermal state condition

$$[a_{\mathbf{k}}^{\dagger}a_{\mathbf{k}} - \tilde{a}_{\mathbf{k}}^{\dagger}\tilde{a}_{\mathbf{k}}]|a\rangle_{th} = 0, \qquad \text{for any} \quad \mathbf{k}, \tag{5.131}$$

holds for any $|a\rangle_{th}$ in H_{th}. In H_{th} we have for example

$$\langle j_{\mu}(x)\rangle_{th} = \langle \tilde{j}_{\mu}(x)\rangle_{th}, \tag{5.132}$$

where $\langle\cdot\rangle_{th}$ denotes matrix elements in H_{th}. We observe that H_{th} is invariant under the dynamics described by $\hat{L}$ (even in the more general case in which some interaction term is present in $\hat{L}$ provided the charge is conserved). We now notice that due to Eq. (5.132) the matrix elements in H_{th} of the TFD Lagrangian Eq. (5.128) (as well as of a more general Lagrangian than the simple one presently considered) are invariant under the simultaneous local gauge transformations of ψ and $\tilde{\psi}$ fields given by Eq. (5.123) and by

$$\tilde{\psi}(x) \quad \rightarrow \quad \exp\left[ig\alpha(x)\right]\tilde{\psi}(x), \tag{5.133}$$

respectively. Thus, under (5.123) and (5.133)

$$\langle \hat{L}\rangle_{th} \quad \rightarrow \quad \langle \hat{L}'\rangle_{th} = \langle \hat{L}\rangle_{th} \quad \text{in} \quad H_{th}. \tag{5.134}$$

In the following, equalities between matrix elements in H_{th}, say $\langle A\rangle_{th} = \langle B\rangle_{th}$, will be denoted as $A \cong B$ and called thermal weak (th-w-) equalities. We thus see that the term $\overline{\tilde{\psi}}\gamma^{\mu}\partial_{\mu}\tilde{\psi}$ plays a crucial role in the th-w-gauge invariance of $\hat{L}$ under (5.123) and (5.133) since it transforms in such a way to compensate the local gauge transformation of the physical kinematical term, i.e.,

$$\overline{\psi}(x)\gamma^{\mu}\partial_{\mu}\tilde{\psi}(x) \rightarrow \overline{\tilde{\psi}}(x)\gamma^{\mu}\partial_{\mu}\tilde{\psi}(x) + g\partial^{\mu}\alpha(x)\tilde{j}_{\mu}(x). \tag{5.135}$$

This motivates us to introduce the vector field A'_{μ} by

$$gj^{\bar{\mu}}(x)A'_{\bar{\mu}}(x) \cong \overline{\tilde{\psi}}(x)\gamma^{\bar{\mu}}\partial_{\bar{\mu}}\tilde{\psi}(x), \qquad \bar{\mu} = 0,1,2,3. \tag{5.136}$$

The bar over μ means, here and in the following, no summation. Due to Eq. (5.135) A'_{μ} thus introduced transforms as

$$A'_{\mu}(x) \rightarrow A'_{\mu}(x) + \partial_{\mu}\alpha(x), \tag{5.137}$$

when (5.123) and (5.133) are implemented. This suggests to us to assume that A'_μ can be identified with the conventional $U(1)$ gauge vector field.

We shall now confirm that this identification is consistent with TFD when we restrict ourselves to the thermal-weak-equalities (th-w-equalities), i.e., to matrix elements in H_{th}. To this aim let us first observe that matrix elements of physical observables, which in TFD are not functions of the tilde-field, but solely of the $\psi(x)$ field, are not changed by the introduction of the assumption (5.136), and therefore also their statistical averages, whose computation is the main motivation of TFD, are not changed; the reason for this is that the position (5.136) does not change the thermal vacuum structure. Next, we have to show that the conservation laws implied by the TFD scheme are also preserved as th-w-equalities when the position (5.136) is adopted. In the simple case of Eq. (5.128) here considered, we have the current conservation laws:

$$\partial^\mu j_\mu(x) = 0, \qquad \partial^\mu \tilde{j}_\mu(x) = 0. \tag{5.138}$$

From the Dirac equation for the ψ field coupled to A'_μ,

$$\gamma^\mu \partial_\mu \psi(x) = ig\gamma^\mu \psi(x) A'_\mu(x), \tag{5.139}$$

we have,

$$\overline{\psi}(x)\gamma^\mu \partial_\mu \psi(x) = ig\overline{\psi}(x)\gamma^\mu \psi(x) A'_\mu(x) \cong \overline{\tilde{\psi}}(x)\gamma^\mu \partial_\mu \tilde{\psi}(x), \tag{5.140}$$

where the position (5.136) has been used. Since

$$\langle \overline{\psi}\gamma^\mu \partial_\mu \psi \rangle_{th} = \langle \overline{\tilde{\psi}}\gamma^\mu \partial_\mu \tilde{\psi} \rangle_{th}, \tag{5.141}$$

from (5.140) we obtain

$$\overline{\psi}(x)\gamma^\mu \partial_\mu \psi(x) \cong ig\overline{\psi}(x)\gamma^\mu \psi(x) A'_\mu(x). \tag{5.142}$$

Using $\gamma_\mu^\dagger = \gamma_\mu$, the adjoint of (5.142) is

$$\overline{\psi}(x)\gamma^\mu \overleftarrow{\partial}_\mu \psi(x) \cong -ig\overline{\psi}(x)\gamma^\mu \psi(x) A'_\mu(x), \tag{5.143}$$

i.e., adding Eqs. (5.142) and (5.143),

$$\partial^\mu j_\mu(x) \cong 0. \tag{5.144}$$

Similarly, from the th-w-equality in (5.140) and its adjoint we get

$$\partial^\mu \tilde{j}_\mu(x) \cong 0. \tag{5.145}$$

Of course, we could derive Eq. (5.144) from $\partial^\mu j_\mu = 0$, which directly follows from (5.139). Eq. (5.145) is then obtained by use of Eq. (5.132). Such a derivation, however, does not involve (5.136) and thus does not give us

information on its consistency. We observe that subtracting term by term
Eqs. (5.142) and (5.143) we obtain the th-w-equality $igj^\mu A'_\mu \propto g_{\mu\nu}T^{\mu\nu}$
where $T^{\mu\nu}$ is the energy-momentum tensor [343] of the matter field ψ.

In the case A'_μ does not represent just an external field but is indeed a
dynamical field, from the assumption that A'_μ may be identified with the
$U(1)$ gauge vector field, we have

$$\partial^\nu F'_{\mu\nu}(x) = -gj_\mu(x), \tag{5.146}$$

which means that also the kinematical term $-\frac{1}{4}F^{\mu\nu}F_{\mu\nu}$ has to be added to
the Lagrangian (5.127) in the well-known fashion. Then, one more route
may be followed to show that the conservation laws implied by TFD are
preserved as th-w-equality when the position (5.136) is adopted. By multiplying both sides of Eq. (5.146) by A'_μ and by using (5.136) we have

$$(\partial^\nu F'_{\bar\mu\nu}(x))A'^{\bar\mu}(x) = -gj_{\bar\mu}(x)A'^{\bar\mu}(x) \cong -\overline{\tilde\psi}(x)\gamma_{\bar\mu}\partial^{\bar\mu}\tilde\psi(x). \tag{5.147}$$

Under gauge transformation (with $\partial_\mu\alpha(x) \neq 0$) this gives

$$(\partial^\nu F'_{\bar\mu\nu}(x))\partial^{\bar\mu}\alpha(x) \cong -g\tilde{j}_{\bar\mu}(x)\partial^{\bar\mu}\alpha(x), \tag{5.148}$$

i.e.,

$$\partial^\nu F'_{\mu\nu}(x) \cong -g\tilde{j}_\mu(x). \tag{5.149}$$

Use of Eq. (5.132) also gives

$$\partial^\nu F'_{\mu\nu}(x) \cong -gj_\mu(x). \tag{5.150}$$

From Eqs. (5.149) and (5.150) we thus obtain the TFD conservation laws
Eqs. (5.138) restricted to H_{th}, namely Eqs. (5.144) and (5.145). Note that
Eq. (5.144) also holds "strongly", since it can be directly derived from
Eq. (5.146). Then Eqs. (5.144) and (5.145) may be again derived by using
also Eq. (5.132). Eq. (5.136), however, is not used in this case.

By chosing the Lorentz gauge, from Eq. (5.150) we also obtain

$$\partial^2 A'_\mu(x) \cong gj_\mu(x), \qquad \partial^\mu A'_\mu(x) \cong 0. \tag{5.151}$$

We further observe that under our assumptions observables are invariant
under gauge transformations. The fact that Eqs. (5.139) and (5.150) hold
in H_{th} suggests to us to adopt in H_{th} the following Lagrangian:

$$\hat{L}_g = -1/4F'^{\mu\nu}F'_{\mu\nu} - \overline{\psi}\gamma^\mu\partial_\mu\psi + ig\overline{\psi}\gamma^\mu\psi A'_\mu, \quad \text{in} \quad H_{th}. \tag{5.152}$$

In conclusion the TFD Lagrangian (5.128) has been substituted in H_{th} by
the Lagrangian (5.152) which clearly shows the intrinsic gauge properties

of TFD. We remark that the tilde-kinematical term is replaced, in a th-w-sense, by the gauge field-current coupling. This suggests that the gauge field A'_μ may play the role of "thermal reservoir" analogous to the one of the tilde-system in the standard TFD. In this respect, it is interesting to observe that Eq. (5.149) relates variations of the gauge field tensor $F'_{\mu\nu}$ to the "reservoir" current $\tilde{j}_\mu$.

Our conclusions also hold true in the case of a massive Dirac field. In such a case in Eq. (5.152) the terms $-m\tilde{\bar{\psi}}\tilde{\psi}$ and $-m\bar{\psi}\psi$ will be added to $\hat{L}_g$ (note that, however, no mass term is introduced for $A'_\mu(x)$). Finally, we observe that in the case of an interaction term in the starting Lagrangian (5.128) $\hat{L}_{tot} = \hat{L} + \hat{L}_I$, $\hat{L}_I = L_I - \tilde{L}_I$, the above discussion and assumptions still hold provided H_{th} is an invariant subspace under the dynamics described by $\hat{L}_{tot}$.

The case of the $U(1)$ local gauge can be generalized to a non-Abelian compact Lie group G. For simplicity consider $SU(2)$. Let ψ and $\tilde{\psi}$ denote two spinor doublet fields transforming under $SU(2)$ as

$$\psi(x) \rightarrow \exp\left[ig\,\boldsymbol{\alpha}(x) \cdot \mathbf{t}\,\right]\psi(x), \tag{5.153a}$$

$$\tilde{\psi}(x) \rightarrow \exp\left[ig\,\boldsymbol{\alpha}(x) \cdot \mathbf{t}\,\right]\tilde{\psi}(x), \quad \mathbf{t} = \boldsymbol{\tau}/2. \tag{5.153b}$$

Then the above discussion for $U(1)$ is generalized to $SU(2)$ provided the position (5.136) is replaced by

$$g\,\mathbf{j}^{\bar{\mu}} \cdot \mathbf{A}'_{\bar{\mu}} \cong \tilde{\bar{\psi}}\gamma^{\bar{\mu}}\partial_{\bar{\mu}}\tilde{\psi}, \tag{5.154}$$

where $\mathbf{j}_\mu = i\bar{\psi}\gamma_\mu\mathbf{t}\psi$. Due to (5.153) the gauge field $\mathbf{A}'_\mu(x)$ transforms as

$$\mathbf{A}'_\mu(x) \rightarrow \mathbf{A}'_\mu(x) + \partial_\mu\boldsymbol{\alpha}(x) + g\mathbf{A}_\mu(x) \wedge \boldsymbol{\alpha}(x), \tag{5.155}$$

for infinitesimal transformations, which is indeed the usual infinitesimal transformation property for $SU(2)$ Yang–Mills gauge field.

Our discussion can be extended to complex scalar field ϕ or else to the non-relativistic Schrödinger case. For the scalar field the position (5.136) is now replaced by:

$$g[(\partial_{\bar{\mu}}\phi)^\dagger\phi - \phi^\dagger(\partial_{\bar{\mu}}\phi)]A'^{\bar{\mu}} - g^2\phi^\dagger\phi A'_{\bar{\mu}}A'^{\bar{\mu}} \cong (\partial_{\bar{\mu}}\tilde{\phi})^\dagger(\partial^{\bar{\mu}}\tilde{\phi}), \tag{5.156}$$

where $\bar{\mu} = 0, 1, 2, 3$. In the Schrödinger case a position similar to Eq. (5.156) is assumed, provided ϕ denotes the wave-function and $\partial_{\bar{\mu}}$ represents only the spatial derivatives ($\bar{\mu} = 1, 2, 3$).

In conclusion, the tilde-kinematical term in TFD Lagrangian can be replaced by the minimal coupling of the system field with a gauge vector field in the subspace H_{th} of states with even number of tilde and non-tilde particles. This is possible due to the local gauge invariance of TFD in H_{th}. Therefore, in H_{th} finite temperature QFT has the structure of a gauge theory, the gauge vector field playing the role of the "reservoir" response to the system dynamics. It is an interesting question to ask if, on the converse, any gauge theory also describes thermal effects in some subspace of states.

As already mentioned, much work has been devoted to non-equilibrium TFD (see [34–38, 58, 153, 185, 227, 333, 405, 406, 408, 615, 617, 622] and references quoted therein). The gauge structure of TFD for dissipative and non-homogeneous systems has been discussed in [148]. The covariant derivative may be conveniently introduced in order to recover the invariance of the Lagrangian under spacetime-dependent Bogoliubov transformations. In the case of non-homogeneous systems, although canonical commutation relations can be recovered, there are unsolved problems with the non-locality of the gauge field [148]. In TFD, a time-dependent quasiparticle picture has been also considered in terms of the coupling to an external (classical) gauge field [319, 320].

5.7   Boson condensation at finite temperature

In Chapters 3 and 4 we have seen that in theories with spontaneous breakdown of symmetry the value acquired by the order parameter characterizes the physical phase of the system. Changes in the order parameter, considered as a function of the temperature and/or of the time, are thus associated to changes of the system phase, i.e., to the process of phase transition. In the present Section, we consider the temperature effects on homogeneous and non-homogeneous boson condensation. We consider the problem first in the framework of the variational approach and then in the TFD formalism. Here we assume that the instability of the quasiparticle states due to thermal effects is negligible in the range of temperature we are interested in.

The variational method and the boson condensate

We consider, for simplicity, the Heisenberg real scalar field $\phi(x)$, $x \equiv (\mathbf{x}, t)$, at $T \neq 0$ with Lagrangian

$$\mathcal{L} = \frac{1}{2}\partial^\mu \phi(x)\partial_\mu \phi(x) - \frac{1}{2}\mu^2\phi^2(x) - \frac{1}{4}\lambda\phi^4(x), \qquad \lambda > 0. \qquad (5.157)$$

We use $g^{00} = -g^{ii} = 1$ and $c = \hbar = 1$. The equation of motion and the Hamiltonian are

$$(\partial^2 + \mu^2)\phi(x) = -\lambda\phi^3(x)\,, \tag{5.158}$$

$$H = \int d^{D-1}x \left[\frac{1}{2}(\partial_0\phi(x))^2 + \frac{1}{2}(\nabla\phi(x))^2 + \frac{1}{2}\mu^2\phi^2(x) + \frac{1}{4}\lambda\phi^4(x)\right]\,, \tag{5.159}$$

respectively. D is the spacetime dimensionality. In the variational approach, the free energy functional is given by the Bogoliubov inequality [237] (see Appendix P)

$$F \le F_1 = F_0 + \langle H - H_0\rangle_0\,, \tag{5.160}$$

where $F = -k_B T \ln \text{Tr}[\exp(-\beta H)]$, with $\beta = \frac{1}{k_B T}$, is the free energy and the symbol $\langle\ldots\rangle_0$ denotes the statistical average: $\langle A\rangle_0 = \frac{1}{Z}\text{Tr}[\exp(-\beta H_0)A]$, with $Z = \text{Tr}[\exp(-\beta H_0)]$. H_0 is the trial Hamiltonian. One wants to separate in the interacting system an effective free excitation against an effective vacuum background, which can be regarded as the quasi-classical temperature-dependent field [438]. The quantum free excitation can be interpreted as the effective free particle with Hamiltonian H_0. The variational approach selects among the set of trial Hamiltonians H_0 the one which can be interpreted as the effective free field Hamiltonian. The trial free field Lagrangian is

$$\mathcal{L}_0 = \frac{1}{2}\partial_\mu\rho(x)\partial^\mu\rho(x) - \frac{1}{4}\mu_0^2\rho^2(x)\,, \tag{5.161}$$

where μ_0 is a variational parameter which in general depends on x and on β and has to be consistently computed as a function of temperature and other parameters as μ and λ. The equation for $\rho(x)$ is

$$(\partial^2 + \mu_0^2)\rho(x) = 0\,. \tag{5.162}$$

The field ρ is introduced as

$$\rho(x) = \frac{1}{(2\pi)^{(D-1)/2}} \int \frac{d^{D-1}k}{(2k_0)^{1/2}} \left[c_\mathbf{k}\, e^{-ikx} + c_\mathbf{k}^\dagger\, e^{-ikx}\right]\,, \tag{5.163}$$

where the creation and annihilation operators $c_\mathbf{k}$ and $c_\mathbf{k}^\dagger$ satisfy the usual bosonic commutation relations, and $kx \equiv k_\mu x^\mu = \mathbf{k}^2 + \mu_0^2$. The vacuum state is $|0\rangle$, $c_\mathbf{k}|0\rangle = 0$ and we assume that

$$\langle 0|\phi(x)|0\rangle = v(x)\,. \tag{5.164}$$

In the variational approach it is then postulated that

$$\phi(x) = U^\dagger(v)\rho(x)U(v)\,, \tag{5.165}$$

where

$$U(v) = \exp\left(-\int d^{D-1}k \left[v_{\mathbf{k}}^{*} c_{\mathbf{k}} - v_{\mathbf{k}} c_{\mathbf{k}}^{\dagger}\right]\right). \tag{5.166}$$

We have

$$\phi(x) = \rho(x) + v(x), \quad \langle 0|\rho(x)|0\rangle = 0. \tag{5.167}$$

Note that Eq. (5.166) introduces the coherent state representation of the condensate of the field $\phi(x)$ in the state $|0\rangle$, the c-number $v(x)$ describing the boson condensate. In full generality, $v(x)$ can be spacetime-dependent and then it describes non-homogeneous boson condensate. Here we consider homogeneous condensate, thus in the following v does not depend on spacetime.

We observe that $\langle \rho \rangle_0 = \langle \rho^3 \rangle_0 = 0$ and that Wick theorem gives $\langle \rho^4 \rangle_0 = 3\langle \rho^2 \rangle_0 = 0$. We can introduce normal ordering to remove divergencies by replacing $\langle \rho^2 \rangle_0$ with $\langle : \rho^2 : \rangle_0$, or by more sophisticated renormalization procedures. We then use Eq. (5.167) in (5.158) and (5.160). We find

$$\left[-\partial^2 - (\mu^2 + 3\lambda v^2)\right]\rho = \lambda\rho^3 + 3\lambda\rho^2 v + \lambda v^3 + \mu^2 v, \tag{5.168}$$

and

$$\frac{\partial F_0}{\partial \mu_0^2} = \frac{1}{2}\int d^{D-1}x \, \langle \rho^2 \rangle_0. \tag{5.169}$$

The variational equations

$$\frac{\partial F_1}{\partial \mu_0^2} = 0, \qquad \frac{\partial F_1}{\partial v} = 0, \tag{5.170}$$

then give

$$\mu_0^2 = \mu^2 + 3\lambda v^2 + 3\lambda\langle \rho^2 \rangle_0, \tag{5.171a}$$

$$\mu^2 v + \lambda v^3 + 3\lambda v\langle \rho^2 \rangle_0 = 0, \tag{5.171b}$$

respectively (integration on the volume has been done and put equal to one). The second of Eqs. (5.171) gives $v = 0$, or

$$\mu^2 + \lambda v^2 + 3\lambda\langle \rho^2 \rangle_0 = 0. \tag{5.172}$$

Eq. (5.172) implies $\mu^2 < 0$ consistently with the condition of spontaneous breakdown of symmetry Eq. (5.164). When $v \neq 0$, Eq. (5.172) shows that v is temperature-dependent $v = v(\beta)$ and from Eqs. (5.171) we get

$$\mu_0^2(\beta) = 2\lambda v^2(\beta). \tag{5.173}$$

Thus, the value of the variational parameter $\mu_0(\beta)$ has been obtained and temperature dependence is introduced through the statistical average $\langle \rho^2 \rangle_0$:

$$\langle : \rho^2 : \rangle_0 = \frac{1}{(2\pi)^{D-1}} \int \frac{dk}{k_0} \frac{1}{e^{\beta k_0} - 1} . \tag{5.174}$$

Moreover, we also have

$$v(\beta) = \left[\frac{1}{\lambda} \left(-\mu^2 - 3\lambda \langle : \rho^2 : \rangle_0 \right) \right]^{\frac{1}{2}} , \tag{5.175}$$

i.e., $v^2(\beta) \to 0$, as T approaches the critical temperature T_C such that

$$|\mu^2| = 3\lambda \langle : \rho^2 : \rangle_0 |_{T=T_C} . \tag{5.176}$$

At the dimensions D which allow symmetry breaking, we have symmetry restoration at T_C. The factor $\frac{1}{\lambda}$ in Eq. (5.175) expresses the nonperturbative character of the boson condensate: this cannot be obtained as a perturbation around $\lambda \approx 0$ solutions.

In the following we follow the alternative formalism of TFD.

The TFD formalism

We start by considering the expectation value in the thermal state $|0(\beta)\rangle$

$$\langle 0(\beta)|\phi(x)|0(\beta)\rangle = v(x, \beta) , \tag{5.177}$$

with

$$\phi(x) = \rho(x) + v(x, \beta) , \tag{5.178a}$$

$$\langle 0(\beta)|\rho(x)|0(\beta)\rangle = 0 . \tag{5.178b}$$

We assume that $v(x, \beta)$ is spacetime-independent $v = v(\beta)$. Use of Eq. (5.178) in Eq. (5.158) again gives Eq. (5.168). We have

$$\rho^2 = : \rho^2 :_\beta + \langle : \rho^2 : \rangle_0 , \tag{5.179a}$$

$$\rho^3 = : \rho^3 :_\beta + 3\rho \langle : \rho^2 : \rangle_0 , \tag{5.179b}$$

where $: :_\beta$ denotes normal ordering with respect to $|0(\beta)\rangle$, as introduced in Section 5.3. We also have

$$\langle 0(\beta)| : \rho^2 : |0(\beta)\rangle = \langle : \rho^2 : \rangle_0 . \tag{5.180}$$

Taking the expectation value of both members of Eq. (5.168) in the state $|0(\beta)\rangle$, we obtain

$$\mu^2 v(\beta) + \lambda v^3(\beta) + 3\lambda v(\beta)\langle : \rho^2 : \rangle_0 = 0 , \tag{5.181}$$

which is similar to the second of the Eqs. (5.171), giving $v(\beta) = 0$, or

$$v^2(\beta) = \frac{1}{\lambda} \left(-\mu^2 - 3\lambda \langle : \rho^2 : \rangle_0 \right), \tag{5.182}$$

i.e., $v^2(\beta) \to 0$, as T approaches the critical temperature T_C such that the r.h.s. of Eq. (5.182) vanishes: at T_C we have symmetry restoration.

When T is such that $v^2(\beta) \neq 0$, use of Eq. (5.182) in Eq. (5.168) gives:

$$\left(-\partial^2 - 2\lambda v^2(\beta) \right) \rho(x) = \lambda : \rho^3(x) :_\beta + 3\lambda : \rho^2(x) :_\beta v(\beta). \tag{5.183}$$

The associated free field equation is

$$\left(-\partial^2 - 2\lambda v^2(\beta) \right) \rho_{in,\beta}(x) = 0, \tag{5.184}$$

where $\rho_{in,\beta}$ denotes the quasiparticle field in the $|0(\beta)\rangle$ representation: $(\rho_{in,\beta})^{(+)}|0(\beta)\rangle = 0$. Note that the instability of the quasiparticle at non-zero temperature is considered to be negligible in the domain of temperature we are interested in, i.e., we assume $T \ll T_i$, where T_i is the temperature above which the instability of the quasiparticle cannot be neglected.

The comparison with the variational method shows that μ_0 is found to be the same in both formalisms ($\mu_0^2 = 2\lambda v^2(\beta)$). In the TFD formalism μ_0 is found in a self-consistent way: $v(\beta)$ depends on the choice of the state $|0(\beta)\rangle$ (Eq. (5.177)) and $|0(\beta)\rangle$ is the vacuum for $\rho_{in,\beta}$, whose mass depends on $v(\beta)$. In the variational approach μ_0 is obtained from the extremum conditions (5.170). In other words, the extremum conditions in the variational approach are equivalent to selecting the temperature-dependent vacuum in the TFD approach.

5.7.1 *Free energy and classical energy*

Let us now assume real spacetime-dependent $v(x, \beta)$. Eq. (5.160) gives

$$F_1(v(x, \beta)) = \int d^{D-1}x \left[(\partial_0 v(x, \beta))^2 - \mathcal{L}_{eff} \left(v(x, \beta), \partial_\mu v(x, \beta) \right) \right], \tag{5.185}$$

where

$$\mathcal{L}_{eff} = \frac{1}{2} \partial^\mu v \partial_\mu v - U_{eff}(v), \tag{5.186a}$$

$$U_{eff}(v) = U_{cl}(v) + F_0(\mu_0^2) - \frac{3}{4}\lambda \langle \rho^2 \rangle_0^2, \tag{5.186b}$$

$$U_{cl}(v) = \frac{1}{2}\mu^2 v^2(x, \beta) + \frac{1}{4}\lambda v^4(x, \beta). \tag{5.186c}$$

Eq. (5.185) and $U_{eff}(v)$ are the generalized Ginzburg–Landau (GL) functional and potential, respectively. An important remark is that the free energy F_1 plays the role of the energy for the c-number field $v(x, \beta)$ whose Lagrangian is $\mathcal{L}_{eff}$: in this way the dynamics for the quantum field ϕ manifests as the dynamics for the classical field $v(x, \beta)$ describing the non-homogeneous boson condensate. Note also that $U_{eff}(v)$ includes not only the classical terms but also the quantum and the thermal contributions. The Euler–Lagrange equation of motion deduced from $\mathcal{L}_{eff}$ is:

$$\left(-\partial^2 + m^2(\beta)\right) v(x, \beta) = \lambda v^3(x, \beta), \tag{5.187}$$

where

$$m^2(\beta) = |\mu^2| - 3\lambda \langle \rho^2 \rangle_0, \tag{5.188}$$

and therefore m^2 is temperature-dependent. Note that Eq. (5.187) is the same as the variational equation $\frac{\partial F_1}{\partial v} = \frac{\partial U_{eff}}{\partial v} = 0$. The variational equation $\frac{\partial F_1}{\partial \mu_0^2} = 0$ gives the variational mass parameter $\mu_0^2(\beta)$, which appears to play the role of an effective potential for the ρ field (cf. Eq. (5.162))

$$\left(-\partial^2 - \mu_0^2(\beta)\right)\rho = 0. \tag{5.189}$$

Eq. (5.187) describes the evolution of the order parameter, namely the evolution of the system in the "space of the representations of the canonical commutation relations", each representation being labeled by $v(x, \beta)$, and therefore the evolution through the different phases of the system (phase transitions). Eq. (5.188) is a consistency relation between the "inner" dynamics described by $\mathcal{L}$ (Eq. (5.157)) and the dynamics in the space of the representations described by $\mathcal{L}_{eff}$ (Eq. (5.186)). In the following we will study the trajectories in such a space.

Notice that the "mass term" in Eq. (5.187) has the "wrong" sign with respect to a Klein–Gordon equation for a physical particle field: $-m^2$ is interpreted as the GL chemical potential and we recall that the "wrong" sign of the mass term signals spontaneous symmetry breaking. Temperature dependence in the mass term is usually introduced by hand. Here, through Eq. (5.188) we see the microscopic origin of the temperature dependence of the chemical potential in the GL potential.

An analysis similar to the one presented here can be made also for the $T = 0$ case [438], where the Bogoliubov inequality considered above reduces to the well-known Ritz method. In such a case, U_{eff} corresponds to the Weinberg-Coleman potential [167].

In Chapters 7 and 8 we will consider solutions for $v(x, \beta)$ corresponding, in different models and dimensions D, to several extended objects, such as the kink, the vortex, the monopole, etc.

The critical or Ginzburg regime

We will now study non-homogeneous boson condensation occurring in the process of non-instantaneous phase transitions characterized by time-dependent order parameter and leading to the formation of correlated domains. Defects or extended objects may be thought of as the normal symmetric phase region trapped in between correlated (ordered) domains. We consider the transition processes occurring in a finite span of time (see, for example [27, 73, 327]) in which new phenomena, as the formation of exponentially decaying defects, are expected to occur. In these processes the transition starts at the critical temperature T_C and, after a certain lapse of time, the maximally stable configuration is attained at the so-called "Ginzburg temperature" T_G ($T_G < T_C$). Between T_C and T_G the system is said to be in the critical regime [27, 73]. Phenomenological descriptions and numerical simulations have been made and the formation of long strings in such a regime has been investigated [27, 73, 327]. Our aim is to study the microscopic dynamics leading to the formation of non-homogeneous structures in the critical region. We want to model the time dependence of $m^2(\beta)$ during the critical regime evolution, i.e., for transitions lasting a finite time interval where the formation of extended condensation domains is allowed.

Our attention is focused on the harmonic limit which is obtained by considering only the (linear) l.h.s. of Eq. (5.187). Of course, this introduces a strong constraint and it does not accurately describe the behavior of the system. However, it gives enough reliable information on the critical regime behavior [27, 73]. We notice that in such an approximation $\mu_0^2 = -m^2$.

Let us consider the expansion of the v-field into partial waves

$$v(x, \beta) = \sum_{\mathbf{k}} \left[u_{\mathbf{k}}(t, \beta) e^{i\mathbf{k}\cdot\mathbf{x}} + u_{\mathbf{k}}^{\dagger}(t, \beta) e^{-i\mathbf{k}\cdot\mathbf{x}} \right]. \qquad (5.190)$$

Then, in the harmonic potential approximation, the Ginzburg–Landau (GL) equation (5.187) gives for each k-mode ($k \equiv \sqrt{\mathbf{k}^2}$):

$$\ddot{u}_{\mathbf{k}}(t) + (\mathbf{k}^2 - m^2)u_{\mathbf{k}} = 0, \qquad (5.191)$$

i.e., it leads to the equations for the parametric oscillator modes $u_{\mathbf{k}}$ [519] (see also [14, 19]). We denote the frequency of the k-mode by

$$M_k(t) = \sqrt{\mathbf{k}^2 - m^2(\beta, t)}. \qquad (5.192)$$

which is required to be real for each k. In full generality, we assume that m^2 as well as β may depend on time. We remark that the reality condition on $M_k(t)$ turns out to be a condition on the k-modes propagation. The reality condition in fact is satisfied provided at each t, during the critical regime time interval, it is

$$\mathbf{k}^2 \geq m^2(\beta, t), \qquad (5.193)$$

for each k-mode. For shortness we are using the notation $\beta \equiv \beta(t)$. Let $t = 0$ and $t = \tau$ denote the times at which the critical regime starts and ends, respectively. For a given $\mathbf{k}$, Eq. (5.193) will hold up to a time $t = \tau_k$ after which m^2 is larger than $\mathbf{k}^2$. Such a τ_k represents the maximal propagation time of that k mode. The value of τ_k is given when the explicit form of m^2 is assigned. In the following we consider two possibilities for the time dependence of $m(\beta, t)$. The first one can lead to large correlation domains, and thus possibly excludes defect formation. The second one does not allow very large domains and therefore it may allow defect formation.

Our first model choice is

$$a) \qquad m^2(\beta, t) = m_0^2 \left(e^{2f(\beta(t), t)} - 1 \right), \qquad (5.194)$$

with $t = 0$ assumed to correspond to the minimum of m^2 and $f(0) = 0$.

Although at the transition temperature infinite correlation length is allowed, the corresponding mode has only a limited time for propagating. So the formation of domains, i.e., the 'effective causal horizon' [371–374,539,630,672] can be inside the system (domain formation) or outside (single domain), according to whether the time occurring for reaching the boundaries of the system is longer or shorter than the allowed propagation time. So, the dimensions to which the domains can expand depend on the rate between the speed at which the correlation can propagate, at a certain time, and the correlation length $\lambda_k \propto (m(\beta, t))^{-1}$, at that time. In the present case, each k-mode can propagate for a span of time $0 \leq t \leq \tau_k$. From Eqs. (5.193) and (5.194) we obtain:

$$f(\tau_k) = \ln \left(\frac{\sqrt{k^2 + m_0^2}}{m_0} \right) \propto \ln \left(\frac{\mathcal{E}_k}{\mathcal{E}_0} \right), \qquad (5.195)$$

where $\mathcal{E}_k$ and $\mathcal{E}_0$ are the k-mode energies for non-zero and zero k, respectively. The equilibrium time at which $T = T_G$ is $\tau \geq \tau_k$ for any k.

A second possibility to model m^2 is:

$$b) \qquad m^2(\beta, t) = m_0^2 \, e^{2f(\beta(t), t)}. \qquad (5.196)$$

In this case, a cut-off exists for the correlation length, $L \propto m_0^{-1}$ and the propagation time is implicitly given by:

$$f(\beta(\tau_k), \tau_k) = \ln \left(\frac{k}{m_0} \right) \propto \ln \left(\frac{L}{\xi} \right), \qquad (5.197)$$

$f(\beta(\tau_k), \tau_k)$ resembling the commonly called string tension [672]. In Eq. (5.197), ξ is the correlation length corresponding to the k-mode propagation. In this case the reality condition acts as an intrinsic infrared cut-off since small k values are excluded, due to Eq. (5.193), by the non-zero minimum value of m^2. This means that infinitely long wave-lengths are actually precluded, i.e., only domains of finite size can be obtained. Finally (phase) transitions through different vacuum states (which would be unitarily inequivalent vacua in the infinite volume limit) at a given t are possible. This is consistent with the fact that the system is indeed in the middle of a phase transition process (it is in the critical regime) [15]. At the end of the critical regime the correlation may extend over domains of linear size of the order of $\tilde{\lambda}_k \propto (m(\beta_G, \tau))^{-1}$.

In conclusion we see that the model choice $a)$ differs from the model choice $b)$ in the fact that the case $a)$ allows the formation of large correlation domains (infrared modes are allowed); in the case $b)$, on the contrary, only finite-size domains can be formed. Due to the fact that defects may be thought of as the normal symmetric phase region trapped in between correlated (ordered) domains, we see that model $a)$ may exclude defect formation if, as mentioned above, k-mode life-times are longer than the ones needed to reach the system boundaries.

We may then specify the function f by assuming that it is positive and a possible analytic expression for $f(\beta(t), t)$ is

$$f(\beta(t), t) = \frac{at}{bt^2 + c}, \tag{5.198}$$

where a, b, c are (positive) parameters chosen so as to guarantee the correct dimensions. T decreases from T_C to T_G as time grows from $t = 0$ to $t = \tau$; $f(\beta(t), t)$ is positive in the critical region (and even for $t > \tau$, i.e., in the spontaneous breakdown of symmetry region). The equilibrium time-scale is given by $\tau^2 = \frac{c}{b}$. In order to obtain this result, the variations in time of $\beta(t)$ have to be assumed small, according to the picture of a slow (non-instantaneous) transition (slow transitions are those for which τ is large; for fixed parameter a, large τ also means large τ_Q, see below).

The study of the behavior of $m^2(\beta, t)$ corresponding to (5.198) shows [17] that the maximum of $m^2(\beta)$ has to be identified with the minimum of the potential $\frac{\partial U_{eff}}{\partial v} = 0 = -m(\beta)^2 v(x, \beta) + \lambda v^3(x, \beta)$ (cf. Eq. (5.188)), so it corresponds to $T = T_G$. Moreover, for $T > T_C$ (the "negative time" region) f turns out to be negative.

Finally, the present treatment allows us to recover the known results on the number of defects, i.e., $n_{def} \propto (\tau_0/\tau_Q)^{1/2}$ (see [539], for a review).

In fact, recalling that the equilibrium time τ is given by $\tau = \sqrt{c/b}$ it is possible to introduce the time-scales: $\tau_Q = c/a\lambda$ and $\tau_0 = a\lambda/b$, with λ arbitrary constant and a, b, c introduced in Eq. (5.198). Thus $\tau = \sqrt{\tau_Q \tau_0}$. We now observe that, at the first order approximation, one has

$$e^{g(x)} - g(x) \approx e^{g(x_0)} - g(x_0), \qquad \text{if} \qquad \frac{\partial g(x)}{\partial x}\Big|_{x_0} = 0, \qquad (5.199)$$

so, in our case, $e^{2f(\tau)} = 2f(\tau) + e^{2f(0)}$. The number of defects can be finally obtained. In the case $a)$ it is:

$$n_{def} \propto m^2(\beta_G, \tau) \approx 2m_0^2 f(\tau) = m_0^2 \, \tau/\lambda\tau_Q \propto \sqrt{(\tau_0/\tau_Q)}. \qquad (5.200)$$

Similarly, in the case $b)$:

$$n_{def} \propto m^2(\beta_G, \tau) \approx m_0^2(2f(\tau) + 1) \approx m_0^2 \, \tau/\lambda\tau_Q, \qquad (5.201)$$

for large τ (slow transitions). This agrees with the result obtained by Zurek [672].

Size and life-time of correlated domains

The analysis of the size and life-time of these domains leads to an interesting spectrum of possibilities. Let us write $M_k^2(t)$ as

$$M_k^2(\Lambda_k(t)) = M_k^2(0) \, e^{-2\Lambda_k(t)}, \qquad (5.202)$$

$$e^{-2\Lambda_k(t)} = e^{2f(\beta(t),t)} \, \frac{\sinh(f(\tau_k) - f(\beta(t),t))}{\sinh f(\tau_k)}, \qquad (5.203)$$

where $f(0) = 0$ has been used. Eq. (5.203) shows that $\Lambda_k(t) \geq 0$ for $0 \leq t \leq \tau_k$, $\Lambda_k(0) = 0$ and $\Lambda_k(\tau_k) = \infty$. Since $M_k(\Lambda_k(\tau_k)) = 0$ we see that $\Lambda_k(t)$ acts as a life-time, say with $\Lambda_k(t) \propto t_k$, for the k-mode: each k-mode "lives" with a proper time t_k, i.e., it is born when t_k is zero and it dies for $t_k \to \infty$. Only the modes satisfying the reality condition are present at a certain time t, the other ones have been decayed. In this way the causal horizon sets up. Eqs. (5.202) and (5.203) show that modes with larger k have a longer life with reference to time t.

Since, on the other hand, longer wave-lengths correspond to smaller k, we see that domains with a specific spectrum of k-modes components may coexist, some of them disappearing before, some other ones persisting longer in dependence of the number in the spectrum of the smaller or larger k components, respectively. In general, the boundaries of larger size domains are thus expected to be less persistent than those of smaller size domains. This fits with the observation [27, 73, 327] that "critical regime has little effect over the small scale dynamics", thus allowing the survival of localized defects (such as vortex strings).

5.8 Trajectories in the space of representations

In this Section we discuss the chaotic behavior, under certain conditions, of the trajectories in the space $\mathcal{H}$ of the unitarily inequivalent representations of the CCR.

Let us recall some of the features of the $SU(1,1)$ group structure (see, e.g., [519] and Appendix C).

$SU(1,1)$ realized on $\mathbb{C} \times \mathbb{C}$ consists of all unimodular 2×2 matrices leaving invariant the hermitian form $|z_1|^2 - |z_2|^2$, $z_i \in \mathbb{C}, i = 1, 2$. The complex z plane is foliated under the group action into three orbits: $X_+ = \{z : |z| < 1\}$, $X_- = \{z : |z| > 1\}$ and $X_0 = \{z : |z| = 1\}$.

The unit circle $X_+ = \{\zeta : |\zeta| < 1\}$, $\zeta \equiv e^{i\phi} \tanh\theta$, is isomorphic to the upper sheet of the hyperboloid which is the set $\mathbf{H}$ of pseudo-Euclidean bounded (unit norm) vectors $\mathbf{n} : \mathbf{n} \cdot \mathbf{n} = 1$. $\mathbf{H}$ is a Kählerian manifold with metrics

$$ds^2 = 4 \frac{\partial^2 F}{\partial \zeta \partial \bar{\zeta}} d\zeta \cdot d\bar{\zeta} .$$ (5.204)

The Kählerian potential is

$$F \equiv -\ln(1 - |\zeta|^2) .$$ (5.205)

The metrics is invariant under the group action [519].

The Kählerian manifold $\mathbf{H}$ is known to have a symplectic structure. It may thus be considered as the phase space for the classical dynamics generated by the group action [519].

The $SU(1,1)$ generalized coherent states are recognized to be "points" in $\mathbf{H}$ and transitions among these points induced by the group action are therefore classical trajectories [519] in $\mathbf{H}$ (a similar situation occurs [519] in the $SU(2)$ (fermion) case).

Summarizing, the space of the unitarily inequivalent representations of the CCR, which in the boson case is the space of the $SU(1,1)$ generalized coherent states, is a Kählerian manifold, $\mathcal{H} \equiv \{\mathcal{H}_\theta, \ \forall \theta\} \approx \mathbf{H}$; it has a symplectic structure and a classical dynamics is established on it by the $SU(1,1)$ action (generated by $\mathcal{G}$ or, equivalently, by p_θ: $\mathcal{H}_\theta \to \mathcal{H}_{\theta'}$, cf. Section 5.4). Variations of the θ–parameter induce transitions through the representations $\mathcal{H}_\theta = \{|0(\theta)\rangle\}$, i.e., through the physical phases of the system, the system order parameter being dependent on θ. These transitions are described as trajectories through the "points" in $\mathcal{H}$. One may then assume time-dependent θ: $\theta = \theta(t)$. For example, this is the case of dissipative systems and of non-equilibrium thermal field theories where $\theta_k = \theta_k(\beta(t))$, with $\beta(t) = \frac{1}{k_B T(t)}$.

The role played by the Kählerian potential in the motion over $\mathcal{H}$, e.g., in the transitions $\mathcal{H}_\theta \to \mathcal{H}_{\theta'}$, i.e., $|0(\theta)\rangle \to |0(\theta')\rangle$, is expressed by

$$\langle 0(\theta)|0(\theta')\rangle = e^{-\frac{V}{2(2\pi)^3}\int d^3 k F_{\mathbf{k}}(\theta,\theta')}, \tag{5.206}$$

where $F_{\mathbf{k}}(\theta,\theta')$ is given by Eq. (5.205) with $|\zeta_{\mathbf{k}}|^2 = \tanh^2(\theta_k - \theta'_{\mathbf{k}})$. In [195,438] the result that the group action induces classical trajectories in $\mathcal{H}$ has been obtained on the ground of more phenomenological considerations.

With reference to the discussion presented in Section 5.2, we may say that on the (classical) trajectories in $\mathcal{H}$ it is $x_+ = x_- = x_{classical}$, i.e., on these trajectories the quantum effects accounted for by the y doubled coordinate is fully shielded by the thermal bath. In [576] it has been indeed observed that the y freedom contributes to the imaginary part of the action which becomes negligible in the classical regime, but is relevant for the quantum dynamics, namely in each of the "points" in $\mathcal{H}$ (i.e., in each of the spaces $\mathcal{H}_\theta$, for each θ) through which the trajectory goes as θ changes. Upon "freezing" the action of $G(\theta)$ (i.e., upon "freezing" the "motion" through the unitarily inequivalent representations) the quantum features of $\mathcal{H}_\theta$, at given θ, become manifest.

For any $\theta(t) = \{\theta_k(t), \forall \mathbf{k}\}$ we have

$$_\theta\langle 0(t)|0(t)\rangle_\theta = 1, \quad \forall t, \tag{5.207}$$

where $|0(t)\rangle_\theta \equiv |0(\theta(t))\rangle$ is used. We will now restrict the discussion to the case in which, for any k, $\theta_k(t)$ is a growing function of time and

$$\theta(t) \neq \theta(t'), \ \forall t \neq t', \qquad \text{and} \qquad \theta(t) \neq \theta'(t'), \ \forall t,t'. \tag{5.208}$$

Under such conditions the trajectories in $\mathcal{H}$ satisfy the requirements for chaotic behavior in classical non-linear dynamics [326]:

i) the trajectories are bounded and each trajectory does not intersect itself.

ii) trajectories specified by different initial conditions do not intersect.

iii) trajectories of different initial conditions are diverging trajectories.

Let $t_0 = 0$ be the initial time. The "initial condition" of the trajectory is then specified by the $\theta(0)$-set, $\theta(0) = \{\theta_k(0), \forall \mathbf{k}\}$. One obtains

$$_\theta\langle 0(t)|0(t')\rangle_\theta \xrightarrow[V\to\infty]{} 0, \ \forall t,t', \ \text{with } t \neq t', \tag{5.209}$$

provided $\int d^3 k \, \ln \cosh(\theta_k(t) - \theta_k(t'))$ is finite and positive for any $t \neq t'$.

Eq. (5.209) expresses the unitary inequivalence of the states $|0(t)\rangle_\theta$ (and of the associated Hilbert spaces $\{|0(t)\rangle_\theta\}$) at different time values $t \neq t'$ in the infinite volume limit.

The trajectories are bounded in the sense of Eq. (5.207), which shows that the "length" (the norm) of the "position vectors" (the state vectors at time t) in $\mathcal{H}$ is finite (and equal to one) for each t. Eq. (5.207) rests on the invariance of the form $|z_1|^2 - |z_2|^2$, $z_i \in \mathbb{C}$, $i = 1, 2$ and we also recall that the manifold of points representing the coherent states $|0(t)\rangle_\theta$ for any t is isomorphic to the product of circles of radius $r_\mathbf{k}^2 = \tanh^2(\theta_k(t))$ for any $\mathbf{k}$.

Eq. (5.209) expresses the fact that the trajectory does not cross itself as time evolves (it is not a periodic trajectory): the "points" $\{|0(t)\rangle_\theta\}$ and $\{|0(t')\rangle_\theta\}$ through which the trajectory goes, for any t and t', with $t \neq t'$, after the initial time $t_0 = 0$, never coincide. The requirement $i)$ is thus satisfied.

In the infinite volume limit, we also have

$$_\theta\langle 0(t)|0(t')\rangle_{\theta'} \underset{V\to\infty}{\longrightarrow} 0 \ \forall\, t, t' \quad, \ \forall\, \theta \neq \theta'. \tag{5.210}$$

Under the assumption (5.208), Eq. (5.210) is true also for $t = t'$. The meaning of Eqs. (5.210) is that trajectories specified by different initial conditions $\theta(0) \neq \theta'(0)$ never cross each other. The requirement $ii)$ is thus satisfied.

In order to study how the "distance" between trajectories in the space $\mathcal{H}$ behaves as time evolves, consider two trajectories of slightly different initial conditions, say $\theta'(0) = \theta(0) + \delta\theta$, with small $\delta\theta$. A difference between the states $|0(t)\rangle_\theta$ and $|0(t)\rangle_{\theta'}$ is the one between the respective expectation values of the number operator $a_\mathbf{k}^\dagger a_\mathbf{k}$. For any k at any given t, it is

$$\Delta\mathcal{N}_{a_\mathbf{k}}(t) \equiv \mathcal{N}'_{a_\mathbf{k}}\big(\theta'(t)\big) - \mathcal{N}_{a_\mathbf{k}}\big(\theta(t)\big)$$
$$= {}_{\theta'}\langle 0(t)|a_\mathbf{k}^\dagger a_\mathbf{k}|0(t)\rangle_{\theta'} - {}_\theta\langle 0(t)|a_\mathbf{k}^\dagger a_\mathbf{k}|0(t)\rangle_\theta$$
$$= \sinh^2\theta'_k(t) - \sinh^2\theta_k(t) = \sinh\big(2\theta_k(t)\big)\delta\theta_k(t), \tag{5.211}$$

where $\delta\theta_k(t) \equiv \theta'_\mathbf{k}(t) - \theta_k(t)$ is assumed to be greater than zero, and the last equality holds for "small" $\delta\theta_k(t)$ for any k at any given t. By assuming that $\frac{\partial\delta\theta_k}{\partial t}$ has negligible variations in time, the time-derivative gives

$$\frac{\partial}{\partial t}\Delta\mathcal{N}_{a_\mathbf{k}}(t) = 2\frac{\partial\theta_k(t)}{\partial t}\cosh\big(2\theta_k(t)\big)\delta\theta_k. \tag{5.212}$$

This shows that, provided $\theta_k(t)$ is a growing function of t, small variations in the initial conditions lead to growing in time $\Delta\mathcal{N}_{a_\mathbf{k}}(t)$, namely to diverging trajectories as time evolves.

In the assumed hypothesis, at enough large t the divergence is dominated by $\exp(2\theta_k(t))$. For each k, the quantity $2\theta_k(t)$ could thus be thought to play the role similar to the one of the Lyapunov exponent.

Since $\sum_{\mathbf{k}} E_{\mathbf{k}} \dot{\mathcal{N}}_{a_{\mathbf{k}}} dt = \frac{1}{\beta} d\mathcal{S}_a$, where $E_{\mathbf{k}}$ is the energy of the mode $a_{\mathbf{k}}$ and $d\mathcal{S}_a$ is the entropy variation associated to the modes a (cf. Eq. (5.119)) [153], the divergence of trajectories of different initial conditions may be expressed in terms of differences in the variations of the entropy (cf. Eqs. (5.211) and (5.212)):

$$\Delta \sum_{\mathbf{k}} E_{\mathbf{k}} \dot{\mathcal{N}}_{a_{\mathbf{k}}}(t) dt = \frac{1}{\beta} \left(d\mathcal{S}'_a - d\mathcal{S}_a \right) . \qquad (5.213)$$

The discussion above thus also shows that the requirement iii) is satisfied. The conclusion is that trajectories in the $\mathcal{H}$ space exhibit, under the condition (5.208) and with $\theta(t)$ a growing function of time, properties typical of the chaotic behavior in classical non-linear dynamics. The Kählerian manifold of the unitarily inequivalent representations in QFT thus appears as a *classical blanket* [633] covering the quantum dynamical evolution going on in each representation.

Let us close this Chapter with a final comment. We have seen in Section 5.4 that the q-deformation parameter acts as a label indexing the unitarily inequivalent representations, i.e., the physical phases of the system. When such a parameter does not depend on temperature (although in most cases it does, as we have seen) changes in such a parameter induce transitions through the inequivalent representations and therefore through the system physical phases. In such cases, one usually speaks of quantum phase transitions [545], namely of transitions not induced by changes in the system temperature.

Chapter 6

Selected topics in thermal field theory

6.1 Introduction

In the previous Chapter the algebraic operator approach to thermal QFT has been considered. There, the guiding principle was the doubling of the Hilbert space. It is, however, often convenient to consider the alternative approach based on functional integrals. Though these are less suitable tools for the discussion of inequivalent representations, canonical transformations and algebra deformations, they have an immense wealth of (both perturbative and non-perturbative) analytical techniques and numerical recipes that make them extremely valuable in the finite temperature context. In particular, the analysis of phase transitions or finite-temperature instantons can be conveniently done by use of functional integral semiclassical methods. In this connection, *thermal* effective action analysis plays a central role. Several duality transformations for topological defects, such as the duality between magnetic vortices and the Ginzburg–Landau theory of Bardeen–Cooper–Shrieffer superconductor [381, 382], or the (finite-temperature) duality between the sine-Gordon soliton and the fundamental fermion of the massive Thirring model [165, 293], are conveniently formulated in the language of functional integrals.

In order to tackle both dynamical and statistical features, as well as to stay comparatively close to TFD, we will concentrate in this Chapter on the Keldysh–Schwinger closed time-path formalism (CTP) of finite temperature QFT. For simplicity, we will consider zero chemical potential μ, i.e., we will assume that particle number is not conserved. Inclusion of μ can be often achieved by formally shifting the energy spectrum by μ, and the corresponding modifications (e.g., for KMS condition or free thermal Green's functions) can be easily obtained.

235

This Chapter has two aims. One is to introduce some essentials of the finite-temperature functional techniques that will be useful in the following. The other one is to discuss the $\eta - \xi$ spacetime, which provides a geometric background unifying the various thermal field theory approaches, namely Thermo Field Dynamics and thermal functional methods.

6.2   The Gell-Mann–Low formula and the closed time-path formalism

In order to set up the diagrammatic formalism at thermal equilibrium, we start by considering the Heisenberg field equations

$$\dot{\phi}(x) = i[H, \phi(x)], \tag{6.1a}$$

$$\dot{\pi}(x) = i[H, \pi(x)]. \tag{6.1b}$$

Here π is the momentum conjugate to the Heisenberg real scalar field operator ϕ and H is the Hamiltonian in the Heisenberg picture: $H = \int d^3x\, \mathcal{H}(x)$. Assuming that the Heisenberg and interaction pictures coincide at some time t_i, the formal solution of (6.1) is [169]:

$$\phi(x) = Z_\phi^{1/2} \Lambda^{-1}(t)\, \phi_{in}(x)\, \Lambda(t), \tag{6.2a}$$

$$\pi(x) = Z_\pi^{1/2} \Lambda^{-1}(t)\, \pi_{in}(x)\, \Lambda(t), \tag{6.2b}$$

$$\Lambda(t) = e^{i(t-t_i)H_{in}^0}\, e^{-i(t-t_i)H}, \tag{6.2c}$$

$$\Lambda(t_2)\, \Lambda^{-1}(t_1) = U(t_2; t_1) = T\left[\exp\left(-i\int_{t_1}^{t_2} d^4x\, \mathcal{H}_{in}^I(x)\right)\right]. \tag{6.2d}$$

T is the time-ordering symbol and Z_ϕ, Z_π are the wave function renormalizations (usually $\pi \propto \partial_t \phi$, and so $Z_\phi = Z_\pi$). Eqs. (6.2) must be understood in a weak sense, i.e., valid for each matrix element separately. If not, we would obtain the canonical commutator between ϕ and π being equal to $iZ_\phi \delta^3(\mathbf{x} - \mathbf{y})$ and thus canonical quantization would require that $Z_\phi = 1$. On the other hand, non-perturbative considerations (e.g., the Källen–Lehmann representation [343]) require $Z_\phi < 1$. The solution of this problem is well known [286, 308, 343]: the Hilbert spaces for ϕ and ϕ_{in} are unitarily inequivalent and the wave function renormalizations Z_ϕ and/or Z_π are then "indicators" of how much the unitarity is violated. This is nothing but the Haag theorem [114, 272, 307, 308, 557, 558, 569, 579] discussed in previous Chapters.

The interaction picture evolution operator $U(t_2, t_1)$ is [491]

$$U(t_2, t_1) = T^* \left[\exp \left(i \int_{t_1}^{t_2} d^4x \, \mathcal{L}_{in}^I(x) \right) \right]. \qquad (6.3)$$

where the T^*-product (or covariant product) [343,345,491] is defined in such a way that for fields in the interaction picture it is simply the T-product with all the derivatives pulled out of the T-ordering symbol. For free fields without derivatives: $T^*[\phi_{in}(x_1) \dots \phi_{in}(x_n)] = T[\phi_{in}(x_1) \dots \phi_{in}(x_n)]$. $\mathcal{L}_{in}^I$ is the interaction Lagrangian density in the interaction picture. Eq. (6.3) is valid even when the derivatives of fields are present in $\mathcal{L}^I$ (i.e., $\mathcal{H}^I \neq -\mathcal{L}^I$).

By use of (6.3), we can rewrite Eq. (6.2) for the Heisenberg field $\phi(x)$ in the following form

$$\begin{aligned} \phi(x) &= Z_\phi^{1/2} \, U(t_i; t) \, \phi_{in}(x) \, U^{-1}(t_i; t) \\ &= Z_\phi^{1/2} \, U(t_i; t_f) U(t_f; t) \, \phi_{in}(x) \, U(t; t_i) \\ &= Z_\phi^{1/2} \, T_C \left[\phi_{in}(x) \exp \left(-i \int_C d^4x \, \mathcal{H}_{in}^I(x) \right) \right] \\ &= Z_\phi^{1/2} \, T_C^* \left[\phi_{in}(x) \exp \left(i \int_C d^4x \, \mathcal{L}_{in}^I(x) \right) \right]. \end{aligned} \qquad (6.4)$$

Here C denotes a closed-time (or Schwinger) contour, running from t_i to a later time t_f, and back again (see Fig. 6.1). T_C denotes the corresponding time-path ordering symbol (analogously for the T_C^* ordering). In the limit $t_i \to -\infty$, ϕ_{in} turns out to be the usual in-(or asymptotic) field. Since t_f is an arbitrary time, we set $t_f = +\infty$. Eq. (6.4) may be viewed as the Haag expansion (the dynamical map) of the Heisenberg field ϕ. Generalization of Eq. (6.4) to more fields is straightforward. So, for instance, for the time ordered product of n Heisenberg fields we may write

$$\begin{aligned} &T[\phi(x_1) \dots \phi(x_n)] \\ &= Z_\phi^{n/2} T_C^* \left[\phi_{in}(x_1) \dots \phi_{in}(x_n) \exp \left(i \int_C d^4x \, \mathcal{L}_{in}^I(x) \right) \right]. \end{aligned} \qquad (6.5)$$

When we consider the *vacuum expectation value* of Eq. (6.5) with respect to *in*-field vacuum $|0, in\rangle \equiv |0\rangle$, we have at our disposal two equivalent representations for the zero temperature Green's functions. The first one is simply

$$\begin{aligned} &\langle 0 | T(\phi(x_1) \dots \phi(x_n)) | 0 \rangle \\ &= Z_\phi^{n/2} \langle 0 | T_C^* \left[\phi_{in}(x_1) \dots \phi_{in}(x_n) \exp \left(i \int_C d^4x \, \mathcal{L}_{in}^I(x) \right) \right] | 0 \rangle. \end{aligned} \qquad (6.6)$$

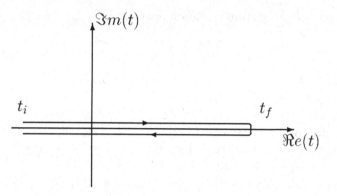

Fig. 6.1   The closed time-path.

The second representation is obtained under the condition that

$$T^* \left[\exp \left(i \int_{-\infty}^{\infty} d^4 x \, \mathcal{L}_{in}^I(x) \right) \right] |0\rangle = \alpha |0\rangle, \tag{6.7}$$

which is nothing but the vacuum stability condition: the lowest energy state is preserved during the interaction process. Here α is an eigenvalue (basically a phase factor) of the evolution operator $U(\infty, -\infty)$ (also known as S matrix) corresponding to the eigenstate $|0\rangle$. Then we can write

$$\langle 0 | T_C^* \left[\cdots e^{i \int_C d^4 x \, \mathcal{L}_{in}^I(x)} \right] |0\rangle$$

$$= \langle 0 | T^* \left[\cdots e^{i \int_{-\infty}^{\infty} d^4 x \, \mathcal{L}_{in}^I(x)} \right] \left(T^* \left[e^{i \int_{-\infty}^{\infty} d^4 x \, \mathcal{L}_{in}^I(x)} \right] \right)^{-1} |0\rangle$$

$$= \langle 0 | T^* \left[\cdots e^{i \int_{-\infty}^{\infty} d^4 x \, \mathcal{L}_{in}^I(x)} \right] \alpha^{-1} |0\rangle$$

$$= \frac{\langle 0 | T^* \left[\cdots \exp \left(i \int_{-\infty}^{\infty} d^4 x \, \mathcal{L}_{in}^I(x) \right) \right] |0\rangle}{\langle 0 | T^* \left[\exp \left(i \int_{-\infty}^{\infty} d^4 x \, \mathcal{L}_{in}^I(x) \right) \right] |0\rangle}. \tag{6.8}$$

With this we can recast (6.6) into the form

$$\langle 0 | T(\phi(x_1) \ldots \phi(x_n)) |0\rangle$$

$$= Z_\phi^{n/2} \frac{\langle 0 | T^* \left[\phi_{in}(x_1) \ldots \phi_{in}(x_n) \exp \left(i \int_{-\infty}^{\infty} d^4 x \, \mathcal{L}_{in}^I(x) \right) \right] |0\rangle}{\langle 0 | T^* \left[\exp \left(i \int_{-\infty}^{\infty} d^4 x \, \mathcal{L}_{in}^I(x) \right) \right] |0\rangle}, \tag{6.9}$$

which is the well-known Gell-Mann–Low formula [275] for Green's functions. Notice, however, that it holds *only* for vacuum expectation values.

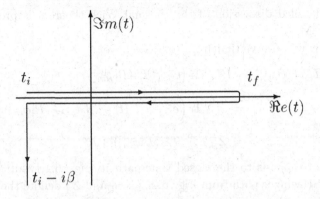

Fig. 6.2   Keldysh–Schwinger thermal path.

Indeed, the passage from (6.6) to the Gell-Mann–Low formula can be established only provided Eq. (6.7) holds. In more general cases, where expectation values are taken with respect to a state $|\Psi\rangle \neq |0\rangle$, the Green's function cannot be written as in Eq. (6.9) and the T_C^* prescription is obligatory. In particular, for mixed states, where $\rho \neq |0\rangle\langle 0|$, the vacuum expectation value is replaced by the statistical average $\text{Tr}\,(\rho\ldots) \equiv \langle\ldots\rangle$ and

$$\langle T(P[\phi])\rangle = \left\langle T_C^* \left[P_r[\phi_{in}] \exp\left(i \int_C d^4 x\, \mathcal{L}_{in}^I(x) \right) \right] \right\rangle , \qquad (6.10)$$

where $P[\ldots]$ is an arbitrary (generally composite) polynomial in ϕ, and the subscript r suggests that the corresponding renormalization factors are included. An important special case arises for systems in thermodynamical equilibrium. There, the statistical properties are described by the canonical density matrix (for simplicity we omit from present consideration grand-canonical ensembles). As $\rho \propto e^{-\beta H}$, the density matrix is basically the generator of the (imaginary) time translations. Using Eq. (6.2) we have

$$e^{-\beta H} = e^{-\beta H_{in}^0} U(t_i - i\beta, t_i) . \qquad (6.11)$$

From this it is evident that one may exchange the *full* density matrix for the density matrix of a corresponding free system provided one adds to the path C a vertical part running from $t_i - i\beta$ to t_i (see Fig. 6.2). The advantage of this rather formal step is that the free density matrix is Gaussian and correspondingly $\langle T_C[\phi_{in}(x_1)\ldots\phi_{in}(x_n)]\rangle$ is 0 for n odd and a symmetrized product of the (free) two-point Green's function if n is even. This is nothing but the thermal (or thermodynamical) Wick theorem which provides the basis for a perturbation calculus in QFT at finite temperatures.

A more detailed discussion of the thermal Wick theorem is presented in Appendix M.

Finally, we observe that from (6.9) we have

$$\langle T_C^*(P[\phi_{in}]) \rangle = Z_\beta^{-1} \mathrm{Tr} \left(e^{-\beta H} T_C^*(P[\phi_{in}]) \right)$$

$$= Z_\beta^{-1} \mathrm{Tr} \left(e^{-\beta H_{in}^0} U(t_i - i\beta, t_i) T_C^*(P[\phi_{in}]) \right)$$

$$= Z_\beta^{-1} Z_\beta^0 \langle T_{C'}^*(P[\phi_{in}]) \rangle_0 . \tag{6.12}$$

Here C corresponds to the closed time-path in Fig. 6.1, while C' is the Keldysh–Schwinger path from Fig. 6.2. $\langle \ldots \rangle_0$ and Z_β^0 denote the *free* field theory expectation value and partition function, respectively.

6.3   The functional integral approach

In this Section we shall present a functional integral approach to the calculation of Green's functions at finite temperature.

Functional integral representations are in general useful tools providing a bookkeeping method for generating various perturbation expansions. They provide a valuable instrument for quantizing non-Abelian gauge theories or implementing spontaneous breakdown of symmetry. Also, many results in QFT can be derived in a compact way through formal manipulations with functional integrals.

Benefits of functional integrals at zero temperature are to a large extent shared by their finite-temperature counterparts. The structure of the latter, however, also exhibit novel features such as the matrix structure of propagators and self-energies, periodic (or antiperiodic) boundary conditions for fields, non-unique analytic extension from imaginary to real times, etc. In the following we introduce the functional integral representation of the finite-temperature Green's functions and the associated generating functionals. The latter are typical instruments in treatment of phase transitions [381, 382, 668] (see also Appendix O).

6.3.1   *Generating functionals for Green's functions*

We consider the bosonic multiplet field[1] $\hat{\phi} = \{\hat{\phi}_1, \ldots, \hat{\phi}_N\}$ in a heat bath at temperature $T = \beta^{-1}$. The key objects used at finite temperature are

---

[1]To avoid possible confusion, we insert in this Section the "hat" symbol over all operators.

contour-ordered thermal Green's functions

$$D^{(c)}_{i_1,\ldots,i_n}(x_1,\ldots,x_n) = \left\langle T_C \left[\hat{\phi}_{i_1}(x_1)\ldots\hat{\phi}_{i_n}(x_n) \right] \right\rangle, \quad i_k = 1,\ldots,n. \quad (6.13)$$

Fields $\hat{\phi}_i$ are bare Heisenberg fields. Due to the permutation symmetry of Eq. (6.13) one can combine the entire hierarchy $\{D^{(c)}_{i_1,\ldots,i_n}; n \in \mathbb{N}^+\}$ into one generating functional. For simplicity, we consider a single real scalar field $\hat{\phi}(x)$ and a single c-number source term (Schwinger source) $J(x)$. The generating functional for Green's functions can then be written as

$$\begin{aligned}
Z_\beta[J] &= Z_\beta[0] \sum_{n=0}^{\infty} \frac{i^n}{n!} \int_C \prod_{i=1}^{n} d^4x_i \; J(x_1)\ldots J(x_n) \; D^{(c)}(x_1,\ldots,x_n) \\
&= Z_\beta[0] \left\langle T_C \left[\exp\left(i \int_C d^4x \; J(x)\hat{\phi}(x) \right) \right] \right\rangle.
\end{aligned} \quad (6.14)$$

By using Eq. (6.10), we can rewrite this in terms of free fields as

$$Z_\beta[J] = Z_\beta[0] \left\langle T_C^* \left[\exp\left(i \int_C d^4x \left(\hat{\mathcal{L}}^I_{in}(x) + J(x)\hat{\phi}_{in}(x) \right) \right) \right] \right\rangle. \quad (6.15)$$

It is convenient to keep the arbitrary normalization factor $Z_\beta[0]$ unspecified.

It follows from Eq. (6.15) that thermal Green's functions can be obtained by functionally differentiating $Z_\beta[J]$ with respect to $J(x)$, e.g.,

$$D^{(c)}(x_1,\ldots,x_n) = \frac{1}{Z_\beta[0]} \frac{(-i)^n \; \delta^n}{\delta J(x_1)\ldots J(x_n)} Z_\beta[J] \bigg|_{J=0}. \quad (6.16)$$

The generating functional (6.15) is an analogue of the moment generating function (or characteristic function) used in mathematical statistics. Eq. (6.16) shows that Eq. (6.15) can be formally rewritten as

$$\begin{aligned}
Z_\beta[J] &= Z_\beta[0] \exp\left(i \int_C d^4y \; \mathcal{L}^I\left(-i \frac{\delta}{\delta J(y)} \right) \right) \\
&\quad \times \left\langle T_C \left[\exp\left(i \int_C d^4x \; J(x)\hat{\phi}_{in}(x) \right) \right] \right\rangle \bigg|_{J=0} \\
&= \exp\left(i \int_{C'} d^4y \; \mathcal{L}^I\left(-i \frac{\delta}{\delta J(y)} \right) \right) Z^0_\beta[J] \bigg|_{J=0}. \quad (6.17)
\end{aligned}$$

Here $Z^0_\beta[J]$ is the generating functional for a free theory (i.e., $\mathcal{L}^I = 0$). The key observation now is that

$$\begin{aligned}
Z^0_\beta[J] &= Z^0_\beta[0] \left\langle T_{C'} \left[\exp\left(i \int_{C'} d^4x \; J(x)\hat{\phi}_{in}(x) \right) \right] \right\rangle_0 \\
&= Z^0_\beta[0] \exp\left(-\frac{i}{2} \int_{C'} d^4x \, d^4y \; J(x) D^{(c)}_F(x,y) J(y) \right), \quad (6.18)
\end{aligned}$$

where

$$iD_F^{(c)}(x,y) = D^{(c)}(x,y) = \left\langle T_C\left[\hat{\phi}_{in}(x)\hat{\phi}_{in}(y)\right]\right\rangle_0 \qquad (6.19)$$

is the two-point thermal Green's function for free fields (also known as thermal propagator).

To prove the identity (6.18) it suffices to compare the coefficients in the corresponding Taylor expansions of J. The l.h.s. of (6.18) has the k-th coefficient in the series

$$\frac{1}{Z_\beta^0[0]} \left.\frac{(-i)^k\,\delta^k Z_\beta^0[J]}{\delta J(x_1)\cdots\delta J(x_k)}\right|_{J=0} = \left\langle T_C\left[\hat{\phi}_{in}(x_1)\ldots\hat{\phi}_{in}(x_k)\right]\right\rangle_0$$

$$= \begin{cases} 0, & \text{if } k \text{ odd} \\ \sum_{\text{distinct pairings}} iD_F^{(c)}(x_{i_1},x_{i_2})\cdots iD_F^{(c)}(x_{i_{k-1}},x_{i_k}), & \text{if } k \text{ even.} \end{cases} \qquad (6.20)$$

Here $\sum_{\text{distinct pairings}}$ denotes the sum over all $(2m-1)!! = (2m)!/2^m m!$ ways of breaking $(1,2,\ldots,k=2m)$ into m pairs. The second identity in (6.20) represents nothing but the thermal Wick theorem (cf. Appendix M).

On the other hand, the fact that

$$\frac{(-i)^k\delta^k}{\delta J(x_1)\cdots\delta J(x_k)} \exp\left(-\frac{i}{2}\int_C d^4x d^4y\; J(x)D_F^{(c)}(x,y)J(y)\right)\Bigg|_{J=0}$$

$$= \begin{cases} 0, & \text{if } k \text{ odd} \\ \sum_{\text{distinct pairings}} iD_F^{(c)}(x_{i_1},x_{i_2})\cdots iD_F^{(c)}(x_{i_{k-1}},x_{i_k}), & \text{if } k \text{ even,} \end{cases} \qquad (6.21)$$

is obvious by mere inspection. Note that since the field-time argument lives on the Schwinger time-path, one can omit the vertical part in the Keldysh–Schwinger ordering [225, 412]. Note also that $Z_\beta[0] = \text{Tr}(e^{-\beta H}) \neq 1$. Only at zero temperature $e^{-\beta\hat{H}} \to \mathbb{1}$ (for normally ordered Hamiltonians) and hence $Z_{\beta\to\infty}[0] = 1$.

We close this discussion with the following remark; let us decompose $D_F^{(c)}(\mathbf{x},t;\mathbf{y},0)$ into "retarded" and "advanced" parts as

$$D_F^{(c)}(\mathbf{x},t;\mathbf{y},0) = \theta_C(t)\,D^>(\mathbf{x},t;\mathbf{y},0) + \theta_C(-t)\,D^<(\mathbf{x},t;\mathbf{y},0). \qquad (6.22)$$

Here

$$iD^>(\mathbf{x},t;\mathbf{y},0) = \left\langle\hat{\phi}_{in}(\mathbf{x},t)\hat{\phi}_{in}(\mathbf{y},0)\right\rangle_0 = iD^<(\mathbf{y},0;\mathbf{x},t), \qquad (6.23)$$

are thermal Wightman's functions.

The cyclic property of the trace together with the identity $e^{-\beta H}\hat{\phi}(t,\mathbf{x})e^{\beta H} = \hat{\phi}(t+i\beta,\mathbf{x})$ then yields

$$D^>(\mathbf{x},t;\mathbf{y},0) = D^<(\mathbf{x},t+i\beta;\mathbf{y},0) = D^>(\mathbf{y},0;\mathbf{x},t+i\beta). \qquad (6.24)$$

Consequently, the propagator $D_F^{(c)}$ itself fulfils a periodicity condition between its two domains of analyticity. Indeed, consider $0 < \Im m(t) < \beta$ then $\theta_C(-t) = \theta_C(t - i\beta) = 1$ and correspondingly

$$D_F^{(c)}(\mathbf{x}, t; \mathbf{y}, 0) = D_F^{(c)}(\mathbf{x}, t - i\beta; \mathbf{y}, 0). \qquad (6.25)$$

The relations (6.24) and (6.25) for the thermal propagator are known as Kubo–Martin–Schwinger (KMS) periodicity conditions for boson fields.

It should be stressed that the KMS condition (6.24) not only results from the explicit form of the canonical distribution $\rho \propto e^{-\beta H}$ but it is, in fact, satisfied *only* in the case when the density matrix is the canonical one. This can be seen as follows; assume that yet another density matrix $\rho' \neq \rho$ fulfills the KMS condition and consider a pair of arbitrary operators say A and B that are defined in the Hilbert space in which "Tr" is defined (in our case $\mathcal{H}_F$). The KMS condition at $t = 0$ then acquires the form (spatial variables are suppressed)

$$\mathrm{Tr}(\rho' A(0) B(0)) = \mathrm{Tr}(\rho' B(0) A(i\beta)) = \mathrm{Tr}(\rho' B(0) e^{-\beta H} A(0) e^{\beta H}). \qquad (6.26)$$

Since B is arbitrary we can take $B = |\psi\rangle\langle\psi|$ where $|\psi\rangle \in \mathcal{H}_F$. Since $|\psi\rangle$ is allowed to run over all $\mathcal{H}_F$ we obtain the (weak) operator identity

$$\rho' A(0) = e^{-\beta H} A(0) e^{\beta H} \rho' \quad \Leftrightarrow \quad A(0) e^{\beta H} \rho' = e^{\beta H} \rho' A(0). \qquad (6.27)$$

The latter indicates that the operator $e^{\beta H} \rho'$ commutes with any A defined on $\mathcal{H}_F$ and hence according to Schur's lemma $e^{\beta H} \rho' \propto \mathbb{1}$. From the normalization condition $\mathrm{Tr}(\rho') = 1$ we thus have that $\rho' = e^{-\beta H}/\mathrm{Tr}(e^{-\beta H})$. So ρ' cannot be different from ρ and so it must coincide with the canonical Gibbs distribution.

The KMS condition (6.24) was originally expressed by Kubo [391] and Martin and Schwinger [445] as a boundary condition on the analytic behavior of thermal Green functions. The subject was further developed and deepened in the framework of *algebraic QFT* in [309, 365], in particular, a close connection with the Tomita–Takesaki theory [123] makes the KMS condition a key instrument in studying von Neumann algebras.

The functional integral and its measure

In order to establish contact with functional integrals, let us consider the identity for Fresnel integrals $(a \in \mathbb{R})$

$$\int_{-\infty}^{\infty} \frac{dx}{\sqrt{2\pi}} \exp\left(i \frac{a}{2} x^2\right) = \frac{1}{\sqrt{|a|}} \begin{cases} e^{i\pi/4}, & a > 0, \\ e^{-i\pi/4}, & a < 0, \end{cases} \qquad (6.28)$$

which implies that

$$\int_{-\infty}^{\infty} \prod_{i=1}^{N} dc_i \; \exp\left(\frac{i}{2} \sum_{n,m} c_n \mathbb{A}_{nm} c_m\right) = \prod_{i=1}^{N} \sqrt{\frac{2\pi}{|\lambda_i|}} \; e^{i\pi \, \text{sign}(\lambda_i)/4}$$

$$= \left|\det\left(\frac{\mathbb{A}}{2\pi}\right)\right|^{-\frac{1}{2}} e^{i\eta\pi/4}. \qquad (6.29)$$

Here $\mathbb{A}_{mn}$ is a real, symmetric (hence diagonalizable) $N \times N$ matrix with eigenvalues $\{\lambda_i; \; i = 1, \cdots, N\}$. The index $\eta = \sum_i^n \text{sign}(\lambda_i) \cdot 1$ is referred to as the Morse or Maslov index (or the Keller–Maslov index). The latter is important mostly in the context of transition amplitudes in Quantum Mechanics. For typical applications in QFT it turns out that it has a lesser importance and is typically omitted (for exceptions see, e.g., [175, 176]). For this reason we drop the exponential factor on the r.h.s. of (6.24) and replace $|\det(\ldots)|$ with $\det(\ldots)$.

For the connection to fields, we observe that any real function $\phi(x)$ can be expanded in terms of some real orthonormal basis $\{v_n(x); n \in \mathbb{N}\}$, i.e., $\phi(x) = \sum_n c_n v_n(x)$ with c_n's being the real expansion coefficients. So in particular, we can write

$$\int_\Sigma d^4x d^4y \; \phi(x) A(x,y) \phi(y) = \sum_{n,m} c_n \mathbb{A}_{nm} c_m, \qquad (6.30)$$

with

$$\mathbb{A}_{nm} = \int_\Sigma d^4x d^4y \; v_n(x) A(x,y) v_m(y). \qquad (6.31)$$

Here Σ denotes some spacetime region. The basis $\{v_n(x)\}$ must be chosen in such a way that it conforms with the prescribed behavior of ϕ on the boundary of Σ. We are interested in real $A(x,y)$. Since in Eq. (6.29) only the symmetric part of $A(x,y)$ survives, we assume, without any harm, that $A(x,y)$ is symmetric. Then, both $A(x,y)$ and $\mathbb{A}$ are diagonalizable, i.e., there are polar bases $\{u_n(x); n \in \mathbb{N}\}$ and $\{u_m^{(n)}; n, m \in \mathbb{N}\}$ such that

$$\int_\Sigma d^4x \; A(x,y) u_n(y) = \lambda_n u_n(y) \quad \text{and} \quad \sum_k \mathbb{A}_{m,k} u_k^{(n)} = \lambda_n u_m^{(n)}, \quad (6.32)$$

with

$$u_k^{(n)} = \int_\Sigma d^4x \; u_n(x) v_k(x) \quad \text{and} \quad u_n(x) = \sum_k u_k^{(n)} v_k(x). \quad (6.33)$$

Let us now discretize the spacetime points in Σ, so that Σ is spanned by N points x_i. Then any $\{\phi(x_i), i \in N\}$ can be expanded into N base functions $v_n(x_i)$ only. Then, we can formulate the integral measure as

$$\prod_{i=1}^{N} d\phi(x_i) = \prod_{n=1}^{N} dc_n \, \mathcal{J}^{(N)}, \tag{6.34}$$

with the Jacobian

$$\mathcal{J}^{(N)} = \det \begin{bmatrix} v_1(x_1) \; v_1(x_2) \cdots & \cdots \\ v_2(x_1) \; v_2(x_2) & \\ \vdots & \ddots \\ \vdots & v_n(x_n) \end{bmatrix}. \tag{6.35}$$

Note that, due to the orthonormality of the base system $\{v_n(x)\}$, we have that $\mathcal{J}^{(N)} \to 1$ in the large N limit (modulo x-independent multiplicative constant). Truncation of the base system elements changes the infinite-dimensional matrix $\mathbb{A}$ to finite-dimensional $N \times N$ matrix $\mathbb{A}^{(N)}$. Recalling Eq. (6.29), this allows us to define the functional integral over ϕ as

$$\int \mathcal{D}\phi \, \exp \left(\frac{i}{2} \int_{\Sigma} d^4x d^4y \; \phi(x) A(x,y)\phi(y) \right)$$

$$\equiv \lim_{N \to \infty} \int \prod_{i=1}^{N} dc_i \, \exp \left(\frac{i}{2} \sum_{n,m} c_n \mathbb{A}_{nm}^{(N)} c_m \right) = \lim_{N \to \infty} \left[\det \left(\frac{\mathbb{A}^{(N)}}{2\pi} \right) \right]^{-\frac{1}{2}}$$

$$= \mathcal{N}' \left[\det \left(A(x,y) \right) \right]^{-\frac{1}{2}}, \tag{6.36}$$

where we have assimilated all irrelevant constants in front of the functional determinant into the overall factor $\mathcal{N}'$. The functional "measure" is then formally defined as

$$\mathcal{D}\phi = \lim_{N \to \infty} \prod_{i=1}^{N} d\phi(x_i) \left(\mathcal{J}^{(N)} \right)^{-1}, \tag{6.37}$$

and it is implicitly assumed that the limit appears in front of integrals. In fact, the limiting procedure and integrations are not mutually interchangeable operations [286,541,668]. The method of truncating to a finite number of modes N and taking the limit $N \to \infty$ to define functional determinants is referred to as a regularization procedure.

Let us now choose $\Sigma = C \times \mathbb{R}^3$ (C is the Keldysh–Schwinger time-path) and $[A(x,y)]^{-1} = D_F^{(c)}(x,y)$; we then obtain from the KMS condition

(6.24), from Eqs. (6.32) and (6.33) that $\phi(x)$ appearing in the functional integral (6.36) is restricted by the boundary condition $\phi(t_i - i\beta, \mathbf{x}) = \phi(t_i, \mathbf{x})$, known as the single-field KMS condition. The free partition function can now be written as

$$Z_\beta^0[J] = \left[\det D_F^{(c)}\right]^{\frac{1}{2}} \exp\left(-\frac{i}{2}\int_C d^4x d^4y \; J(x) D_F^{(c)}(x,y) J(y)\right)$$

$$= \mathcal{N}\int \mathcal{D}\phi \; \exp\left(\frac{i}{2}\int_C d^4y d^4x \; [\phi + J D_F^{(c)}](D_F^{(c)})^{-1}[\phi + D_F^{(c)}J]\right)$$

$$\times \exp\left(-\frac{i}{2}\int_C d^4x d^4y \; J(x) D_F^{(c)}(x,y) J(y)\right), \tag{6.38}$$

where we have used the translational invariance of the measure under the shift of the field ϕ to $\phi' = (\phi + D_F^{(c)} J)$ and the fact that also ϕ' fulfills the KMS boundary condition $\phi'(t_i - i\beta, \mathbf{x}) = \phi'(t_i, \mathbf{x})$. So finally we find

$$Z_\beta^0[J] = \mathcal{N}\int \mathcal{D}\phi \; \exp\left(\frac{i}{2}\int_C d^4y d^4x \; \phi(x)(D_F^{(c)})^{-1}(x,y)\phi(y)\right)$$

$$\times \exp\left(+i\int_C d^4x \; \phi J\right)$$

$$= \mathcal{N}\int \mathcal{D}\phi \; \exp\left(iS_0[\phi,\beta] + i\int_C d^4x \; \phi(x)J(x)\right). \tag{6.39}$$

Here the $[\det D_F^{(c)}]$ term (i.e., a constant term) and $Z_\beta^0[0]$ were included in the multiplicative pre-factor $\mathcal{N}$.

6.3.2   *The Feynman–Matthews–Salam formula*

Consider now the field theory with an interaction. The functional integral representation of $Z_\beta[J]$ is obtained by use of the last line of Eq. (6.17):

$$Z_\beta[J] = \mathcal{N}\exp\left(i\int_C d^4y \; \mathcal{L}^I\left(-i\frac{\delta}{\delta J(y)}\right)\right)$$

$$\times \int \mathcal{D}\phi \; \exp\left(iS_0[\phi,\beta] + i\int_C d^4x \; \phi(x)J(x)\right)$$

$$= \mathcal{N}\int \mathcal{D}\phi \; \exp\left(iS[\phi,\beta] + i\int_C d^4x \; \phi(x)J(x)\right). \tag{6.40}$$

Since $\mathcal{N}$ is J independent and Green's functions are generated through the ratio $Z_\beta[J]/Z_\beta[0]$, we can eliminate $\mathcal{N}$ in favor of $Z_\beta[0]$ and write

$$\frac{Z_\beta[J]}{Z_\beta[0]} = \frac{\int \mathcal{D}\phi \; \exp\left(iS[\phi,\beta] + i\int_C d^4x \; \phi(x)J(x)\right)}{\int \mathcal{D}\phi \; \exp\left(iS[\phi,\beta]\right)}. \tag{6.41}$$

So far we have considered only real scalar fields. Extension to complex scalar fields is obtained by means of an analogue of Eq. (6.29), namely

$$
\int_{-\infty}^{\infty} \prod_{i=1}^{N} dc_i dc_i^* \, \exp\left(i \sum_{n,m} c_n^* \mathbb{A}_{nm} c_m \right) = \prod_{i=1}^{N} \frac{2\pi}{|\lambda_i|} \, e^{i\pi \, \text{sign}(\lambda_i)/2}
$$

$$
= \left[\det\left(\frac{|\mathbb{A}|}{2\pi} \right) \right]^{-1} e^{i\eta\pi/2}. \quad (6.42)
$$

Then the following result naturally emerges from what we have done so far:

$$
Z_\beta^0[J] = \mathcal{N} \int \mathcal{D}\phi \mathcal{D}\phi^* \, \exp\left(i \int_C d^4y d^4x \, \phi^* (D_F^{(c)})^{-1} \phi \right)
$$

$$
\times \exp\left(i \int_C d^4x \, \phi J + i \int_C d^4x \, \phi^* J^* \right)
$$

$$
= \mathcal{N} \int \mathcal{D}\phi \mathcal{D}\phi^* \, \exp\left(i S_0[\phi,\beta] + i \int_C d^4x \, \phi J + i \int_C d^4x \, \phi^* J^* \right). \quad (6.43)
$$

The factor $\mathcal{N}$ can again be eliminated in favor of $Z_\beta[0]$ and consequently obtain

$$
\frac{Z_\beta[J]}{Z_\beta[0]} = \frac{\int \mathcal{D}\phi \mathcal{D}\phi^* \, \exp\left(i S[\phi,\phi^*,\beta] + i \int_C d^4x \, \phi J + i \int_C d^4x \, \phi^* J^* \right)}{\int \mathcal{D}\phi \mathcal{D}\phi^* \, \exp\left(i S[\phi,\phi^*,\beta] \right)}. \quad (6.44)
$$

By differentiations with respect to J or J^* we obtain from Eq. (6.41) (and Eq. (6.44)) the following results

$$
\langle T(P[\hat{\phi}]) \rangle = \frac{\int \mathcal{D}\phi \, P_r[\phi] \, \exp\left(i S[\phi,\beta] \right)}{\int \mathcal{D}\phi \, \exp\left(i S[\phi,\beta] \right)}, \quad (6.45)
$$

for a real scalar field,

$$
\langle T(P[\hat{\phi}]) \rangle = \frac{\int \mathcal{D}\phi \mathcal{D}\phi^* \, P_r[\phi] \, \exp\left(i S[\phi,\phi^*,\beta] \right)}{\int \mathcal{D}\phi \mathcal{D}\phi^* \, \exp\left(i S[\phi,\phi^*,\beta] \right)}, \quad (6.46a)
$$

$$
\langle T(P[\hat{\phi}^*]) \rangle = \frac{\int \mathcal{D}\phi \mathcal{D}\phi^* \, P_r[\phi^*] \, \exp\left(i S[\phi,\phi^*,\beta] \right)}{\int \mathcal{D}\phi \mathcal{D}\phi^* \, \exp\left(i S[\phi,\phi^*,\beta] \right)}, \quad (6.46b)
$$

for a complex scalar field. Here $P[\ldots]$ is an arbitrary polynomial in $\hat{\phi}$ (or $\hat{\phi}^*$), and the subscript r suggests that the wave function renormalization factors are included. A simple formula akin to Eqs. (6.45)–(6.46) holding for $P[\hat{\phi},\hat{\phi}^*]$ does not exist in general. This is related to the ordering problem; operators $\hat{\phi}$ and $\hat{\phi}^*$ do not commute at equal times. In particular, the r.h.s. would depend on the order of $\hat{\phi}$ and $\hat{\phi}^*$ at equal time, while the l.h.s.

would not (due to the c-number nature of fields in the functional integral). This ordering issue can be, however, unambiguously resolved in coherent state functional integrals. Eqs. (6.45) and (6.46) are referred to as thermal Feynman–Matthews–Salam formulas [408, 461].

For most applications (though not all [224,225,412]) it turns out that the vertical part of the contour C can be neglected. It is then usual practice to label the field (source) with the time argument on the upper branch of C as ϕ_+ (J_+) and that with the time argument on the lower branch of C as ϕ_- (J_-). Introducing the metric $(\sigma_3)_{\alpha\beta}$ (σ_3 is the Pauli matrix and $\alpha, \beta = \{+, -\}$) we can write e.g., $J_+\phi_+ - J_-\phi_-$ as $J_\alpha(\sigma_3)^{\alpha\beta}\phi_\beta$. For the raised and lowered indices we simply read: $\phi_+ = \phi^+$ and $\phi_- = -\phi^-$ (similarly for J_α). With this convention $Z^0_\beta[J]$ is given by

$$Z^0_\beta[J] = \mathcal{N} \int \mathcal{D}\phi \, \exp\left(\frac{i}{2} \int d^4y d^4x \, \phi_\alpha(x)(D_F^{-1})^{\alpha\beta}(x, y)\phi_\beta(y)\right)$$

$$\times \exp\left(+i \int d^4x \, \phi_\alpha(x)J^\alpha(x)\right)$$

$$= Z^0_\beta[0] \, \exp\left(-\frac{i}{2} \int d^4x d^4y \, J_\alpha(x)D_F^{\alpha\beta}(x, y)J_\beta(y)\right). \qquad (6.47)$$

Here we have used

$$\int d^4y \, (D_F^{-1})_{\alpha\gamma}(x, y)D_F^{\gamma\beta}(y, z) = (\sigma_3)^\beta_\alpha \delta^4(x - z). \qquad (6.48)$$

When both x and y are on the $-\infty$ to ∞ part of C (i.e., upper branch) then the T_C ordering is just ordinary chronological ordering T; this corresponds to the D_F^{11} element of $D_F^{\alpha\beta}$. When both time arguments are on the ∞ to $-\infty$ part of C (i.e., lower branch), then T_C reduces to anti-chronological ordering $\bar{T}$; this corresponds to the D_F^{22} element. The off-diagonal elements of $D_F^{\alpha\beta}$ then correspond to x_0 being on one part of C and y_0 on the other. One can thus write the free thermal Green's function as (cf. Section 5.3.1)

$$D_F^{\alpha\beta}(x, y) = \begin{bmatrix} \langle T[\hat{\phi}_{in}(x)\hat{\phi}_{in}(y)]\rangle_0 & \langle \hat{\phi}_{in}(y)\hat{\phi}_{in}(x)\rangle_0 \\ \langle \hat{\phi}_{in}(x)\hat{\phi}_{in}(y)\rangle_0 & \langle \bar{T}[\hat{\phi}_{in}(x)\hat{\phi}_{in}(y)]\rangle_0 \end{bmatrix}. \qquad (6.49)$$

Due to the trace in the definition of $\langle \ldots \rangle_0$ and due to the fact that the Hamiltonian $\hat{H}$ and the total 4-momentum $\hat{P}^\mu$ commute, there is an invariance under translation in each matrix element of $D_F^{\alpha\beta}$, i.e., $D_F^{\alpha\beta}$ depends only on the difference $x - y$. For a single scalar field the matrix structure of the propagator can be easily calculated, and its Fourier transform reads

$$D_F^{\alpha\beta}(k) = \begin{bmatrix} \frac{i}{k^2-m^2+i\epsilon} & 2\pi\delta^-(k^2-m^2) \\ 2\pi\delta^+(k^2-m^2) & \frac{-i}{k^2-m^2-i\epsilon} \end{bmatrix}$$

$$+ 2\pi\delta(k^2-m^2)n(k^0)\begin{bmatrix} 1 & 1 \\ 1 & 1 \end{bmatrix}. \tag{6.50}$$

Note that in addition to the standard zero-temperature contribution, we have a term proportional to the occupation number $n(k_0)$. It has been known already since the seminal work of Feynman [236] that QFT-free Green's functions describe on the *first* quantized level an amplitude of transition for an associated relativistic particle (in this case the Klein–Gordon particle). In this view $D_F^{\alpha\beta}(x,y)$ gives the amplitude of transition that a particle is detected at the spacetime point x provided it was inserted into the system at the point y. The first matrix term in (6.50) corresponds to the particle propagating from y to x without presence of a heat bath, precisely as in the original ($T = 0$) closed time-path formalism (cf. Fig. 6.1). There is, however, a possibility that the detected particle comes from the heat bath; this is accounted for by the second (thermal) term. Representation (6.50) for the free thermal Green's function is known as Mills' representation [475].

It is possible to extend the above considerations to fermion field theories. Despite the complications related with the anticommuting nature of the fermion fields, the final formulas for the functional integrals (6.45) and (6.46) are the same with the proviso that the single-field KMS condition (i.e., boundary condition for functional integration) is antiperiodic. As for the matrix propagator (6.50), there is a rather similar matrix propagator for a free Fermi field, but with the Fermi–Dirac distribution replacing the Bose–Einstein one.

6.3.3 *More on generating functionals*

The generating functional $Z_\beta[J]$ serves as an important bookkeeping device for generating higher order thermal Green's functions. However, in order to systematize the actual perturbation expansions, it turns out that it is more convenient to introduce the connected Green's functions and vertex (or proper) functions. In analogy to the zero-temperature case, the connected thermal n-point Green's function $G_{\alpha_1\ldots\alpha_n}^c$ describe the sum of all thermal diagrams with n external lines that do not have topologically disconnected sub-diagrams. Sub-indices $\alpha_1, \alpha_2, \ldots$ correspond to thermal indices at the end of external lines. The thermal n-point vertex function $\Gamma^{(n)}_{\alpha_1...\alpha_n}$ corresponds to the sum of all 1PI (one-particle irreducible) thermal diagrams with n external lines bearing the thermal indices from α_1 to α_n.

By 1PI thermal diagram we mean a connected diagram that cannot separate into two parts by cutting a single line. Conventionally, 1PI thermal diagrams are evaluated with no thermal propagators on external lines.

The generating functional for connected Green's functions is defined as

$$W_\beta[J] = \sum_{n=1}^{\infty} \frac{1}{n!} \int_C \prod_{i=1}^{n} d^4 x_i \, J(x_1)\ldots J(x_n) \, G^c(x_1,\ldots,x_n). \qquad (6.51)$$

Here the thermal indices are implicit in the time contour C. In contrast to the $Z_\beta[J]$ case, the generating functional $W_\beta[J]$ is expanded without i factors, and the sum is taken from "1".

There exists a simple relationship between Z_β and W_β. Consider indeed $(-i\delta Z_\beta[J]/\delta J(y))/Z_\beta[0]$. The actual effect of this is that we strip one leg of the external source J. As a result one obtains a sum of diagrams that can be factorized into a product of two sums. The first sum contains diagrams with a source $J(y)$ missing at one external leg, the second sum comprises diagrams all having external legs terminating at sources. The first sum does not have, by construction, any disconnected subdiagrams and hence it must coincide with $\delta W_\beta[J]/\delta J(y)$. The second sum is nothing but $Z_\beta[J]/Z_\beta[0]$. So finally we can write

$$-i\frac{\delta Z_\beta[J]}{\delta J(y)} = Z_\beta[J] \frac{\delta W_\beta[J]}{\delta J(y)}, \qquad (6.52)$$

with the boundary condition $W_\beta[0] = 0$. The solution is then

$$Z_\beta[J] = Z_\beta[0] \exp(iW_\beta[J]). \qquad (6.53)$$

Connected Green's functions are also known as *cumulants*.

The first functional derivative of $W_\beta[J]$ defines the thermal expectation value of the Heisenberg field $\hat{\phi}(x)$ in the presence of the source $J(x)$:

$$\phi_{c,\alpha}(x) \equiv \frac{\delta W_\beta[J]}{\delta J^\alpha(x)} = -\frac{i}{Z_\beta[J]} \frac{\delta Z_\beta[J]}{\delta J^\alpha(x)}$$

$$= \frac{Z_\beta[0]}{Z_\beta[J]} \left\langle T_C \left[\hat{\phi}_\alpha(x) \exp\left(i \int_C d^4 x \, J(x)\hat{\phi}(x) \right) \right] \right\rangle. \qquad (6.54)$$

Here $\alpha = \{+, -\}$. The index c indicates that $\phi_{c,\alpha}$ is essentially classical in its nature.

In the theory of phase transitions, to deal with these thermal-vacuum degrees of freedom a systematic method has been developed by considering the (quantum) effective action, $\Gamma_\beta^{\text{eff}}$, which is is defined as the functional Legendre transform of $W_\beta[J]$:

$$\Gamma_\beta^{\text{eff}}[\phi_c] = W_\beta[J] - \int_C d^4x \, J(x)\phi_c(x). \tag{6.55}$$

Here J is expressed in terms of ϕ_c: $J(x) \equiv J(x, \phi_c(x))$. We note that

$$\frac{\delta\Gamma_\beta^{\text{eff}}[\phi_c]}{\delta\phi_c(x)} = \frac{\delta W_\beta[J]}{\delta\phi_c(x)} - J(x) - \int_C d^4y \, \frac{\delta J(y)}{\delta\phi_c(x)} \phi_c(y), \tag{6.56a}$$

$$\frac{\delta W_\beta[J]}{\delta\phi_c(x)} = \int_C d^4y \, \frac{\delta W_\beta[J]}{\delta J(y)} \frac{\delta J(y)}{\delta\phi_c(x)} = \int_C d^4y \, \phi_c(y) \frac{\delta J(y)}{\delta\phi_c(x)}. \tag{6.56b}$$

Substituting the latter equation into the former, we have

$$\frac{\delta\Gamma_\beta^{\text{eff}}[\phi_c]}{\delta\phi_{c,\alpha}(x)} = -J^\alpha(x). \tag{6.57}$$

In particular, when we set $J = 0$ we obtain from Eq. (6.54) that $\phi_{c,\alpha} = \langle\hat\phi_\alpha\rangle$. Due to the cyclic property of the trace entering the definition of thermal averages, $\phi_{c,\alpha}$ becomes time-independent. In many bosonic systems, such as in superfluid helium ^4He, $\langle\hat\phi\rangle$ directly corresponds to the (finite-temperature) order parameter. The order parameter is thus the solution of the equation

$$\left. \frac{\delta\Gamma_\beta^{\text{eff}}[\phi_c]}{\delta\phi_c(x)} \right|_{\phi_c=\langle\hat\phi\rangle} = 0. \tag{6.58}$$

Near the critical temperature Eq. (6.58) turns out to be the so called Landau–Ginzburg equation (see also Appendix O).

To introduce the generating functional of 1PI thermal vertex functions we assume, following [163], the existence of a functional $\Gamma_\beta^{1\text{PI}}[\phi]$ such that

$$\exp\left(\frac{i}{a} W_\beta[J]\right) = \int \mathcal{D}\phi \, \exp\left[\frac{i}{a}\left(\Gamma_\beta^{1\text{PI}}[\phi] + \int_C d^4x \, \phi(x)J(x)\right)\right], \tag{6.59}$$

to the lowest order in a. Here, for simplicity, the normalization constant $\mathcal{N}$ is assimilated into the measure. The point of stationary phase $\phi_c(x)$ is at

$$\frac{\delta\Gamma_\beta^{1\text{PI}}[\phi]}{\delta\phi_\alpha(x)} = -J^\alpha(x), \tag{6.60}$$

and hence

$$W_\beta[J] = \Gamma_\beta^{1\text{PI}}[\phi_c] + \int_C d^4x \, \phi_c(x)J(x), \tag{6.61}$$

which is the inverse of the Legendre transform (6.55). Using Eq. (6.60), we can expand $\Gamma_\beta[\phi_c]$ as

$$\Gamma_\beta^{1PI}[\phi_c] = \sum_{n=1}^{\infty} \frac{1}{n!} \int_C \prod_{i=1}^{n} d^4 x_i \, \phi_c(x_1) \dots \phi_c(x_n) \, \Gamma^n(x_1, \dots, x_n). \quad (6.62)$$

The limit for $a \to 0$ selects, in the r.h.s. of (6.59), the tree level graphs of a theory whose action is $\Gamma_\beta^{1PI}[\phi]$ (similarly, the $\hbar \to 0$ limit picks up at $T = 0$ the tree level of the theory defined by $S[\phi]$). A vertex in one of these tree graphs is some Γ^n, defined in Eq. (6.62). On the other hand, $W_\beta[J]$ is the sum of all connected thermal Green's functions of the original theory. Since the connected Green's functions can be broken down into 1PI components, $\Gamma^n(x_1, \dots, x_n)$ must coincide with the thermal n-point vertex function. Note also that a formal comparison between Legendre transform (6.55) and (6.61) shows that we can identify Γ_β^{1PI} with $\Gamma_\beta^{\text{eff}}$.

In conclusion, we arrive at the following important relations:

$$\frac{1}{Z_\beta} \frac{\delta^n Z_\beta}{\delta J(x_1) \dots \delta J(x_n)} \bigg|_{J=0} = i^n \, D^{(c)}(x_1, \dots, x_n), \quad (6.63a)$$

$$\frac{\delta^n W_\beta}{\delta J(x_1) \dots \delta J(x_n)} \bigg|_{J=0} = (-1)^{n+1} \, G^c(x_1, \dots, x_n), \quad (6.63b)$$

$$\frac{\delta \phi_c(x)}{\delta J(y)} \bigg|_{J=0} = -G^c(x, y), \quad (6.63c)$$

$$\frac{\delta^n \Gamma_\beta^{1PI}}{\delta \phi(x_1) \dots \phi(x_n)} \bigg|_{\phi=\phi_c} = \Gamma^{(n)}(x_1 \dots x_n). \quad (6.63d)$$

6.4   The effective action and the Schwinger–Dyson equations

The effective action formalism was introduced by Schwinger [561, 562], Goldstone, Salam and Weinberg [291], and Jona-Lasinio [360] to account for the spontaneous symmetry breaking (SSB) phenomenon.

The Schwinger–Dyson equations [211, 559, 560] (SDE) were originally constructed with the motivation that they could provide some information about the complete Green's functions outside the scope of perturbative theory. We shall now derive the finite-temperature SDE with the help of functional integrals.

The $(T \neq 0)$ generating functional of Green's functions $Z_\beta[J]$ is

$$Z_\beta[J] = \int \mathcal{D}\phi \exp\left(i(S[\phi, \beta] + \int_C d^4x \, J(x)\phi(x))\right) \quad \text{(6.64a)}$$

$$S[\phi, \beta] = \int_C d^4x \, \mathcal{L}(x), \quad \text{(6.64b)}$$

where $\int_C d^4x = \int_C dx_0 \int_{\mathbb{R}^3} d\mathbf{x}$ with the subscript C indicating that time runs along some contour in the complex plane. In the real-time formalism, which we adopt throughout, the most natural version is the so-called Keldysh–Schwinger one [367, 560], which is represented by the contour depicted in Fig. 6.2. Within the functional integral formalism, the c-number fields are further restricted by the periodic boundary condition — KMS condition [391, 445], which for bosonic fields reads:

$$\phi(t_i - i\beta, \mathbf{x}) = \phi(t_i, \mathbf{x}). \quad \text{(6.65)}$$

The zero-temperature generating functional may be recovered from Eq. (6.64) if one integrates over the close-time-path (no vertical parts) and omits the KMS condition (6.65). The l.h.s. of Eq. (6.64) is independent of ϕ and thus it is invariant under infinitesimal point transformation

$$\phi(x) \rightarrow \phi(x) + \varepsilon f(x) \equiv \phi'(x), \quad \varepsilon \ll 1, \quad \text{(6.66)}$$

where $f(x)$ is an arbitrary (ϕ-independent) function which fulfils the periodic boundary condition

$$f(t_i - i\beta, \mathbf{x}) = f(t_i, \mathbf{x}). \quad \text{(6.67)}$$

Eq. (6.66) implies that the functional Jacobian is one, i.e., $\mathcal{D}\phi = \mathcal{D}\phi'$. The invariance of $Z_\beta[J]$ under infinitesimal field transformation (6.66) gives

$$\begin{aligned}
Z_\beta[J] &= \int \mathcal{D}\phi' \, e^{iS[\phi'-\varepsilon f]+i\int_C d^4x \, J(x)\phi'(x)-i\varepsilon\int_C d^4x \, J(x)f(x)} \\
&= -i\varepsilon \int \mathcal{D}\phi' \left\{ \int_C d^4x \left(\frac{\delta S}{\delta \phi'}(x) + J(x) \right) f(x) \right\} e^{iS[\phi']+i\int_C J\phi'} \\
&\quad + Z_\beta[J] + \mathcal{O}(\varepsilon^2),
\end{aligned} \quad \text{(6.68)}$$

which means that

$$\begin{aligned}
0 &= \int \mathcal{D}\phi \left\{ \int_C d^4x \left(\frac{\delta S}{\delta \phi}(x) + J(x) \right) f(x) \right\} e^{iS[\phi]+i\int_C J\phi} \\
&= \int_C d^4x \left\langle \frac{\delta S[\phi]}{\delta \phi}(x) + J(x) \right\rangle f(x).
\end{aligned} \quad \text{(6.69)}$$

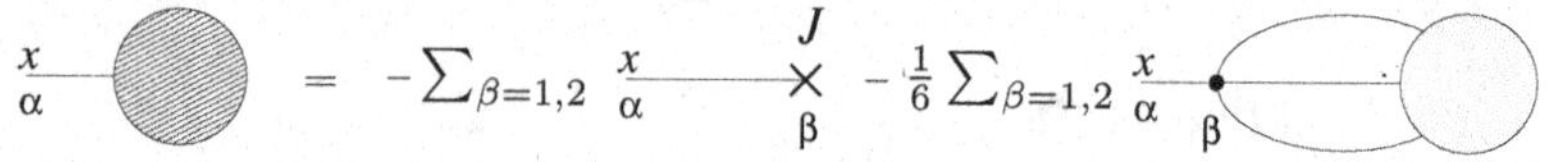

Fig. 6.3 The graphical representation of Eq. (6.76). Hatched blob refers to the (full) 1-point amputated Green's function, the dotted blob refers to the (full) 3-point amputated Green's function, the cross denotes the source J, while the heavy dot without coordinate indicates that the vertex must be integrated over all possible positions.

Since Eq. (6.69) is true for any $f(x)$ fulfilling the condition (6.67), it is

$$0 = \left\langle \frac{\delta S[\phi]}{\delta \phi}(x) + J(x) \right\rangle = \left(\frac{\delta S}{\delta \phi}\left[\frac{\delta}{i\delta J(x)} \right] + J(x) \right) Z[J]. \quad (6.70)$$

A more useful form of the solution (6.70) is obtained by using the identity

$$F\left[-\frac{i}{Z}\frac{\delta}{\delta J} Z \right] \Psi = F\left[\left(\phi - i\frac{\delta}{\delta J} \right) \right] \Psi, \quad (6.71)$$

which is valid for any analytic function $F[\ldots]$ and any test function Ψ. This identity follows from the commutation relation

$$-i\frac{\delta}{\delta J} Z\Psi = Z\left(\phi - i\frac{\delta}{\delta J} \right)\Psi. \quad (6.72)$$

Using Eq. (6.71) and setting the test function $\Psi = \mathbb{1}$, Eq. (6.70) becomes

$$-J(x) = \frac{\delta S}{\delta \phi}\left[\phi(x) - i\frac{\delta}{\delta J(x)} \right] \mathbb{1}$$

$$= \frac{\delta S}{\delta \phi}\left[\phi(x) + i\int_C d^4z\, D^c(x,z)\frac{\delta}{\delta\phi(z)} \right] \mathbb{1}. \quad (6.73)$$

Eq. (6.73), together with its derivatives, represents the functional version of the SDE.

As an illustration of the SDE, we consider the $\lambda\phi^4$ theory. Following the prescription (6.73) we find (no sum over α!):

$$-J_\alpha(x) = -(\partial^2 + m_0^2)\phi_\alpha(x) + \frac{\lambda_0}{3!} D^{(3)}_{\alpha\alpha\alpha}(x,x,x). \quad (6.74)$$

If we now use the identity:

$$(\partial_x^2 + m_0^2) D_F^{\alpha\beta}(x,y) = -(\sigma_3)^{\alpha\beta}\delta(x-y), \quad (6.75)$$

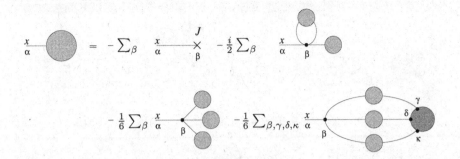

Fig. 6.4 The graphical representation of Eq. (6.76) in terms of connected Green's functions. Here the double-hatched blob describes the 3-point vertex function and the hatched blobs refer to the connected (truncated) Green's functions.

we may invert the differential operator in Eq. (6.74) and write

$$- \int d^4y \, J_\beta(y) \, D_F^{\beta\alpha}(y,x)$$

$$= \phi^\alpha(x) + \frac{\lambda_0}{3!} \int d^4y \, D_F^{\alpha\beta}(x,y) \, D_{\beta\beta\beta}^{(3)}(y,y,y) \,. \qquad (6.76)$$

The graphical representation of Eq. (6.76) is depicted in Fig. 6.3. The corresponding graphical representation in terms of connected Green's functions can be obtained from the second equality in (6.73). Using the identity

$$\frac{\delta D_{\alpha\beta}^c(x,y)}{\delta\phi_\gamma(z)} = - \int dy_1^4 \, dy_2^4 \, D_{\alpha\delta}^c(x,y_1)\Gamma^{(3)\,\delta\kappa\gamma}(y_1,y_2,z) \, D_{\kappa\beta}^c(y_2,y) \,,$$

we find

$$J_\alpha(x) = (\partial^2 + m_0^2)\phi_\alpha(x) + \frac{\lambda_0}{3!}\Big\{ (\phi_\alpha(x))^3 + i3\phi_\alpha(x) \, D_{\alpha\alpha}^c(x,x)$$

$$- \int d^4y \, d^4w \, d^4z \, D_{\alpha\beta}^c(x,y) \, D_{\alpha\gamma}^c(x,w) \, \Gamma^{(3)\,\beta\gamma\delta}(y,w,z) \, D_{\delta\alpha}^c(z,x) \Big\} \,, \qquad (6.77)$$

which has its diagrammatic counterpart depicted in Fig. 6.4.

To obtain the SDE for the 2-point (connected) Green's functions, we differentiate Eq. (6.74) with respect to $J_\beta(y)$. This yields

$$(\partial_x^2 + m_0^2) \, D_{\alpha\beta}^c(x,y) \, - \, \frac{\lambda_0}{3!} D_{\alpha\alpha\alpha\beta}^{(4)}(x,x,x,y) \, + \, \frac{\lambda_0 i}{3!} \, \phi_\beta(y) \, D_{\alpha\alpha\alpha}^{(3)}(x,x,x)$$

$$= -\,(\sigma_3)_{\alpha\beta}\delta(x-y) \,, \qquad (6.78)$$

or, after inverting the differential operator

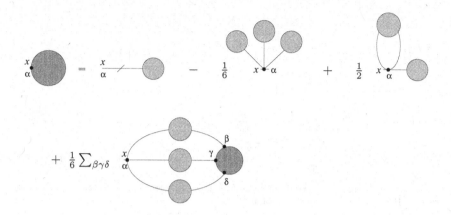

Fig. 6.5   The graphical representation of the SDE for the 1-point vertex function in $\lambda\phi^4$ theory. Here the slash stands for the inverse free thermal propagator.

$$D^c_{\alpha\beta}(x,y) = D_{F\,\alpha\beta}(x,y) \;-\; \frac{\lambda_0}{3!} \int d^4z\, D_F{}^{\gamma}{}_\alpha(x,z)\, D^{(4)}_{\gamma\gamma\gamma\beta}(z,z,z,y)$$

$$+ \frac{\lambda_0 i}{3!} \int d^4z\, D_F{}^{\gamma}{}_\alpha(x,z)\, D^{(3)}_{\gamma\gamma\gamma}(z,z,z)\, \phi_\beta(y)\,. \qquad (6.79)$$

As in the case of 1-point Green's function, Eq. (6.79) can be formulated in terms of connected Green's functions and vertex functions.

The functional relations (6.73) can also be used to generate the SDE for the 1PI Green's functions. These may be obtained by applying various powers of derivatives with respect to ϕ on Eq. (6.73). So, for instance, for the 1-point vertex function, i.e.,

$$\Gamma^{(1)}_\alpha(x) \;=\; \frac{\delta\Gamma}{\delta\phi^\alpha(x)} \;=\; -J_\alpha(x)\,, \qquad (6.80)$$

one may verify by mere inspection that the graphical representation for the $\lambda\phi^4$ theory (cf. Eq. (6.74)) is that of Fig. 6.5.

The 2- and 4-point vertex functions $\Gamma^{(2)}$ and $\Gamma^{(4)}$ turn out to be indispensable tools in renormalisation prescriptions.

Finally, when the source term in Eq. (6.73) is written with the help of Eq. (6.80), we obtain the remarkable identity:

$$\frac{\delta\Gamma[\phi]}{\delta\phi^\alpha(x)} \;=\; \frac{\delta S}{\delta\phi^\alpha(x)}\left[\phi(x) + i\int_C d^4z\, D^c(x,z)\frac{\delta}{\delta\phi(z)}\right]\mathbb{1}\,. \qquad (6.81)$$

Eq. (6.81) provides the reason why the generating functional for 1PI Green's function $\Gamma[\phi]$ is called the *effective action*. If the derivatives are dropped, the effective action reduces to the classical action S. The role of the derivatives is to take into account both quantum and thermal fluctuations. At $T = 0$ (no thermal fluctuations) the derivative terms generate quantum corrections in terms of loops. In fact, effective action and classical action are equal in the leading order in the saddle-point (or WKB) approximation.

6.5 Imaginary-time formalism

So far we have been discussing two key approaches to thermal quantum field theory, namely Thermo Field Dynamics and real (or Minkowskian) time-path formalism. Because in both cases the time argument in Green's functions is real, these methods are commonly known as real-time methods. An important advantage of the real-time methods lies in the fact that Green's functions can be formulated directly as functions of energy and time. In this Section we briefly mention another (historically the oldest) approach, namely the imaginary-time (also Matsubara or Euclidean) formalism of thermal QFT.

As seen in Section 6.2, in the real time-path formalism the time argument runs along the contour C (Fig. 6.2), i.e., from an arbitrary initial time $t_i \in \mathbb{R}$ down to $t_i - i\beta$. We also know that the contour C must pass through the time arguments of all Green's functions under consideration. Since the 2-point thermal Green's function is defined as (cf. Appendix M)

$$D^{(c)}(x,y) = \theta_C(x_0 - y_0)D_W(x,y) + \theta_C(y_0 - x_0)D_W(y,x), \quad (6.82)$$

and the thermal Wightman's function $D_W(x,y) \equiv D^>(x,y)$ has the spectral decomposition (n and m label the eigenstates of H)

$$
\begin{aligned}
D_W(x,y) &= (Z_\beta)^{-1}\text{Tr}\left(e^{-\beta H}\phi(x)\phi(y)\right) \\
&= (Z_\beta)^{-1}\text{Tr}\left(e^{-\beta H}e^{iHx_0}\phi(0,\mathbf{x})e^{-iHx_0}e^{iHy_0}\phi(0,\mathbf{y})e^{-iHy_0}\right) \quad (6.83) \\
&= (Z_\beta)^{-1}\sum_{n,m} e^{iE_n(x_0-y_0+i\beta)}\langle n|\phi(0,\mathbf{x})|m\rangle e^{-iE_m(x_0-y_0)}\langle m|\phi(0,\mathbf{y})|n\rangle,
\end{aligned}
$$

there is a restriction on the domain of definition of $D^{(c)}(x,y)$, namely[2]

$$-\beta \leq \Im m(x_0 - y_0) \leq 0 \quad \text{for} \quad \theta_C(y_0 - x_0) = 0, \quad (6.84)$$

---

[2]The exponentials are assumed to dominate the convergence of the sum for large E_n.

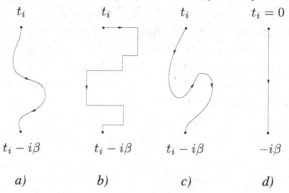

Fig. 6.6   Complex time paths. Paths $a)$, $b)$ and $d)$ are allowed, while $c)$ is not. Path $d)$ gives rise to imaginary-time formalism.

and

$$0 \leq \Im m(x_0 - y_0) \leq \beta \quad \text{for} \quad \theta_C(x_0 - y_0) = 0. \qquad (6.85)$$

These restrictions imply that the time argument along C in $D^{(c)}$ must have a *non-increasing* imaginary part, which is sufficient also for the existence of all higher-order thermal Green's functions [475]. The restrictions on C allow for many complex-time-paths among which the Keldysh–Schwinger one is only a particular choice, cf. Fig. 6.6. In the imaginary-time formalism the choice of the contour is the simplest one, namely a straight line running entirely along the imaginary axis and having t_i set to 0. Since explicit physical time is not involved, the Matsubara formalism is convenient only for evaluation of *static* thermodynamic properties, i.e., partition function and thermodynamical potentials. In this case the contour ordering, say for 2-point Green's function, is simply an ordering in "imaginary time":

$$\bar{D}_F(x, y) = \theta(-\Im m(t)) D_F^{21}(x, y) + \theta(\Im m(t)) D_F^{12}(x, y), \qquad (6.86)$$

with $t = x^0 - y^0 \in \Im m(\mathbb{R})$. Since x^0 and y^0 are integrated from 0 to $i\beta$, we need $\bar{D}_F(x, y)$ for $\Im m(t)$ in the range $[-\beta, \beta]$. In this interval $\bar{D}_F(t, \mathbf{x})$ can be expanded in a Fourier series

$$\bar{D}_F(t, \mathbf{x}) = \frac{i}{\beta} \sum_{n=-\infty}^{\infty} D_n(\mathbf{x}) \, e^{\omega_n t}. \qquad (6.87)$$

Here $\omega_n = n\pi/\beta$ is the so-called Matsubara frequency [446]. Since the KMS condition

$$D_F^{21}(t, \mathbf{x}) = D_F^{12}(t - i\beta, \mathbf{x}), \qquad (6.88)$$

implies that $\bar{D}_F(t, \mathbf{x}) = \bar{D}_F(t + i\beta, \mathbf{x})$, only *even* values of n appear in the sum (6.87). Similarly, for fermions one can use the KMS condition

$$S_F^{21}(t, \mathbf{x}) = -S_F^{12}(t - i\beta, \mathbf{x}), \tag{6.89}$$

which implies that $\bar{S}_F(t, \mathbf{x}) = -\bar{S}_F(t + i\beta, \mathbf{x})$, and so only *odd* values of n contribute to the corresponding Fourier expansion.

By employing the continuous three-dimensional Fourier transform in (6.87) and inverting the Fourier summation over n, we obtain

$$D_n(\mathbf{k}) = \int_0^{-i\beta} dt \, e^{-\omega_n t} \, D_F^{21}(t, \mathbf{k}) = \frac{1}{\omega_n^2 + \mathbf{k}^2 + m^2}, \tag{6.90}$$

which is the ordinary (i.e., $T = 0$) Feynman propagator with $k^0 = i\omega_n$. So, the Feynman rules are just like the $T = 0$ ones, except that the energy-conserving δ-function at each vertex is replaced with the Kronecker delta which imposes conservation of the discrete energy and the conventional loop integration is replaced with the energy (or Matsubara) summation according to prescription

$$\int \frac{d^4 k}{(2\pi)^4} \cdots \rightarrow \frac{i}{\beta} \sum_{n=-\infty}^{\infty} \int \frac{d\mathbf{k}}{(2\pi)^3} \cdots. \tag{6.91}$$

Note that computation of the higher order Matsubara frequency sums leads to a proliferation of the so-called vertex-ordered diagrams [212].

Let us now turn to the generating functionals. One can formally pass from the representation Eq. (6.40) (respective Eq. (6.43)) to a Matsubara formalism though the replacement

$$\int_C d^4 x \cdots \rightarrow \int_0^{-i\beta} dx_0 \int_{\mathbb{R}^3} d\mathbf{x} \cdots. \tag{6.92}$$

(C represents here the Keldysh–Schwinger contour.) For convenience we may parametrize the time-integration interval by an imaginary-time variable $x_0 = -i\tau$, where τ runs from 0 to β. Introducing Euclidean version of the field $\phi_E(\tau, \mathbf{x}) = \phi(x)$, the source $J_E(\tau, \mathbf{x}) = J(x)$ and the Lagrangian $\mathcal{L}_E(\phi_E, J_E) = -\mathcal{L}(\phi_E, J_E)$, we can write

$$S[\phi, J, \beta] = \int_0^{-i\beta} d^4 x \, \mathcal{L}(\phi, J, \beta) = -i \int_0^\beta d\tau \int_{\mathbb{R}^3} d\mathbf{x} \, \mathcal{L}(\phi_E, J_E, \beta)$$

$$= i \int_0^\beta d\tau \int_{\mathbb{R}^3} d\mathbf{x} \, \mathcal{L}_E(\phi_E, J_E, \beta) \equiv i S_E[\phi_E, J_E, \beta]. \tag{6.93}$$

This introduces the Euclidean action. Note that for the Euclidean fields the KMS boundary condition Eq. (6.65) reads $\phi_E(\beta, \mathbf{x}) = \phi_E(0, \mathbf{x})$. The corresponding Matsubara version of the thermal generating functional for Green's functions reads

$$Z_{E,\beta}[J] = \mathcal{N} \int_{\beta-\text{periodic}} \mathcal{D}\phi \, \exp\left(-\int_0^\beta d\tau \int_{\mathbb{R}^3} d\mathbf{x}\, \mathcal{L}_E(\phi, J, \beta)\right)$$

$$= \mathcal{N} \int_{\beta-\text{periodic}} \mathcal{D}\phi \, \exp\left[-\left(S_E[\phi,\beta] - \int_0^\beta d^4x\, \phi(x)J(x)\right)\right]. \quad (6.94)$$

In the integration we have dropped the sub-index E in ϕ_E since the functional integral is invariant under relabeling the integration variable. We have also simplified the notation for the space temperature integration.

By analogy with the CTP formalism we define the Matsubara version of the generating functional $W_{E,\beta}$ for connected Green's functions as

$$Z_{E,\beta}[J] = Z_{E,\beta}[0] \exp(W_{E,\beta}[J]). \quad (6.95)$$

We discuss Matsubara functional integrals in phase transitions in Appendix O.

6.6 Geometric background for thermal field theories

It has been observed that if one wishes to work with real-time propagators at finite temperature, a naive use of the imaginary-time propagator analytically continued to real-time results in ambiguities [204]. Such ambiguities appear in higher-point Green's functions, e.g., in three-point thermal Green's functions [46, 223, 265]. Ambiguities were also reported in the β-function calculations at the one-loop level [120, 406].

As already pointed out in previous Sections, real-time formalisms are generally more suitable to the study of transition processes or linear responses since no analytic continuations are required to reach the physical region, while the imaginary time formalism is tailored for (equilibrium) thermodynamic calculations. Real- and imaginary-time formalisms agree in calculations of self-energies and thermodynamical potentials.

An interesting question is whether such different formalisms have some roots in common and if their features can be understood in a deeper way so that they appear unified. A clue to an answer may be found in the well-known discovery of Hawking [315] that temperature arises in a quantum theory as a result of a non-trivial background endowed with event-horizon(s), in this case a black-hole spacetime. Rindler spacetime, the spacetime of an accelerated observer, was also shown to exhibit thermal features [78, 538]. This is known as the Unruh (or Davies–Unruh) effect.

On this basis, a flat background with a non-trivial structure which exhibits thermal features has been constructed, the so-called η-ξ space-time [302, 303]. Fields and states in this spacetime look everywhere as if they were immersed in a thermal bath contained in a Minkowski background.

In this Section, we consider some aspects of field quantization in η-ξ spacetime, showing that it acts as a geometric background for thermal field theories. The discussion is limited to the case of a scalar boson field, for fermion fields and further discussion of the properties of η-ξ spacetime, see [302–304].

6.6.1 *The η-ξ spacetime*

The η-ξ spacetime [302, 303] is a four-dimensional complex manifold defined by the line element

$$ds^2 = \frac{-d\eta^2 + d\xi^2}{\alpha^2 \left(\xi^2 - \eta^2\right)} + dy^2 + dz^2 \,, \tag{6.96}$$

where $\alpha \equiv 2\pi/\beta$ is a real constant and $(\eta, \xi, y, z) \in \mathbb{C}^4$. The symbol $\xi^\mu \equiv (\eta, \xi, y, z)$ denotes as a whole the set of η-ξ coordinates. For simplicity, we drop the index μ when no confusion arises.

Euclidean section

The Euclidean section of η-ξ spacetime is obtained by assuming that $(\sigma, \xi, y, z) \in \mathbb{R}^4$ where $\eta \equiv i\sigma$. In this section the metric is (cf. Eq. (6.96))

$$ds^2 = \frac{d\sigma^2 + d\xi^2}{\alpha^2 \left(\sigma^2 + \xi^2\right)} + dy^2 + dz^2 \,. \tag{6.97}$$

By use of the transformation

$$\sigma = (1/\alpha) \exp\left(\alpha x\right) \sin\left(\alpha\tau\right) \,, \tag{6.98a}$$

$$\xi = (1/\alpha) \exp\left(\alpha x\right) \cos\left(\alpha\tau\right) \,, \tag{6.98b}$$

the metric becomes that of the cylindrical Euclidean flat spacetime,

$$ds^2 = d\tau^2 + dx^2 + dy^2 + dz^2 \,, \tag{6.99}$$

where the time τ has a periodic structure, i.e. $\tau \equiv \tau + \beta$.

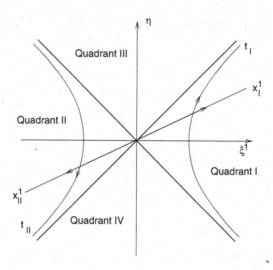

Fig. 6.7 Lorentzian section of η-ξ spacetime: the solid lines represent the singularities at $\xi^2 - \eta^2 = 0$. On the straight lines time is constant, while on the hyperbolas the Minkowski coordinate x is constant. Note that time flows in opposite directions in regions I and II.

Lorentzian section

In the Lorentzian section the metric is given by Eq. (6.96) where $(\eta, \xi, y, z) \in \mathbb{R}^4$. This metric is singular on the two hyperplanes $\eta = \pm \xi$ which are called the "event-horizons". They divide η-ξ spacetime into four regions denoted by R_I, R_{II}, R_{III} and R_{IV} (see Fig. 6.7).

In regions R_I R_{II}, one can define the following sets of coordinates

$$\text{in } R_I: \quad \begin{cases} \eta = +(1/\alpha) \exp{(\alpha x_I)} \sinh{(\alpha t_I)}, \\ \xi = +(1/\alpha) \exp{(\alpha x_I)} \cosh{(\alpha t_I)}, \end{cases} \tag{6.100a}$$

$$\text{in } R_{II}: \quad \begin{cases} \eta = -(1/\alpha) \exp{(\alpha x_{II})} \sinh{(\alpha t_{II})}, \\ \xi = -(1/\alpha) \exp{(\alpha x_{II})} \cosh{(\alpha t_{II})}. \end{cases} \tag{6.100b}$$

Similar transformations can be defined to cover R_{III} and R_{IV} (see [302]). The metric Eq. (6.96) becomes the usual Minkowski metric in these regions and in the new coordinates,

$$ds^2 = -dt^2_{I,II} + dx^2_{I,II} + dy^2 + dz^2. \tag{6.101}$$

Regions I to IV are thus nothing but copies of the Minkowski spacetime glued together along the "event-horizons" making up the Lorentzian section of η-ξ spacetime.

Note that, although Eqs. (6.100) are formally Rindler transformations [538], the $(t_{I,II}, x_{I,II}, y, z)$ coordinates should not be confused with Rindler coordinates since in Eqs. (6.100) the role of the inertial and non-inertial coordinates are actually reversed with respect to the Rindler case [302,538].

A careful analysis [92] of the analytical properties of the transformations given in Eqs. (6.100) leads to the following fundamental relation valid in $R_I \cup R_{II}$:

$$t_{II}(\xi) = t_I(\xi) + i\beta/2, \tag{6.102a}$$
$$x_{II}(\xi) = x_I(\xi). \tag{6.102b}$$

Extended Lorentzian section

An interesting class of complex sections of η-ξ spacetime is obtained [92] from the above Lorentzian section by shifting the Minkowski time coordinates in the imaginary direction only in region R_{II},

$$\text{in } R_I \cup R_{III} \cup R_{IV}: \quad t_q \to t_{q_\delta} = t_q', $$
$$\text{in } R_{II}: \quad t_{II} \to t_{II_\delta} = t_{II} + i\beta\delta, \tag{6.103}$$

where $\delta \in [-1/2, 1/2]$. Let R_{II_δ} denote the set of these sections, which we will call "extended Lorentzian section". In R_{II_δ} the η-ξ coordinates become complex variables and are transformed according to $(\eta, \xi) \to (\eta_\delta, \xi_\delta)$ where, from Eq. (6.103),

$$\eta_\delta = -(1/\alpha) \exp(\alpha x_{II}) \sinh[\alpha(t_{II} + i\beta\delta)], \tag{6.104a}$$
$$\xi_\delta = -(1/\alpha) \exp(\alpha x_{II}) \cosh[\alpha(t_{II} + i\beta\delta)]. \tag{6.104b}$$

In terms of the real η-ξ variables, we have

$$\eta_\delta = +\eta \cos(2\pi\delta) + i\xi \sin(2\pi\delta), \tag{6.105a}$$
$$i\xi_\delta = -\eta \sin(2\pi\delta) + i\xi \cos(2\pi\delta). \tag{6.105b}$$

The time shift thus induces a rotation in the $(\eta, i\xi)$ plane of R_{II}. In terms of the rotated coordinates the metric becomes

$$ds^2 = \frac{-d\eta_\delta^2 + d\xi_\delta^2}{\alpha^2(\xi_\delta^2 - \eta_\delta^2)} + dy^2 + dz^2, \tag{6.106}$$

and is unchanged by the time shift, which is thus an isometry of the four-dimensional complex η-ξ spacetime. The analogues of Eqs. (6.102) are now

$$t_{II_\delta}(\xi_\delta) = t_I(\xi_\delta) + i\beta\left(\tfrac{1}{2} + \delta\right), \tag{6.107a}$$
$$x_{II_\delta}(\xi_\delta) = x_I(\xi_\delta). \tag{6.107b}$$

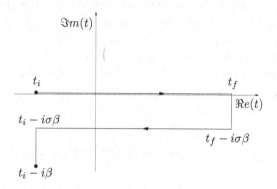

Fig. 6.8   Complex time-path with the parameter $\sigma \neq 0$.

6.6.2   *Fields in η-ξ spacetime*

Let us now consider a free scalar field in the above defined η-ξ spacetime (for the case of a fermion field see [304]). We denote by $\Phi(\xi)$ the "global" scalar field in η-ξ coordinates satisfying the Klein–Gordon equation

$$(D_\mu D^\mu + m^2)\Phi(\xi) = 0\,, \qquad (6.108)$$

where D_μ denotes the covariant derivative and

$$D_\mu D^\mu \;=\; \frac{1}{\sqrt{g}}\,\partial_\mu \sqrt{g}\, g^{\mu\nu}\partial_\nu\,, \qquad (6.109)$$

is the Laplace–Beltrami operator with $g = |\det g_{\alpha\beta}|$. The scalar product of two fields is given by

$$(\Phi_1, \Phi_2) = -i \int_\Sigma d\Sigma \sqrt{g}\,\Phi_1(\xi) n^\nu \overset{\leftrightarrow}{\partial_\nu} \Phi_2^*(\xi)\,, \qquad (6.110)$$

where Σ is any space-like surface and n^ν an orthonormal vector to this surface. The covariant Lagrangian for Φ including the source J is given by

$$\mathcal{L}[\Phi, \mathrm{J}] = \sqrt{g}\left(\frac{1}{2}\,g^{\mu\nu}\partial_\mu\Phi\partial_\nu\Phi + \frac{m^2}{2}\,\Phi^2 - V(\Phi) - \mathrm{J}\Phi\right). \qquad (6.111)$$

Euclidean section

In the Euclidean section we have $\Phi = \Phi(\sigma, \xi, y, z)$, and the field in the t-x coordinates defined in Eqs. (6.98) shall be denoted by $\phi = \phi(\tau, x, y, z)$. These two fields are related by

$$\phi(\tau, x, y, z) \;=\; \Phi(\sigma(\tau, x), \xi(\tau, x), y, z). \qquad (6.112)$$

The periodic nature of τ and the requirement of single valuedness imply

$$\phi(\tau, x, y, z) = \phi(\tau + \beta, x, y, z). \tag{6.113}$$

The generating functional for the Green's functions in the Euclidean section of η-ξ spacetime is [302]

$$Z_E[J] = \mathcal{N} \int \mathcal{D}\Phi \exp\left\{-\int d\sigma d\xi dy dz \, \mathcal{L}_{\sigma,\xi}[\Phi, J]\right\}, \tag{6.114}$$

where

$$\mathcal{L}_{\sigma,\xi}[\Phi, J] = \frac{1}{2}\left[(\partial_\sigma \Phi)^2 + (\partial_\xi \Phi)^2\right]$$
$$+ \frac{1}{\alpha^2(\sigma^2 + \xi^2)}\left\{\frac{1}{2}(\boldsymbol{\nabla}_\perp \Phi)^2 + \frac{m^2}{2}\Phi^2 + V(\Phi) - J\Phi\right\}. \tag{6.115}$$

We now perform the change of coordinates given in Eqs. (6.98) in the generating functional. In the τ-x coordinates, the sum over fields is taken over all periodic fields satisfying the periodic constraint in Eq. (6.113),

$$Z_E[J] = \mathcal{N} \int\limits_{\beta-\text{periodic}} \mathcal{D}\phi \exp\left\{-\int_0^\beta d\tau \int_{\mathbb{R}^3} dx dy dz \, \mathcal{L}_{\tau,x}[\phi, J]\right\}, \tag{6.116}$$

where $J(\tau, x, y, z) = J(\sigma, \xi, y, z)$ and

$$\mathcal{L}_{\tau,x}[\phi, J] = \frac{1}{2}\left[(\partial_\tau \phi)^2 + (\boldsymbol{\nabla}\phi)^2 + m^2 \phi^2\right] + V(\phi) - J\phi. \tag{6.117}$$

By differentiating $Z_E[J]$ with respect to the source J we obtain the Matsubara propagator (see Section 6.5).

Lorentzian section

In the Lorentzian section we have four different regions, each of them being a complete Minkowski spacetime. Since we are interested only in regions R_I and R_{II}, we shall consider the quantum field over these two regions only. Our aim is to find an expansion for the global field Φ in the joining $R_I \cup R_{II}$.

We start by defining the "local" fields $\phi^I(x_I)$ and $\phi^{II}(x_{II})$ by

$$\Phi(\xi) = \begin{cases} \phi^I(x_I(\xi)), & \text{when } \xi \in R_I, \\ \phi^{II}(x_{II}(\xi)), & \text{when } \xi \in R_{II}. \end{cases} \tag{6.118}$$

They have support in R_I and R_{II} respectively. By choosing the particular surface $\eta = a\xi$ where a is a constant satisfying $-1 < a < 1$, one shows from Eq. (6.110) that the global scalar product is given by

$$(\Phi_1, \Phi_2) = \langle \phi_1^I, \phi_2^I \rangle + \langle \phi_1^{II}, \phi_2^{II} \rangle, \tag{6.119}$$

where $\langle\,,\,\rangle$ is the local scalar product in Minkowski spacetime

$$\langle\phi_1,\phi_2\rangle = -i\int_{\mathbb{R}^3} d^3x\,\phi_1(x)\,\overset{\leftrightarrow}{\partial_t}\,\phi_2^*(x). \qquad (6.120)$$

In η-ξ spacetime covered by t-x coordinates given in Eq. (6.100), the solutions of the Klein–Gordon equation are local plane waves restricted to a given region. They are given by

$$u_{\mathbf{k}}(x_I) = (4\pi\omega_{\mathbf{k}})^{-\frac{1}{2}}\,e^{i(-\omega_{\mathbf{k}}\,t_I+\mathbf{k}\cdot\mathbf{x}_I)}, \qquad (6.121a)$$

$$v_{\mathbf{k}}(x_{II}) = (4\pi\omega_{\mathbf{k}})^{-\frac{1}{2}}\,e^{i(+\omega_{\mathbf{k}}\,t_{II}+\mathbf{k}\cdot\mathbf{x}_{II})}, \qquad (6.121b)$$

where $\omega_{\mathbf{k}} = \sqrt{\mathbf{k}^2 + m^2}$. From these Minkowski modes, one defines the two wave functions $U_{\mathbf{k}}(\xi)$ and $V_{\mathbf{k}}(\xi)$ with support in R_I and R_{II} respectively by

$$U_{\mathbf{k}}(\xi) = \begin{cases} u_{\mathbf{k}}(x_I(\xi)), & \text{when } \xi \in R_I, \\ 0, & \text{when } \xi \in R_{II}, \end{cases} \qquad (6.122a)$$

$$V_{\mathbf{k}}(\xi) = \begin{cases} 0, & \text{when } \xi \in R_I, \\ v_{\mathbf{k}}(x_{II}(\xi)), & \text{when } \xi \in R_{II}. \end{cases} \qquad (6.122b)$$

Their power spectrum with respect to the momenta conjugated to $\xi^{\pm} \equiv \eta \pm \xi$ contains negative contributions, which are furthermore not bounded from below. Consequently, the sets of functions $\{U_{\mathbf{k}}(\xi), U_{-\mathbf{k}}^*(\xi)\}_{\mathbf{k}\in\mathbb{R}^3}$ and $\{V_{\mathbf{k}}(\xi), V_{-\mathbf{k}}^*(\xi)\}_{\mathbf{k}\in\mathbb{R}^3}$ defined on R_I and R_{II} respectively are both over-complete since the same energy contribution (i.e. momentum contribution conjugate to η) can appear twice in these sets. In other words, the energy spectrum of $U_{\mathbf{k}}$ and $U_{-\mathbf{k}}^*$ overlap, and so do the ones of $V_{\mathbf{k}}$ and $V_{-\mathbf{k}}^*$. Thus, these sets cannot be used as a basis in their respective regions, and the joining of these sets is clearly not a basis in $R_I \cup R_{II}$.

To construct a basis in $R_I \cup R_{II}$, we could solve the Klein–Gordon equations in η-ξ coordinates to obtain the field modes in these coordinates. However, the Bogoliubov transformations resulting from this basis choice are rather complicated. So, instead, we shall construct from the wave functions $u_{\mathbf{k}}(x_I(\xi))$ and $v_{-\mathbf{k}}^*(x_{II}(\xi))$ basis elements having positive energy spectrum.

We shall demand these basis elements to be analytical functions in the lower complex planes of ξ^+ and ξ^-, so that their spectrum contains only positive contributions of the momenta conjugate with respect to ξ^+ and ξ^-. As a consequence, they shall have positive energy spectra.

We thus extend analytically the two wave functions $u_{\mathbf{k}}(x_I(\xi))$ and $v_{-\mathbf{k}}^*(x_{II}(\xi))$ in the lower complex planes of ξ^+ and ξ^- (the cut in the complex planes is given by $\mathbb{R}^- + i\epsilon$). By using Eqs. (6.102) we get directly

$$u_{\mathbf{k}}(x_I(\xi)) = e^{-\frac{\beta}{2}\omega_{\mathbf{k}}} v^*_{-\mathbf{k}}(x_{II}(\xi)), \tag{6.123a}$$

$$v_{\mathbf{k}}(x_{II}(\xi)) = e^{-\frac{\beta}{2}\omega_{\mathbf{k}}} u^*_{-\mathbf{k}}(x_I(\xi)). \tag{6.123b}$$

The expressions on the r.h.s. and l.h.s. of these last equations are analytic continuations of each other. In this way we are led to introduce the two normalized linear combinations

$$F_{\mathbf{k}}(\xi) = (1 - f_{\mathbf{k}})^{-\frac{1}{2}} \left[U_{\mathbf{k}}(\xi) + f_{\mathbf{k}}^{\frac{1}{2}} V^*_{-\mathbf{k}}(\xi) \right], \tag{6.124a}$$

$$\widetilde{F}_{\mathbf{k}}(\xi) = (1 - f_{\mathbf{k}})^{-\frac{1}{2}} \left[V_{\mathbf{k}}(\xi) + f_{\mathbf{k}}^{\frac{1}{2}} U^*_{-\mathbf{k}}(\xi) \right], \tag{6.124b}$$

where $f_{\mathbf{k}} = e^{-\beta\omega_{\mathbf{k}}}$, and where $U_{\mathbf{k}}(\xi)$ and $V_{\mathbf{k}}(\xi)$ are defined in Eq. (6.122). These wave functions are still solutions of the Klein–Gordon equation. They are analytical in $R_I \cup R_{II}$ and in particular at the origin $\xi^+ = \xi^- = 0$. Since they are analytical complex functions in the lower complex planes of ξ^+ and ξ^-, their spectrum has only positive energy contributions. The set $\{F_{\mathbf{k}}, F^*_{-\mathbf{k}}, \widetilde{F}_{\mathbf{k}}, \widetilde{F}^*_{-\mathbf{k}}\}_{\mathbf{k}\in\mathbb{R}^3}$ is thus complete but not over-complete over the joining $R_I \cup R_{II}$. Furthermore it is an orthogonal set since

$$(F_{\mathbf{k}}, F_{\mathbf{p}}) = (\widetilde{F}^*_{\mathbf{k}}, \widetilde{F}^*_{\mathbf{p}}) = +\delta^3(\mathbf{k} - \mathbf{p}), \tag{6.125a}$$

$$(F^*_{\mathbf{k}}, F^*_{\mathbf{p}}) = (\widetilde{F}_{\mathbf{k}}, \widetilde{F}_{\mathbf{p}}) = -\delta^3(\mathbf{k} - \mathbf{p}), \tag{6.125b}$$

with all the other scalar products vanishing.

The local scalar fields can be expanded in the Minkowski modes given in Eq. (6.121),

$$\phi^I(x_I) = \int d^3k \left[a^I_{\mathbf{k}} u_{\mathbf{k}}(x_I) + a^{I\dagger}_{\mathbf{k}} u^*_{\mathbf{k}}(x_I) \right], \tag{6.126a}$$

$$\phi^{II}(x_{II}) = \int d^3k \left[a^{II}_{\mathbf{k}} v_{\mathbf{k}}(x_{II}) + a^{II\dagger}_{\mathbf{k}} v^*_{\mathbf{k}}(x_{II}) \right]. \tag{6.126b}$$

The global scalar field can be expanded in terms of the "global" modes given in Eqs. (6.124) as

$$\Phi(\xi) = \int d^3k \left[b_{\mathbf{k}} F_{\mathbf{k}}(\xi) + b^\dagger_{\mathbf{k}} F^*_{-\mathbf{k}}(\xi) + \widetilde{b}_{\mathbf{k}} \widetilde{F}_{\mathbf{k}}(\xi) + \widetilde{b}^\dagger_{\mathbf{k}} \widetilde{F}^*_{-\mathbf{k}}(\xi) \right]. \tag{6.127}$$

These expansions define the local and global creation and annihilation operators, which are related by Bogoliubov transformations. To obtain these, we recall the definition (6.118) relating local and global fields and use the field expansions (6.126) and (6.127). We obtain

$$b_{\mathbf{k}} = a^I_{\mathbf{k}} \cosh\theta_{\mathbf{k}} - a^{II\dagger}_{\mathbf{k}} \sinh\theta_{\mathbf{k}}, \tag{6.128a}$$

$$\widetilde{b}_{\mathbf{k}} = a^{II}_{\mathbf{k}} \cosh\theta_{\mathbf{k}} - a^{I\dagger}_{\mathbf{k}} \sinh\theta_{\mathbf{k}}, \tag{6.128b}$$

where $\sinh^2\theta_{\mathbf{k}} = n(\omega_{\mathbf{k}}) = (e^{\beta\omega_{\mathbf{k}}} - 1)^{-1}$.

This is nothing but the thermal Bogoliubov transformation introduced in Section 5.3. A similar treatment for the extended Lorentzian section [92], gives the non-hermitian Bogoliubov transformation of Section 5.3.2.

The connection of TFD with the geometrical picture of η-ξ spacetime is immediate once we make the identification

$$\begin{pmatrix} \phi \\ \tilde{\phi} \end{pmatrix} \equiv \begin{pmatrix} \phi^I \\ \phi^{II} \end{pmatrix} \tag{6.129}$$

under the constraint $\delta = \sigma - 1/2$. Then the Bogoliubov transformations in Eqs. (6.128) and those in Eqs. (5.36) are identical.

Real time formalism: Closed time-path method

Let us now consider the extended Lorentzian section of η-ξ spacetime. The generating functional is given by [302]

$$Z[J] = \mathcal{N} \int \mathcal{D}\Phi \exp \left\{ i \int d\eta_\delta d\xi_\delta dy dz \, \mathcal{L}_{\eta_\delta, \xi_\delta}[\Phi, J] \right\}, \tag{6.130}$$

where

$$\mathcal{L}_{\eta_\delta, \xi_\delta}[\Phi, J] = \frac{1}{2} \left[\left(\partial_{\eta_\delta} \Phi \right)^2 - \left(\partial_{\xi_\delta} \Phi \right)^2 \right]$$

$$+ \frac{1}{\alpha^2 \left(\eta_\delta^2 - \xi_\delta^2 \right)} \left\{ -\frac{1}{2} (\boldsymbol{\nabla}_\perp \Phi)^2 - \frac{m^2}{2} \Phi^2 - V(\Phi) + J\Phi \right\}. \tag{6.131}$$

Since we shall be interested only in the propagators whose spacetime arguments belong only to the joining $R_I \cup R_{II_\delta}$, we can set the source to zero in regions R_{III} and R_{IV}: $J(x) = 0$ when $x \in R_{III} \cup R_{IV}$. Then Eq. (6.130) becomes

$$Z[J] = \mathcal{N} \int \mathcal{D}\Phi \exp \left\{ i \int_{R_I \cup R_{II_\delta}} d\eta_\delta d\xi_\delta dy dz \, \mathcal{L}_{\eta_\delta, \xi_\delta}[\Phi, J] \right\}. \tag{6.132}$$

We express the fields in regions R_I and R_{II_δ} in terms of the local Minkowskian coordinates by using the transformations given in Eq. (6.104):

$$Z[J] = \mathcal{N} \int \mathcal{D}\phi \exp \left\{ i \int dt_I dx_I dy dz \, \mathcal{L}_{t,x}[\phi, J] \right.$$

$$\left. + i \int dt_{II_\delta} dx_{II_\delta} dy dz \, \mathcal{L}_{t,x}[\phi, J] \right\}, \tag{6.133}$$

where the integration is taken over the Minkowski spacetime, ϕ is the local field and where

$$\mathcal{L}_{t,x}[\phi, J] = \frac{1}{2} \left[(\partial_t \phi)^2 - (\boldsymbol{\nabla}\phi)^2 - m^2 \phi^2 \right] - V(\phi) + J\phi. \tag{6.134}$$

We use the relations (6.107) and obtain

$$Z[J] = \mathcal{N} \int \mathcal{D}\phi \exp\left\{ i \int_C d^4x \left[\mathcal{L}_{t,x}[\phi, J](t, x) - \mathcal{L}_{t,x}[\phi, J](t + i\beta\delta, x) \right] \right\}.$$

(6.135)

In the last step we have dropped the subscript I and taken into account that the time direction in R_{II_δ} is opposite to the one in R_I, resulting in a minus sign in the second integration.

The generating functional in the CTP formalism, cf. Eq. (6.40), is

$$Z_\beta[J] = \mathcal{N}' \int \mathcal{D}\phi \exp\left\{ i \int_C d^4x \, \mathcal{L}_{t,x}[\phi, J] \right\},$$

(6.136)

where the time-path C is shown in Fig. 6.8. The contribution from the vertical parts of the contour may be included in the normalization factor $\mathcal{N}'$ and neglected when calculating the real-time Green's functions [411]. The generating functionals in Eqs. (6.135) and (6.136) can then be identified provided $\delta = \sigma - 1/2$.

We thus see that the type of time-path in the CTP formalism is related directly to the "rotation angle" between the two regions R_I and R_{II_δ} of η-ξ spacetime. In the free field case, from the above generating functionals we obtain the well-known thermal matrix propagator [174, 408, 411]:

$$D_{11}(k) = \frac{i}{k^2 - m^2 + i\epsilon} + 2\pi \, n(k_0) \, \delta(k^2 - m^2),$$

(6.137a)

$$D_{22}(k) = D_{11}^*(k),$$

(6.137b)

$$D_{12}(k) = e^{\sigma\beta k_0} \left[n(k_0) + \theta(-k_0) \right] 2\pi \, \delta(k^2 - m^2),$$

(6.137c)

$$D_{21}(k) = e^{-\sigma\beta k_0} \left[n(k_0) + \theta(k_0) \right] 2\pi \, \delta(k^2 - m^2),$$

(6.137d)

where $n(k_0) = (e^{\beta|k_0|} - 1)^{-1}$. The parameter σ appears explicitly only in the off-diagonal components of the matrix propagator, which, for $\sigma = 0$ reduces to the one in Eq. (6.50).

From the above discussion we can infer the equivalence of QFT in η-ξ spacetime with TFTs when considering a scalar field. In several sections of η-ξ spacetime QFT naturally reproduces the known formalisms of TFTs.

In the Euclidean section of η-ξ spacetime, QFT corresponds to the imaginary-time formalism and the Green's functions are the Matsubara Green's functions. On the other hand, in the extended Lorentzian section, QFT reproduces the two known formalisms of TFT with real time, namely the CTP formalism and TFD. Furthermore, the parameter δ of the extended Lorentzian section can be related with the parameter σ appearing in these formalisms. The most general thermal matrix propagator is thus obtained in the framework of η-ξ spacetime. The geometric structure of η-ξ spacetime plays also a crucial role in obtaining the matrix real-time propagator from the Matsubara one [670].

A final comment is about the *tilde-conjugation* in the context of η-ξ spacetime. The tilde-conjugation rules (cf. Section 5.3) are postulated in TFD in order to connect the physical and the tilde operators. Due to the geometrical structure of η-ξ spacetime, these rules can be there regarded simply as coordinate transformations [671].

Appendix M

Thermal Wick theorem

There is a close similarity between thermal perturbative calculations and the usual zero-temperature perturbation theory. In fact, at finite temperature the perturbation calculus reduces to drawing the thermal diagrams, that are in many respects identical with the ordinary Feynman diagrams. The only difference is that instead of the internal lines in the ordinary Feynman diagrams representing vacuum expectation values, in thermal diagrams they are represented by non-interacting thermal averages. This behavior is possible due to thermal Wick theorem.

It is remarkable that the actual form of the Wick theorem in thermal QFT is just the same as for the usual zero temperature form, with the only proviso that one considers expectation values with respect to the canonical ensemble of free fields (non-interacting thermal averages), instead of vacuum expectation values, i.e.,

$$\langle T_C[\phi_{in}(x_1)\phi_{in}(x_2)\cdots\phi_{in}(x_n)]\rangle$$

$$= \begin{cases} 0, & \text{if } n \text{ odd} \\ \sum_{\text{distinct pairings}} iD_F^{(c)}(x_{i_1}, x_{i_2})\cdots iD_F^{(c)}(x_{i_{k-1}}, x_{i_n}), & \text{if } n \text{ even}, \end{cases} \quad \text{(M.1)}$$

where $\sum_{\text{distinct pairings}}$ denotes the sum over all $(2m-1)!! = (2m)!/2^m m!$ ways of breaking $(1, 2, \ldots, n = 2m)$ into m pairs, and

$$iD_F^{(c)}(x, y) = \langle T_C[\phi_{in}(x)\phi_{in}(y)]\rangle. \quad \text{(M.2)}$$

In order to derive (M.1) we have used the identity

$$e^{-\beta H_{in}^0} \phi_{in}(x) e^{\beta H_{in}^0} = \phi_{in}(t - i\beta, \mathbf{x}). \quad \text{(M.3)}$$

By using the cyclic property of the trace we obtain, for any operator X,

$$\langle[\phi_{in}(x), X]\rangle = (Z_\beta^0)^{-1}\text{Tr}\left(e^{-\beta H_{in}^0}(\phi_{in}(x)X - X\phi_{in}(x))\right)$$

$$= (Z_\beta^0)^{-1}\text{Tr}\left(e^{-\beta H_{in}^0}(\phi_{in}(x) - e^{\beta H_{in}^0}\phi_{in}(x)e^{-\beta H_{in}^0})X\right)$$

$$= (Z_\beta^0)^{-1}\text{Tr}\left(e^{-\beta H_{in}^0}(\phi_{in}(x) - \phi_{in}(t + i\beta, \mathbf{x}))X\right)$$

$$= \left(1 - e^{i\beta\partial_t}\right)\langle\phi_{in}(x)X\rangle. \tag{M.4}$$

Then, the Wick theorem for thermal Wightman's functions follows. Indeed,

$$\langle\prod_{i=1}^{n}\phi_{in}(x_i))\rangle = \langle\phi_{in}(x_1)\prod_{i=2}^{n}\phi_{in}(x_i)\rangle$$

$$= (1 - \exp(i\beta\partial_{t_1}))^{-1}\langle[\phi_{in}(x_1), \prod_{i=2}^{n}\phi_{in}(x_i)]\rangle$$

$$= \sum_{l=2}^{n}(1 - \exp(i\beta\partial_{t_1}))^{-1}[\phi_{in}(x_1), \phi_{in}(x_l)]\langle\prod_{i=2,i\neq l}^{n}\phi_{in}(x_i)\rangle$$

$$= \sum_{l=2}^{n}(1 - \exp(i\beta\partial_{t_1}))^{-1}\langle[\phi_{in}(x_1), \phi_{in}(x_l)]\rangle\langle\prod_{i=2,i\neq l}^{n}\phi_{in}(x_i)\rangle$$

$$= \sum_{l=2}^{n}\langle\phi_{in}(x_1)\phi_{in}(x_l)\rangle\langle\prod_{i=2,i\neq l}^{n}\phi_{in}(x_i)\rangle. \tag{M.5}$$

On the 4th line of Eq. (M.5) we used the fact that $[\phi_{in}(x), \phi_{in}(y)]$ is a c-number and thus

$$[\phi_{in}(x), \phi_{in}(y)] = \langle[\phi_{in}(x), \phi_{in}(y)]\rangle. \tag{M.6}$$

By induction the result (M.5) can be rewritten as

$$\langle\phi_{in}(x_1)\phi_{in}(x_2)\ldots\phi_{in}(x_n)\rangle$$

$$= \begin{cases} 0, & \text{if } n \text{ odd} \\ \sum_{\text{distinct pairings}} D_W(x_{i_1}, x_{i_2})\cdots D_W(x_{i_{k-1}}, x_{i_n}), & \text{if } n \text{ even}, \end{cases} \tag{M.7}$$

with

$$D_W(x, y) = \langle\phi_{in}(x)\phi_{in}(y)\rangle \tag{M.8}$$

representing 2-point thermal Wightman's function. In deriving (M.7) we have used the fact that for free fields

$$\langle\phi_{in}(x)\rangle = (Z_\beta^0)^{-1}\text{Tr}\left(e^{-\beta H_{in}^0}\phi_{in}(x)\right) = (Z_\beta^0)^{-1}\text{Tr}\left(\mathcal{P}e^{-\beta H_{in}^0}\mathcal{P}^{-1}\phi_{in}(x)\right)$$

$$= (Z_\beta^0)^{-1}\text{Tr}\left(e^{-\beta H_{in}^0}\mathcal{P}^{-1}\phi_{in}(x)\mathcal{P}\right) = -\langle\phi_{in}(x)\rangle = 0, \tag{M.9}$$

where $\mathcal{P}$ denotes the field sign reversal operator (or $\mathbb{Z}_2$ operator). For instance, for a free real scalar field one has $\mathcal{P} = e^{i\pi \sum_k a_k^\dagger a_k}$ and then $\mathcal{P}\phi_{in}(x)\mathcal{P}^{-1} = -\phi_{in}(x)$. On the r.h.s. of (M.7), in each D_W, x is to the left/right of y, if this is the case in the original n-point thermal Wightman's function.

The passage from thermal Wightman's functions to thermal Green's functions is obtained via the relation

$$\langle T_C[\phi_{in}(x_1)\phi_{in}(x_2)\ldots\phi_{in}(x_n)]\rangle$$

$$= \sum_P \theta_C(t_{p_1}, t_{p_2}, \cdots, t_{p_n})\langle \phi_{in}(x_{p_1})\phi_{in}(x_{p_2})\ldots\phi_{in}(x_{p_n})\rangle. \quad \text{(M.10)}$$

Here $\sum_P$ denotes the sum over permutations of time arguments and $\theta_C(t_{p_1}, \cdots, t_{p_n})$ represents the n-point step function on the oriented time contour C, i.e.,

$$\theta_C(t_1, t_2, \cdots, t_n) = \begin{cases} 1, & \text{if } t_1 > t_2 > \cdots > t_n \text{ along } C \\ 0, & \text{otherwise}. \end{cases} \quad \text{(M.11)}$$

The multi-point step function (M.11) has an explicit realization in terms of 2-point contour step functions θ_C

$$\theta_C(t_1, t_2, \cdots, t_n) = \prod_{i=1}^{n-1} \theta_C(t_i - t_{i+1}). \quad \text{(M.12)}$$

By inspection, we see that (M.7) together with (M.10) directly imply the thermal Wick theorem for Green's functions (M.1).

The one discussed above is the weak (expectation values) version of thermal Wick theorem. A discussion of the strong (operatorial) version of the same theorem, which is based on the concept of "thermal normal ordering (or product)" (see Section 5.3), can be found in [226].

Appendix N

Coherent state functional integrals

In this Appendix we discuss the functional integral representation of coherent states, both for the case of Glauber coherent states (see Appendix B) and for the case of generalized coherent states (see Appendix C).

N.1 Glauber coherent states

We start by constructing the path-integral representation for the transition amplitude $\langle z_f, t_f | z_i, t_i \rangle$ for the case of one degree of freedom. The Heisenberg-picture resolution of unity reads (cf. Eq. (B.8))

$$\mathbb{1} = \int \frac{dz dz^*}{2\pi i} \, e^{-zz^*} |z, t\rangle\langle z, t|. \tag{N.1}$$

Let us now partition the time interval $[t_i, t_f]$ into $N + 1$ equidistant pieces Δt (see Fig. N.1) by writing $t_f - t_i = (N + 1)\Delta t$. Consequently we have:

$$\begin{aligned}
\langle z_f, t_f | z_i, t_i \rangle &= \left(\int \prod_{k=1}^{N} \frac{dz_k dz_k^*}{2\pi i} \right) \langle z_f, t_f | z_N, t_f - \Delta t \rangle \, e^{-z_N^* z_N} \\
&\quad \times \langle z_N, t_f - \Delta t | z_{N-1}, t_f - 2\Delta t \rangle \, e^{-z_{N-1}^* z_{N-1}} \\
&\quad \times \langle z_{N-1}, t_f - 2\Delta t | z_{N-2}, t_f - 3\Delta t \rangle \, e^{-z_{N-2}^* z_{N-2}} \\
&\qquad \vdots \\
&\quad \times \langle z_1, t_i + \Delta t | z_i, t_i \rangle.
\end{aligned} \tag{N.2}$$

It is convenient to set $t_0 = t_i$ and $t_{N+1} = t_f$.

The infinitesimal time-transition amplitudes can be written in terms of the Schrödinger picture base vectors as

$$\langle z_j, t_j | z_{j-1}, t_{j-1} \rangle = \langle z_j | T \left[\exp\left(-i \int_{t_{j-1}}^{t_j} dt \, H(t) \right) \right] |z_{j-1}\rangle, \tag{N.3}$$

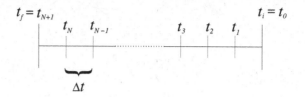

Fig. N.1   Sliced time interval.

where the time ordering prescription T must be enforced when H is explicitly time-dependent (e.g., when it includes time-dependent external fields). We remind that the Schrödinger-picture *base* vectors are time-independent in contrast with Schrödinger-picture *state* vectors that are, of course, time-dependent (the opposite holds for Heisenberg picture). Using the expansion

$$T\left[\exp\left(-i\int_{t_{j-1}}^{t_j} dt\, H(t)\right)\right]$$

$$= 1 - i\int_{t_{j-1}}^{t_j} dt_1\, H(t_1) - \int_{t_{j-1}}^{t_j} dt_1\, H(t_1)\int_{t_{j-1}}^{t_1} dt_2\, H(t_2)$$

$$+ i\int_{t_{j-1}}^{t_j} dt_1\, H(t_1)\int_{t_{j-1}}^{t_1} dt_2\, H(t_2)\int_{t_{j-1}}^{t_2} dt_3\, H(t_3)\ldots, \qquad (N.4)$$

for a very small Δt we can write

$$\langle z_j, t_j | z_{j-1}, t_{j-1}\rangle \simeq \langle z_j | \left(1 - i\int_{t_{j-1}}^{t_j} dt\, H(t)\right) |z_{j-1}\rangle$$

$$\simeq \langle z_j | z_{j-1}\rangle \left(1 - i\, H(z_j^*, z_{j-1}, t_{j-1})\Delta t\right)$$

$$\simeq \exp\left(z_j^* z_{j-1} - iH(z_j^*, z_{j-1}, t_{j-1})\Delta t\right). \qquad (N.5)$$

In the latter we have defined

$$H(z_j^*, z_{j-1}, t_{j-1}) \equiv \frac{\langle z_j | H(a^\dagger, a, t) | z_{j-1}\rangle}{\langle z_j | z_{j-1}\rangle}, \qquad (N.6)$$

and $t_j = t_i + j\Delta t$. Inserting (N.5) into (N.2) we obtain the exponent that is composed of two parts; the sum of Hamiltonians and the sum

$$z_{N+1}^* z_N - z_N^* z_N + z_N^* z_{N-1} + \ldots + z_2^* z_1 - z_1^* z_1 + z_1^* z_0$$

$$= z_{N+1}^* z_{N+1} - \sum_{k=1}^{N+1} z_k^* \left(\frac{z_k - z_{k-1}}{\Delta t}\right)\Delta t. \qquad (N.7)$$

In the continuous time limit (i.e., when $N \to \infty$) we have

$$\exp((\text{N.7})) \;\to\; \exp\left(z_f^* z_f - \int_{t_i}^{t_f} dt\, z^*(t)\dot{z}(t)\right),$$

and

$$\exp\left(-i \sum_{k=0}^{N} H(z_{k+1}^*, z_k, t_k)\Delta t\right) \;\to\; \exp\left(-i \int_{t_i}^{t_f} dt\, H(z^*(t), z(t), t)\right).$$

Adopting the formal notation

$$\lim_{N\to\infty} \prod_{k=1}^{N} \left[\int \frac{dz_k^* dz_k}{2\pi i}\right] = \int_{z(t_i)=z_i}^{z^*(t_f)=z_f^*} \mathcal{D}z^* \mathcal{D}z, \qquad (\text{N.8})$$

the path-integral representation of the amplitude (N.2) is written as

$$\langle z_f, t_f | z_i, t_i \rangle = \int_{z(t_i)=z_i}^{z^*(t_f)=z_f^*} \mathcal{D}z^* \mathcal{D}z\; e^{z_f^* z_f} \exp\left(i \int_{t_i}^{t_f} dt\, L\right). \qquad (\text{N.9})$$

Here $L = iz^*(t)\dot{z}(t) - H(z^*(t), z(t), t)$.

Eq. (N.7) can be expressed in more symmetric form. This can be done by realizing that

$$(\text{N.7}) = \frac{1}{2} \sum_{k=1}^{N+1} z_{k-1} \left(\frac{z_k^* - z_{k-1}^*}{\Delta t}\right) \Delta t - \frac{1}{2} \sum_{k=1}^{N+1} z_k^* \left(\frac{z_k - z_{k-1}}{\Delta t}\right) \Delta t$$

$$+ \frac{1}{2}\, z_{N+1}^* z_{N+1} + \frac{1}{2}\, z_0^* z_0\,. \qquad (\text{N.10})$$

The corresponding exponent goes in the large N limit to

$$\exp\left(\frac{1}{2}(z_f^* z_f + z_i^* z_i)\right) \exp\left(-\frac{1}{2} \int_{t_i}^{t_f} dt\, (z^*(t)\dot{z}(t) - z(t)\dot{z}^*(t))\right).$$

With this, the transition amplitude can be cast into the form

$$\langle z_f, t_f | z_i, t_i \rangle = \int_{z(t_i)=z_i}^{z^*(t_f)=z_f^*} \mathcal{D}z^* \mathcal{D}z\; \exp\left(\int_{t_i}^{t_f} dt\, \left(\frac{1}{2}(\dot{z}^* z - z^* \dot{z}) - iH\right)\right)$$

$$\times \exp\left(\frac{1}{2}(z_f^* z_f + z_i^* z_i)\right), \qquad (\text{N.11})$$

which often serves as a starting point for discussions concerning the classical or semiclassical limits of quantum theory, see [379, 380].

Note that the factors $e^{\frac{1}{2}(z_f^* z_f + z_i^* z_i)}$ in Eq. (N.11) and $e^{z_f^* z_f}$ in Eq. (N.9) do not have an analogue in the phase-space path integrals. Their appearance reflects the specificity of boundary conditions: we fix $z(t_i)$ and $z^*(t_f)$ but $z(t_f)$ and $z^*(t_i)$ are arbitrary.

Extension to many degrees of freedom

We generalize the previous discussion for single-mode coherent states to the case of an infinite number of degrees of freedom. We start with the Weyl–Heisenberg algebra for n modes, denoted by W_n. The single-mode coherent state has a natural generalization in W_n, namely

$$|[z]\rangle = \otimes_{l=1}^{n} |z_l\rangle = \otimes_{l=1}^{n} e^{z_l a_l^\dagger} |0\rangle = \exp \sum_{l=1}^{n} z_l a_l^\dagger |0\rangle, \qquad (\text{N.12})$$

with

$$a_m |[z]\rangle = z_m |[z]\rangle, \qquad m = 1, \ldots, n. \qquad (\text{N.13})$$

The corresponding normalized coherent state then reads

$$|[z]) = \otimes_{l=1}^{n} |z_l) = \otimes_{l=1}^{n} e^{-|z|^2/2} e^{z_l a_l^\dagger} |0\rangle = \exp \sum_{l=1}^{n} (z_l^* a - z_l a_l^\dagger)|0\rangle, \qquad (\text{N.14})$$

and the resolution of the unity has the form

$$\mathbb{1} = \int [dz dz^*] \, |[z]\rangle\langle[z]|, \qquad (\text{N.15})$$

with

$$[dz dz^*] = \prod_{l=1}^{n} \left(\frac{dz_l dz_l^* \, e^{-|z_l|^2}}{2\pi i} \right), \qquad \text{and} \qquad |[z]\rangle\langle[z]| = \otimes_{l=1}^{n} |z_l\rangle\langle z_l|.$$

In the Heisenberg picture Eq. (N.15) becomes

$$\mathbb{1} = \int [dz dz^*] \, |[z], t\rangle\langle[z], t|, \qquad (\text{N,16})$$

with

$$|[z], t\rangle\langle[z], t| = \otimes_{l=1}^{n} |z_l, t\rangle\langle z_l, t|.$$

The passage to QFT is established by performing the limit $n \to \infty$ (i.e., going to W_∞), and replacing in Eqs. (N.12)–(N.16)

$$\sqrt{V 2\omega_{\mathbf{k}}} \, a_{\mathbf{k}} \to a(\mathbf{k}), \qquad \frac{1}{V} \sum_{\mathbf{k}} \to \int \frac{d^3 k}{(2\pi^3)}, \qquad V \delta_{\mathbf{kk}'} \to \delta(\mathbf{k} - \mathbf{k}'),$$

where V is the volume of the system.

The Heisenberg-picture coherent states for a scalar field ψ then has the form

$$|\psi, t\rangle = \exp \left(\int d^3 \mathbf{x} \, \hat{\psi}^\dagger(\mathbf{x}, t) \psi(\mathbf{x}) \right) |0, t\rangle, \qquad (\text{N.17a})$$

$$\langle \psi, t| = \langle 0, t| \, \exp \left(\int d^3 \mathbf{x} \, \hat{\psi}(\mathbf{x}, t) \psi^*(\mathbf{x}) \right), \qquad (\text{N.17b})$$

where

$$\hat{\psi}(\mathbf{x},t)|\psi,t\rangle \;=\; \psi(\mathbf{x})|\psi,t\rangle \quad \text{and} \quad \langle\psi,t|\hat{\psi}^\dagger(\mathbf{x},t) \;=\; \psi^*(\mathbf{x})\langle\psi,t|\,, \quad \text{(N.18)}$$

and

$$\mathbb{1} \;=\; \int \mathcal{D}\psi\mathcal{D}\psi^* \; |\psi,t\rangle \; e^{-\int d^3\mathbf{x}\,\psi^*(\mathbf{x})\psi(\mathbf{x})} \; \langle\psi,t|\,. \quad \text{(N.19)}$$

By a comparison with Eq. (N.9), we see that the functional integral representation of the transition amplitude $\langle\psi_f^*,t_f|\psi_i,t_i\rangle$ can be written as

$$\langle\psi_f,t_f|\psi_i,t_i\rangle \;=\; \int \mathcal{D}\psi\mathcal{D}\psi^* \; \exp\left(i\int_{t_i}^{t_f} dt\, L(\psi,\psi^*,\dot{\psi},t)\right)$$

$$\times\; \exp\left(\int d^3\mathbf{x}\,\psi^*(\mathbf{x},t_f)\psi(\mathbf{x},t_f)\right), \quad \text{(N.20)}$$

where the Lagrangian has the form (cf. (N.9))

$$L(\psi,\psi^*,\dot{\psi},t) \;=\; \int d^3\mathbf{x}\left(\psi^*(\mathbf{x},t)i\frac{\partial}{\partial t}\psi(\mathbf{x},t) - H(\psi^*,\psi,t)\right). \quad \text{(N.21)}$$

The main applications of the canonical coherent state functional integrals are in tackling classical or semiclassical limits of QFT systems. Examples are low-energy/soft-momenta behavior (i.e., infra-red behavior) of QFT gauge theories, or a semiclassical treatment of collective phenomena (e.g., solitons), or thermal statistical physics.

We do not consider here the case of fermions, which requires the use of Grassmann variables [504].

N.2 Generalized coherent states

For the case of generalized coherent states discussed in Appendix C, we follow the derivation that leads us to Eq. (N.2) (in the case of Glauber coherent states), and obtain

$$\langle 0(\mathbf{x}_f),t_f|0(\mathbf{x}_i),t_i\rangle \;=\; \lim_{N\to\infty}\left(\int\prod_{k=1}^{N} c\,dx_k\right)\;\langle 0(\mathbf{x}_f),t_f|0(\mathbf{x}_N),t'-\Delta t\rangle$$

$$\times\; \langle 0(\mathbf{x}_N),t'-\Delta t|0(\mathbf{x}_{N-1}),t'-2\Delta t\rangle$$

$$\times\; \langle 0(\mathbf{x}_{N-1}),t'-2\Delta t|0(\mathbf{x}_{N-2}),t'-3\Delta t\rangle$$

$$\vdots$$

$$\times\; \langle 0(\mathbf{x}_1),t+\Delta t|0(\mathbf{x}_i),t_i\rangle\,, \quad \text{(N.22)}$$

where we used the resolution of the unity Eq. (C.10) and the integration is over the corresponding coset space (cf. Appendix C).

As we have seen in Section N.1, the infinitesimal-time transition amplitude can be written as

$$\langle 0(\mathbf{x}_k), t_k | 0(\mathbf{x}_{k1}), t_{k-1} \rangle \simeq \langle 0(\mathbf{x}_k)| \left(1 - i \int_{t_{k-1}}^{t_k} dt \hat{H}(t) \right) | 0(\mathbf{x}_{k-1}) \rangle$$

$$\simeq \langle 0(\mathbf{x}_k) | 0(\mathbf{x}_{k-1}) \rangle \left(1 - i \Delta t H(\mathbf{x}_k, \mathbf{x}_{k-1}, t_k^*) \right)$$

$$\simeq \langle 0(\mathbf{x}_k) | 0(\mathbf{x}_{k-1}) \rangle \exp \left(-i \int_{t_{k-1}}^{t_k} dt H(\mathbf{x}, \dot{\mathbf{x}}, t) \right), \quad \text{(N.23)}$$

where

$$H(\mathbf{x}_k, \mathbf{x}_{k-1}, t_k) = \frac{\langle 0(\mathbf{x}_k) | \hat{H}(t_k) | 0(\mathbf{x}_{k-1}) \rangle}{\langle 0(\mathbf{x}_k) | 0(\mathbf{x}_{k-1}) \rangle}, \quad \text{(N.24)}$$

is the normalized matrix element of the Hamiltonian. At the same time, one may write (for square integrable states)

$$\langle 0(\mathbf{x}_k) | 0(\mathbf{x}_{k-1}) \rangle \simeq 1 - \langle 0(\mathbf{x}_k) | \{ |0(\mathbf{x}_k)\rangle - |0(\mathbf{x}_{k-1})\rangle \}$$

$$\simeq \exp \left(-\Delta t \frac{\langle 0(\mathbf{x}_k) | \{ |0(\mathbf{x}_k)\rangle - |0(\mathbf{x}_{k-1})\rangle \}}{\Delta t} \right)$$

$$\simeq \exp \left(- \int_{t_{k-1}}^{t_k} \langle 0(\mathbf{x}) | \frac{d}{dt} | 0(\mathbf{x}) \rangle \, dt \right). \quad \text{(N.25)}$$

Thus the transition amplitude can be cast into the form

$$\langle 0(\mathbf{x}_f), t_f | 0(\mathbf{x}_i), t_i \rangle$$

$$= \lim_{N \to \infty} \left(\int \prod_{k=1}^{N} c \, d\mathbf{x}_k \right) \exp \left(i \int_{t_i}^{t_f} dt \left[i \langle 0(\mathbf{x}) | \frac{d}{dt} | 0(\mathbf{x}) \rangle - H(\mathbf{x}, \dot{\mathbf{x}}, t) \right] \right)$$

$$= \int_{\mathbf{x}(t_i)=\mathbf{x}_i}^{\mathbf{x}(t_f)=\mathbf{x}_f} \mathcal{D}\mu(\mathbf{x}) \exp \left(i \int_{t_i}^{t_f} dt \left[i \langle 0(\mathbf{x}) | \frac{d}{dt} | 0(\mathbf{x}) \rangle - H(\mathbf{x}, \dot{\mathbf{x}}, t) \right] \right). \quad \text{(N.26)}$$

We note that by using the square integrability of generalized coherent states we have

$$\langle 0(\mathbf{x}) | \frac{d}{dt} | 0(\mathbf{x}) \rangle = -\frac{d}{dt} \{ \langle 0(\mathbf{x}) | \} | 0(\mathbf{x}) \rangle = -(\langle 0(\mathbf{x}) | \frac{d}{dt} | 0(\mathbf{x}) \rangle)^*, \quad \text{(N.27)}$$

that is $\langle 0(\mathbf{x}) | d/dt | 0(\mathbf{x}) \rangle$ is a purely imaginary term. There is no general criterion for an explicit form of this term, but when the particular group representation is specified, then the term can be found quite easily. There is, however, a strict connection of this therm with Berry phase. To see it we write

$$i \int_{t_i}^{t_f} \langle 0(\mathbf{x})| \frac{d}{dt} |0(\mathbf{x})\rangle \, dt \; = \; i \int_{t_i}^{t_f} \langle 0(\mathbf{x})|\boldsymbol{\nabla}_{\mathbf{x}} \, 0(\mathbf{x})\rangle \cdot d\mathbf{x}. \qquad \text{(N.28)}$$

In particular, when $|0(\mathbf{x})\rangle$ are eigenstates of the Hamiltonian (as, for instance, in the non-linear σ model where $|0(\mathbf{x})\rangle$ describe the degenerate ground state) and when $\mathbf{x}(t)$ traverses in the interval $t_f - t_i$ a closed path in the $\mathbf{x}$ space, then (N.28) corresponds to the formula [72] for the Berry phase.

SU(2) coherent states

$SU(2)$ coherent states are also known as spin or Bloch coherent states and the associated path-integral transition amplitude representation follows from Eq. (N.26) [378, 396, 519]:

$$
\begin{aligned}
\langle 0(\xi_f), t_f | 0(\xi_i), t_i \rangle \; &= \; \lim_{N \to \infty} \int \cdots \int \prod_{k=1}^{N} d\mu(\xi_k^*, \xi_k) \\
&\quad \times \exp\left(i \sum_{l=0}^{N} \Delta t \left[\frac{i}{\Delta t} \langle 0(\xi_l)|\Delta|0(\xi_l)\rangle - H(\xi_l^*, \xi_{l-1}, t_l) \right] \right) \\
&= \int_{\xi(t_i)=\xi_i}^{\xi^*(t_f)=\xi_f^*} \mathcal{D}\mu(\xi^*, \xi) \exp\left(i \int_{t_i}^{t_f} dt \left[i \, \langle 0(\xi)| \frac{d}{dt} |0(\xi)\rangle - H(\xi^*, \xi, t) \right] \right) \\
&= \int_{\xi(t_i)=\xi_i}^{\xi^*(t_f)=\xi_f^*} \mathcal{D}\mu(\xi^*, \xi) \exp\left(i \int_{t_i}^{t_f} dt \left[i \frac{j(\xi^* \dot{\xi} - \dot{\xi}^* \xi)}{(1+|\xi|^2)} - H(\xi^*, \xi, t) \right] \right).
\end{aligned}
$$

$$\text{(N.29)}$$

Here

$$d\mu(\xi_k^*, \xi_k) \; \equiv \; \frac{d\xi_k d\xi_k^*}{(1+|\xi_k|^2)^2} \quad \text{and} \quad H(\xi_l^*, \xi_{l-1}, t_l) \; \equiv \; \frac{\langle 0(\xi_l)|H(t_l)|0(\xi_{l-1})\rangle}{\langle 0(\xi_l)|0(\xi_{l-1})\rangle}.$$

Use was also made of the fact that up to the order $\Delta \xi_l = \xi_l - \xi_{l-1}$ one has

$$\langle 0(\xi_l)|\Delta|0(\xi_l)\rangle = \langle 0(\xi_l)|\{|0(\xi_l)\rangle - |0(\xi_{l-1})\rangle\} = \frac{j(\xi_l^* \Delta \xi_l - \xi_l \Delta \xi_l^*)}{1+|\xi_l|^2}. \qquad \text{(N.30)}$$

SU(1,1) coherent states

For the path integral for $SU(1,1)$ coherent states [280] we have:

$$
\langle 0(\zeta_f), t_f | 0(\zeta_i), t_i \rangle = \lim_{N \to \infty} \int \cdots \int \prod_{k=1}^{N} d\mu(\zeta_k^*, \zeta_k)
$$

$$
\times \exp\left(i \sum_{l=0}^{N} \Delta t \left[\frac{i}{\Delta t} \langle 0(\zeta_l) | \Delta | 0(\zeta_l) \rangle - H(\zeta_l^*, \zeta_{l-1}, t_l) \right] \right)
$$

$$
= \int_{\zeta(t_i)=\zeta_i}^{\zeta^*(t_f)=\zeta_f^*} \mathcal{D}\mu(\zeta^*, \zeta) \exp\left(i \int_{t_i}^{t_f} dt \left[i \langle 0(\zeta) | \frac{d}{dt} | 0(\zeta) \rangle - H(\zeta^*, \zeta, t) \right] \right)
$$

$$
= \int_{\zeta(t_i)=\zeta_i}^{\zeta^*(t_f)=\zeta_f^*} \mathcal{D}\mu(\zeta^*, \zeta) \exp\left(i \int_{t_i}^{t_f} dt \left[i \frac{|j|(\zeta^* \dot\zeta - \dot\zeta^* \zeta)}{(1 - |\zeta|^2)} - H(\zeta^*, \zeta, t) \right] \right),
$$

$$(N.31)$$

where the resolution of the unity Eq. (C.38) has been used and

$$
d\mu(\zeta_k^*, \zeta_k) \equiv \frac{d\zeta_k d\zeta_k^*}{(1 - |\zeta_k|^2)^2} \quad \text{and} \quad H(\zeta_k^*, \zeta_{k-1}, t_k) \equiv \frac{\langle 0(\zeta_k) | H(t_k) | 0(\zeta_{k-1}) \rangle}{\langle 0(\zeta_k) | 0(\zeta_{k-1}) \rangle}.
$$

We have also made use of the fact that, up to order of $\Delta \zeta_k = \zeta_k - \zeta_{k-1}$,

$$
\langle 0(\zeta_k) | \Delta | 0(\zeta_k) \rangle = \langle 0(\zeta_k) | \{ | 0(\zeta_k) \rangle - | 0(\zeta_{k-1}) \rangle \} = \frac{|j|(\zeta_k^* \Delta \zeta_k - \zeta_k \Delta \zeta_k^*)}{1 - |\zeta_k|^2}.
$$

$$(N.32)$$

Generalization to field theory can now proceed by formally exchanging the coset space variables $\zeta^a(t)$ ($a = 1, \ldots, \dim G/H$) with the coset space fields $\phi^a(\mathbf{x}, t)$, which provide the mapping from d-dimensional physical space (e.g., spacetime) to $\mathcal{M} = G/H$,

$$
\phi^a(\mathbf{x}, t): \ \mathbb{R}^d \to \mathcal{M}. \tag{N.33}
$$

The space $\mathcal{M}$ is called the *target space*.

Non-linear σ models

Since functional integrals for generalized coherent states are naturally expressed in terms of coset space fields, they are well suited to describe the dynamics of Goldstone bosons. From Chapters 3 and 4 (cf. also Appendix I) we know that Goldstone fields take values in the target space which is a coset of G/H (G is the symmetry group of the normal phase and H of the broken phase).

Massless field theories where the target space is the group coset space G/H are commonly known as non-linear σ models (or G/H-σ models). With a suitable choice of the Hamiltonian $H(\mathbf{x}, \dot{\mathbf{x}}, t)$ (cf. Eq. (N.26)), the generalized coherent state functional integrals describe low-energy effective field theories — non-linear σ models, in which only Goldstone bosons, including their mutual interactions, are retained.

Let us consider the case of the $SU(2)$ functional integral. We consider small fluctuations around the ground state in the Heisenberg model of ferromagnets. In the long-wavelength limit we obtain the Landau–Lifshitz non-linear σ model which describes the dynamics of corresponding Goldstones — the magnons.

We write the action in the path integral (N.29) in terms of the unit-vector parameters $\mathbf{n}(t)$. The first term can be written as (cf. Appendix C)

$$i\frac{j(\xi^* d\xi - d\xi^* \xi)}{(1 + |\xi|^2)} = -2j \sin^2(\theta/2)\, d\varphi = -\frac{j}{r(z+r)}\,(x\, dy - y\, dx)$$
$$= \mathbf{A}_B(\mathbf{x}) \cdot d\mathbf{x}, \tag{N.34}$$

where the vector potential (Berry's connection) is

$$\mathbf{A}_B(\mathbf{x}) = -\frac{j}{r(z+r)}\,(-y, x, 0). \tag{N.35}$$

Since the vector $\mathbf{x}$ should sweep the surface of $\mathcal{S}^2$ we have that $\mathbf{x} = \mathbf{n}$ ($\mathbf{n}^2 = 1$). The first term in the action in Eq. (N.29) thus reads

$$i\int_{t_i}^{t_f} dt\, \frac{j(\xi^* \dot{\xi} - \dot{\xi}^* \xi)}{(1 + |\xi|^2)} = \int_{t_i}^{t_f} \mathbf{A}_B(\mathbf{n}) \cdot \frac{d\mathbf{n}}{dt}\, dt = \int_{\Sigma} \mathbf{B}_B \cdot d\boldsymbol{\sigma}. \tag{N.36}$$

With Σ denoting the area of $\mathcal{S}^2$ bounded by a closed loop traversed by $\mathbf{n}(t)$. Berry's magnetic induction $\mathbf{B}_B$ has the explicit form

$$\mathbf{B}_B(\mathbf{x}) = \boldsymbol{\nabla} \wedge \mathbf{A}_B(\mathbf{x}) = \frac{j}{r^3}\,\mathbf{x} = \frac{j}{r^2}\,\mathbf{n} = j\mathbf{n}, \tag{N.37}$$

and thus

$$\int_{\mathcal{S}^2} \mathbf{B}_B \cdot d\boldsymbol{\sigma} = 4\pi j. \tag{N.38}$$

Eq. (N.37), together with Eq. (N.38), shows that there is a monopole of the magnetic charge j located in the origin of the target space. From Eqs. (N.36) and (N.37) it follows

$$i \int_{t_i}^{t_f} dt \, \frac{(\xi^* \dot{\xi} - \dot{\xi}^* \xi)}{(1 + |\xi|^2)} \;=\; \int_0^1 du \int_{t_i}^{t_f} dt \, \mathbf{n}(t, u) \cdot [\partial_t \mathbf{n}(t, u) \wedge \partial_u \mathbf{n}(t, u)]$$

$$\equiv S_{WZ}[\mathbf{n}], \qquad\qquad\qquad (N.39)$$

where $\mathbf{n}(t, u)$ is an arbitrary extension of $\mathbf{n}(t)$ into the spherical rectangle defined by the limits of integration and fulfilling conditions: $\mathbf{n}(t, 0) = \mathbf{n}(t)$, $\mathbf{n}(t, 1) = (1, 0, 0)$, and $\mathbf{n}(t_i, u) = \mathbf{n}(t_f, u)$. The $S_{WZ}[\mathbf{n}]$ is a special member of a wide class of actions known as Wess–Zumino actions which were introduced in [663]. Eq. (N.39) then demonstrates a typical situation (ubiquitous in effective theories) where the Berry phase gives rise to the Wess–Zumino action.

We now turn to many-spin systems — lattice of spins. We will consider first the Hamiltonian $H(\xi^*, \xi, t)$. To this end we consider the Hamiltonian for the ferromagnetic Heisenberg model, i.e.,

$$\hat{H}(\mathbf{J}) \;=\; K \sum_{(\mathbf{x}, \mathbf{x}')} \hat{\mathbf{J}}(\mathbf{x}) \cdot \hat{\mathbf{J}}(\mathbf{x}'), \qquad\qquad (N.40)$$

where $K = -|K|$ is the exchange coupling and $(\mathbf{r}, \mathbf{r}')$ are pairs of neighboring lattice sites. According to the definition of $H(\xi_k^*, \xi_{k-1}, t)$ we have

$$H(\xi_k^*, \xi_{k-1}, t) \;=\; H(\mathbf{n}_k, \mathbf{n}_{k-1}) \;=\; \frac{\langle 0(\mathbf{n}_k) | \hat{H}(\mathbf{J}) | 0(\mathbf{n}_{k-1}) \rangle}{\langle 0(\mathbf{n}_k) | (\mathbf{n}_{k-1}) \rangle}$$

$$\approx \langle 0(\mathbf{n}_k) | \hat{H}(\mathbf{J}) | 0(\mathbf{n}_k) \rangle + \mathcal{O}(\Delta t). \qquad (N.41)$$

By taking advantage of the identity $\langle 0(\mathbf{n}_k) | \hat{\mathbf{J}}(\mathbf{x}) | 0(\mathbf{n}_k) \rangle = j \mathbf{n}_k(\mathbf{x})$ (cf. Appendix C), we obtain

$$H(\mathbf{n}_k, \mathbf{n}_{k-1}) \;\approx\; -|K| j^2 \sum_{(\mathbf{x}, \mathbf{x}')} \mathbf{n}_k(\mathbf{x}) \cdot \mathbf{n}_k(\mathbf{x}'), \qquad (N.42)$$

so that action in the path integral (N.29) reads

$$S[\mathbf{n}] \;=\; j \sum_{\mathbf{x}} S_{WZ}[\mathbf{n}(\mathbf{x})] + |K| j^2 \sum_k \Delta t \sum_{(\mathbf{x}, \mathbf{x}')} \mathbf{n}_k(\mathbf{x}) \cdot \mathbf{n}_k(\mathbf{x}'). \quad (N.43)$$

Here the first sum runs over all the sites of the lattice and thus represents the sum of the Wess–Zumino terms of individual spins. Note that the time derivative enters only through the Wess–Zumino term.

For definiteness sake, we consider the D-dimensional hypercubical lattice and restrict $\sum_{(\mathbf{x},\mathbf{x}')}$ to nearest neighbors only. We can write

$$\sum_{(\mathbf{x},\mathbf{x}')_{\mathrm{nn}}} \mathbf{n}_k(\mathbf{x}) \cdot \mathbf{n}_k(\mathbf{x}') = -\frac{1}{2} \sum_{(\mathbf{x},\mathbf{x}')_{\mathrm{nn}}} [\mathbf{n}_k(\mathbf{x}) - \mathbf{n}_k(\mathbf{x}')]^2 + \mathrm{const..} \quad (\mathrm{N.44})$$

Consider now the long-wavelength limit, in which $\mathbf{n}_k(\mathbf{x})$ are smooth functions of $\mathbf{x}$. By denoting the lattice spacing a and, taking the $N \to \infty$ (i.e., continuous-time) limit, we obtain an effective theory described by the action

$$S[\mathbf{n}] = \frac{j}{a^D} \int_{\mathbb{R}^D} d^D\mathbf{x}\, S_{WZ}[\mathbf{n}(\mathbf{x})]$$

$$- \frac{j^2|K|}{2a^{D-2}} \int_{t_i}^{t_f} dt \int_{\mathbb{R}^D} d^D\mathbf{x}\, \partial_i \mathbf{n}(\mathbf{x},t) \cdot \partial_i \mathbf{n}(\mathbf{x},t). \quad (\mathrm{N.45})$$

Here we have dropped the irrelevant constant appearing in Eq. (N.44). The non-trivial measure $\mathcal{D}\mu(\mathbf{n})$ in the functional integral is rewritten as $\mathcal{D}\mu(\mathbf{n})\delta[\mathbf{n}^2 - 1]$, where the integration variables $\mathbf{n}$ are not any more restricted to a target space $\mathcal{S}^2$. The functional δ-function can be represented via functional Fourier transform as

$$\delta[\mathbf{n}^2 - 1] = \lim_{N \to \infty} \prod_{i=1}^{N} \delta(\mathbf{n}^2(\mathbf{x}_i, t_i) - 1)$$

$$= \int \mathcal{D}\lambda \exp\left(i \int_{t_i}^{t_f} dt \int_{\mathbb{R}^D} d^D\mathbf{x}\, \lambda(\mathbf{x},t)(\mathbf{n}^2(\mathbf{x},t) - 1) \right), \quad (\mathrm{N.46})$$

which then leads to a new *total* action

$$S_{\mathrm{tot}}[\mathbf{n}] = S[\mathbf{n}] + \int_{t_i}^{t_f} dt \int_{\mathbb{R}^D} d^D\mathbf{x}\, \lambda(\mathbf{x},t)(\mathbf{n}^2(\mathbf{x},t) - 1). \quad (\mathrm{N.47})$$

We now look at the classical equation of motion whose solution represents the dominant field configuration in the semiclassical approach to quantum ferromagnetism. The variation $\delta S_{\mathrm{tot}}[\mathbf{n}] = 0$ implies the equations

$$j(\mathbf{n} \wedge \partial_t \mathbf{n}) + 2a^D \lambda \mathbf{n} = -a^2|K|j^2 \boldsymbol{\nabla}^2 \mathbf{n} \quad \text{and} \quad \mathbf{n}^2 = 1. \quad (\mathrm{N.48})$$

Here we have used

$$\delta S_{WZ}[\mathbf{n}(\mathbf{x})] = \int_0^1 du \int_{t_i}^{t_f} dt\, \partial_u\{\delta\mathbf{n}(\mathbf{x},t,u) \cdot [\mathbf{n}(\mathbf{x},t,u) \wedge \partial_t \mathbf{n}(\mathbf{x},t,u)]\}$$

$$+ 3 \int_0^1 du \int_{t_i}^{t_f} dt\, \delta\mathbf{n}(\mathbf{x},t,u) \cdot [\partial_t \mathbf{n}(\mathbf{x},t,\tau) \wedge \partial_u \mathbf{n}(\mathbf{x},t,u)]$$

$$= \int_{t_i}^{t_f} dt\, \delta\mathbf{n}(\mathbf{x},t) \cdot [\mathbf{n}(\mathbf{x},t) \wedge \partial_t \mathbf{n}(\mathbf{x},t)]. \quad (\mathrm{N.49})$$

The term on the second line is zero because $\mathbf{n}\delta\mathbf{n} = 0$ (on mass-shell solutions must fulfill the condition $\mathbf{n}^2 = 1$) and $\partial_t\mathbf{n} \wedge \partial_u\mathbf{n}$ is parallel to $\mathbf{n}$. Using the identity $\mathbf{n} \cdot (\mathbf{n} \wedge \partial_t\mathbf{n}) = 0$, λ is eliminated from Eq. (N.48). We obtain

$$\lambda = -\frac{|K|j^2}{2a^{D-2}}\, \mathbf{n} \cdot \boldsymbol{\nabla}^2 \mathbf{n}, \tag{N.50}$$

which leads to the equation of motion

$$\partial_t\mathbf{n} = a^2|K|j\,(\mathbf{n} \wedge \boldsymbol{\nabla}^2\mathbf{n}). \tag{N.51}$$

In deriving Eq. (N.51), which is known as the Landau–Lifshitz equation for quantum ferromagnet [403], we have utilized the identity $\mathbf{n} \wedge (\mathbf{n} \wedge \partial_t\mathbf{n}) = -\partial_t\mathbf{n}$. It essentially describes the dynamics of ferromagnetic spin waves. To see a leading dispersion behavior, we go to the linear regime and assume that the spins are aligned along the third axis around which they fluctuate (precess). So n_3 changes with t and $\mathbf{x}$ much slower that $n_{1,2}$. By defining, $\mathbf{n} = (\pi_1, \pi_2, \sigma)$ $(\boldsymbol{\pi}^2 + \sigma^2 = 1)$, omitting derivatives of σ and setting $\sigma \approx 1$ we linearize the Landau–Lifshitz equation and obtain

$$\partial_t\pi_1 \approx -a^2|K|j\boldsymbol{\nabla}^2\pi_2 \quad \text{and} \quad \partial_t\pi_2 \approx a^2|K|j\boldsymbol{\nabla}^2\pi_1. \tag{N.52}$$

Fourier transform of (N.52) yields the dispersion relation $\omega(\mathbf{k}) \propto \mathbf{k}^2$. The modes that obey this dispersion are *ferromagnetic magnons*. These are true (non-relativistic) Nambu–Goldstone modes.

In the large j limit the $SU(2)$ functional integral is dominated by the stationary points of $S_{\text{tot}}[\mathbf{n}]$ (i.e., by solution of (N.51)). This yields the semiclassical representation of the $SU(2)$ functional integral for Heisenberg ferromagnet. Semiclassical result can be arranged as power series in $1/j$. This expansion is known as the Holstein–Primakoff expansion [331].

There exists similar analysis for anti-ferromagnets [243]. Here the classical lowest energy configuration is described by the Néel state where the neighboring lattice spins flip the sign, i.e., $\mathbf{n}(l) \to (-1)^l\mathbf{n}(l)$. Consequently the dispersion relation of spin waves have the linear (relativistic) form $\omega(\mathbf{k}) \propto |\mathbf{k}|$. This linear and gapless dispersion describes the NG modes, which are called *anti-ferromagnetic magnons*. In anti-ferromagnets Berry phase does not play a dynamical role because in the Néel state the Wess–Zumino term reduces to a topological charge [243].

Appendix O

Imaginary-time formalism and phase transitions

In this Appendix we briefly discuss the application of imaginary-time formalism to phase transitions. We note that in the Matsubara formalism the effective action is more relevant with respect to the generating functional $W_{E,\beta}$. This is related mainly to the central role it plays in phase transitions. Indeed, in the Matsubara formalism, it is straightforward to tackle the subtleties concerning the convexity of the effective action in the ordered phase. This is because for the Matsubara functional integral one can easily prove that $W_\beta[J]$ is a *convex* function in J, i.e., $\forall\,\lambda \in [0,1]$.

$$W_{E,\beta}[\lambda J_1 + (1-\lambda)J_2] \;\leq\; \lambda W_{E,\beta}[J_1] \;+\; (1-\lambda)W_{E,\beta}[J_2]\,. \qquad (O.1)$$

This can be formally proven by setting

$$a_m = \frac{\mathcal{N}}{Z_{E,\beta}[J_m]} \exp\left(-S[\phi,\beta] + \int_0^\beta d^4x\,\phi(x)J_m(x)\right), \qquad (O.2)$$

and using the Jensen inequality [3]

$$a_1^\lambda a_2^{1-\lambda} \;\leq\; \lambda a_1 + (1-\lambda)a_2, \qquad (a_m > 0)\,. \qquad (O.3)$$

After functionally integrating both sides of Eq. (O.3), we obtain the convexity condition (O.1). For those sources $J(x)$ where $W_{E,\beta}[J]$ is twice differentiable, the convexity can also be deduced from the fact that

$$\frac{\delta^2 W_{E,\beta}[J]}{\delta J(x)\delta J(y)} \;=\; \langle(\phi(x) - \langle\phi(x)\rangle_J)(\phi(y) - \langle\phi(y)\rangle_J)\rangle_J\,, \qquad (O.4)$$

is the covariance matrix which is always positive semi-definite. The subscript J indicates that the averaging is performed with the weight function $Z_{E,\beta}^{-1}[J]e^{-S_E[\phi,J,\beta]}$ at non-zero current.

We now define the effective action $\Gamma_{E,\beta}^{\text{eff}}$ as the functional Legendre–Fenchel transform (LFT) of $W_{E,\beta}[J]$, i.e.,

$$\Gamma_{E,\beta}^{\text{eff}}[\Phi] \;=\; \max_J\left[\int_0^\beta d^4x\,J(x)\Phi(x) - W_{E,\beta}[J]\right] \;\equiv\; \mathfrak{L}W_{E,\beta}[J]\,. \qquad (O.5)$$

Because $W_{E,\beta}[J]$ is a convex function in J, the above relation defines J at a given β as an *invertible* function of Φ, i.e., $J(x) \equiv J(\Phi(x), \beta)$. We may then write

$$\Gamma_{E,\beta}^{\text{eff}}[\Phi] = \int_0^\beta d^4x\, J(\Phi(x), \beta)\Phi(x) - W_{E,\beta}[J(\Phi)], \qquad (\text{O.6})$$

which was also our starting point in the CTP formalism (cf. Eq. (6.55)). Note that when $W_{E,\beta}[J]$ is differentiable then from Eq. (O.5) follows that

$$\Phi(x) = \frac{\delta W_{E,\beta}[J]}{\delta J(x)} = \langle \phi(x) \rangle_J \equiv \phi_{E,c}(x). \qquad (\text{O.7})$$

To define the generating functional of 1PI thermal diagrams, we need only appeal to the Euclidean version of Coleman prescription (6.59) which now gives

$$W_{E,\beta}[J] = \max_\Phi \left[\int_0^\beta d^4x J(x)\Phi(x) - \Gamma_{E,\beta}^{\text{1PI}}[\Phi] \right] \equiv \mathfrak{L}\Gamma_{E,\beta}^{\text{1PI}}[\phi_{E,c}], \qquad (\text{O.8})$$

where

$$\Gamma_{E,\beta}^{\text{1PI}}[\phi_{E,c}] = \sum_{n=1}^\infty \frac{1}{n!} \int_0^\beta \prod_{i=1}^n d^4x_i\, \phi(x_1)\ldots\phi(x_n)\Gamma_E^n(x_1,\ldots,x_n)|_{\phi=\phi_{E,c}}. \qquad (\text{O.9})$$

Here $\Gamma_E^n(x_1,\ldots,x_n)$ are Matsubara n-point vertex functions.

Substituting Eq. (O.8) into Eq. (O.5), we obtain that $\Gamma_{E,\beta}^{\text{eff}} = (\mathfrak{L} \circ \mathfrak{L})\Gamma_{E,\beta}^{\text{1PI}}$. At this stage we should stress that the Legendre–Fenchel transform is not necessary *self-inverse* (involutive), that is to say, $\mathfrak{L} \circ \mathfrak{L}$ need not necessarily be equal to $\mathbb{1}$. In particular, $\mathfrak{L} \circ \mathfrak{L} = \mathbb{1}$ only on concave functions [542]. In fact, $\mathfrak{L} \circ \mathfrak{L}$ produces the *convex envelope* of the function on which it acts. By the convex envelope of a function, say $f(x)$, we mean the function that is identical with $f(x)$ everywhere except for non-convex parts. The non-convex parts of $f(x)$ are replaced by a straight line (or hyperplanes) connecting the neighboring convex branches. In this sense $(\mathfrak{L} \circ \mathfrak{L})f$ is the largest convex function fulfilling $(\mathfrak{L} \circ \mathfrak{L})f \leq f$. Existence of (strictly) non-convex regions translates at the level of $\mathfrak{L}f$ to non-differentiable points. As an illustration, let us consider a non-convex $\Gamma_{E,\beta}^{\text{1PI}}$, see Fig. O.1. The $\mathfrak{L}\Gamma_{E,\beta}^{\text{1PI}}$ produces $W_{E,\beta}$ which we know is convex. The double LFT of $\Gamma_{E,\beta}^{\text{1PI}}$, i.e., $\mathfrak{L}W_{E,\beta} = \Gamma_{E,\beta}^{\text{eff}}$ is the convex envelope of $\Gamma_{E,\beta}^{\text{1PI}}$. The LFT is uniquely invertible only between $W_{E,\beta}$ and $\Gamma_{E,\beta}^{\text{eff}}$ as only in this case both functionals are convex. Consequently $\Gamma_{E,\beta}^{\text{1PI}} \neq \Gamma_{E,\beta}^{\text{eff}}$.

In view of the above discussion, it is important to stress that the general transform which arises in (thermal) QFT is the LFT, not the Legendre

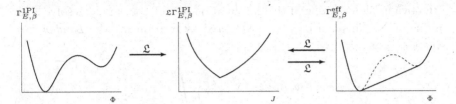

Fig. O.1 Structure of the Legendre–Fenchel transform for non-convex $\Gamma_{E,\beta}^{\text{1PI}}$. Note that $\Gamma_{E,\beta}^{\text{eff}} = \mathfrak{L}W_{E,\beta} = (\mathfrak{L} \circ \mathfrak{L})\Gamma_{E,\beta}^{\text{1PI}}$ represents a convex envelope of $\Gamma_{E,\beta}^{\text{1PI}}$.

---

transform. This is because of two facts: firstly $W_{E,\beta}$ is convex, and secondly the stationary phase (or steepest descent) method used by Coleman gives directly LFT between $W_{E,\beta}$ and $\Gamma_{E,\beta}^{\text{1PI}}$. Use of the Legendre transform in this context is strictly speaking erroneous because the Legendre transform is defined only for differentiable convex functions. In this sense LFT is a generalization of the Legendre transform, which reduces to the Legendre transform only when applied to convex differentiable functions. This will have important implications in systems with SSB as we shall see in the following.

Let us close with two more observations. In Matsubara formulation $\Gamma_{E,\beta}^{\text{eff}}$ has a simple physical meaning, namely it corresponds to a dimensionless Helmholtz free energy of sourceless system. In order to see this we recall that the Helmholtz free energy is defined as

$$F[J,\beta] = -T \log Z_{E,\beta}[J] = \langle H \rangle_J - T\langle S \rangle_J, \qquad (\text{O.10})$$

where S denotes the entropy of the systems and $\langle \cdots \rangle_J$ is evaluated with respect to the full Hamiltonian $H[\phi, J]$. So in particular $\beta F[J, \beta] = -W_{E,\beta}[J] - \log Z_{E,\beta}[0]$. By dropping the unimportant constant factor we have (cf. Eq. (O.6))

$$\Gamma_{E,\beta}^{\text{eff}}[\Phi] = \int_0^\beta d^4x\, J(\Phi(x), \beta)\Phi(x) + \beta F[J(\Phi), \beta]. \qquad (\text{O.11})$$

By setting $J = 0$ the Φ will correspond to the order parameter $\phi \equiv \langle \hat\phi \rangle$ (cf. Section 6.3.3) and thus $\Gamma_{E,\beta}^{\text{eff}}[\phi] = \beta F[\phi, \beta]$.

Although $\Gamma_{E,\beta}^{\text{eff}}$ can describe the phase transition exactly it does not possess the intuitive picture in which phase transitions can be viewed in terms of competing minima of a continuously varying, smooth function. For this purpose would seem $\Gamma_{E,\beta}^{\text{1PI}}$ more suitable. In fact, from (O.10) follows that

$$\beta F[J,\beta] = -W_{E,\beta}[J] \simeq \Gamma_{E,\beta}^{\text{1PI}}[\Phi] - \int_0^\beta d^4x\, J\Phi. \qquad (\text{O.12})$$

An obvious shortcoming of this formulation is that one cannot resolve J in terms of Φ. This can be indirectly cured by introducing the *Landau free energy* F_L as

$$\beta F_L[J, \Phi, \beta] = \Gamma_{E,\beta}^{\text{1PI}}[\Phi] - \int_0^\beta d^4x \, J\Phi. \qquad (\text{O}.13)$$

$\beta F_L[J, \Phi, \beta]$ should not be confused with $-W_{E,\beta}$ as it depends both on J and Φ as independent variables. Landau free energy is useful in exploring the phase structure of the whole theory, including both thermal and quantum effects, since the field expectation value (or at $J = 0$ the order parameter) can be obtained by minimizing $F_L[J, \Phi, \beta]$. The possible phases may be identified by searching for non-trivial minima. The actual phase is represented by the absolute minimum, while the other minima should correspond to metastable states. This is the key concept that is taken over in the Landau–Ginzburg treatment of phase transitions (see below).

In conclusion, it may be noted that the doubling of degrees of freedom in the real-time-path approach cannot be achieved by simply analytically continuing in the imaginary-time approach the complex time from imaginary to real axis. In Section 6.6 we show that both real and imaginary-time approaches can be, however, conveniently unified through the geometric background method where the mutual analytic continuations find their proper mathematical framework.

O.1   Landau–Ginzburg treatment

The idea that $\Gamma_\beta^{\text{1PI}}$ summarizes (both in the real and imaginary-time formalism) the joint effect of quantum and thermal fluctuations that superpose on the top the classical theory turns out to be important concept in dealing with phase transitions. To see the effect of the fluctuations explicitly one can try to recast $\Gamma_\beta^{\text{1PI}}[\phi]$ into a form that looks like a generalization of the classical action $S[\phi, \beta]$. Because of the non-local nature of $\Gamma_\beta^{\text{1PI}}$ (cf. (6.62) and (O.9)) one can bring $\Gamma_\beta^{\text{1PI}}$ only into a quasi-local form. This is achieved by expanding ϕ_c in Eqs. (6.62) and (O.9) about the point x_1 that is common to each integrand, i.e.,

$$\begin{aligned}
\phi_c(x_i) = \; & \phi_c(x_1) + (x_i - x_1)^\mu \partial_\mu \phi_c(x_1) \\
& + \tfrac{1}{2}(x_i - x_1)^\mu (x_i - x_1)^\nu \partial_\mu \partial_\nu \phi_c(x_1) + \cdots.
\end{aligned} \qquad (\text{O}.14)$$

Integrating over $x_2, x_3 \ldots$ and collecting the derivatives of ϕ_c we can cast $\Gamma_{E,\beta}^{1PI}$ into the form[3]

$$\Gamma_{E,\beta}^{1PI}[\phi_c] = \int_0^\beta d^4x \left[V_\beta(\phi_c) + \tfrac{1}{2} Z_\beta(\phi_c)(\partial_\mu \phi_c)^2 + \cdots \right], \qquad (O.15)$$

where we have relabeled x_1 as x. The "$\cdots$" in the integrand denote terms that contain four and more derivatives. Because of the proliferation of the higher order derivatives, the expansion (O.15) is known as the *gradient expansion*. Terms $V_\beta(\phi_c)$, $Z(\phi_c)$, etc., are ordinary functions (not functionals) in ϕ_c. The term without any derivatives – $V_\beta(\phi_c)$, defines the *effective potential*. The structure of $V_\beta(\phi_c)$ can be put into more manageable form by taking the Fourier transform of Γ_E^n, i.e.,

$$\Gamma_E^n(x_1, \ldots, x_n) = \int \frac{d^4k_1}{(2\pi)^4} \cdots \frac{d^4k_n}{(2\pi)^4} (2\pi)^4 \, \delta^4(k_1 + \cdots + k_n)$$

$$\times \; e^{i(k_1 x_1 + \cdots + k_n x_n)} \, \tilde{\Gamma}_E^n(k_1, \ldots, k_n). \qquad (O.16)$$

Here δ-function reflects the energy conservation in each vertex. For simplicity we have also assimilated the discrete summation over Matsubara frequencies into an integration measure. We now insert the result (O.16) into (O.9) and compare it with (O.15). This directly gives that

$$V_\beta(\phi_c) = \sum_{n=1}^\infty \frac{1}{n!} \tilde{\Gamma}_E^n(0, 0, \cdots, 0)(\phi_c(x))^n. \qquad (O.17)$$

The latter simply says that the vertex functions at zero momenta act as (T and $\hbar$ dependent) "couplings" in an effective potential. Effective potential generally have a different structure of stationary points (stable or unstable minima) than the classical potential. This is, for instance, nicely demonstrated in the celebrated Coleman–Weinberg mechanism [163, 167]. In fact, the radiative and thermal corrections can both create new minima as well as erase the old classical ones. In practice they can even spoil the breakdown of symmetry or generate a breakdown of symmetry even if it was not existent at classical level. In the latter case one speaks [476] about *dynamical breakdown of symmetry*.

The key observation in the framework of $\Gamma_{E,\beta}^{1PI}$ is that its essential behavior near the critical temperature $T_c = \beta_c^{-1}$ in second order phase transitions can be grasped via the Landau–Ginzburg approximation [282, 381]. In the Landau–Ginzburg approach one directly works with the order parameter

---

[3]For a definiteness we illustrate the concept with the Matsubara 1PI generating functional. This is also most typical framework used in the literature.

$\phi \equiv \langle \hat{\phi} \rangle$ (see Eq. (6.58)) rather than $\phi_c = \langle \hat{\phi} \rangle_J$. According to the note after Eq. (6.57), the order-parameter field can be considered as independent of time and temperature. By expanding (O.15) in ϕ we can write

$$\Gamma^{1PI}_{E,\beta}[\phi]_{T \approx T_c} = \beta F_L[0, \phi, \beta]_{T \approx T_c} \simeq \Gamma_{\beta, LG}[\phi]$$

$$\equiv \Gamma_{\beta, LG}[0] + \beta_c \int_{\mathbb{R}^3} d^3x \, [a(T_c)(\nabla \phi)^2 + b(T)\phi^2 + c(T_c)\phi^4] . \quad (O.18)$$

The expansion has been truncated after the fourth term in ϕ and the second-order gradient term, i.e., ϕ is supposed to be a slowly varying function [282]. The coefficients appearing in $\Gamma_{\beta, LG}[\phi]$ can be identified as follows

$$\beta a(T) = \frac{1}{2} Z_\beta(0) , \quad (O.19a)$$

$$\beta b(T) = \frac{1}{2} \tilde{\Gamma}^2_E(0, 0) , \quad (O.19b)$$

$$\beta c(T) = \frac{1}{4!} \tilde{\Gamma}^4_E(0, 0, 0, 0) . \quad (O.19c)$$

To have a phase transition, the coefficient $b(T)$ must flip sign at T_c, and to justify the truncation, the coefficients $a(T_c)$ and $c(T_c)$ must be positive, otherwise $\Gamma_{\beta, LG}[\phi]$ can be minimized by $|\phi| \to \infty$, whereas we wish to describe how the order parameter rises from zero and has a finite value as the coupling constants are varied through T_c to lower temperatures. We may observe that $\Gamma_{\beta, LG}[\phi]$ has a unique minimum in the phase where $b(T) > 0$ (Wigner phase), while when $b(T) < 0$ (Goldstone phase) it acquires the double-well potential with degenerate minima at $\phi = \pm\sqrt{-b/2c}$. By minimizing $\Gamma_{\beta, LG}[\phi]$ with respect to order-parameter fluctuations we obtain that ϕ near T_c fulfills the Landau–Ginzburg equation

$$a(T_c)\nabla^2 \phi = b(T)\phi + 2c(T_c)\phi^3 . \quad (O.20)$$

We finally remark that first-order phase transitions can be described by an extended Landau–Ginzburg treatment, for example by including a sixth-order term in ϕ with a positive coefficient and considering $c(T_c) \leq 0$ (see, e.g., [403]).

Appendix P

Proof of Bogoliubov inequality

Here we prove the Bogoliubov inequality $F \leq F_0 + \langle H - H_0 \rangle_0$, used in Section 5.7. In the actual proof we employ the functional integrals. To this end it is convenient to reformulate (L2) and (L3) in terms of a Hamiltonian and corresponding phase-space variables ϕ and π. So, in particular, for the action we can write

$$S[\phi, \beta] = \int_0^{-i\beta} dx_0 \int_{\mathbb{R}^3} d\mathbf{x}\, \mathcal{L}(\phi, \beta)$$

$$\mapsto \quad S[\phi, \pi, \beta] = \int_0^{-i\beta} dx_0 \int_{\mathbb{R}^3} d\mathbf{x}\, [\pi \partial_0 \phi - \mathcal{H}(\phi, \pi, \beta)] . \quad \text{(P.1)}$$

Let us now parametrize the time-integration by an imaginary-time variable $x_0 = -i\tau$ ($\tau \in [0, \beta]$) and introduce the Euclidean fields $\phi_E(\tau, \mathbf{x}) = \phi(x)$ and

$$\pi_E(\tau, \mathbf{x}) = \frac{\partial \mathcal{L}_E(\phi_E)}{\partial \dot{\phi}_E} = -\frac{\partial \mathcal{L}(\phi_E)}{\partial \dot{\phi}_E} = -\pi(x). \quad \text{(P.2)}$$

With these we obtain

$$S[\phi, \pi, \beta] = -i \int_0^\beta d\tau \int_{\mathbb{R}^3} d\mathbf{x} \left[i\pi_E \dot{\phi}_E - \mathcal{H}(\phi_E, \pi_E, \beta) \right]$$

$$\equiv -i \int_0^\beta d\tau \left[i\pi_E \dot{\phi}_E - H(\phi_E, \pi_E, \beta) \right] . \quad \text{(P.3)}$$

Here the assumption was made that $\mathcal{H}(\phi_E, -\pi_E, \beta) = \mathcal{H}(\phi_E, \pi_E, \beta)$. This is usually true in systems without gauge symmetry.[4] In addition, the

---

[4]In general cases it is the phase-space formulation with a Hamiltonian that is more fundamental. The usual configuration-space formulation is just a derived formulation after the functional integration over π's is performed [384].

293

configuration-space measure goes over into phase-space measure according to

$$\int_{\beta-\text{periodic}} \mathcal{D}\phi_E \;\mapsto\; \int_{\beta-\text{periodic}} \mathcal{D}\phi_E \mathcal{D}\pi_E \,. \tag{P.4}$$

Here the β-periodicity condition is applied only in the ϕ_E integration. Similarly as in ordinary integral calculus we can relabel the integration variables and call them simply ϕ and π. In passing, the reader may note that a similar phase-space structure was already obtained in coherent-state functional integrals, see, for instance, (N.20) and (N.21).

Exploiting the connection between Helmholtz free energy and Euclidean generating functional, we may write the partition function as (cf. Eq. (O.10))

$$e^{-\beta F_0} = \int_{\beta-\text{periodic}} \mathcal{D}\phi \mathcal{D}\pi \, \exp\left(\int_0^\beta d\tau [i\dot\phi\pi - H_0(\phi,\pi)] \right) , \tag{P.5a}$$

$$e^{-\beta F} = \int_{\beta-\text{periodic}} \mathcal{D}\phi \mathcal{D}\pi \, \exp\left(\int_0^\beta d\tau [i\dot\phi\pi - H(\phi,\pi)] \right) , \tag{P.5b}$$

and

$$\langle H - H_0 \rangle_0 = e^{\beta F_0} \int_{\beta-\text{periodic}} \mathcal{D}\phi \mathcal{D}\pi \, (H - H_0) \exp\left(\int_0^\beta d\tau [i\dot\phi\pi - H_0(\phi,\pi)] \right) . \tag{P.6}$$

The normalization constant $\mathcal{N}$ was for simplicity assimilated into a functional measure.

To proceed we will assume that the functional measures (i.e., what is here loosely denoted as $\mathcal{D}\phi\mathcal{D}\pi$) for both $e^{-\beta F_0}$ and $e^{-\beta F}$ are identical. This is generally not the case when both dynamics possess, for instance, different gauge symmetries. In such cases the inequality cannot be proven with the present method, and in fact, it can be argued that it does not hold in general.

In the next step we use Jensen (or convexity) inequality [3]

$$a_1^\lambda a_2^{1-\lambda} \leq \lambda a_1 + (1 - \lambda)a_2 \,, \tag{P.7}$$

which is valid for all $a_{1,2} \geq 0$ and all real $\lambda \in [0,1]$. By setting

$$a_1 = \exp\left(\int_0^\beta d\tau [i\dot\phi\pi - H(\phi,\pi)] \right) , \tag{P.8a}$$

$$a_2 = \exp\left(\int_0^\beta d\tau [i\dot\phi\pi - H_0(\phi,\pi)] \right) , \tag{P.8b}$$

we obtain, after a functional integration of (P.7), the inequality

$$e^{-\beta F_\lambda} - \lambda e^{-\beta F} - (1-\lambda)e^{-\beta F_0} \leq 0. \tag{P.9}$$

Here F_λ is defined as

$$e^{-\beta F_\lambda} = \int_{\beta-\text{periodic}} \mathcal{D}\phi \mathcal{D}\pi \exp\left(\int_0^\beta d\tau [i\dot{\phi}\pi - (H_0 + \lambda(H - H_0))]\right). \tag{P.10}$$

Since (P.9) holds for all $\lambda \in [0,1]$, it holds also for $\lambda \ll 1$ (for such λ's is, due to convexity, the inequality close to identity). To this end we expand $e^{-\beta F_\lambda}$ to order $\mathcal{O}(\lambda)$, so that

$$e^{-\beta F_\lambda} \simeq e^{-\beta F_0 - \lambda\beta F_\lambda'|_{\lambda=0}} \simeq e^{-\beta F_0}(1 - \lambda\beta F_\lambda'|_{\lambda=0})$$

$$= e^{-\beta F_0}[1 - \lambda\beta\langle H - H_0\rangle_0]. \tag{P.11}$$

By inserting this into (P.9) we obtain

$$e^{-\beta F_0}[1 - \beta\langle H - H_0\rangle_0] \leq e^{-\beta F}. \tag{P.12}$$

Exponential Jensen inequality

$$e^{-\beta\langle H - H_0\rangle_0} \leq [1 - \beta\langle H - H_0\rangle_0], \tag{P.13}$$

then implies that

$$e^{-\beta F_0}e^{-\beta\langle H - H_0\rangle_0} \leq e^{-\beta F_0}[1 - \beta\langle H - H_0\rangle_0] \leq e^{-\beta F}. \tag{P.14}$$

Taking logarithm on both sides, and using the fact that $\log(\dots)$ is a monotonic function of its argument, we finally obtain

$$F \leq F_0 + \langle H - H_0\rangle_0. \tag{P.15}$$

Chapter 7

Topological defects as non-homogeneous condensates. I

7.1 Introduction

In this and in the following Chapter, we study explicit examples of "extended objects" of quantum origin (with or without topological singularities), which behave classically. Vortices in superconductors and superfluids, domain wall in ferromagnets, and other soliton-like objects in many other physical systems are extended objects of this type. An introduction to solitons is presented in Chapter 10.

The study of many-body physics and elementary particle physics shows that, at a basic level, Nature is ruled by quantum dynamical laws. Many phenomena are observed where quanta coexist and interact with extended objects which show a classical behavior. Systems such as superconductors, superfluids, crystals, ferromagnets are described by quantum microscopic dynamics from which most of their macroscopic behaviors are derived. As mentioned in Chapters 3 and 4, these and other systems, characterized by a certain degree of ordering in their fundamental state, appear as *macroscopic quantum systems*, in the sense that some of their macroscopic observable properties cannot be explained without recourse to the underlying quantum dynamics. Even for structures at cosmological scale, the question of their dynamical origin from elementary components requires an answer consistent with quantum dynamical laws [371–373]. Quantum theory thus appears not to be confined to microscopic phenomena. The question then arises of how quantum dynamics generates not only the observed macroscopic properties such as ordered patterns and coherent behaviors, but also the variety of soliton-like *defects* mentioned above. Thus, we are faced also with the question of the quantum origin of the macroscopically behaving extended objects and of their interaction with quanta [619]. In other words, how it happens that, out of the microscopic scale of the quantum elementary components, the macroscopic scale characterizing those systems dynamically emerges together with the classically behaving solitons [617]. We have already analyzed such a problem in Section 4.3.2, where defect formation in the process of symmetry breaking phase transitions has been considered in terms of non-homogeneous boson condensation. There, we have seen the role played by boson transformation functions with topological singularities and observed that topologically non-trivial defects appear in systems presenting an ordered state.

In the present and in the following Chapter we consider examples of topological defects as non-homogeneous condensates [102]. In this Chapter, we will consider a number of soliton-like extended objects, such as the $\lambda\phi^4$ kink, the sine-Gordon soliton, the soliton solution of non-linear Schrödinger equation, including the Davydov soliton which has been applied to model biological processes. In the following Chapter, we will consider topological defects in gauge theories. We will study the explicit form of the boson condensation function $f(x)$ for the various cases under considerations and show how the classical solitons are recovered in appropriate limits from the QFT framework. Moreover, we will study the effects of finite temperature by means of formalisms introduced in previous Chapters, such as Thermo Field Dynamics and Closed Time-Path.

7.2   Quantum field dynamics and classical soliton solutions

In the following Sections we will consider various examples of classical soliton solutions arising from the underlying quantum dynamics via the boson transformation method. The specific form and singularities of $f(x)$ are the ones consistent with the topological properties of the specific solution of the classical field equation which we want to study. We show that the extended objects described by the classical soliton solutions are obtained as non-homogeneous boson condensation of the quantum fields: they appear as the macroscopic envelopes of localized boson condensates. Such a condensation is controlled, under convenient boundary conditions, by the quantum dynamics. The boson transformation (7.8) is thus determined by the internal consistency of the dynamics.

In the case $\psi(x)$ is a fermion field, one considers products of an even number of $\psi(x)$ fields, as done, e.g., in the chiral gauge model discussed in Section 3.8.2, and then one considers the boson condensate of NG boson fields. For example, in the case of superconductivity [422], one considers the order parameter $\Delta(x) = \langle 0|\psi_\downarrow \psi_\uparrow|0\rangle$ for the spin-down and spin-up Heisenberg electron field, the NG bosons being bound states of two electrons. The Gor'kov equation [294, 295] for $\Delta(x)$ is a well-known example of the classical Euler–Lagrange equation [458, 459] (for small $\Delta(x)$ the Gor'kov equation becomes the Ginzburg–Landau equation).

7.2.1 *The dynamical map and the boson transformation*

For our subsequent discussion, it is useful to summarize some aspects of QFT formalism presented in previous Chapters. For simplicity, in the following we only consider the case of one boson Heisenberg field $\psi(x)$ and one asymptotic field $\varphi(x)$.

The dynamics is described in terms of the Heisenberg field, which satisfy equal-time canonical commutation relations and the Heisenberg field equation:

$$\Lambda(\partial)\,\psi(x) = J[\psi(x)]\,. \tag{7.1}$$

$\Lambda(\partial)$ is a differential operator and J is some functional of ψ. They are specific to the model dynamics describing the interaction. The observable level is described in terms of the asymptotic (in- and/or out-) field, $\varphi(x)$ satisfying the field equation

$$\Lambda(\partial)\,\varphi(x) = 0\,. \tag{7.2}$$

Eq. (7.1) can be recast in the integral form (Yang–Feldman equation):

$$\psi(x) = \varphi(x) + \Lambda^{-1}(\partial)\,J[\psi(x)]\,. \tag{7.3}$$

The symbol $\Lambda^{-1}(\partial)$ formally denotes the $\varphi(x)$ field Green's function, whose precise form is specified by the boundary conditions. Eq. (7.3) can be solved by iteration, thus giving an expression for the Heisenberg field $\psi(x)$ in terms of powers of the $\varphi(x)$ field; this is the Haag expansion (or "dynamical map" [199, 617, 619], cf. Section 2.2), which might be formally written as

$$\psi(x) = \Psi\,[x; \varphi]\,. \tag{7.4}$$

We recall that such an expression is valid in a weak sense, i.e., for the matrix elements only (cf. Chapter 2). This implies that Eq. (7.4) is not unique, since in general different sets of asymptotic fields (and the corresponding Hilbert spaces) can be used in its construction. Let us indeed consider a c–number function $f(x)$, solution of

$$\Lambda(\partial)\,f(x) = 0\,, \tag{7.5}$$

where $\Lambda(\partial)$ is the same as in Eq. (7.2). Then the corresponding Yang–Feldman equation takes the form

$$\psi^f(x) = \varphi(x) + f(x) + \Lambda^{-1}(\partial)J[\psi^f(x)]. \qquad (7.6)$$

The latter gives rise to a *different* Haag expansion for a field $\psi^f(x)$ still satisfying the Heisenberg equation (7.1):

$$\psi^f(x) = \Psi^f[x; \varphi + f]. \qquad (7.7)$$

The difference between the two solutions ψ and ψ^f is only in the boundary conditions. An important point is that the expansion Eq. (7.7) is obtained from that in Eq. (7.4), by the spacetime-dependent translation (*boson transformation*):

$$\varphi(x) \rightarrow \varphi(x) + f(x). \qquad (7.8)$$

Eqs. (7.6)–(7.8) express the *boson transformation theorem* [619] (cf. Section 4.3): the dynamics embodied in Eq. (7.1), contains an internal freedom, represented by the possible choices of the function $f(x)$, satisfying the free field equation (7.2). Eq. (7.8) is a canonical transformation since it leaves invariant the canonical commutation relations.

The vacuum expectation value of Eq. (7.6) gives

$$\phi^f(x) \equiv \langle 0|\psi^f(x)|0\rangle = f(x) + \langle 0|\left[\Lambda^{-1}(\partial)J[\psi^f(x)]\right]|0\rangle. \qquad (7.9)$$

The classical solution is obtained by means of the classical or Born approximation, which consists in taking $\langle 0|J[\psi^f]|0\rangle = J[\phi^f]$, i.e., neglecting all contractions of the physical fields. In this limit, $\phi_{cl}^f(x) \equiv \lim_{\hbar \to 0} \phi^f(x)$ is the solution of the classical Euler–Lagrange equation:

$$\Lambda(\partial)\,\phi_{cl}^f(x) = J[\phi_{cl}^f(x)]. \qquad (7.10)$$

The Yang–Feldman equation (7.6) describes not only the equations for $\phi_{cl}^f(x)$ in the classical approximation, i.e., Eq. (7.10), but also, at higher orders in $\hbar$, the dynamics of one or more quantum physical particles in the potential generated by the macroscopic object $\phi^f(x)$ [619].

To see this, we rewrite the dynamical map (7.7) by expanding $\psi^f(x)$ around $\phi^f(x)$. Using the relation

$$\frac{\delta J^f(x)}{\delta f(y)} = \int d^4z \frac{\delta J^f(x)}{\delta \phi^f(z)} \frac{\delta \phi^f(z)}{\delta f(y)}, \qquad (7.11)$$

we obtain [619]:

$$\psi^f(x) = \phi^f(x) + \int d^4y\,\varphi(y)\frac{\delta}{\delta f(y)}\phi^f(x)$$

$$+ \frac{1}{2} : \int d^4y_1\,d^4y_2\,\varphi(y_1)\,\varphi(y_2)\frac{\delta}{\delta f(y_1)}\frac{\delta}{\delta f(y_2)}\phi^f(x) : + \ldots$$

$$\equiv \phi^f(x) + \psi_f^{(1)}(x) + \frac{1}{2} : \psi_f^{(2)}(x) : + \ldots, \qquad (7.12)$$

with

$$\Lambda(\partial)\,\phi^f(x) \,=\, J[\phi^f(x)]\,, \tag{7.13}$$

$$\Lambda(\partial)\,\psi_f^{(1)}(x) \,-\, \int d^4y\, V^f(x,y)\,\psi_f^{(1)}(y) \,=\, 0\,, \tag{7.14}$$

where $V^f(x,y) \,=\, \frac{\delta}{\delta\phi^f(y)}\,\langle 0|J[\psi^f(x)]|0\rangle$ is the (self-consistent) potential induced by the macroscopic object which is self-consistently created in the quantum field system.

Eq. (7.14) describes a quantum particle under the influence of the potential $V^f(x,y)$. It is possible to derive a complete hierarchy of such equations, describing, at higher order, scattering of two or more particles in the potential of the macroscopic object. For example, the equation of next order with respect to Eq. (7.14) is:

$$\Lambda(\partial)\,\psi_f^{(2)}(x) \,-\, \int d^4y\, V^f(x,y)\,\psi_f^{(2)}(y) \,=$$

$$=\, \int d^4y_1\, d^4y_2\, V^f(x,y_1,y_2)\,\psi_f^{(1)}(y_1)\,\psi_f^{(1)}(y_2)\,, \tag{7.15}$$

with $V^f(x,y_1,y_2) \,=\, \frac{\delta}{\delta\phi^f(y_1)}\,\frac{\delta}{\delta\phi^f(y_2)}\,\langle 0|J[\psi^f(x)]|0\rangle$.

For a more detailed description of these aspects, see [617, 619].

The class of solutions of Eq. (7.5), which lead to non-trivial (i.e., carrying a non-zero topological charge) solutions of Eq. (7.10), are those which have some sort of singularity with respect to Fourier transform. These can be either *divergent singularities* or *topological singularities*. The former are associated to divergences of $f(x)$ for $|\mathbf{x}| = \infty$, at least in some direction. The latter arise when $f(x)$ is not single-valued, i.e., is path-dependent.[1] In both cases, the macroscopic object described by the order parameter will carry a non-zero topological charge (cf. Sections 3.5, 4.3, 4.3.1).

The computational strategy is the following (see Fig. 7.1): the first step consists in writing down the dynamical map(s) for the Heisenberg operator(s). Then one introduces the boson transformation function(s) f controlling the choice of the Hilbert space. The next step is to determine f, say $\tilde{f}$, corresponding to a particular soliton solution by taking the classical limit $\phi_{cl}^f \equiv \lim_{\hbar\to 0}\phi^f$ of the order parameter $\phi^f \equiv \langle 0|\psi^f|0\rangle$. The function $\tilde{f}$ is then used to obtain the Heisenberg field operator in the chosen soliton sector: $\psi^{\tilde{f}}$. At this point there are various possibilities: $i)$ calculate quantum corrections to the order parameter, by taking higher orders in the $\hbar$ expansion of $\langle 0|\psi^{\tilde{f}}|0\rangle$; $ii)$ study finite temperature effects on the order parameter,

---

[1] An interesting study on multivalued fields and topological defects can be found in [383].

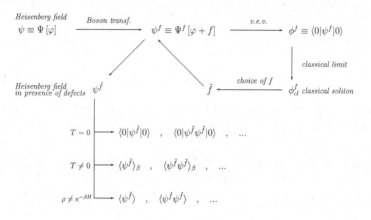

Fig. 7.1　Computational strategy

by considering $\langle \psi^{\tilde{f}} \rangle_\beta$, where $\langle \ldots \rangle_\beta$ stands for thermal average; *iii)* calculate Green's functions, such as $\langle T[\psi^{\tilde{f}} \psi^{\tilde{f}}] \rangle$, in the presence of defects, both at zero and finite temperature. Another possibility, not considered here, is the study of the non-equilibrium phase transition properties of systems containing defects.

Finally, we remark that the *c*-number (classical) field $\phi^f(x)$ is obtained without need of performing the $\hbar \to 0$ limit. Since a large number of bosons is condensed, we may have

$$\frac{\hbar \Delta n}{\hbar n} = \frac{\Delta n}{n} \ll 1 , \tag{7.16}$$

i.e., quantum fluctuations $\hbar \Delta n$ can be much less than the total quantum number $\hbar n$ [182, 619]. The point is that the inequality (7.16) does not depend on $\hbar$. Therefore, classical behavior is possible even for non-vanishing $\hbar$, which allows the co-existence of quanta and classically (macroscopically) behaving systems (extended objects) and their interaction. We have already commented on the interaction of quanta with classical fields and currents of quantum origin in Chapter 4, see in particular Sections 4.3.1 and 4.4.2.

7.2.2　*The quantum coordinate*

The appearance of a localized macroscopic object breaks the translational invariance of the system (and in particular that of the ground state): this is reflected by the appearance of zero energy modes, i.e., the Nambu–Goldstone (NG) modes associated with the symmetry breaking.

The dynamics of the quasiparticles in the presence of the extended object may admit different solutions: *scattered waves, bound states* and *zero energy modes*. The first ones are recovered by solving by iteration Eq. (7.14) and have a continuous spectrum; the second ones are discrete energy solutions of Eq. (7.14) and represent bound states of φ-quanta and the extended object. The discrete energy solutions may also include zero energy solutions, describing the NG modes created by the breakdown of translational symmetry.

These solutions are easily shown to exist when we consider Eq. (7.14), for simplicity, at tree level and for a static extended object (i.e., $f(x) = f(\mathbf{x})$):

$$\Lambda(-i\omega, \nabla) u_\omega(\mathbf{x}) - \int d^3 \mathbf{y} \frac{\delta J[\phi^f(\mathbf{x})]}{\delta \phi^f(\mathbf{y})} u_\omega(\mathbf{y}) = 0, \qquad (7.17)$$

where $u_\omega(\mathbf{x})$ denotes the wave function of the solution of energy ω. By applying ∇ on Eq. (7.13), we obtain [619] in the static case, $\phi^f(x) = \phi^f(\mathbf{x})$),

$$\Lambda(0, \nabla) \nabla \phi^f(\mathbf{x}) - \int d^3 \mathbf{y} \frac{\delta J[\phi^f(\mathbf{x})]}{\delta \phi^f(\mathbf{y})} \nabla \phi^f(\mathbf{x}) = 0, \qquad (7.18)$$

i.e., $\nabla \phi^f(\mathbf{x})$ is a solution of Eq. (7.17) with zero energy.

The necessity of taking into account all the different solutions of Eq. (7.17) comes when considering the commutation relations for the $\psi(x)$ field. In the tree approximation, the only contribution to the δ function in the canonical commutator $[\psi(x), \partial_t \psi(y)]_{t_x = t_y} = \delta^3(\mathbf{x} - \mathbf{y})$ comes from the $\psi^{(1)}(x)$ component.

The fact that Eq. (7.17) admits both continuous and discrete energy solutions implies that a completeness relation can be obtained only by considering a combination of all the wave functions related to these modes. More explicitly, denoting by $\chi^s(x)$ and $\chi^b(x)$ the scattered waves and the bound states solutions of Eq. (7.17), they are expanded as follows:

$$\chi^s(x) = \int d^3 \mathbf{k} \frac{\hbar}{\sqrt{2\omega_k}} \left[u_{\mathbf{k}}(x) \alpha_{\mathbf{k}} e^{-i\omega_k t} + u_{\mathbf{k}}^*(x) \alpha_{\mathbf{k}}^\dagger e^{i\omega_k t} \right], \quad (7.19)$$

$$\chi^b(x) = \sum_{i>3} \frac{\hbar}{\sqrt{2\omega_i}} \left[u_i(x) \alpha_i e^{-i\omega_i t} + u_i^*(x) \alpha_i^\dagger e^{i\omega_i t} \right], \qquad (7.20)$$

where the operators $\alpha_{\mathbf{k}}$ and α_i are independent. In Eq. (7.20) the first three modes (not included into the sum) are those associated with the three directions for spatial translations, given by $\nabla \phi^f(\mathbf{x})$. Since these modes have zero energy, they cannot be written in the form (7.20) and must be considered separately.

Denoting by β_i, $i = 1, 2, 3$, the annihilation operators for these modes, we can write

$$\mathbf{q} = \frac{1}{\sqrt{2\omega}} \left(\boldsymbol{\beta} + \boldsymbol{\beta}^\dagger \right) , \tag{7.21}$$

$$\mathbf{p} = -i\sqrt{\frac{\omega}{2}} \left(\boldsymbol{\beta} - \boldsymbol{\beta}^\dagger \right) , \tag{7.22}$$

with $[q_i, p_j] = i\delta_{ij}$. The quantum mechanical operator $\mathbf{q}$ is called the *quantum coordinate* and it commutes with $\alpha_{\mathbf{k}}$ and α_i.

The expression for $\psi_f^{(1)}(x)$ is then:

$$\psi_f^{(1)}(x) = \varphi(\mathbf{x}) - (\mathbf{q} \cdot \nabla)\phi^f(\mathbf{x}), \tag{7.23}$$

where $\varphi(x) = \chi^s(x) + \chi^b(x)$. The dynamical map now reads

$$\psi^f(x) = \phi^f(x) - (\mathbf{q} \cdot \nabla)\phi^f(\mathbf{x}) + \varphi(x) + \ldots$$
$$= \phi^f(\mathbf{x} - \mathbf{q}) + \varphi(x) + \ldots , \tag{7.24}$$

which shows how the quantum coordinate appears in the dynamical map in the combination $(\mathbf{x} - \mathbf{q})$ with the spatial coordinate. The quantum coordinate $\mathbf{q}$ and the conjugate momentum $\mathbf{p}$ describe respectively quantum fluctuations of the position and momentum of the soliton.

The appearance of the quantum coordinate modifies the structure of the Hilbert space, since it is now the direct product of the Hilbert space for the quantum coordinate and of the Fock space for the excitation modes. The order parameter is then defined more properly as

$$\phi^f(x) = \langle 0|\psi^f(x)|0\rangle\big|_{\mathbf{q}=\mathbf{p}=0} . \tag{7.25}$$

For a systematic discussion of quantum coordinates and their relation to collective coordinates (cf. Chapter 10) see [617, 619]. A further issue connected with quantum coordinates is the one of the so-called *emergent symmetries*, i.e., those symmetries associated with the macroscopic objects and which do not exist in the basic Lagrangian. An example of this is the Abrikosov lattice formed by vortices in superconductors. Here we will not discuss emergent symmetries, see [617, 619].

7.3   The $\lambda\phi^4$ kink solution

We consider a real scalar Heisenberg field ψ with quartic interaction term in $(1 + 1)$ dimensions. The Lagrangian (throughout we adopt the Minkowski metric with signature $(+-)$) is:

$$\mathcal{L} = \frac{1}{2} (\partial_\mu \psi(x))^2 - \frac{1}{2} \mu^2 \psi^2(x) - \frac{\lambda}{4} \psi^4(x). \tag{7.26}$$

For simplicity we do not consider here the renormalization problem; see [95, 617, 619] for a discussion of this issue. The Heisenberg equation is

$$(\partial^2 + \mu^2)\psi(x) = -\lambda\psi^3(x),$$ (7.27)

where $\partial^2 \equiv \partial_t^2 - \partial_{x_1}^2$ and $x = (x_0, x_1)$. We symbolically write

$$\Lambda(\partial)\psi(x) = J[\psi(x)],$$ (7.28)

where $\Lambda(\partial) \equiv -\partial^2 - \mu^2$ and $J[\psi(x)] \equiv \lambda\psi^3(x)$. In the case $\mu^2 < 0$, it is known that (7.26) admits at a classical level kink solutions [163, 271, 532]. We shall not consider the case $\mu^2 > 0$, since it does not possess the kink solutions. The quantum theory is completely specified when the Hilbert space on which the field operators have to be realized is assigned. We choose to work in the Hilbert space for the in-fields and denote by $|0\rangle$ the vacuum for such fields. We assume that the $\psi(x)$ field has non-vanishing expectation value in the vacuum $|0\rangle$:

$$\langle 0|\psi(x)|0\rangle = v,$$ (7.29)

and define the Heisenberg field operator $\rho(x)$ by the relation

$$\psi(x) = v + \rho(x).$$ (7.30)

As a consequence of the parametrization (7.29) we obtain

$$\mathcal{L} = \frac{1}{2}(\partial_\mu\rho)^2 - \frac{1}{2}m^2\rho^2 - \frac{g^2}{8}\rho^4 - \frac{1}{2}mg\rho^3 + \frac{m^4}{8g^2},$$ (7.31)

and the Heisenberg equation for the field ρ is:

$$(\partial^2 + m^2)\rho(x) = -\frac{3}{2}mg\rho^2(x) - \frac{1}{2}g^2\rho^3(x),$$ (7.32)

with $g = \sqrt{2\lambda}$ and $m^2 = 2\lambda v^2 = -2\mu^2 > 0$. The in-field (or quasiparticle field) $\rho_{in}(x)$ solves the free field equation

$$(\partial^2 + m^2)\rho_{in}(x) = 0.$$ (7.33)

The Heisenberg field $\psi(x)$ operates on the state $|0\rangle$ when expressed in terms of $\rho_{in}(x)$ fields. We thus search for the dynamical map

$$\psi(x) = \Psi[x; \rho_{in}],$$ (7.34)

which, as we know, has to be understood as a weak relation, i.e.,

$$\langle a|\psi(x)|b\rangle = \langle a|\Psi[x; \rho_{in}]|b\rangle.$$ (7.35)

Here, $|a\rangle$ and $|b\rangle$ are any two states belonging to the Hilbert space for $\rho_{in}(x)$. $\Psi[x; \rho_{in}]$ represents a functional of normal products of $\rho_{in}(x)$ determined by the dynamics, which, in the tree approximation, reads

$$\rho(x) = \rho_{in}(x) + \frac{3}{2}mg(-i) \int d^2y\Delta(x-y) : [\rho_{in}(y)]^2 :$$

$$+ \frac{1}{2}g^2(-i) \int d^2y\Delta(x-y) : [\rho_{in}(y)]^3 :$$

$$+ \frac{9}{2}m^2g^2(-i)^2 \int d^2yd^2z\Delta(x-y)\Delta(y-z) : \rho_{in}(y)[\rho_{in}(z)]^2 :$$

$$+ \dots . \tag{7.36}$$

$\Delta(x-y)$ is the Green's function:

$$(\partial^2 + m^2)\Delta(x-y) = -i\delta^{(2)}(x-y) . \tag{7.37}$$

We now write

$$\rho(x) = \sum_{n=1}^{\infty} \rho^{(n)}(x) , \tag{7.38}$$

where n denotes the order of the normal products. In the tree approximation we obtain the recurrence relation

$$\rho^{(n)}(x) = -\frac{3}{2}mg(-i) \int d^2y\Delta(x-y) \sum_{i+j=n} : \rho^{(i)}(y)\rho^{(j)}(y) : \tag{7.39}$$

$$+ \frac{1}{2}g^2(-i) \int d^2y\Delta(x-y) \sum_{i+j+k=n} : \rho^{(i)}(y)\rho^{(j)}(y)\rho^{(k)}(y) : .$$

Now we perform the boson transformation

$$\rho_{in}(x) \rightarrow \rho_{in}(x) + f(x) , \tag{7.40}$$

where $f(x)$ is a c-number function solution of the in-field Eq. (7.33):

$$(\partial^2 + \mu^2)f(x) = 0 . \tag{7.41}$$

We denote the boson-transformed field operators by ψ^f and ρ_f and define

$$\phi^f(x) \equiv \langle 0|\psi^f(x)|0\rangle , \tag{7.42}$$

i.e.,

$$\phi^f(x) = v + \langle 0|\rho_f(x)|0\rangle , \tag{7.43}$$

and put

$$\phi_0^f(x) = \lim_{\hbar \to 0} \phi^f(x) . \tag{7.44}$$

The boson transformation theorem implies that $\psi^f(x) = \psi(x; \phi_{in} + f)$ also satisfies the Heisenberg field equation as $\psi(x)$ does (cf. Eq. (7.28))

$$\Lambda(\partial)\psi^f(x) = J[\psi^f(x)]. \tag{7.45}$$

The $\hbar \to 0$ limit in Eq. (7.44) indicates that the tree approximation is used, i.e., we neglect the contributions $O(\hbar)$ of the order of $\hbar$ due to the contraction of fields which create loop diagrams. In other words, one writes all products of normal products as normal products without contractions. Therefore, when considering the expectation values of the non-linear functional J of the ψ field, we have

$$\langle 0|J[\psi]|0\rangle = J[\langle 0|\psi|0\rangle] + O(\hbar), \tag{7.46}$$

and thus

$$\phi_0^f(x) = f(x) + [\Lambda(\partial)]^{-1} J[\phi_0^f(x)], \tag{7.47}$$

which gives the classical Euler–Lagrange equation

$$\Lambda(\partial)\phi_0^f(x) = J[\phi_0^f(x)]. \tag{7.48}$$

The c-number field $\phi^f(x)$ describes a classically behaving object even when $\hbar \neq 0$. We now consider the static case since we are interested in the static kink solution. Non-static configurations can then be generated by boosting static solutions with a Lorentz transformation. We denote the space coordinates $x_1, y_1, \ldots$ simply by $x, y, \ldots$. Then the function $f(x)$ has to be a solution of the equation

$$\left(\frac{d^2}{dx^2} - m^2\right) f(x) = 0. \tag{7.49}$$

We obtain

$$\phi_0^f(x) = v + f(x) + \frac{3}{2}mg \int dy K(x-y)f^2(y) + \frac{1}{2}g^2 \int dy K(x-y)f^3(y)$$
$$+ \frac{9}{2}m^2g^2 \int dy dz K(x-y)K(y-z)f(y)f^2(z) + \ldots, \tag{7.50}$$

where $K(x-y)$ is the Green's function defined by:

$$\left(\frac{d^2}{dx^2} - m^2\right) K(x-y) = \delta(x-y). \tag{7.51}$$

The recurrence relation Eq. (7.39) now becomes

$$\rho_f^{(n)}(x) = \frac{3}{2}mg \int dy K(x-y) : \sum_{i+j=n} \rho_f^{(i)}(y)\rho_f^{(j)}(y) :$$
$$+ \frac{1}{2}g^2 \int dy K(x-y) : \sum_{i+j+k=n} \rho_f^{(i)}(y)\rho_f^{(j)}(y)\rho_f^{(k)}(y) : . \tag{7.52}$$

We choose $f(x) = Ae^{-mx}$ as a solution of Eq. (7.49). We do not choose as $f(x)$ a superposition of e^{-mx} and e^{+mx} since it would be divergent both as $x \to -\infty$ and $x \to +\infty$. Then we choose the Green's function K such that $K(x - y) = 0$ for $x > y$:

$$K(x) = -\theta(-x)\frac{1}{m}\sinh mx. \tag{7.53}$$

The choice of $f(x)$ and the Green's function $K(x)$ are conditioned by the convergence of the integrals in Eq. (7.50). It is, for instance, clear that Feynman causal Green's function would produce exponential divergencies. By induction one can prove from Eq. (7.52) that $\rho_f^{(1)} = f(x)$, $\rho_f^{(n)} = C_n(e^{-mx})^n$ and that $C_1 = A$, and for $n \geq 2$

$$C_n = \frac{1}{m^2(n^2 - 1)}\left(\frac{3}{2}mg\sum_{i+j=n}C_iC_j + \frac{1}{2}g^2\sum_{i+j+k=n}C_iC_jC_k\right). \tag{7.54}$$

This and use of $gv = m$ give $C_n = 2v\left(\frac{A}{2v}\right)^n$, which leads to

$$\phi_0^f(x) = v + 2v\sum_{n=1}^{\infty}\left(\frac{A}{2v}e^{-mx}\right)^n = v\frac{1 + \frac{A}{2v}e^{-mx}}{1 - \frac{A}{2v}e^{-mx}}. \tag{7.55}$$

By choosing $A = -2ve^{ma}$, the kink soliton is obtained in the form

$$\phi_0^f(x) = v\tanh\left[\frac{m}{2}(x - a)\right], \tag{7.56}$$

i.e., the kink solution of the classical Euler–Lagrange equation (7.48):

$$(\partial^2 + \mu^2)\phi_0^f(x) = -\lambda[\phi_0^f]^3(x). \tag{7.57}$$

The antikink solution is similarly obtained for $A = -2ve^{-ma}$ and $f(x) = Ae^{+mx}$, or, alternatively, for $\langle 0|\psi(x)|0\rangle = -v$.

When $A = 2ve^{ma}$ we obtain another solution of (7.57):

$$\phi_0^f(x) = v\coth\left[\frac{m}{2}(x - a)\right]. \tag{7.58}$$

Eqs. (7.56) and (7.58) are the only static solutions of (7.57) satisfying the condition $\phi_0^f(x) \to v$ for $x \to \infty$ (see Section 10.3.2).

For static soliton-like solutions created by non-homogeneous boson condensation, the equation to be solved by $f(x)$ has the general form

$$(\nabla - m^2)f(x) = 0. \tag{7.59}$$

For $m = 0$ this admits the trivial solution $f(x) = constant$. Disregarding such a solution, for $m \geq 0$, Eq. (7.59) does not admit any solution which is Fourier transformable. This implies that static extended objects created by boson condensation carry certain singularities which prohibit the Fourier transform of $f(x)$. Thus $f(x)$ has either a divergent singularity, i.e., $f(x)$ diverges at $|\mathbf{x}| \to \infty$ at least in certain direction of $|\mathbf{x}|$ (as in the case of the kink discussed above) or a topological singularity, i.e., $f(x)$ is not single-valued. We refer to Section 4.3.1 for a discussion of non-homogeneous boson condensation in the case of $f(x)$ not single-valued. There we have seen that such a case is compatible only with the condensation of massless bosons (such as the NG bosons). An example of extended object where $f(x)$ carries topological singularity is the vortex solution, which will be discussed later.

As already observed in Section 7.2.1, we remark that $\phi^f(x)$ contains quantum effects, although it describes a macroscopic object. In other words, $\phi^f(x)$ is the soliton solution with quantum corrections given by loop diagrams [619].

7.3.1 *The kink solution and temperature effects*

In Section 5.7 we have seen how to treat in the TFD formalism and in the Ginzburg–Landau functional method at finite temperature the occurrence of a non-zero vacuum expectation value $v(\beta)$ of a real scalar field $\psi(x) = \rho(x) + v(\beta)$ (for uniformity of notation here we denote by $\psi(x)$ the scalar quantum field)

$$\langle 0(\beta)|\psi(x)|0(\beta)\rangle = v(\beta) . \tag{7.60}$$

In the present Section we consider the case of dimensions $D = 2$, i.e., $1+1$ dimensions, and the kink solution in the tree approximation.

We start from Eqs. (5.183) obtained in the TFD formalism and in the variational method, here rewritten for the reader's convenience:

$$(\partial^2 - \mu_0^2(\beta))\rho(x) =: \lambda\rho^3(x) :_\beta +3\lambda v(\beta) : \rho^2(x) :_\beta , \tag{7.61}$$

where, according to Eq. (5.182),

$$\mu_0^2(\beta) = 2\lambda v^2(\beta) = 2(-\mu^2 - 3\lambda\langle: \rho^2 :\rangle_0) . \tag{7.62}$$

The solution of (7.61) is formally given by the Yang–Feldman relation

$$\rho(x) = \rho_{in,\beta}(x) + (\partial^2 - \mu_0^2)^{-1} \left(: \lambda\rho^3(x) :_\beta +3\lambda v(\beta) : \rho^2(x) :_\beta\right) , \tag{7.63}$$

where $\rho_{in,\beta}(x)$ is the quasiparticle field in the β-representation $\{|0(\beta)\rangle\}$ solution of Eq. (5.184), i.e.:

$$(\partial^2 - \mu_0^2(\beta))\rho_{in,\beta}(x) = 0 . \tag{7.64}$$

At $T = 0$ the soliton solution is obtained in the tree approximation as the macroscopic envelope of the localized Bose–Einstein condensation. When $T \neq 0$ we assume that the instability of the quasiparticle state due to thermal fluctuations can be neglected provided we work at a temperature low enough with respect to the critical temperature T_C where instability is dominant. The soliton is then obtained similarly as in the case $T = 0$, provided the temperature is carefully kept below T_C. As $T \to T_C$ the soliton solution may vanish due to large thermal fluctuations. We proceed therefore as follows. The non-homogeneous boson condensation of the $\rho_{in,\beta}$ field is induced by the translation

$$\rho_{in,\beta}(x) \to \rho_{in,\beta}(x) + f_\beta(x) . \tag{7.65}$$

This is an invariant translation provided the condensation function $f_\beta(x)$ is a solution of the field equation

$$(\partial^2 - \mu_0^2(\beta)) f_\beta(x) = 0 . \tag{7.66}$$

We are interested in obtaining the static kink solution, thus we consider the static case where $f_\beta(x) = const. \, e^{-\mu_0(\beta)x_1}$. Then

$$\psi(x) \to \psi^f(x) = \rho^f(x) + v(\beta) , \tag{7.67a}$$

$$\rho(x) \to \rho^f(x) = F\left[\rho_{in,\beta}(x) + f_\beta(x)\right] , \tag{7.67b}$$

and

$$\langle 0(\beta)|\psi^f(x)|0(\beta)\rangle = v(\beta) + \langle 0(\beta)|\rho^f(x)|0(\beta)\rangle \equiv v^f(x, \beta) . \tag{7.68}$$

Note that $f_\beta(x)$ is self-consistently determined under the condition (7.66) through (7.61), (7.63) and (7.64). The static kink solution

$$v_0^f(x_1, \beta) = \lim_{\hbar \to 0} v^f(x_1, \beta) , \tag{7.69}$$

is obtained as solution of the classical Euler–Lagrange equation derived from (5.186)

$$\left(\nabla + |\mu^2| - 3\lambda \langle : \rho^2 : \rangle_0\right) v_0^f(x_1, \beta) = \lambda \left[v_0^f(x_1, \beta)\right]^3 . \tag{7.70}$$

The kink solution of Eq. (7.70) centered at $x_1 = a$ is thus

$$v_0^f(x_1, \beta) = v(\beta) \tanh \sqrt{\frac{\lambda}{2}} v(\beta) (x_1 - a) . \tag{7.71}$$

Eq. (7.71) shows that the "order parameter" $v_0^f(x_1, \beta)$ (the kink solution) depends on β through $\langle : \rho^2 : \rangle_0$ (cf. Eq. (7.62)). From (7.62), we see that $v^2(\beta) \to 0$ as T approaches the critical temperature T_C such that

$$|\mu^2| = 3\lambda \langle : \rho^2 : \rangle_0|_{T=T_C} . \tag{7.72}$$

Therefore, in the same limit $(T \to T_C)$ the kink solution and the associated topological charge $v_0^f(+\infty, \beta) - v_0^f(-\infty, \beta) \to 0$ go to zero [438, 637].

The boson condensation function $f_\beta(x) = const. \, e^{-\mu_0(\beta)x_1}$ plays the role of "form factor" and the density of condensed bosons is proportional to

$$|f_\beta(x)|^2 = const. \, e^{-2\mu_0(\beta)x_1} = e^{-2\mu_0(\beta)\,(x_1-a)} . \qquad (7.73)$$

Such a number is maximal near the kink center $x_1 = a$ and decreases over the size $\xi_\beta = \frac{2}{\mu_0(\beta)}$.

The meaning of the transformation (7.65) is clearly expressed by Eq. (7.68) which shows that the f_β-translation breaks the homogeneity of the otherwise constant in space order parameter $v(\beta)$. On the other hand, the mass $\mu_0(\beta) = (2\lambda)^{\frac{1}{2}} v(\beta)$ of the "constituent" fields ρ_{in} fixes the kink size $\xi_\beta \propto \frac{2}{\mu_0} = \frac{\sqrt{2}}{\sqrt{\lambda} v(\beta)}$. Since in our picture the kink is made out of the coherent condensation of the ρ_{in} particles, which are confined around the kink center, we see that there is a deconfinement at $T \to T_C$.

It is interesting to note that in the $T \to 0$ limit the kink size ξ_0 is $\xi_0 \propto \frac{\sqrt{2}}{\sqrt{\lambda} v} < \frac{\sqrt{2}}{\sqrt{\lambda} v(\beta)} = \xi_\beta$, since $v^2(\beta) = v^2 - 3\langle: \rho^2 :\rangle_0 < v^2$. At $T = 0$ the kink equation and the kink solution are in fact

$$(\nabla + |\mu^2|) \, v_0^f(x_1) = \lambda [v_0^f(x_1)]^3 \, , \qquad \mu^2 = -\lambda v^2 \, , \qquad (7.74)$$

and

$$v_0^f(x_1) = v \tanh \sqrt{\frac{\lambda}{2}} v \, (x_1 - a) \, , \qquad (7.75)$$

respectively, according to the assumption $\langle 0|\psi(x)|0\rangle = v \neq 0$ at $T = 0$.

As T starts to be different from zero, the thermal Bose–Einstein condensate $\langle: \rho^2(x) :\rangle_0$ develops, acting as a potential term (or thermal mass [411]), for the quantum field $\rho(x)$, cf. Eqs. (7.61) and (7.62): in this way quantum dynamics is affected by temperature effects. Similarly, such temperature effects manifest at classical level as potential term for the classical field v_0^f (cf. Eq. (7.70) with Eq. (7.74)). It is such a potential term which actually controls the size of the kinks.

Notice that in Eq. (7.64) $\mu_0^2(\beta)$ also acts as a potential for the $\rho_{in,\beta}(x)$ field. Only in the limit $v(x, \beta) \to const.$ the $\rho_{in,\beta}(x)$ field may be considered as a free field; e.g., far from the kink core. In general, however, the free field description breaks down for $v \neq const$. At $T > T_C$, μ_0 becomes imaginary (cf. Eq. (7.62)) and ρ_{in} becomes unstable. Their condensation disappears and we have symmetry restoration (cf. (7.62) and (7.71)).

For $D = 4$ dimensions the soliton-like solution has been discussed in the literature, see, e.g., [597], and in thin-wall approximation [425] the spherical symmetric solution has the same form as the kink solution. Also in these cases, the temperature effects may lead to symmetry restoration in a way very similar to the one presented above for the kink in $D = 2$.

7.3.2 *The kink solution: closed time-path approach*

In this Section we present an alternative derivation of the kink solution considered above, by means of the closed time-path approach, introduced in Section 6.2. In this formalism, some aspects of the boson condensation mechanism will be clearer and the dynamical map are given in a closed form, which allows a straightforward extension to the case of finite temperature. Another application of this formalism will be presented in Section 7.7.

The Heisenberg operator in the presence of a kink

We consider the real scalar field $\psi(x)$ with quartic interaction term in $1+1$ dimensions already considered in Section 7.3. The interaction Lagrangian $\mathcal{L}^I$ for the field ρ introduced in Eq. (7.31) is

$$\mathcal{L}^I[\rho] = -\frac{1}{8}\, g^2\rho^4 - \frac{1}{2}\, mg\rho^3 + \frac{m^4}{8g^2}\,, \tag{7.76}$$

and the dynamical map for the field ψ is written in the form (see Eq. (6.4))

$$\psi(x) = v + T_C^* \left[\rho_{in}(x) \exp\left\{ i \int_C d^2y\, \mathcal{L}^I[\rho_{in}] \right\} \right]. \tag{7.77}$$

Note that the constant factor in $\mathcal{L}^I$ automatically cancels in performing the contour integration. We then consider the boson transformation $\rho_{in}(x) \to \rho_{in}(x) + f(x)$ with $(\partial^2 + m^2)f(x) = 0$ (cf. Eqs. (7.40) and (7.41)). As a result we get the following Haag expansion for the field $\psi^f(x)$

$$
\begin{aligned}
\psi^f(x) &= v + T_C^* \left[(\rho_{in}(x) + f(x)) \exp\left\{ i \int_C d^2y\, \mathcal{L}^I[\rho_{in}(y) + f(z)] \right\} \right] \\
&= v + \left[\frac{\delta}{i\delta J(x)} + f(x) \right] \exp\left\{ i \int_C d^2y\, \mathcal{L}^I \left[\frac{\delta}{i\delta J(y)} + f(y) \right] \right\} \\
&\quad \times T_C \left(\exp i \int_C d^2y\, J(y)\rho_{in}(y) \right)\Big|_{J=0}\,,
\end{aligned}
\tag{7.78}
$$

where we have introduced the c–number source J to perform some formal manipulations. By use of the (operatorial) Wick theorem [343], we get

$$\psi^f(x) = v + \left[\frac{\delta}{i\delta J(x)} + f(x)\right] \exp\left\{i \int_C d^2y\, \mathcal{L}^I \left[\frac{\delta}{i\delta J(y)} + f(y)\right]\right\}$$

$$\times\, :\exp i \int_C d^2y\, J(y)\rho_{in}(y): \exp\left[-\frac{1}{2}\int_C d^2y d^2z\, J(y)\Delta_C(y;z)J(z)\right]\Bigg|_{J=0},$$
$$(7.79)$$

where $\Delta_C(x;y) = \langle 0|T_C(\rho_{in}(x)\rho_{in}(y))|0\rangle$ and $|0\rangle$ is the vacuum for the ρ_{in} field. Once the function f is properly chosen (see below), Eq. (7.79) provides a representation of the Heisenberg operator in the presence of kinks.

In order to determine the function f leading to kink solutions, we consider the vacuum expectation value of the Heisenberg field ψ^f. Here the normal ordered term drops, and we get ($\langle \ldots \rangle \equiv \langle 0|\ldots|0\rangle$)

$$\langle\psi^f(x)\rangle = v + \left[\frac{\delta}{i\delta J(x)} + f(x)\right] \exp\left\{i \int_C d^2y\, \mathcal{L}^I \left[\frac{\delta}{i\delta J(y)} + f(y)\right]\right\}$$

$$\times\, \exp\left[-\frac{1}{2}\int_C d^2y d^2z\, J(y)\Delta_C(y;z)J(z)\right]\Bigg|_{J=0}. \qquad (7.80)$$

By use of the relation

$$F\left[\frac{\delta}{i\delta J}\right] G[J] = G\left[\frac{\delta}{i\delta K}\right] F[K]e^{i\int KJ}\Bigg|_{K=0}, \qquad (7.81)$$

we obtain

$$\langle\psi^f(x)\rangle = v + \exp\left[-\frac{1}{2}\int_C d^2y d^2z\, \Delta_C(y;z)\frac{\delta}{i\delta K(y)}\frac{\delta}{i\delta K(z)}\right]$$

$$\times\, (K(x) + f(x)) \exp\left\{i\int_C d^2y[\mathcal{L}^I[K(y) + f(y)] + K(y)J(y)]\right\}\Bigg|_{K=J=0}$$
$$(7.82)$$

Performing the change of variables $K(x) \to K(x) + f(x)$ and setting to zero the J-term (there are no derivatives with respect to it), we obtain

$$\langle\psi^f(x)\rangle = v + \exp\left[-\frac{1}{2}\int_C d^2y d^2z\, \Delta_C(y;z)\frac{\delta}{i\delta K(y)}\frac{\delta}{i\delta K(z)}\right] K(x)B[K]\Bigg|_{K=f},$$
$$(7.83)$$

with

$$B[K] \equiv \exp\left\{i\int_C d^2y\, \mathcal{L}^I[K(y)]\right\}. \qquad (7.84)$$

We can thus express Eq. (7.83) as a sum of three terms[2]

$$\langle \psi^f(x) \rangle = v + C[K](x)|_{K=f} + D[K](x)|_{K=f} , \tag{7.85a}$$

$$C[K](x) = -\int_C d^2y \Delta_C(x;y) \frac{\delta}{\delta K(y)} \times$$

$$\times \exp\left[-\frac{1}{2}\int_C d^2z d^2y \Delta_C(z;y) \frac{\delta^2}{\delta K(z)\delta K(y)}\right] B[K], \tag{7.85b}$$

$$D[K](x) = K(x) \exp\left\{-\frac{1}{2}\int_C d^2x d^2y \Delta_C(x;y) \frac{\delta^2}{\delta K(x)\delta K(y)}\right\} B[K]. \tag{7.85c}$$

The kink solution in the Born approximation

So far, all results obtained were of a full quantum nature. We now deal with the Born, or classical, approximation of Eqs. (7.85) and for this purpose we reintroduce $\hbar$. Each propagator is then proportional to $\hbar$ whilst B has in the exponent the factor $i\hbar^{-1}$. The Born approximation means that only terms of order $\hbar^0$ in (7.85) must be taken into account. We recall that the particle mass and momentum are obtained by multiplying ω_0 and k by $\hbar$ (we do set $c = 1$). $\lambda = g^2/2$ goes as $[E^1 l^{-3}]$ where E means energy and l means length, so dimensionally $[\lambda] = [\hbar^{-1}l^{-2}]$.

We rewrite some of the previous expressions making the $\hbar$ dependence explicit. In the following we will consider $\Delta_C(\ldots) \propto \hbar^0$. We have

$$C[K](x,\hbar) = -\hbar \int_C d^2y \, \Delta_C(x;y) \frac{\delta}{\delta K(y)} \exp[\hbar a] \exp\left[\frac{1}{\hbar}b\right], \tag{7.86a}$$

$$D[K](x,\hbar) = K(x) \exp[\hbar a] \exp\left[\frac{1}{\hbar}b\right], \tag{7.86b}$$

with

$$a = -\frac{1}{2}\int_C d^2z d^2y \, \Delta_C(z;y) \frac{\delta^2}{\delta K(z)\delta K(y)}, \tag{7.87a}$$

$$b = -\frac{i}{2}\int_C d^2z \left[\frac{g^2}{4} K^4(z) + \omega_0 g \, K^3(z)\right]. \tag{7.87b}$$

---

[2]We use the identity

$$\exp\left[\frac{1}{2}\sum_{ij} \Delta_{ij}\partial_{x_i}\partial_{x_j}\right] x_k B(x) = \left(x_k + \sum_j \Delta_{kj}\partial_{x_j}\right) B(x_l + \sum_j \Delta_{lj}\partial_{x_j}) \mathbb{1} ,$$

where $x_k \to K(x)$ and $B(x) \to B[K]$.

Keeping only the finite terms in the $\hbar \to 0$ limit we get

$$C[K](x, \hbar \to 0) = - \int_C d^2 y \, \Delta_C(x; y) \frac{\delta}{\delta K(y)} \, Res_{\hbar=0} \left(\exp\left[\hbar a \right] \exp\left[\frac{1}{\hbar} b \right] \right)$$

$$= -i \int_{-\infty}^{\infty} d^2 y \, G_R(x, y) \frac{\delta}{\delta K(y)} \sum_{n=0}^{\infty} \frac{1}{n!(n+1)!} a^n b^{n+1} , \qquad (7.88)$$

where $i G_R(x, y) = \theta(x_0 - y_0) \, \Delta(x; y)$ is the (position space) retarded Green's function of the free theory and $\Delta(x; y)$ is the Pauli–Jordan function:

$$\Delta(x_0, x_1; 0, 0) = \langle 0 | [\rho_{in}(x), \rho_{in}(0)] | 0 \rangle$$

$$= \int \frac{d^2 k}{(2\pi)} \delta(k^2 - \omega_0^2) \, \varepsilon(k_0) \, e^{-ikx}$$

$$= -\frac{i}{2} \theta(x_0 - |x_1|) \, J_0 \left(\omega_0 \sqrt{x_0^2 - x_1^2} \right) . \qquad (7.89)$$

The D term gives

$$D(x, \hbar \to 0) = K(x) \, Res_{\hbar=0} \left(\frac{1}{\hbar} \exp\left[\hbar a \right] \exp\left[\frac{1}{\hbar} b \right] \right)$$

$$= K(x) \sum_{n=0}^{\infty} \frac{1}{(n!)^2} a^n b^n . \qquad (7.90)$$

In [95] it was shown that $D[K, \hbar \to 0] = K$. The final result is then

$$\langle \psi_0^f(x) \rangle = v + f(x) - i \int_{-\infty}^{\infty} d^2 y G_R(x, y) \frac{\delta}{\delta K(y)} \sum_{n=0}^{\infty} \frac{1}{n!(n+1)!} a^n b^{n+1} \Bigg|_{K=f}$$

$$= v + \sum_{n=1}^{\infty} P_n[K](x) \Bigg|_{K=f} , \qquad (7.91)$$

where

$$P_1(x) = K(x) ,$$

$$P_n(x) = -\frac{1}{[(n-2)!]^2} \int_{-\infty}^{\infty} d^2 y \, G_R(x, y) \, a^{n-2} \left\{ \left[\frac{3}{2} \omega_0 g \, K^2(y) \right. \right.$$

$$\left. \left. + \frac{1}{2} g^2 \, K^3(y) \right] b^{n-2} \right\} ; \quad n \geq 2 . \qquad (7.92)$$

Using mathematical induction the following recurrence relation is obtained:

$$P_n(x) = -\int_{-\infty}^{\infty} d^2 y \, G_R(x, y) \left[\frac{3}{2} \omega_0 g \sum_{i+j=n} P_i(y) \, P_j(y) \right.$$

$$\left. + \frac{1}{2} g^2 \sum_{i+j+k=n+1} P_i(y) \, P_j(y) \, P_k(y) \right] ; \quad n \geq 2 , \qquad (7.93)$$

where $i, j, k = 1, 2 \ldots$. This may be "diagonalized" to obtain

$$Q_1(x) = K(x),$$

$$Q_2(x) = -\int_{-\infty}^{\infty} d^2y \frac{3}{2v} G_R(x, y) \sum_{i+j=2} Q_i(y) Q_j(y),$$

$$Q_n(x) = -\int_{-\infty}^{\infty} d^2y\, G_R(x, y) \left[\frac{3}{2}\omega_0 g \sum_{i+j=n} Q_i(y) Q_j(y) + \right.$$

$$\left. \frac{1}{2}g^2 \sum_{i+j+k=n} Q_i(y) Q_j(y) Q_k(y) \right] \; ; \quad n \geq 3 . \qquad (7.94)$$

with $Q_n = P_1 P_{n+1} + P_n$. This equation can be solved by the Ansatz [95]:

$$Q_n(x) = A_n f^n(x) = A_n\, e^{\pm \omega_0 n \gamma (x_1 - x_0 u)} , \qquad (7.95)$$

where $\gamma = (1 - u^2)^{-\frac{1}{2}}$ (u will be later interpreted as the velocity of the kink). Plugging this form into the recurrence relation (7.94), we arrive at the following equation for the factor A_n (cf. Eq. (7.54) for C_n)

$$A_n = \frac{1}{(n^2 - 1)} \left\{ \frac{3}{2v} \sum_{i+j=n} A_i A_j + \frac{1}{2v^2} \sum_{i+j+k=n} A_i A_j A_k \right\} . \qquad (7.96)$$

This is a trivial version of Cauchy–Marley equation which has the only (non-zero) fundamental solution, namely $A_n \propto (A_1)^n$. Using the identities:

$$\sum_{i+j=n} 1 = n - 1 \; ; \quad \sum_{i+j+k=n} 1 = \frac{1}{2}(n-1)(n-2) , \qquad (7.97)$$

it might be easily checked that a solution of the recurrence relation (7.96) reads for C_n)

$$A_n = 2v \left(\frac{s}{2v} \right)^n , \qquad (7.98)$$

with s being a real constant. Thus, finally, we have

$$\langle \psi_0^f(x) \rangle = v + 2v \sum_{n=1}^{\infty} \left(\frac{sf(x)}{2v} \right)^n = v\, \mathrm{cth} \left[\frac{1}{2}\mathrm{Ln} \left(\frac{sf(x)}{2v} \right) \right] . \qquad (7.99)$$

Here $\mathrm{Ln}(z) = \ln|z| + i\,\mathrm{arg}z$ is the usual principal value of the logarithm of z. Thus, provided $f(x)$ is an exponential solution of Eq. (7.41), the solution (7.99) fulfils the (classical) Euler–Lagrange equation of motion:

$$(\partial^2 + \mu^2)\langle \psi_0^f(x) \rangle = -\lambda \langle \psi_0^f(x) \rangle^3 , \quad \mu^2 = -\omega_0^2/2 . \qquad (7.100)$$

The latter is nothing but the expectation value of equation (7.27) in the Born approximation. For instance, if we choose $f(x) = e^{-\omega_0\gamma(x_1-x_0u)}$ with $s = -2ve^{\omega_0\gamma a}$, we easily obtain the standard kink solution [176, 290, 524]

$$\langle\psi_0^f(x)\rangle = v \tanh\left[\frac{\omega_0}{2}\gamma((x_1 - a) - x_0u)\right], \qquad (7.101)$$

describing a constantly moving kink of a permanent profile with a center localized at $a + ux_0$. Note that $f(x)$ is the solution of the homogeneous Klein–Gordon equation (7.41); it is not Fourier transformable and it fulfils the initial value condition: $f(x_0 \to -\text{sgn}(u)\infty, x_1) \to 0$.

As the Lagrangian (7.26) is $\mathbb{Z}_2$ invariant we could equally choose $\langle0|\psi(x)|0\rangle = -v$. In this case we would get

$$\langle\psi_0^f(x)\rangle = -v \tanh\left[\frac{\omega_0}{2}\gamma((x_1 - a) - x_0u)\right], \qquad (7.102)$$

which is the antikink solution. Note that it is also obtained by choosing $f(x) = e^{+\omega_0\gamma(x_1-x_0u)}$ and $s = -2ve^{-\omega_0\gamma a}$, provided we keep $\langle0|\psi(x)|0\rangle = v$.

According to the scheme of Fig. 7.1, the above solutions for the shift function f corresponding to kink solutions can now be used in Eq. (7.79) to get the Heisenberg operator in a given kink sector. One can thus calculate Green's functions and higher quantum corrections to ψ^f [95, 96].

The kink solution at finite temperature

We now consider the finite temperature kink solution in the framework of the CTP formalism.

As already discussed in Section 6.2, in thermal equilibrium the most convenient choice for the time-path is the one in Fig. 6.3, to which we refer as the *thermal path*. The crucial observation at finite temperature is that the operatorial Wick theorem still holds (see, e.g., [226] and Appendix M) and consequently Eq. (7.79) retains its validity provided the following substitutions are performed:

$$\Delta_C(x; y) = \langle0|T_C(\rho_{in}(x)\rho_{in}(y))|0\rangle \longrightarrow \Delta_C(x; y, T) = \langle T_C(\rho_{in}(x)\rho_{in}(y))\rangle_\beta$$

$$: \dots : \longrightarrow : \dots :_\beta,$$

where $\langle\dots\rangle_\beta \equiv \text{Tr}\left(e^{-\beta H} \dots\right)/\text{Tr}\left(e^{-\beta H}\right)$ and $\beta = 1/T$. The thermal normal ordering $: \dots :_\beta$ is defined in such a way [226] that $\langle: \dots :_\beta\rangle_\beta = 0$, where the dots stand for a product of $T = 0$ free fields. This ensures that all the formal considerations developed above go through also for finite T.

There is, however, a fundamental difference between the zero temperature case and the finite temperature case, when the classical limit is in question. Indeed, at non-zero temperature the question of $\hbar$ appearance is more delicate than in the zero-temperature case. The whole complication is hidden in the thermal propagator $\Delta_C(x; y, T)$. Whilst at $T = 0$ the latter is directly proportional to $\hbar$, at finite T the situation is very different. To understand this let us make $\hbar$ explicit. The free thermal propagator in the spectral or Mills representation [411, 475] reads

$$\Delta_C(x; y, T) = \hbar \int \frac{d^2k}{(2\pi)^2} e^{-ik(x-y)} \rho(k)[\theta_C(x_0 - y_0) + f_b(\hbar k_0/T)]$$
$$= \Delta_C(x; y) + \Delta_C^T(x; y), \tag{7.103}$$

where $f_b(x) = (e^x - 1)^{-1}$ the Bose–Einstein distribution is. The $T = 0$ part, which is proportional to $\hbar$, reads

$$\Delta_C(x; y) = \hbar \int \frac{d^2k}{(2\pi)^2} e^{-ik(x-y)} \rho(k) \left[\theta_C(x_0 - y_0) - \theta(-k_0)\right]. \tag{7.104}$$

Here $\rho(k) = (2\pi)\varepsilon(k_0)\,\delta(k^2 - \omega_0^2)$, with $\varepsilon(k_0) = \theta(k_0) - \theta(-k_0)$, is the spectral density. The contour step function $\theta_C(x_0 - y_0)$ is 1 if y_0 precedes x_0 along the contour C. One obtains the usual elements of the (free) thermal propagator (cf. Eq. (6.50)):

$$\Delta_{11}(x; y, T) = \hbar \int \frac{d^2k}{(2\pi)^2} e^{-ik(x-y)} \left\{ \frac{i}{k^2 - \omega_0^2 + i\varepsilon} \right.$$
$$\left. + 2\pi\,\delta(k^2 - \omega_0^2) f_b(\hbar|k_0|/T) \right\}, \tag{7.105}$$

$$\Delta_{21}(x; y, T) = \hbar \int \frac{d^2k}{(2\pi)} e^{-ik(x-y)} \left\{ \theta(k_0) + f_b(\hbar|k_0|/T) \right\} \delta(k^2 - \omega_0^2),$$

$$\Delta_{22}(x; y, T) = (\Delta_{11}(x; y, T))^* \quad ; \qquad \Delta_{12}(x; y, T) = (\Delta_{21}(x; y, T))^*,$$

where the particle mass $m = \hbar\omega_0$. From Eqs. (7.105) we see that the thermal part of $\Delta_C(x; T)$ is identical for all matrix elements and in (7.105) it appears $f_b(\hbar|k_0|/T)$ and not $f_b(\hbar k_0/T)$. Note that k in the integration is a wave vector – a reciprocal length – and not a momentum.

Due to the mentioned analogy with the $T = 0$ situation, we may write for the order parameter

$$\langle \psi^f(x) \rangle_\beta = v + C[K](x; T)|_{K=f} + D[K](x; T)|_{K=f}, \tag{7.106}$$

where we took the thermal average of the expression analogous to the one in Eq. (7.79), but with the normal ordering and the propagator replaced with their thermal counterparts. Both $C[K](x; T)$ and $D[K](x; T)$ entering Eq. (7.106) formally coincide with their zero temperature counterparts provided one uses the thermal propagator $\Delta_C(x; y, T)$ instead of $\Delta_C(x; y)$.

A detailed analysis of the classical limit of Eq. (7.106) is given in [95] where it is shown that $D[f](x; T, \hbar \to 0) = f(x)$.

The analysis of the $C[f](x; T)$ term is more complicated and requires an explicit Laurent expansion of $\Delta_C(x, T)$ around $\hbar = 0$. Using the Bernoulli expansion:

$$\frac{x}{e^x - 1} = \sum_{\alpha=0}^{\infty} B_\alpha \frac{x^\alpha}{\alpha!}, \qquad |x| < 2\pi,$$

(B_α are Bernoulli's numbers) we may Laurent expand f_b as

$$f_b(\hbar k_0/T) + \frac{1}{2} = \frac{T}{\hbar k_0} + \frac{1}{12} \frac{\hbar k_0}{T} - \frac{1}{90} \left(\frac{\hbar k_0}{T} \right)^2 + \dots . \tag{7.107}$$

The key point is that the foregoing series converges only for $\hbar |k_0| < 2\pi T$. The leading term in (7.107) gives the *classical* thermal part of the propagator,[3] and the classical approximation is then equivalent to taking the leading term in Laurent expansion (7.107). Higher quantum corrections are due to higher terms in the expansion, but for large $|k_0|$ the expansion does not work, i.e., an expansion in $\hbar$ is unwarranted. Of course, for $\hbar |k_0| \gg T$ the distribution f_b is exponentially small (Wien distribution law) and is dropped in comparison to the (zero point) first term in the integral (7.103), which returns the $T = 0$ approach from the previous subsection. For the high temperature case see [95].

7.4 The sine-Gordon solution

We now present the explicit derivation of the sine-Gordon soliton solutions in the tree approximation following closely [503]. When no confusion arises we will use the symbol x to denote either (x_1, t) or x_1.

Consider the Heisenberg real scalar (boson) field $\psi(x)$ and its motion equation for the $(1 + 1)$-dimensional quantum sine-Gordon model:

$$\partial^2 \psi(x) + \frac{m^3}{g} \sin \left[\frac{g}{m} \psi(x) \right] = 0, \tag{7.108}$$

which we symbolically rewrite as

$$\Lambda(\partial)\psi(x) = J[\psi(x)]. \tag{7.109}$$

The quasiparticle field $\rho_{in}(x)$ solves the free field equation

$$\Lambda(\partial)\rho_{in}(x) = 0. \tag{7.110}$$

---

[3] We can call it Rayleigh–Jeans sector of the thermal propagator as the corresponding distribution function $f(\omega) = T/\omega$ is nothing but Rayleigh–Jeans distribution law.

Let $|0\rangle$ denote the vacuum for such a field and assume that

$$\langle 0|\psi(x)|0\rangle = 0. \qquad (7.111)$$

By expanding the sine function around $\psi = 0$, Eq. (7.108) becomes:

$$(\partial^2 + m^2)\,\psi(x) = \frac{m^3}{g} \sum_{k=1}^{\infty} \frac{(-1)^{k+1}}{(2k+1)!} \left(\frac{g}{m}\psi(x)\right)^{2k+1}. \qquad (7.112)$$

We perform the boson transformation

$$\rho_{in}(x) \to \rho_{in}(x) + f(x), \qquad (7.113)$$

where the c-number function $f(x)$ is a solution of Eq. (7.110). The dynamical map then becomes:

$$\psi^f(x) = \Psi^f[x; \rho_{in} + f], \qquad (7.114)$$

where $\psi^f(x)$ also satisfies Eq. (7.108). The dynamical map can also be written as

$$\psi(x) = \sum_{n \geq 1} \psi^{(n)}(x), \qquad (7.115)$$

where $\psi^{(n)}(x)$ denotes a product of n normal ordered field factors. In the tree approximation ($h \to 0$), use of Eq. (7.115) into Eq. (7.112) gives

$$(\partial^2 + m^2)\psi^{(n)}(x) = \frac{m^3}{g} \sum_{k=1}^{\frac{n-1}{2}} \frac{(-1)^{k+1}}{(2k+1)!} \left(\frac{g}{m}\right)^{2k+1}$$

$$\times \sum_{\substack{i_1,i_2,\ldots,i_{2k+1}\geq 1 \\ i_1+i_2+\cdots+i_{2k+1}=n}} :\psi^{(i_1)}(x)\psi^{(i_2)}(x)\ldots\psi^{(i_{2k+1})}(x): , \qquad (7.116)$$

where n is restricted to be odd integer. By performing the boson transformation (7.113), the order parameter $\phi(x)$ is obtained as

$$\phi_f(x) \equiv \langle 0|\psi^f(x)|0\rangle = \sum_{n \geq 1} \phi_f^{(n)}(x), \qquad (7.117)$$

where n is odd, $\phi_f^{(1)}(x) = f(x)$ and

$$(\partial^2 + m^2)\phi_f^{(n)}(x) = \frac{m^3}{g} \sum_{k=1}^{\frac{n-1}{2}} \frac{(-1)^{k+1}}{(2k+1)!} \left(\frac{g}{m}\right)^{2k+1}$$

$$\times \sum_{\substack{i_1,i_2,\ldots,i_{2k+1}\geq 1 \\ i_1+i_2+\cdots+i_{2k+1}=n}} \phi_f^{(i_1)}(x)\phi_f^{(i_2)}(x)\ldots\phi_f^{(i_{2k+1})}(x). \qquad (7.118)$$

Summation over n of both sides of Eq. (7.118) then leads to the classical Euler–Lagrange equation for the order parameter $\phi_f(x)$:

$$(\partial^2 + m^2)\phi_f(x) = m^2\phi_f(x) - \frac{m^3}{g}\sin\frac{g}{m}\phi_f(x). \qquad (7.119)$$

In the following we simplify the notation by introducing $\phi \equiv \frac{g}{m}\phi_f$ and by rescaling the coordinates as $t \to t/m$ and $x \to x/m$. The differential operator $(\partial^2 + m^2)$ then becomes $(\partial^2 + 1)$. If its inverse is well defined, Eq. (7.119) can be formally solved and (cf. Eq. (7.117))

$$\phi^{(n)}(x) = \sum_{k=1}^{\frac{n-1}{2}} \frac{(-1)^{k+1}}{(2k+1)!}$$

$$\times \sum_{\substack{i_1,i_2,\ldots,i_{2k+1}\geq 1 \\ i_1+i_2+\cdots+i_{2k+1}=n}} (\partial^2 + 1)^{-1}\left[\phi^{(i_1)}(x)\phi^{(i_2)}(x)\ldots\phi^{(i_{2k+1})}(x)\right], \qquad (7.120)$$

which is a recurrence relation. Since $\phi^{(1)}(x,t) = f(x,t)$ is a solution of

$$(\partial^2 + 1)\,f(x,t) = 0, \qquad (7.121)$$

the problem is reduced to the one of the summation of a series in $f(x,t)$. As a result, we can choose the following Ansatz for $f(x,t)$

$$f(x,t) = \exp[X], \qquad X = \gamma x - \beta t + \delta, \qquad (7.122)$$

where γ and β are real constants subject to the condition $\gamma^2 - \beta^2 = 1$. The constant δ is real modulo $i\pi$ to allow for a change of sign of f.

Note that this solution can be obtained by boosting the solution $f(x) = const.\exp[x]$, or $f(x) = const.\exp[-x]$, of the time-independent equation for f. We remark that a superposition of $\exp[x]$ and $\exp[-x]$ would make f divergent at both $x = +\infty$ and $x = -\infty$, but in that case the operator $(\partial^2 + 1)^{-1}$ cannot be defined. Therefore, we restrict ourselves to the use of $\exp[-x]$ only, or $\exp[x]$ only. Such a restriction will not affect the final result, as we will see. By using Eqs. (7.120) and (7.121) we get

$$\phi^{(3)}(x,t) = \frac{1}{1-3^2}\frac{\exp[3X]}{3!}, \qquad (7.123)$$

where we used $(\partial^2 + 1)\exp[\alpha X] = (-\alpha^2 + 1)\exp[\alpha X]$, $\alpha \neq 1$, which shows that $(\partial^2 + 1)^{-1}$ is well defined when operating on $\exp[\alpha X]$, provided $\alpha \neq 1$, but not when operating on a linear combination of $\exp[\alpha X]$ and $\exp[-\alpha X]$, as observed above. The general term is then of the form

$$\phi^{(n)}(x,t) = A_n \exp[nX]. \qquad (7.124)$$

The constants A_n satisfy the recurrence relation

$$(1 - n^2)A_n = \sum_{k=1}^{\frac{n-1}{2}} \frac{(-1)^{k+1}}{(2k+1)!} \sum_{\substack{i_1,i_2,\ldots,i_{2k+1}\geq 1 \\ i_1+i_2+\cdots+i_{2k+1}=n}} A_{i_1} A_{i_2} \ldots A_{i_{2k+1}}, \qquad n \geq 3$$

$$A_1 = 1. \tag{7.125}$$

The generating function $F(z)$ can be introduced, where z is an auxiliary variable such that the coefficient of order n of $F(z)$ is the r.h.s. of Eq. (7.125):

$$F(z) = -\sin\left(A_1 z + A_3 z^3 + \cdots + A_{n-2} z^{n-2}\right). \tag{7.126}$$

From Eq. (7.125) we see that $A_3 = -\frac{1}{3}\left(\frac{1}{4}\right)^2$. This suggests to assume that

$$A_p = \frac{(-1)^{\frac{p-1}{2}}}{p}\left(\frac{1}{4}\right)^{p-1}, \qquad p = 1, 3, \ldots, n-2, \tag{7.127}$$

and we can thus write

$$F(z) = -\sin\left(4\arctan\frac{z}{4} - \varphi\right)$$

$$= -\cos\varphi\sin\left(4\arctan\frac{z}{4}\right) + \sin\varphi\cos\left(4\arctan\frac{z}{4}\right), \tag{7.128}$$

where φ stands for the subtraction of all the terms of order $\geq n$ in the series expansion of $4\arctan(z/4)$. Since we are interested only in the coefficient of z^n, the relevant part of $F(z)$ is

$$\varphi - \sin\left(4\arctan\frac{z}{4}\right) = \varphi - z\frac{1 - (z/4)^2}{(1 + (z/4)^2)^2}. \tag{7.129}$$

By equating the coefficient of z^n of the r.h.s. of Eq. (7.129) with $(1-n^2)A_n$, we get

$$A_n = \frac{(-1)^{\frac{n-1}{2}}}{n}\left(\frac{1}{4}\right)^{n-1}, \qquad n \text{ odd}. \tag{7.130}$$

Now the series (7.117) can be summed up giving

$$\phi(x) = 4\arctan\frac{e^X}{4}. \tag{7.131}$$

This is the well-known single-soliton solution of the classical sine-Gordon equation. The factor $1/4$ can be incorporated in the e^X, which we will do in the following, thus reproducing the usual expression for the sine-Gordon solution [57,126,346,532,563] (see also Section 10.3.1). We recall that quantum corrections have been neglected since the solution has been obtained in the tree approximation. When collective coordinates (or "quantum coordinates") are introduced one can see that, provided the fluctuations of these coordinates are negligible with respect to the size of the soliton, it behaves as a classical object; otherwise quantum features show up. An extensive treatment of these occurrences is presented in [619].

The sine-Gordon single-soliton solution presented above only describes static objects. Time-dependent solutions are described by the N-soliton solutions which account for the collisions and the bound states of static solutions. The rest frame of each of these static solutions can only appear via the boson transformation function f. In [503] it is shown that the N-soliton solution is obtained by considering the function

$$f(x,t) = \sum_{i=1}^{N} f_i(x,t) = \sum_{i=1}^{N} \exp[X_i], \qquad (7.132)$$

where each $X_i = \gamma_i x - \beta_i t + \delta_i$, $\gamma_i^2 - \beta_i^2 = 1$, $i = 1, \ldots, N$, corresponds to a moving reference frame and the velocities β_i/γ_i are all different. The solution for $N = 2$ is obtained [503] as

$$\phi_{(N=2)}(x,t) = 4 \arctan \frac{\exp[X_1] + \exp[X_2]}{1 + a \exp[X_1 + X_1]}, \qquad (7.133a)$$

$$a = \frac{1 - \gamma_1\gamma_2 + \beta_1\beta_2}{1 + \gamma_1\gamma_2 - \beta_1\beta_2}, \qquad (7.133b)$$

which, up to phase factors, is the known 2-soliton solution [328]. Here we are not going to present the details of the derivation of Eq. (7.133), which can be found in [503]. Instead, in the following Section we are going to discuss the coherent state representation of the soliton solution and discuss the "quantum image" of the Bäcklund transformation which connects the $N - 1$-soliton to the N-soliton solution in the classical soliton treatment (see also the Appendix S).

We close this Section with a comment on the equivalence between the sine-Gordon model and the massive Thirring model, i.e., a two-dimensional self-coupled Fermi field model with vector interaction (see, e.g., [1]). The equivalence between the two models was derived in [165]. On the basis of results obtained in the massless Thirring model [377], it can be shown [165] that the perturbation series in the mass parameter μ of the Thirring model is term-by-term identical with the perturbation series in m^4 for the sine-

Gordon model, provided the identification is made

$$\frac{4\pi m^2}{g^2} = 1 + \frac{\lambda}{\pi}, \tag{7.134a}$$

$$\bar{\chi}\gamma_\mu\chi = i\frac{g}{2\pi m}\epsilon_{\mu\nu}\partial_\nu\psi, \tag{7.134b}$$

$$i\mu\kappa\bar{\chi}\chi = -\frac{m^4}{g^2}\cos(\frac{g}{m}\psi), \tag{7.134c}$$

where χ, λ and κ are the fermion field, the coupling constant and a constant dependent on a cut-off, respectively, in the Thirring model. For $\mu = 0, 1$, the γ^μ are the two-dimensional gamma matrices. The soliton solutions thus correspond to states with unit fermion number. It is remarkable that for $\frac{g^2}{m^2} = 4\pi$ the sine-Gordon model is equivalent to a free Fermi field, as implied by Eq. (7.134). In [436] the creation and annihilation operators for the quantum sine-Gordon soliton are obtained and shown to satisfy anticommutation rules and solve the field equations in the fermionic massive Thirring model. Analogous operators were obtained in [200] in the massless Thirring model. However, in this case they do not possess the physical significance as they do in the case of the massive model. These results, also obtained by means of functional integration technique [494], hold at zero temperature and extension to finite temperature can be done [198] in the functional integration approach. It is known that in $(1+1)$-dimensional theories fermionic degrees of freedom may be expressed in terms of bosonic degrees of freedom and vice-versa [1]. The remarkable fact in the equivalence massive Thirring model/sine-Gordon model is the duality symmetry between the two theories expressed by Eq. (7.134), namely the correspondence between the strong coupling regime in one theory and the weak coupling one in the other. We will not comment further on this problem.

7.4.1 *The quantum image of the Bäcklund transformations*

The kink solution and the sine-Gordon solitons are two examples of topologically non-trivial solutions. However, the sine-Gordon equation, contrary to the equation for the kink solution, belongs to the family of the so-called completely integrable Hamiltonian systems [629], which have an infinite number of conservation laws (of conserved functionals). In the frame of classical integrable field theory, one can see (cf. Chapter 10 and Appendix S) that a transformation exists such that a new solution can be obtained from a given one. Such a transformation is called the Bäcklund transformation and it expresses the invariance of the theory under certain symmetry transformations (see, e.g., [629]). In the present Section we show that a "quantum image" of the Bäcklund transformation exists [469, 642] in the boson condensation formalism. Then we discuss the coherent state representation of the soliton solution. We use the sine-Gordon model as an example for our discussion.

In the N-soliton case, one has to use the boson transformation function (cf. Eq. (7.132))

$$f^{(N)}(x) = \sum_{i=1}^{N} f_i(x), \qquad i = 1, \ldots, N, \tag{7.135}$$

where each of the $f_i(x)$ has to be a solution of the equation for ρ_{in}. This means that we need to consider the canonical transformation $\beta^{(N)}$ on the in-field:

$$\rho_{in}(x) \to \rho_{in}^{(N)}(x) = \beta^{(N)} \rho_{in}(x) = \rho_{in}(x) + f^{(N)}(x). \tag{7.136}$$

This can be considered as the chain of N successive β transformations:

$$\rho_{in}(x) \xrightarrow[\beta]{} \rho_{in}(x) + f_1(x) \xrightarrow[\beta]{} \rho_{in}(x) + f_1(x) + f_2(x) \xrightarrow[\beta]{} \ldots, \tag{7.137}$$

so as to induce the transformations

$$\psi(x) \xrightarrow[B]{} \psi^{(1)}(x) \xrightarrow[B]{} \psi^{(2)}(x) \xrightarrow[B]{} \ldots, \tag{7.138}$$

on the Heisenberg field. The N-soliton solution is then given by

$$\phi_c^{(N)}(x) = \langle 0|B^{(N)}\psi(x)|0\rangle. \tag{7.139}$$

In this sense we refer to $\beta^{(N)}$ as the "quantum image" of the Bäcklund transformation. Note that, as in the case of Bäcklund transformation, $\beta^{(N)}$ is a canonical transformation which keeps the Hamiltonian invariant.

Let us now consider the generator of the transformation (7.113) (or β in Eq. (7.137)). It is given by

$$D_{in} = - \int_{-\infty}^{+\infty} dy_1 g(y_1)\partial_t \rho_{in}(y), \tag{7.140}$$

where we have made explicit the notation for the space coordinate x_1 and the temporal one; $g(x_1) \equiv \theta(x_1)f(x_1)$ and $\theta(x_1)$ is the step function. Use of the canonical commutation relations shows indeed that

$$\beta \rho_{in}(x) = e^{-iD_{in}} \rho_{in}(x)e^{iD_{in}}, \tag{7.141}$$

and therefore

$$\langle 0|e^{-iD_{in}} \rho_{in}(x)e^{iD_{in}}|0\rangle = \langle f|\rho_{in}(x)|f\rangle = f(x_1), \tag{7.142}$$

where we have used the notation

$$|f\rangle = e^{iD_{in}}|0\rangle. \tag{7.143}$$

Note that a restriction to a convenient x_1 domain has to be made in order to make Eq. (7.141) well defined. Due to the isometry of the transformation induced by (7.140) it is $\langle f|f\rangle = 1$ and we have

$$\alpha_{in,\mathbf{k}}|f\rangle = \alpha_{\mathbf{k}}|f\rangle, \tag{7.144}$$

where $\alpha_{in,\mathbf{k}}$ is the annihilation operator of the field ρ_{in} and $\alpha_{\mathbf{k}}$ is a c-number related with the Fourier image of the $f(x_1)$ function. The state $|f\rangle$ is thus a coherent state for the ρ_{in} field. Use of the dynamical map and of basic properties of the coherent states [519] gives (cf. Appendices B and C)

$$\langle f|\psi(x)|f\rangle = \Psi[f(x_1)], \quad \langle f|J[\Psi[\rho_{in}]]|f\rangle = J[\Psi[f(x_1)]]. \tag{7.145}$$

Thus $\langle f|\psi(x)|f\rangle$ solves the static Euler–Lagrange equation

$$\Lambda_{static}(\partial)\langle f|\psi(x)|f\rangle = J[\langle f|\psi(x)|f\rangle], \tag{7.146a}$$

$$\langle f|\psi(x)|f\rangle = \phi_c^{(1)}(x_1), \tag{7.146b}$$

where $\Lambda_{static}(\partial) \equiv \partial_x^2 - m^2$. Note that our discussion can be easily extended to the case of the N-soliton solution.

We remark that $|f\rangle$ is not the soliton solution coherent state for $\psi(x)$ (see [29, 129, 604]), i.e., $|\phi_c^{(1)}(x_1)\rangle = e^{iD}|0\rangle$, where

$$D = -\int_{-\infty}^{+\infty} dy\, \phi_c^{(1)}(x_1)\partial_t\psi(y), \tag{7.147}$$

which is indeed different from D_{in}. Since $|0\rangle$ is the vacuum for ρ_{in}, the dynamical map for $\psi(x)$ must be used to express D in terms of ρ_{in}. In this way the non-linearity of the dynamics is taken into account. It must also be observed that in the infinite volume limit ($x_1 \to \pm\infty$), since the soliton solution goes to a constant ($\phi_c^{(1)}(x_1 = \pm\infty) = \text{const}^\pm$), a generator such as the one in Eq. (7.147) cannot be defined for the transformation $\psi(x) \to \psi(x) + \text{const}^\pm$ at $x_1 \to \pm\infty$ and $\text{const}^+ \neq \text{const}^-$, as explained in Chapters 1 and 2. This fact expresses the stability of the soliton against the decay into a number of components ρ_{in}. Indeed, the overlap between the soliton state $|\phi_c^{(1)}(x_1)\rangle$ and the vacuum $|0\rangle$ is given by [604]

$$\langle 0|\phi_c^{(1)}\rangle|_{t=\pm\infty} = \exp\left[-\frac{1}{2}\int dk\, |\tilde\phi_c^{(1)}|^2 \omega_k\right], \tag{7.148}$$

where Eq. (7.147) and the asymptotic field for ψ have been used, $\tilde\phi_c^{(1)}$ is the Fourier image of $\phi_c^{(1)}$ and $\omega_k = (\mathbf{k}^2 + m^2)^{1/2}$. Since $\int dk|\tilde\phi_c^{(1)}|^2\omega_k >$

$\int dk |\tilde{\phi}_c^{(1)}|^2 = \int dx_1 |\phi_c^{(1)}(x_1)|^2 = \infty$ the transition probability is zero. The existence of the conservation law of the topological charge corresponds indeed [604] to the soliton being a stable bound state of infinite number of constituents. We thus conclude that the confinement of the constituents of an extended object finds its origin in the fact that the transformation

$$\rho_{in}(x) \to \rho_{in}(x) + \text{const.}^{\pm} , \qquad (7.149)$$

is not unitarily implementable. In order to implement the transformation (7.149) an infinite amount of energy would be required because it is not an invariant transformation for the ρ_{in} equation. The interpolating character of the soliton solution between degenerate vacua

$$_{2\pi}\langle 0|\psi(x)|0\rangle_{2\pi} = -2\pi \frac{m}{g} \underset{-\infty \leftarrow x_1}{\longleftarrow} \phi_c^{(1)}(x_1) \underset{x_1 \to +\infty}{\longrightarrow} 0 = {}_0\langle 0|\psi(x)|0\rangle_0 , \quad (7.150)$$

equivalently expresses the existence of the conservation law of the topological charge. There does not exist any unitary operator generating the transformation $|0\rangle_{2\pi} \to |0\rangle_0$, or vice-versa, due to their unitarily inequivalence. Nevertheless, the interpolation between the expectation values of ψ in the states $|0\rangle_0$ and $|0\rangle_{2\pi}$, as in (7.150), is characteristic of the non-trivial topology of the soliton $\phi_c^{(1)}(x_1)$. *This is a manifestation of the classical nature of the soliton solution.* At a quantum level such an interpolation would be strictly forbidden, which is the meaning of unitarily inequivalence. The non-trivial mapping between the soliton field values at $x_1 = \pm\infty$ and the inequivalent representations associated to $|0\rangle_0$ and $|0\rangle_{2\pi}$ is an example of homotopy mapping (see Chapter 10).

We remark that the above stability argument fails if the integration in Eq. (7.148) does not extend over an infinite range. Thus for a one-dimensional system of finite length the transition probability (7.148) is not zero and the local deformation (the kink or the sine-Gordon soliton) is not stable against decay in a number of constituent excitations.

7.5   Soliton solutions of the non-linear Schrödinger equation

The two-dimensional non-linear Schrödinger equation (NSE) provides another example of a completely integrable dynamical system [179]. It admits a soliton solution which, contrary to the sine-Gordon soliton, goes to zero at $x_1 \to \pm\infty$. In this Section we discuss some physical systems where NSE is relevant, such as the ferromagnetic chain, kink-like deformations acting as a potential well for the magnon fields, the case of Toda lattice back-reaction potential and ring solitons in molecular aggregates. NSE has been used also to tackle the problem of efficient, non-dissipative energy transfer which is of great interest in many physical systems, including biological systems. For example, in order to explain efficient energy transfer on long muscle fibers, Davydov has proposed a non-linear dynamics of electric dipole excitations in protein molecular chains ruled by NSE [178–180, 180, 193, 194, 645].

7.5.1 *The ferromagnetic chain*

An interesting physical system which, in certain approximations, is described by the NSE equation is the ferromagnetic chain. We discuss such a system in the present Section. Our conclusions are, however, applicable to the soliton solution of the NSE equation in general cases.

One speaks of magnetic chains since magnetic substances exist [187, 578] with chain structures. In these substances the interaction between chains is smaller than the interaction of the neighboring spins within a chain at a temperature slightly higher than the crystal magnetic ordering temperature T_c. In these conditions, although long-range magnetic order does not exist in the one-dimensional system (cf. Section 3.10), one can speak of a magnetic chain, provided there exists a correlation length much greater than the lattice constant along the chain. The generalized Heisenberg spin chain with phonon interaction was studied, e.g., in [394], where real magnetic systems have been obtained by some reduction procedure and have been shown to be close to integrable models in certain limits. Usually semiclassical quantization procedures are used to relate the c-number soliton solution to collective excitations such as magnons [549]. In the following we will introduce the boson creation and annihilation operators, which we now denote ϕ and $\phi^\dagger$, respectively, for the magnon quantum field excitations considered to be the quanta of the spin wave. The soliton appears thus strictly related to the quantum magnon-magnon bound state. In the semiclassical approach of crucial importance is the bosonization of the basic spin variables, usually achieved by introduction of the Holstein–Primakoff boson spin operators. From Sections 4.4.1 and 4.4.3 we know that when spin operators S^+, S^-, S^z, which are functionals of the Heisenberg field operators, are expressed in terms of the (quasiparticle) magnon field operators ϕ, $\phi^\dagger$, they reproduce the Holstein–Primakoff linear representation without any approximation, namely $S_j^+ = \sqrt{2S}\phi_j$, $S_j^- = \sqrt{2S}\phi_j^\dagger$, $S_j^z = S - \phi_j^\dagger\phi_j$, with $j = 1, 2, \ldots$ denoting the spin site in the chain lattice of spacing δ. Here and in the following S denotes the expectation value of S^z. In the present

Section we show that the soliton in a one-dimensional ferromagnetic chain can be described by localized quantum condensation of magnons as an effect of the non-linear magnon-magnon interaction [134]. This interaction will turn out to be dynamically equivalent to the coupling between the deformation of the chain lattice structure and the magnon field [134]. Let us denote by a_j the Heisenberg magnon operators in terms of which the spin operators are expressed. These a_j are in general complicated non-linear functionals of the quasiparticles magnon fields ϕ. Following [530], in the continuum limit (where the discretization implied by the lattice spacing δ may be ignored due to the large number of sites which are considered) the field equation for the Heisenberg magnon field $a(x,t)$ is

$$\left(i\hbar\frac{\partial}{\partial t} + JS\frac{\partial^2}{\partial x^2} - \mathcal{V}\right) a(x,t) = -2(\tilde{J} - J)|a(x,t)|^2 a(x,t)\,, \quad (7.151a)$$

$$\mathcal{V} = \mu\mathcal{H} + 2(\tilde{J} - J)S\,, \qquad (7.151b)$$

which is indeed the non-linear Schrödinger equation (NSE) (of Gross–Pitaevskii-type). In Eq. (7.151) $\mu\mathcal{H}$ is the chemical potential term, with $\mathcal{H}$ the associated energy. $\tilde{J}$ and J are the exchange integrals appearing in the phenomenological Hamiltonian from which the NSE Eq. (7.151) is obtained in the continuum limit and in the biquadratic approximation [530]. Such an Hamiltonian is written as

$$H = -\mu\mathcal{H}\sum_j S_j^z - \frac{1}{4}J\sum_{j,\delta}(S_j^+ S_{j+\delta}^- + S_j^- S_{j+\delta}^+) - \frac{1}{2}\tilde{J}\sum_{j,\delta} S_j^z S_{j+\delta}^z\,. \quad (7.152)$$

The soliton solution of Eq. (7.151) is [352, 550, 551]

$$\alpha(x,t) = \frac{1}{\sqrt{2\ell}}\frac{\exp[i(\eta x + \gamma_1 - \omega t)]}{\cosh[(x - x_0 - vt)/\ell]}\,, \quad (7.153)$$

with

$$\int_{-\infty}^{-\infty} dx|\alpha(x,t)|^2 = 1\,, \quad (7.154)$$

and where γ_1 is the phase constant, v the soliton velocity and

$$\ell = \frac{2JS}{(\tilde{J} - J)} > 0\,, \qquad \eta = \frac{\hbar v}{2JS}\,, \quad (7.155a)$$

$$\hbar\omega \equiv E = \mu\mathcal{H} + 2(\tilde{J} - J)S + \frac{\hbar^2 v^2}{4JS} - \frac{(\tilde{J} - J)^2}{4JS}\,. \quad (7.155b)$$

The quantity ℓ in Eq. (7.155) defines the macroscopic characteristic length.

We now show that the soliton solution Eq. (7.153) can be obtained as the result of the localized boson condensation of the magnon field $\phi(x,t)$ satisfying the free field equation

$$\Lambda(\partial)\phi(x,t) = 0\,, \tag{7.156}$$

with

$$\Lambda(\partial) \equiv \left(i\hbar \frac{\partial}{\partial t} + JS \frac{\partial^2}{\partial x^2} - \mathcal{V} \right). \tag{7.157}$$

The relation between the Heisenberg magnon field $a(x,t)$ and the quasiparticle magnon field is provided by the dynamical map

$$\langle a | a(x,t) | b \rangle = \langle a | F[\phi(x,t)] | b \rangle\,, \tag{7.158}$$

where $|a\rangle$ and $|b\rangle$ are wave packet states of the physical Fock space and F is a functional of normal ordered products of the quasiparticle fields; here we consider only the quasiparticle magnon field since it is the only one relevant to our discussion. As already done in the previous Sections, we then assume that a transformation B for the $a(x,t)$ field exists, which is an invariant transformation for the NSE (7.151):

$$a(x,t) \rightarrow a'(x,t) = Ba(x,t)\,, \tag{7.159a}$$

$$\Lambda(\partial) a'(x,t) = J[a'(x,t)]\,, \tag{7.159b}$$

where $J[a'(x,t)]$ denotes the r.h.s. of the first of Eqs. (7.151) with $a(x,t)$ replaced by $a'(x,t)$. Then, the soliton solution (7.153) is obtained as

$$\alpha(x,t) = \langle 0 | a'(x,t) | 0 \rangle |_{\hbar \to 0}\,. \tag{7.160}$$

By a procedure analogous to the one followed in previous Sections and by using the expansion formula

$$\cosh^{-1} x = 2 \sum_{n=0}^{\infty} (-1)^n e^{-(2n+1)x}\,, \tag{7.161}$$

one can show [134] that a transformation β for the $\phi(x,t)$ field exists, which is a canonical transformation leaving invariant the free field equation (7.156). It induces through the dynamical map (7.158) the B transformation for the $a(x,t)$ field

$$\langle a | Ba(x,t) | b \rangle = \langle a | F[\beta\phi(x,t)] | b \rangle\,. \tag{7.162}$$

In this way one finds that

$$\phi(x,t) \rightarrow \phi'(x,t) = \beta\phi(x,t) = \phi(x,t) + f(x,t)\,, \tag{7.163}$$

with $f(x,t)$ a solution of the equation

$$\Lambda(\partial)f(x,t) = 0\,, \qquad (7.164)$$

e.g.,

$$f(x,t) = \frac{2}{\sqrt{2\ell}} \exp[i(\eta x + \gamma_1 - \omega t)] \exp[-(x - x_0 - vt)/\ell]\,. \qquad (7.165)$$

Eq. (7.163) describes the localized condensation of the magnon field $\phi(x,t)$ out of which the soliton solution emerges as an effect of the non-linear dynamics. The soliton thus appears as the macroscopic envelope of localized quantum condensation of magnons.

As done in the case of the sine-Gordon soliton, we may obtain the coherent state representation of the localized boson condensation. Indeed, in the $v = 0$ reference frame, provided a restriction to a convenient x domain, e.g., to the domain $x \geq 0$, is introduced, the generator of Eq. (7.163) is written as

$$D = - \int_{-\infty}^{\infty} dy[g(y,t)\dot{\phi}(y,t) - \dot{g}(y,t)\phi(y,t)]\,, \qquad (7.166)$$

with

$$g(x,t) = \theta(-x + x_0) \exp[i(\eta x + \gamma_1 - \omega_0 t)] \exp[-(x - x_0)/\ell]\,, \qquad (7.167)$$

and $\omega_0 = \omega|_{v=0}$. Then, by using the canonical commutation relations, Eq. (7.163) is obtained as

$$\phi'(x,t) = e^{-iD}\phi(x,t)e^{iD}\,, \qquad \text{at } v = 0\,, \qquad (7.168)$$

and, by exploiting the coherent state properties, we have from the dynamical map

$$\langle 0|e^{-iD}a(x,t)e^{iD}|0\rangle = F[f_{v=0}(x,t)]\,, \qquad (7.169a)$$

$$\langle 0|e^{-iD}J[a(x,t)]e^{iD}|0\rangle = J\left[F[f_{v=0}(x,t)]\right]\,, \qquad (7.169b)$$

which used in Eq. (7.151) give the c-number NSE equation

$$\Lambda(\partial)F[f(x,t)] = J\left[F[f(x,t)]\right]\,. \qquad (7.170)$$

Here a boost to the original reference frame has been performed. Eq. (7.170) admits the soliton solution $F[f(x,t)] = \alpha(x,t)$ given by Eq. (7.153).

The kink-like deformation of the ferromagnetic chain as a potential well for the magnon field

It is interesting to observe that the localized boson condensation, from which the NSE soliton solution emerges, induces a kink-like deformation excitation which in turn acts as a potential well for the magnon field $a(x,t)$ and travels with velocity v [134, 193]. This recovers the result of [530], to be compared with the Davydov analysis of excitations in molecular chains, see [177–180, 193]. Let us consider the probability density of the magnon field $a(x,t)$ in the soliton state

$$\rho(x,t) = \gamma\langle 0|e^{-iD}a^\dagger(x,t)a(x,t)e^{iD}|0\rangle = \gamma|\alpha(x,t)|^2\,, \qquad (7.171)$$

i.e., after a boost to the original frame,

$$\rho(x,t) = \frac{\gamma}{2\ell \cosh^2[(x-x_0-vt)/\ell]}\,, \qquad (7.172)$$

where γ is a proportionality constant. It is easily seen that $\rho(x,t)$ satisfies the phonon equation

$$\left(\frac{\partial^2}{\partial t^2} - v^2\frac{\partial^2}{\partial x^2}\right)\rho(x,t) = 0\,, \qquad (7.173)$$

which, by using

$$v^2 = v_0^2 - v_0^2(1-s^2)\,, \quad s = \frac{v}{v_0} \ll 1\,, \qquad (7.174)$$

can be rewritten as

$$\left(\frac{\partial^2}{\partial t^2} - v_0^2\frac{\partial^2}{\partial x^2}\right)\beta(x,t) + \frac{K}{M}\frac{\partial}{\partial x}|\alpha(x,t)|^2 = 0\,. \qquad (7.175)$$

v_0 is the sound wave velocity on the chain in the absence of soliton,

$$v_0^2 = \frac{w}{M}\,, \quad K = w(1-s^2)\gamma\,. \qquad (7.176)$$

Here w denotes the elasticity coefficient of the chain, M the molecular mass. $\beta(x,t)$ is defined as

$$-\frac{\partial}{\partial x}\beta(x,t) = \rho(x,t)\,, \qquad (7.177a)$$

$$\beta(x,t) = -\left[\frac{\gamma}{2}\tanh\frac{(x-x_0-vt)}{\ell}\right] + \text{const.} \qquad (7.177b)$$

Note that Eq. (7.177) describes the kink solution (up to an arbitrary constant) discussed in Section 7.3. Eq. (7.170) is now written as

$$\Lambda(\partial)\alpha(x,t) = \frac{2}{\gamma}(\tilde{J}-J)\frac{\partial\beta(x,t)}{\partial x}\alpha(x,t)\,. \qquad (7.178)$$

Eqs. (7.175) and (7.178) describe a system of coupled equations for the phonon and the magnon similar to the ones of [530] where the reader can find an analysis of the soliton stability and velocity. We remark that Eq. (7.178) is also similar to the Davydov Eq. (2.5a) of [177] describing the kink solution traveling on protein chains in biological systems (see also [193]). Thus our present results can be also extended, in their general features, to the discussion of non-linear excitations in biology [193, 194, 196, 287, 358, 645].

Let us now comment on the anisotropy term $\propto (\tilde{J} - J)$ in Eq. (7.151). The non-linear coupling of the field $a(x,t)$ disappears in the isotropic limit $\tilde{J} - J \to 0$. This leads to the disappearance of the soliton solution Eq. (7.153), consistently with its non-linear dynamical origin. Eqs. (7.155) and (7.156) show that the soliton energy is smaller than linear magnon energy by the amount $(\tilde{J} - J)^2/4JS$ in the $v = 0$ rest frame. Moreover, the anisotropic magnon energy increases by $2(\tilde{J} - J)S$, with $\tilde{J} > J$, with respect to the isotropic ($\tilde{J} = J$) magnon energy. Note also that for $v = 0$:

$$\langle 0 | e^{-iD} \phi'^{\dagger}(x,t) \phi'(x,t) e^{iD} | 0 \rangle = \frac{2}{\ell} \exp[-(x - x_0)/\ell], \qquad (7.179)$$

which shows that the condensation of quasimagnons is maximal near the center x_0 of the soliton (localized condensation), whereas it disappears in the limit $\ell \to \infty$. It is interesting that such a localized condensation is a self-sustained dynamical effect since the size of the region over which it extends (twice the macroscopic characteristic length ℓ) is self-consistently controlled by the non-linear coupling constant ($\propto (\tilde{J} - J)$) which is a measure of the anisotropy of the system: as soon the anisotropy is induced (e.g., by external perturbation) a non-linear interaction among magnons appears by which quasimagnon fields are trapped in a one-dimensional bag-like region of size 2ℓ, thus creating the soliton excitation (7.153). By conveniently tuning the anisotropy and the excitation energy one can therefore change the number of solitons on the chain.

Note that as far as the length of the chain is infinite (i.e., in the limit in which the continuum, or infinite number of degrees of freedom, holds) the topological stability of the kink (7.177) is ensured. In the case of chains of finite length, the chain deformation (the kink $\beta(x,t)$ and the associated envelope of localized magnon condensation, i.e., the soliton $\alpha(x,t)$) is unstable and it can decay into a number of free excitations.

7.5.2 *Non-linear Schrödinger equation with Toda lattice back-reaction potential*

In the previous Section we studied the NSE in one space dimension. In its non-linear term it has been used $|\alpha(x,t)|^2 \propto \rho(x,t)$, with $\rho(x,t)$ solving the phonon equation (7.173) and acting as a potential term which can be understood as a back-reaction field to the $\alpha(x,t)$ excitation field. In the present Section we consider another example [334] of NSE where $|\alpha(x,t)|^2$ in the non-linear term is proportional to the soliton solution to the Toda lattice non-linear equation [606–608]. The interest in such a problem comes from the fact that such a system may model a one-dimensional chain of molecules which undergo space displacement due to their excitation. In two space dimensions, a similar molecular system may represent a coherent molecular domain presenting efficient energy transport. Such a two-dimensional system will be studied in the following Section. A potential term in the Dirac equation also originated from a soliton (a kink) and trapping the fermion field will be studied in the subsequent Section.

We consider the molecular excitation field $a(x,t)$, solution of the Heisenberg field equation

$$\left(i\hbar \frac{\partial}{\partial t} + J \frac{\partial^2}{\partial x^2} - \Lambda + V(x,t) \right) a(x,t) = 0 \,, \qquad (7.180)$$

where Λ is the energy of the bottom of the exciton band, $-J$ represents the resonant dipole-dipole interaction between neighboring molecules [178,179] and $V(x,t)$ is the potential arising from the displacement of the excited molecules from their equilibrium positions. The coordinate x refers to a fixed origin on the line. Suppose that we undergo all the formal procedures shown in the previous Sections in order to obtain the coherent state representation of the problem (7.180), or else the boson transformation machinery by introduction of the free field $\phi(x,t)$. We do not repeat these steps here for sake of brevity. We thus assume that a coherent state representation $\alpha(x,t)$ is obtained for the molecular excitation field $a(x,t)$, and use such a representation in the following. The Heisenberg equation (7.180) is then recognized to be the Schrödinger equation for the "wave function" $\alpha(x,t)$ (cf., e.g., the derivation in Eqs. (7.169)–(7.170))

$$\left(i\hbar \frac{\partial}{\partial t} + J \frac{\partial^2}{\partial x^2} - \Lambda + V(x,t) \right) \alpha(x,t) = 0 \,. \qquad (7.181)$$

Let l be the molecular spacing of the aggregate in the absence of excitation: the displacement β_n of a molecule at the site n produces the deformation $l \to l - \rho_n$ with $\rho_n = \beta_{n-1} - \beta_n$. We want to study the back-reaction potential acting on $\alpha(x,t)$ induced by the ρ_n displacement. To this aim we assume that the interaction among the molecules due to the deformation ρ_n is described by an anharmonic potential U of Toda-lattice type [334, 606–608]:

$$U = \Phi(\rho_n) = \frac{w}{\gamma} \left\{ -\rho_n + \frac{1}{\gamma} [e^{\gamma \rho_n} - 1] \right\}, \qquad (7.182a)$$

$$\Phi(\rho_n) = \frac{1}{2} w \rho_n^2 + \frac{1}{6} w \gamma \rho_n^3 + \cdots, \qquad (7.182b)$$

where $\gamma \rho_n \ll 1$. w is the elasticity coefficient of the chain and γ the anharmonicity parameter. The sound velocity is $v_0 = l(\frac{w}{M})^{1/2}$ when $\gamma = 0$ (harmonic interaction potential). M denotes the molecular mass.

The propagation equation for the deformation ρ_n along the chain is

$$M \frac{d^2}{dt^2} \rho_n = \frac{w}{\gamma} [e^{\gamma \rho_{n+1}} + e^{\gamma \rho_{n-1}} - 2e^{\gamma \rho_n}], \qquad (7.183)$$

which has the well-known [177, 179, 180, 606–608] soliton solution

$$\rho_n(t) \simeq \frac{1}{\gamma} \frac{\sinh^2(ql)}{\cosh^2[q(nl - vt)]}, \qquad (7.184)$$

going to zero as $n \to \pm\infty$. q is a parameter (the soliton parameter) upon which the soliton velocity v depends. v is given by

$$v = v_0 \frac{1}{ql} \sinh(ql). \qquad (7.185)$$

Assuming $ql \ll 2\pi$, we can apply the continuum approximation and write Eq. (7.184) as

$$\rho(x,t) \simeq \frac{1}{\gamma} \frac{\sinh^2(ql)}{\cosh^2[q(x - x_0 - vt)]}, \qquad (7.186)$$

where x_0 is the coordinate of the soliton center at the initial time $t = 0$. In such an approximation, the deformation is $\rho(x,t) = -l \frac{\partial}{\partial x} \beta(x,t)$.

We now use in Eq. (7.181) the potential $V(x,t) = \sigma^2 \rho(x,t)$, with $\rho(x,t)$ given by (7.186). The coupling σ^2 is given in Eq. (7.187). The solution of Eq. (7.181) for the lowest energy E is now the soliton

$$\alpha(x,t) = \frac{1}{\gamma^{1/2}} \sinh(ql) \frac{e^{iK(x-x_0) - i\omega t}}{\cosh[q(x - x_0 - vt)]}, \qquad (7.187a)$$

$$K = \frac{\hbar v}{2J}, \qquad E = \hbar\omega - \Lambda - \frac{\hbar^2 v^2}{4J} = -Jq^2, \qquad (7.187b)$$

$$\sigma^2 = \frac{2J\gamma q^2}{\sinh^2(ql)} = \frac{2J\gamma}{l^2} \frac{v_0^2}{v^2}. \qquad (7.187c)$$

The soliton solution (7.187), describing the aggregate molecular excita-tion, does not spread out in time, contrary to the case of a free exciton. From Eqs. (7.186) and (7.187) we see that $\rho(x,t) = |\alpha(x,t)|^2$ so that the po-tential $V(x,t)$ is seen to be induced by the back-reaction displacement field $\rho(x,t)$ and Eq. (7.181) is recognized to be the NSE (of Gross–Pitaevskii-type)

$$\left(i\hbar\frac{\partial}{\partial t} + J\frac{\partial^2}{\partial x^2} - \Lambda\right)\alpha(x,t) = -\sigma^2|\alpha(x,t)|^2\alpha(x,t).\qquad(7.188)$$

We observe that the soliton velocity v, given by Eq. (7.185), is allowed to exceed the sound velocity in the system, $v \gg v_0$, and that $v \to v_0$ and $\rho(x,t) \to 0$ in the limit of $q \to 0$, so that the soliton disappears, which means that the coherent molecular excitation domain on the chain decays to free excitons in that limit. Consequently, the energy efficiently carried by the solitonic molecular excitation also tends to zero: the efficiency of the energy transfer thus drops with decreasing values of q. Note that the size of the coherent excitation domain is

$$\Delta x \simeq \frac{2\pi}{q}.\qquad(7.189)$$

In the discussion presented above we have not considered thermal fluctu-ations which are expected to counteract the cooperative interaction among molecules. Therefore dissipative terms could be introduced in the NSE (7.188) accounting for such thermal effects. On the basis of experiments made on monolayer systems, one observes that the size of the coherent molecular (two-dimensional) domains decreases indeed with increasing tem-perature T. In the one-dimensional case one thus expects that the soliton q parameter is temperature-dependent, e.g., is proportional to T. We do not consider further this problem here. Rather we proceed in the next Section to extend our discussion to the case of highly ordered molecular monolayers called Scheibe aggregates where efficient energy transfer is observed.

7.5.3   *Ring solitons in the Scheibe aggregates*

In this Section we will consider energy transfer in aggregates of highly ordered molecular monolayers, such as those called Scheibe aggregates. For a review on these systems see, e.g., [111]. Such monolayers are made by donor molecules, for example by oxycyanine dyes, doped with fluorescent acceptor molecules, e.g., with thiacyanine dyes, and exhibit a highly efficient transfer of energy from impinging photons via excited host molecules to acceptor guests [477, 478]. The doping level may be as low as a donor to acceptor ratio of 10^4. Systems made by multilayers forming a sandwich type structure, where the acceptor molecules are situated in a layer adjacent to the monolayer containing the donor molecules, are also studied. However, these multilayer systems present a less efficient energy transfer than the one observed in the monolayers mentioned above. In both cases, the efficiency of energy transfer has been observed to decrease with decreasing temperature.

The classical approach which has been used to model the energy transfer in these systems and which reproduces the experimental results is based on the assumption that a coherent exciton may be formed [477, 478], namely it is assumed that a classical analogue of a domain containing a certain number of molecules, all oscillating in phase at the frequency of the exciting radiation, can be dynamically produced. This coherent exciton is then thought to move as a whole unit throughout the layer until it reaches an acceptor site. In agreement with experimental results, an estimate of the coherent exciton mean lifetime, before it is absorbed by an acceptor, is of the order of 10^{-10} s at $300 K$ and its velocity of propagation results as about 10 to 20 times the sound velocity v_0; thus it migrates with a velocity of the order of 10^4 m s^{-1}. The exciton mean life is observed to increase linearly with temperature. The number ν of involved molecules ranges between 10 at room temperature and 150 at 20 K. Theoretical modeling [477] predicts $\nu \propto T^{-1}$ in agreement with experiments.

Delicate points in these approaches are the explanation of the formation of the coherent domain and the explanation of the non-dissipative character observed in the energy transfer mechanism. Both these features might find a natural explanation provided the system is governed by a non-linear dynamics. In [334] it was thus proposed that the coherent exciton domain originates from non-linear dynamical effects in the layer. Prediction of the domain size [334] and mean lifetime of the coherent domain made by using this non-linear model [157, 158] is in good qualitative agreement with experimental measurements [477, 478].

The non-linear model of [334] is inspired by the discussion of the formation of the coherent molecular domain in one-dimensional systems presented in the previous Section and is based on the dynamics of the ring-wave (solitary) solution of the NSE [431]. The general theory, reported in this book, of soliton solutions emerging as macroscopic envelopes of boson condensation and satisfying classical non-linear dynamical equations fills the gap between microscopic quantum dynamics and its macroscopic manifestation, thus giving account, also in the present case, of how the classical behaviors observed in the Scheibe aggregates may be reconciled with the quantum character of the photon-molecule and molecule-molecule interactions out of which they are originated.

Let us start by considering the NSE in two dimensions with radial symmetry [431] modeled on the one-dimensional Eq. (7.188)

$$\left(i\hbar\frac{\partial}{\partial t} + J\frac{\partial^2}{\partial r^2} + J\frac{1}{r}\frac{\partial}{\partial r} - \Lambda\right)\alpha(r,t) = -\sigma^2|\alpha(r,t)|^2\alpha(r,t)\,, \qquad (7.190)$$

where $r^2 = x^2 + y^2$ is the radial coordinate. As in the previous Section the "wave function" $\alpha(r,t)$ denotes the coherent state representation of the molecular excitation field $a(x,t)$. For small values of r, a solution $\alpha(r,t)$ of Eq. (7.190), which remains analytical at $r = 0$, must be of the form

$$\alpha(r,t) = \sum_{n=0}^{\infty} \alpha_n(r,t)\,, \qquad (7.191)$$

where we assume $\alpha_n(t) = 0$ for odd values of n for any t and $\frac{\partial}{\partial t}\alpha(0,t) = 0$. We also require that such a solution approaches to zero as $r \to \infty$ for any t, $\alpha(\infty,t) = 0$. A general solution at $t = 0$ and small r is then found to be

$$\alpha(r,0) \approx \frac{1}{\gamma^{1/2}}\sinh^2(ql)\left\{\frac{e^{iK_0 r}}{\cosh[q(r-r_0)]} + \frac{e^{-iK_0 r}}{\cosh[q(r+r_0)]}\right\}\,, \qquad (7.192)$$

where r_0 approximately denotes the position at which $|\alpha(r,0)|^2$ has its maximum, namely r_0 defines the initial radial coordinate position. K_0 is proportional to the initial time ($t = 0$) radial velocity $v(0)$ (cf. Eq. (7.187)).

Eq. (7.192) represents the so-called ring-wave soliton [431]. The second term in Eq. (7.192) can be neglected in the limit $qr \gg 0$, thus reducing it to the solution (7.187) of the one-dimensional (radial) NSE. Moreover, for small r it is $\frac{\partial^2}{\partial r^2}\alpha(r,t) \simeq \frac{1}{r}\frac{\partial}{\partial r}\alpha(r,t)$, which reduces Eq. (7.192) to the one-dimensional NSE in the $r \to 0$ limit. This justifies the assumption of Eq. (7.192) for the two-dimensional system based on the analogy with the one-dimensional case.

As shown by computer simulations [431], appropriate choices of the wave parameters exhibit the occurrence of three possible regimes: i) diverging ring wave: monotonical increase of $r(t)$ with t; ii) diverging and then converging ring waves with collapse: after reaching a maximum R_{max}, $r(t)$ decreases to zero as time evolves; iii) converging and collapse of ring wave: $r(t)$ decreases to zero as t grows. The analysis of [157, 158] shows that the second regime ii) best fits the experimental results, so that the ring wave may expand up to a maximum radius R_{max}, then it converges towards its center where it collapses.

A simplified picture may be the following. The donors sitting in the circle of initial radius $r(0)$ get excited by the energy transferred to them by the impinging photon(s); their non-linear coupling originates the ring wave exciton, whose non-dissipative evolution to its final collapse in the circle center produces the energy transfer to the acceptor molecule sitting there. Due to its fluorescence nature, the acceptor re-emits the energy absorbed in the ring wave collapse. The net result is a very efficient, non-dissipative harvesting of energy on the domain of maximum radius R_{max}. This mechanism, of obvious great physical and technological interest, has also been referred to in the literature as the "photon funnel" [111, 585]. As already said, the ring wave model is in good agreement with experiments.

From our discussion also emerges the notion of a characteristic range (R_{max}) for the acceptor molecule, within which it efficiently collects energy from the domain of coherently excited donors. Correspondingly, this also gives information on the optimal donor to acceptor ratio N in order to obtain maximal efficiency in the energy transfer. It is indeed observed that for low N ($N \leq 10^3$) the measured fluorescence from the donors is practically null, meaning the high probability that all the energy harvested in the process of excitation of the donors is transferred to the acceptor. When the acceptor density decreases below a certain threshold (with corresponding larger N) the probability of energy losses due to the donor re-emission by fluorescence is higher with consequent lowering of the energy transfer efficiency (see [157, 158] for details).

Temperature effects, radiative and other types of energy losses, which clearly may influence the coherent exciton lifetime and the overall efficiency, are not included in the non-linear model considered above. Moreover, the considered monolayer isotropy and the continuum limit are not always a good approximation in realistic systems, depending on the molecule type and arrangement in the layer. The effects of discreteness have been considered in [53] by using the discrete self-trapping equation formalism (see [214]). Coherent excitation in small aggregates has been considered in [575] and a model based on exciton dynamics without considering the formation of coherent domains has been developed in [59].

7.6 Fermions in topologically non-trivial background fields

It has been shown [347, 350, 584] that whenever the Dirac equation is solved in a topologically non-trivial background field, in addition to the positive and negative energy solutions, related to each other by fermion conjugation, self-conjugate, normalizable, zero-frequency solutions always exist. The background soliton-like field can be the kink [347, 350, 584], the polaron [135, 136] in one spatial dimension, the Abrikosov (Nielsen–Olesen) vortex [191, 498] and the 't Hooft–Polyakov monopole [524, 587] in 2 and 3 spatial dimensions [349], respectively. The presence of the zero mode is essential to establish the occurrence of charge fractionalization mechanism [347, 350, 584]. The one space dimensional fermion-soliton system is particularly interesting since it can be realized in condensed matter physics as a system of electron in the presence of a domain wall. Peculiar experimental effects, which find explanation in the dynamical properties of the fermion-soliton system, have been obtained [135, 136, 350, 584] in polyacetylene.

We also recall that in [567] it has been shown that the non-homogeneous boson transformation together with the fermion conjugation transformation may lead to the fractional charge phenomenon which can be related to the supersymmetry algebra with a non-zero central charge.

By resorting to the discussion of the kink solution presented in previous Section 7.3, here we show that the interpolating character between degenerate vacua of the kink-like potential in a one-dimensional fermionic system requires the existence of the zero-frequency fermion mode with fractional charge in order to recover the x_1-reflection symmetry of the Dirac equation.

We will use the notation $x = (x_0, x_1)$ and consider a spinor field $\psi(x)$ and a real scalar field $\phi(x)$ in a $1+1$-dimensional model with field equations

$$(i\gamma_\mu \partial^\mu - G\phi(x))\psi(x) = 0, \qquad (7.193\text{a})$$

$$(\partial^2 + \mu^2)\phi(x) = -\lambda\phi^3(x). \qquad (7.193\text{b})$$

G is the coupling constant. Eqs. (7.193) are invariant under fermion conjugation $F = F^{-1} = F^\dagger$

$$\psi(x) \to F\psi(x)F = -i\gamma^1\psi^{\dagger T}(x), \qquad \phi(x) \to F\phi(x)F = \phi(x), \qquad (7.194)$$

and discrete chirality $B = B^{-1} = B^\dagger$

$$\psi(x) \to B\psi(x)B = i\gamma^5\psi^\dagger(x), \qquad \phi(x) \to B\phi(x)B = -\phi(x), \qquad (7.195)$$

with two-dimensional gamma matrices: $\gamma^0 = \gamma_0 = \sigma_1$, $\gamma^1 = i\sigma_3 = -\gamma_1$, $\gamma^5 = \gamma^0\gamma^1 = \sigma_2$, and $g_{00} = 1 = -g_{11}$. We observe that there is also an internal phase symmetry $\psi(x) \to e^{-i\theta}\psi(x)$ induced by the fermion number operator $Q = \frac{1}{2}\int dx_1[\psi^\dagger(x), \psi(x)]$.

We assume that discrete chirality is spontaneously broken:

$$\langle 0|\phi(x)|0\rangle = v, \qquad (7.196)$$

and we thus write $\phi(x) = v + \rho(x)$, with $\langle 0|\rho(x)|0\rangle = 0$, so that Eqs. (7.193) become

$$(i\gamma_\mu\partial^\mu - Gv)\psi(x) = G\rho(x)\psi(x)\,, \qquad (7.197a)$$

$$(\partial^2 + m^2)\rho(x) = -J[\rho(x)]\,, \qquad (7.197b)$$

where

$$J[\rho(x)] = \frac{3}{2}mg\rho^2(x) + \frac{1}{2}g^2\rho^3(x), \qquad (7.198a)$$

$$\lambda v^2 = -\mu^2, \quad m^2 = 2\lambda v^2 = g^2 v^2. \qquad (7.198b)$$

The quasiparticle fields $\psi_{in}(x)$ and $\rho_{in}(x)$ satisfy the field equations

$$(i\gamma_\mu\partial^\mu - Gv)\psi_{in}(x) = 0\,, \qquad (7.199)$$

$$(\partial^2 + m^2)\rho_{in}(x) = 0\,, \qquad (7.200)$$

respectively. We remark that the Dirac Eq. (7.199) is not invariant under the substitution $v \to -v$, which occurs when Eq. (7.196) is replaced by

$$\langle 0|\phi(x)|0\rangle = -v\,. \qquad (7.201)$$

Stated in a different way, we can say that the spontaneous breakdown of the discrete chirality (Eq. (7.195)) introduces in the fermion system described by Eq. (7.193) a partition in the set of the $\psi_{in}(x)$ fermion fields: $\psi_{in}(x) \to \{\psi_{in,+}(x), \psi_{in,-}(x)\}$. Provided $v \neq 0$, there is no non-trivial fermion field which is a solution of both Eq. (7.199) and of

$$(i\gamma_\mu\partial^\mu + Gv)\psi_{in}(x) = 0\,. \qquad (7.202)$$

There is no non-trivial intersection between the set of solutions $\psi_{in,+}(x)$ of Eq. (7.199) and the set of solution $\psi_{in,-}(x)$ of Eq. (7.202). We observe that $\psi_{in,+}(x)$ and $\psi_{in,-}(x)$ are related by the discrete chiral transformation

$$\psi_{in,\pm}(x) \to \psi_{in,\mp}(x) = i\gamma^5\psi_{in,\pm}(x)\,, \qquad (7.203)$$

which, however, does not leave Eq. (7.199) (or Eq. (7.202)) invariant.

The problem is then to study the behavior of the fermionic structure when *both* the conditions Eq. (7.196) and Eq. (7.201) are imposed on the system by the non-trivial topology of the kink potential $\phi_c(x)$, which indeed interpolates between $|0\rangle_+$ and $|0\rangle_-$ in the limits $x_1 \to +\infty$ and $x_1 \to -\infty$, respectively:

$$_-\langle 0|\phi(x)|0\rangle_- = -v \underset{-\infty \leftarrow x_1}{\longleftarrow} \phi_c(x) \underset{x_1 \to +\infty}{\longrightarrow} +v = {}_+\langle 0|\psi(x)|0\rangle_+\,. \qquad (7.204)$$

The states $|0\rangle_-$ and $|0\rangle_+$ denote two degenerate unitarily inequivalent vacua of the theory. Eq. (7.204) shows that $x_1 \to -\infty$ in $\phi_c(x)$ induces on $\phi(x)$ the B transformation of Eq. (7.195).

It is useful to recall that, provided $f(x)$, with conveniently selected divergencies, is the solution of $(\partial_1^2 - m^2)f(x) = 0$, the transformation

$$\rho_{in}(x) \to \rho_{in}(x) + f(x) \tag{7.205}$$

may generate non-homogeneous boson condensate whose envelope is the kink $\phi_c(x)$. By denoting with D_{in} the generator of such a transformation and by $|f\rangle$ the coherent state $|f\rangle = \exp(iD_{in})|0\rangle$, we may exploit the coherent state representation for the kink and consider the realization of the dynamical equations (7.193) obtained by taking their matrix elements between the states $\langle 0_b, 0_{\psi_{in}}, f|$ and $|f, 0_{\psi_{in}}, 0_b, \rangle$, where 0_i denotes the vacuum for the i-type particle. We have

$$(i\gamma_\mu \partial^\mu - G\phi^f(x))\psi^f(x) = 0, \tag{7.206}$$

$$(\partial^2 + \mu^2)\phi^f(x) = -\lambda \left(\phi^f(x)\right)^3. \tag{7.207}$$

In the following, $\phi^f(x_1)$ denotes the kink solution $\phi_c(x_1)$ of the classical equation (7.207) in the static case, $\phi^f(x_1) \equiv \phi_c(x_1)$:

$$\phi^f(x_1) = v \tanh\left(\frac{1}{2}mx_1\right). \tag{7.208}$$

Eq. (7.206) then admits the zero-frequency normalizable solution

$$\psi_0^f(x_1) = N \exp\left(-G \int_0^{x_1} \phi^f(y_1)dy_1\right) \begin{pmatrix} 1 \\ 0 \end{pmatrix}. \tag{7.209}$$

By using this equation and the fact that

$$\psi^f(x_1) = \langle 0_b, 0_{\psi_{in}}, 0_{\rho_{in}}|e^{-iD_{in}}\psi[\psi_{in}(x), \rho_{in}(x), b]e^{iD_{in}}|0_{\rho_{in}}, 0_{\psi_{in}}, b\rangle, \tag{7.210}$$

we obtain

$$\psi(x) = \psi_{in}(x) + \;: N \exp\left(-G \int_0^{x_1} \phi[\rho_{in}(y_1)]dy_1\right) : b \begin{pmatrix} 1 \\ 0 \end{pmatrix} + \dots, \tag{7.211}$$

where the equality is understood as a relation between matrix elements, the dots denote higher normal ordered products of ψ_{in} and ρ_{in} and ϕ is a functional of ρ_{in} only, since we assume that the ψ_{in} field is not a source of ϕ. The boson transformation theorem now ensures us that by inducing the field transformation (7.205) in the field $\psi(x)$ as given by Eq. (7.211) and in the field $\phi[\rho_{in}]$ we obtain, respectively, fields which are again solutions of the field equations (7.193). The conclusion is then that as soon as the kink has been created by the boson condensation induced by (7.205), the dynamics described by Eqs. (7.193) requires the existence of a fermionic zero-frequency mode $b \begin{pmatrix} 1 \\ 0 \end{pmatrix}$ (of course we exclude the trivial solution $\langle f|\psi|f, b\rangle = 0$). We note that, as shown by the dynamical map (7.211), the b mode participates to matrix elements only in the presence of the soliton state $|f\rangle$. Indeed, as one approaches spatial infinity the zero-frequency mode wave-function (7.209) goes to zero (it is localized around the center $x_1 = 0$ of the kink): the zero mode is trapped by the soliton which acts as a potential well in Eq. (7.206). As soon as the asymptotic regions at $(\pm)$-infinity are approached the dynamics falls into just one of the two sectors $+v$ or $-v$.

Let us now write

$$\psi^f(x_0, x_1) = \exp[iK(x_1) - i\omega x_0] \begin{pmatrix} \psi_1^f \\ \psi_2^f \end{pmatrix} . \tag{7.212}$$

We remark that a solution of Eq. (7.206), i.e.,

$$\omega \psi_2^f - iK'(x_1)\psi_1^f - G\phi^f(x_1)\psi_1^f = 0 , \tag{7.213a}$$

$$\omega \psi_1^f + iK'(x_1)\psi_2^f - G\phi^f(x_1)\psi_2^f = 0 , \tag{7.213b}$$

must also be a solution of

$$\omega \psi_1^f + iK'(-x_1)\psi_2^f - G\phi^f(-x_1)\psi_2^f = 0 , \tag{7.214a}$$

$$\omega \psi_2^f - iK'(-x_1)\psi_1^f - G\phi^f(-x_1)\psi_1^f = 0 , \tag{7.214b}$$

since the kink interpolates between $+v$ or $-v$ as $x_1 \to \pm\infty$. In these equations K' denotes derivative of K with respect to x_1. Eqs. (7.214) are obtained indeed from Eqs. (7.213) by parity transformation, i.e., letting $x_1 \to -x_1$ and $\psi^f(x_0, x_1) \to \gamma^0 \psi^f(x_0, -x_1)$.

Now it can be shown that, by subtracting Eqs. (7.214) from Eqs. (7.213), the only non-trivial, normalizable solution of Eqs. (7.213) *and* Eqs. (7.214) is the zero-frequency solution $\psi_0^f(x_1)$ given by Eq. (7.209). This is the only fermion solution for which the invariance under discrete chirality transformation is preserved even in the limits of $\phi^f \to \pm v$ at $x_1 \to \pm\infty$. The zero-frequency solution would never appear in the absence of the kink.

We finally remark that since the fermion zero mode $b \begin{pmatrix} 1 \\ 0 \end{pmatrix}$ is a self-conjugate field (cf. Eqs. (7.194) and (7.211)) and the fermion conjugation symmetry (7.194) is unbroken, we have $0 = \langle f|b^\dagger b|f\rangle = \langle f|bb^\dagger|f\rangle$, which gives $\langle f|\{b, b^\dagger\}|f\rangle = 0$ in contradiction with $\{b, b^\dagger\} = 1$. This contradiction is solved by noticing that $\{b^\dagger, b^\dagger\} = 0 = \{b, b\}$ and by introducing [347] two distinct soliton states $|f, +\rangle$ and $|f, -\rangle$ with $F|f, \pm\rangle = |f, \mp\rangle$ and

$$b|f, +\rangle = |f, -\rangle, \quad b^\dagger|f, +\rangle = 0, \tag{7.215a}$$

$$b^\dagger|f, -\rangle = |f, +\rangle, \quad b|f, -\rangle = 0. \tag{7.215b}$$

We note that $|f, +\rangle$ and $|f, -\rangle$ are degenerate states since the presence or not of the fermion mode b in these states does not involve energy. One can now derive that the fermion zero mode carries the fractional charge $Q = \pm\frac{1}{2}$ (see also [347, 350, 584]). Note that in the asymptotic regions $x_1 \to \pm\infty$ the fermions of masses $\pm Gv$, respectively, are non-fractional charge fermions.

7.7 Superfluid vortices

As a final example we study, in the CTP formalism, vortices in a $(3 + 1)$-d relativistic scalar model for superfluidity. We consider the following Lagrangian invariant under global phase invariance [617]:

$$\mathcal{L} = \partial_\mu \psi^\dagger(x) \partial^\mu \psi(x) - \mu^2 \psi^\dagger(x)\psi(x) - \frac{\lambda}{4}|\psi^\dagger(x)\psi(x)|^2. \tag{7.216}$$

The equation of motion for the complex boson field $\psi(x)$ reads

$$\left(\partial^2 + \mu^2 + \frac{\lambda}{2}|\psi(x)|^2\right)\psi(x) = 0. \tag{7.217}$$

ψ (and $\psi^\dagger$) satisfy usual canonical commutation relations. We assume symmetry breaking, i.e., $\mu^2 < 0$ and $\langle 0|\psi(x)|0\rangle = \langle \psi(x)\rangle = v = \sqrt{-2\mu^2/\lambda}$ and parametrize the field as $\psi(x) \equiv (\rho(x) + v)e^{i\chi(x)}$, where both ρ and χ are hermitian and have zero vacuum expectation value (vev) (this can be unambiguously prescribed since $\chi(x)$ and $\rho(x)$ commute). $|0\rangle$ is the vacuum state for the asymptotic (quasiparticle) field. We put $g = \sqrt{\lambda}$ and $m^2 = \lambda v^2 > 0$ and rewrite the Lagrangian (7.216) as

$$\mathcal{L} = (\rho + v)^2(\partial_\mu \chi)^2 + (\partial_\mu \rho)^2 - m^2\rho^2 - gm\rho^3 - \frac{g^2}{4}\rho^4. \tag{7.218}$$

The conjugate momenta are introduced as $\pi_\rho = 2\dot\rho$ and $\pi_\chi = 2(\rho + v)^2\dot\chi$. The commutation relations are

$$[\rho(x_0, \mathbf{x}), \dot\rho(x_0, \mathbf{y})] = \frac{i}{2}\delta(\mathbf{x} - \mathbf{y}), \tag{7.219a}$$

$$[\chi(x_0, \mathbf{x}), (\rho(x_0, \mathbf{y}) + v)^2\dot\chi(x_0, \mathbf{y})] = \delta(\mathbf{x} - \mathbf{y}). \tag{7.219b}$$

The (unrenormalized) Hamiltonian density reads

$$\mathcal{H} = \frac{1}{4}\pi_\rho^2 + (\nabla\rho)^2 + \frac{1}{4(\rho+v)^2}\pi_\chi^2 + (\rho+v)^2(\nabla\chi)^2$$

$$+ m^2\rho^2 + gm\,\rho^3 + \frac{g^2}{4}\rho^4\,, \tag{7.220}$$

where both m and g are unrenormalized mass and coupling respectively. We subtract $\langle 0|\mathcal{H}|0\rangle$ from $\mathcal{H}$. Such a redefinition will render all matrix elements of $\mathcal{H}$ finite (even if $\mathcal{H}$ is not the free field Hamiltonian density).

The equations of motion for the Heisenberg operators ρ and χ, following from the Heisenberg equations: $\dot{\pi}_\rho = i\,[H,\pi_\rho]$ and $\dot{\pi}_\chi = i\,[H,\pi_\chi]$, are

$$\left[\partial^2 - (\partial_\mu\chi)^2 + m^2\right]\rho + \frac{3}{2}m\,g\,\rho^2 + \frac{1}{2}g\,\rho^3 = v\,(\partial_\mu\chi)^2\,, \tag{7.221a}$$

$$\partial_\mu\left[(\rho+v)^2\partial^\mu\chi\right] = 0\,. \tag{7.221b}$$

The asymptotic fields solve the following equations:

$$\left[\partial^2 - :(\partial_\mu\chi_{in})^2: +m^2\right](\rho_{in}+v) = m^2v\,, \tag{7.222a}$$

$$\partial_\mu\left[:(\rho_{in}+v)^2:\partial^\mu\chi_{in}\right] = 0\,. \tag{7.222b}$$

Note that the above equations are not linear in the asymptotic field. Nevertheless, we can apply the boson condensation machinery in order to arrive to the vortex solutions.

The interacting Lagrangian density is

$$\mathcal{L}^I = -gm\rho^3 - \frac{g^2}{4}\rho^4\,, \tag{7.223}$$

and the Haag expansion for the Heisenberg field operator is written as (see Eq. (6.4)):

$$\psi(x) \equiv (\rho(x)+v)e^{i\chi(x)}$$

$$= T_C\left\{(\rho_{in}(x)+v)e^{i\chi_{in}(x)}\exp\left[-i\int_C d^4y\mathcal{L}_{in}^I(y)\right]\right\}\,. \tag{7.224}$$

As usual, the solutions of the asymptotic equations (7.222) are not unique. Indeed, we may define the following shifted fields:

$$\rho_{in}(x) \rightarrow \rho_{in}^f(x) = \rho_{in}(x) + f(x)\,, \tag{7.225a}$$

$$\chi_{in}(x) \rightarrow \chi_{in}^g(x) = \chi_{in}(x) + g(x)\,, \tag{7.225b}$$

satisfying Eqs. (7.222):

$$\left[\partial^2 - :(\partial_\mu\chi_{in}^g)^2: +m^2\right](\rho_{in}^f+v) = 0\,, \tag{7.226a}$$

$$\partial_\mu\left[:(\rho_{in}^f+v)^2:\partial^\mu\chi_{in}^g\right] = 0\,. \tag{7.226b}$$

The c-number functions $h(x) \equiv f(x) + v(x)$ and $g(x)$ are constrained to solve the following equations (obtained by taking the vev of Eqs. (7.226)):

$$\left[\partial^2 - \langle : (\partial_\mu \chi_{in})^2 : \rangle - (\partial_\mu g)^2 + m^2\right](f + v) = m^2 v, \qquad (7.227a)$$

$$\partial_\mu \left[\left(\langle : \rho_{in}^2 : \rangle + (f + v)^2\right) \partial^\mu g\right] = 0, \qquad (7.227b)$$

where $\langle \ldots \rangle$ stands for vev.

The Haag expansion for ψ in terms of the new asymptotic fields reads

$$\psi^{f,g}(x) = T_C \left\{ (\rho_{in}^f(x) + v) e^{i\chi_{in}^g(x)} e^{-i \int_C d^4 y \mathcal{L}_{in}^{I\,f}(y)} \right\}, \qquad (7.228)$$

where the index f indicates that the asymptotic fields entering the Lagrangian are the shifted ones. The field $\psi^{f,g}(x)$ satisfies the same Heisenberg equations as the field $\psi(x)$. A particular choice of any set of solutions of Eqs. (7.226), can lead to the description of the corresponding physical situation. Let us now consider the order parameter given by

$$\langle 0| \psi^{f,g}(x) |0\rangle \equiv e^{ig(x)} F_f(x) \qquad (7.229)$$

$$= e^{ig(x)} \langle 0| T_C \left\{ (\rho_{in}^f(x) + v) e^{i\chi_{in}(x)} e^{-i \int_C d^4 y \mathcal{L}_{in}^{I\,f}(y)} \right\} |0\rangle.$$

By taking the vacuum expectation value of the Heisenberg equations for $\psi^{f,g}(x)$, in the tree approximation we obtain the vortex equations [381]

$$\left[\partial^2 - (\partial_\mu g(x))^2 + m^2 + \lambda F_f^2(x)\right] F_f(x) = 0, \qquad (7.230)$$

$$\partial_\mu \left[F_f^2(x) \partial^\mu g(x)\right] = 0, \qquad (7.231)$$

where $F_f(x) \equiv F[f, x]$ is a functional of $f(x)$. Note that the dependence on $g(x)$ is contained in the phase only. The static vortex along the third axis is obtained by taking $F_f(x)$ time-independent and with a radial dependence only. Eq. (7.231) then reduces to the Laplace equation which admit as a solution the polar angle, i.e.,

$$g(x) = n\theta(x) = n \arctan\left(\frac{x_2}{x_1}\right), \qquad (7.232)$$

where the integer n guarantees the single valuedness of $\langle 0| \psi^{f,g}(x) |0\rangle$.

Substituting Eq. (7.232) into Eq. (7.230), we find that $F_f(r)$ fulfils the following (static) equation:

$$\left[\partial_r^2 + \frac{1}{r}\partial_r - \frac{n^2}{r^2} - m^2\right] F_f(r) = \lambda F_f^3(r). \qquad (7.233)$$

One expects that the order parameter will be constant far away from the vortex core. Thus for large r the solution of Eq. (7.233) must asymptotically approach a (finite) constant value a_0. So for large r we may write

$$F(r) = a_0 + \frac{a_1}{r} + \frac{a_2}{r^2} + \dots . \tag{7.234}$$

Inserting this expansion into (7.233) we may identify the coefficients a_i. It is simple to check that the asymptotic behavior of $F_f(r)$ reads

$$F_f(r \to \infty) = v - \frac{n^2}{2m\sqrt{\lambda}\, r^2} + \dots . \tag{7.235}$$

The short distance behavior of $F_f(r)$ is dominated by the first three terms on the l.h.s. of Eq. (7.233). As a result, Eq. (7.233) reduces to the Bessel equation. Because the order parameter must vanish on the normal phase (i.e., inside the vortex) $F_f(r)$ must fulfil the boundary condition $F_f(0) = 0$. The latter indicates that the Bessel equation has a solution of the form

$$F_f(r \to 0) \propto J_n(r) \approx r^n . \tag{7.236}$$

Let us now consider Eqs. (7.227) for the shift functions. Since we want to consider a system containing a vortex, we know that $g(x)$ has to be of the form given in Eq. (7.232). Thus, for the static case with $h(x)$ depending only on the radial coordinate, we get the following equation for $f(r)$:

$$\left[\partial_r^2 + \frac{1}{r}\partial_r - \frac{n^2}{r^2} - m^2\right] f(r) = v\frac{n^2}{r^2} . \tag{7.237}$$

It admits a general solution in terms of modified Bessel functions I and K. The boundary conditions are such that the solution vanishes for $r \to \infty$. The solution of the associated homogeneous equation satisfying this condition is $K_n(mr)$. The Green's function, satisfying the same boundary conditions and being finite in $r = 0$, is given by $G(r_1, r_2) = I_n(mr_<)K_n(mr_>)$, where $r_{>,<} \equiv \pm|r - r_1|$. Solutions for f and g are then given by

$$\tilde{g}(x) = n\,\theta(x) = n \arctan\left(\frac{x_2}{x_1}\right) , \tag{7.238a}$$

$$\tilde{f}(r) = K_n(mr) + vn^2 \int \frac{dr_1}{r_1^2} I_n(mr_<)K_n(mr_>) , \tag{7.238b}$$

and the Heisenberg operator in the presence of a vortex is:

$$\psi^{\tilde{f},\tilde{g}}(x) = e^{in\theta(x)} T_C\left\{\left(\rho_{in}(x) + \tilde{f}(x) + v\right) e^{i\chi_{in}(x)} \exp\left[-i \int_C d^4 y \mathcal{L}_{in}^{I\,\tilde{f}}(y)\right]\right\} . \tag{7.239}$$

By taking the vev of (7.239) in tree approximation, we get the vortex solution

$$\langle 0|\psi^{vor}(x)|0\rangle \simeq (K_n(mr(x)) + v)\, e^{in\theta(x)}\,. \qquad (7.240)$$

One of the advantages of the formalism here presented is in the fact that it is straightforward (at least in principle) to consider Green's functions of the ψ field in the presence of vortices, since we have obtained the Heisenberg field operator (7.239) which "know" about the appropriate boundary conditions.

In the simplest approximation, valid when we are very far from the vortex core, we can use $f = 0$. Then an approximate relation between the two Heisenberg field operators in the presence and absence of vortex is

$$\psi^\theta(x) \simeq e^{in\theta(x)}\, \psi(x)\,. \qquad (7.241)$$

Let us consider the two-point function in the presence of a vortex (x and y are far from the vortex core). We have:

$$G^\theta(x,y) \equiv \langle 0|\psi^\theta(x)\psi^{\theta*}(y)|0\rangle \simeq e^{in[\theta(x)-\theta(y)]}\, G(x - y)\,. \qquad (7.242)$$

We then have a relation between the (full) two-point Green's function $G^\theta(x,y)$ in presence of a vortex in terms of the (full) two-point Green's function $G(x - y)$ in absence of vortices and the angle function $\theta(x)$.

Such a result presents some analogies with the Bohm–Aharonov effect [11]. In this last case, the presence of the magnetic flux confined into the solenoid, induces a change in the two-point Green's function of the electrons propagating around it [555] and the relation between the Green's function with and without magnetic flux is formally identical to Eq. (7.242).

Indeed, let us consider the Green's function for electrons in the presence of a non-zero vector potential. If we denote by G_0^c the Green's function in the covering space for $\mathbf{A} = 0$ and by $G_\mathbf{A}^c$ the one for $\mathbf{A} = \nabla\Omega \neq 0$, with Ω the 'angle' gauge function, then we have

$$G_0^c(x,y) = e^{ie[\Omega(x)-\Omega(y)]}\, G_\mathbf{A}^c(x,y)\,. \qquad (7.243)$$

On the other hand, comparing with Eq. (7.242), we note that the phase $\theta(x)$ is related to the superfluid velocity as $\mathbf{v}_s \sim \nabla\theta(x)$ and $\mathbf{v}_s$ can be interpreted as the magnetic field.

The Bohm–Aharonov potential induced by a magnetic flux equal to n times the elementary magnetic flux is

$$\mathbf{A} = \nabla\Omega \quad, \quad \oint d\mathbf{s} \cdot \mathbf{A} = -2\pi n/e\,. \qquad (7.244)$$

In the case of a vortex line of strength n we also have, beside Eq. (7.242):

$$\mathbf{J} = n\nabla\theta \quad, \quad \oint d\mathbf{s} \cdot \mathbf{J} = 2\pi n\,. \qquad (7.245)$$

Thus the correspondence reads $n\theta(x) \leftrightarrow -e\Omega(x)$ and $\mathbf{J} \leftrightarrow -e\mathbf{A}$.

Chapter 8

Topological defects as non-homogeneous condensates. II

8.1 Introduction

We have seen in the previous Chapter how extended objects are generated in QFT by means of localized (non-homogeneous) condensation of quanta, controlled by suitable shifts of the associated asymptotic fields (boson transformation). We have studied in detail the case of a scalar field theory, both in $1 + 1$ and $3 + 1$ dimensions, with the appearance of kinks and vortices, respectively.

In the present Chapter we treat the case of topological defects in theories with gauge fields and SSB. In this case, as seen in Chapter 3, Goldstone bosons disappear from the spectrum of the observable fields and their role is to characterize the vacuum structure. In this respect, the significance of the singularities (such as topological singularities) of the boson transformation function for the Goldstone fields acquires particular relevance, since singular boson transformations cannot be "gauged away" and topologically non-trivial solitonic structures may appear. Furthermore, in Chapters 3 and 4 we have seen that the gauge field acquires a mass (the Higgs mechanism) and classical field equations (Maxwell equations) emerge out of the NG boson condensation. Macroscopic fields and currents have been derived in terms of boson condensation and the interaction of quanta with classically behaving extended objects has been included in the S-matrix.

In this Chapter we consider the vortex in $U(1)$ gauge theory, the $SU(2)$ instanton solution, the monopole and the sphaleron solutions. Temperature effects on soliton solutions will also be considered. In our analysis we adopt the covariant operator formalism for the quantization of non-Abelian gauge theories. Details of this approach can be found in [392, 488].

8.2   Vortices in $U(1)$ local gauge theory

In order to discuss the vortex solution [191, 498] we will first briefly summarize the discussion presented in Sections 3.8 and 4.3. The boson condensation function is related to Green's functions and it leads to specific solutions such as the static vortex, the straight infinitely long vortex and a circular loop.

The vortex solutions are objects of special interest since they appear in many condensed matter systems such as the vortex in superconductivity [4, 282] or the filamentary structure in biological systems [194]. The vortex solution also provides an example of bosonic string [296] and cosmic string [371] (cf. also Chapter 10).

We thus return to the Goldstone–Higgs-type model considered in Section 3.8. There we have seen that in the broken phase, the local gauge transformations of the Heisenberg fields

$$\phi_H(x) \rightarrow e^{ie_0\lambda(x)}\phi_H(x)\,, \tag{8.1a}$$

$$A_H^\mu(x) \rightarrow A_H^\mu(x) + \partial^\mu\lambda(x)\,, \tag{8.1b}$$

$$B_H(x) \rightarrow B_H(x)\,, \tag{8.1c}$$

are induced by the in-field transformations

$$\chi_{in}(x) \rightarrow \chi_{in}(x) + e_0\tilde{v}\,Z_\chi^{-\frac{1}{2}}\lambda(x)\,, \tag{8.2a}$$

$$b_{in}(x) \rightarrow b_{in}(x) + e_0\tilde{v}\,Z_\chi^{-\frac{1}{2}}\lambda(x)\,, \tag{8.2b}$$

$$\rho_{in}(x) \rightarrow \rho_{in}(x)\,, \tag{8.2c}$$

$$U_{in}^\mu(x) \rightarrow U_{in}^\mu(x)\,. \tag{8.2d}$$

The notation here is the same as in Section 3.8. Z_χ denotes the wavefunction renormalization constant and $\tilde{v} = \langle 0|\phi_H(x)|0\rangle$. As usual, we use the indexes "H" and "in" for the Heisenberg and the asymptotic or quasiparticle fields, respectively. The global transformation $\phi_H(x) \rightarrow e^{i\theta}\phi_H(x)$ is induced by

$$\chi_{in}(x) \rightarrow \chi_{in}(x) + \theta\,\tilde{v}\,Z_\chi^{-\frac{1}{2}}f(x)\,, \tag{8.3a}$$

$$b_{in}(x) \rightarrow b_{in}(x)\,, \tag{8.3b}$$

$$\rho_{in}(x) \rightarrow \rho_{in}(x)\,, \tag{8.3c}$$

$$U_{in}^\mu(x) \rightarrow U_{in}^\mu(x)\,, \tag{8.3d}$$

with $\partial^2 f(x) = 0$ and the limit $f(x) \to 1$ to be performed at the end of the computation. Under the above in-field transformations the in-field equations and the S matrix are invariant and B_H is changed by an irrelevant c-number (in the limit $f \to 1$).

Consider now the boson transformation $\chi_{in}(x) \to \chi_{in}(x) + f(x)$, where $f(x)$ is a c-number function which is solution of the $\chi_{in}(x)$ motion equation. We have seen that the boson transformation theorem can be proven to be valid also when a gauge field is present. In such a case, any spacetime dependence of the ϵ-term can be eliminated by a gauge transformation when $f(x)$ is a regular (i.e., Fourier transformable) function and the only effect is the appearance of a phase factor in the order parameter: $\tilde{v}(x) = e^{icf(x)}\tilde{v}$, with c being a constant (cf. Section 4.3). The conclusion is that when a gauge field is present, the boson transformation with regular $f(x)$ is equivalent to a gauge transformation.

On the other hand, we have also noted the difference with the case of a theory with global invariance only (cf. Eq. (4.84)). There, non-singular boson transformations of the NG fields can produce non-trivial physical effects (like linear flow in superfluidity).

Let us consider now the case of singular $f(x)$ and see how specific vortex solutions are generated. By resorting to the results of Sections 4.3.1 and 4.3.2, we have [392, 452, 619, 650, 651]

$$\partial_\mu f(x) = 2\pi \int d^4x' \, G^\dagger_{\mu\nu}(x')\partial^\nu_x K(x - x'), \qquad (8.4)$$

$$K(x - x') = -\frac{1}{(2\pi)^4} \int d^4p \, e^{-ip(x-x')} \frac{1}{p^2 + i\epsilon}, \qquad (8.5)$$

where the Green's function K satisfies $\partial^2 K(x - x') = \delta(x - x')$ and $G^\dagger_{\mu\nu}$ is given by Eq. (4.85). Upon contour integration, Eq. (8.4) gives

$$f(x) = 2\pi \int dx^\mu \int d^4x' \, G^\dagger_{\mu\nu}(x')\partial^\nu_x K(x - x'), \qquad (8.6)$$

which is indeed solution of $\partial^2 f(x) = 0$. The massive vector potential is obtained from the classical (Proca) Eq. (4.69)

$$a_\mu(x) = -\frac{m_V^2}{e} \int d^4x' \, \Delta_c(x - x')\partial'_\mu f(x'), \qquad (8.7)$$

$$\Delta_c(x - x') = \frac{1}{(2\pi)^4} \int d^4p \, e^{-ip(x-x')} \frac{1}{p^2 - m_V^2 + i\epsilon}. \qquad (8.8)$$

The electromagnetic tensor and the vacuum current are [452, 619, 650, 651]

$$F_{\mu\nu}(x) = \partial_\mu a_\nu(x) - \partial_\nu a_\mu(x) = 2\pi \frac{m_V^2}{e} \int d^4x' \, \Delta_c(x-x') G_{\mu\nu}^\dagger(x'), \quad (8.9)$$

$$j_\mu(x) = \frac{m_V^2}{e} \partial^2 \int d^4x' \, \Delta_c(x-x') \partial_\mu' f(x')$$

$$= -2\pi \frac{m_V^2}{e} \int d^4x' \, \Delta_c(x-x') \partial_{x'}^\nu G_{\nu\mu}^\dagger(x'), \quad (8.10)$$

respectively, and satisfy $\partial^\mu F_{\mu\nu}(x) = -j_\nu(x)$.

The above preliminaries are sufficient to consider specific vortex solutions. The line singularity for the vortex solution can be parametrized by a single line parameter σ and by the time parameter τ. We consider the following cases.

The *static vortex* solution. This is obtained by setting $y_0(\tau, \sigma) = \tau$ and $\mathbf{y}(\tau, \sigma) = \mathbf{y}(\sigma)$, with $\mathbf{y}$ denoting the line coordinate. $G_{\mu\nu}^\dagger$ is non-zero only on the line at $\mathbf{y}$ (we can consider more lines but here we limit ourselves to one line, for simplicity). Thus, we have:

$$G_{0i}(x) = \int d\sigma \frac{dy_i(\sigma)}{d\sigma} \delta^3[\mathbf{x} - \mathbf{y}(\sigma)], \qquad G_{ij}(x) = 0, \quad (8.11)$$

$$G_{ij}^\dagger(x) = -\epsilon_{ijk} G_{0k}(x), \qquad G_{0i}^\dagger(x) = 0. \quad (8.12)$$

Eq. (8.9) shows that these vortices are purely *magnetic*. Eq. (8.4) gives

$$\partial_0 f(x) = 0,$$

$$\partial_i f(x) = \frac{1}{(2\pi)^2} \int d\sigma \, \epsilon_{ijk} \frac{dy_k(\sigma)}{d\sigma} \partial_j^x \int d^3p \frac{e^{i\mathbf{p}\cdot(\mathbf{x}-\mathbf{y}(\sigma))}}{\mathbf{p}^2}, \quad (8.13)$$

i.e., by using the identity $(2\pi)^{-2} \int d^3p \frac{e^{i\mathbf{p}\cdot\mathbf{x}}}{\mathbf{p}^2} = \frac{1}{2|\mathbf{x}|}$,

$$\nabla f(x) = -\frac{1}{2} \int d\sigma \frac{d\mathbf{y}_k(\sigma)}{d\sigma} \wedge \nabla_x \frac{1}{|\mathbf{x} - \mathbf{y}(\sigma)|}. \quad (8.14)$$

Note that $\nabla^2 f(x) = 0$ is satisfied.

The *straight infinitely long vortex*. It is specified by $y_i(\sigma) = \sigma \delta_{i3}$ with $-\infty < \sigma < \infty$. The only non-vanishing component of $G^{\mu\nu}(x)$ are $G^{03}(x) = G_{12}^\dagger(x) = \delta(x_1)\delta(x_2)$. Eq. (8.14) gives [452, 619, 650, 651]

$$\frac{\partial}{\partial x_1} f(x) = \frac{1}{2} \int d\sigma \frac{\partial}{\partial x_2} [x_1^2 + x_2^2 + (x_3 - \sigma)^2]^{-\frac{1}{2}} = -\frac{x_2}{x_1^2 + x_2^2}, \quad (8.15)$$

$$\frac{\partial}{\partial x_2} f(x) = \frac{x_1}{x_1^2 + x_2^2}, \qquad \frac{\partial}{\partial x_3} f(x) = 0, \quad (8.16)$$

which give

$$f(x) = \arctan\left(\frac{x_2}{x_1}\right) = \theta(x).\tag{8.17}$$

Use of these results gives a_μ, $F_{\mu\nu}$ and the vacuum current. The only non-zero components of these fields are a_1, a_2, $F_{12} = -H_3$, j_1 and j_2.

The condition (4.90) can be shown to be violated if the line singularity has isolated end points inside the system [619]. Thus the consistency with the continuity equation (4.90) implies that either the string is infinite, or that it forms a closed loop. If there is more than one string, the end points of different strings can be connected in a vertex. Eq. (4.90) results then in a condition for the relative string tensions ν_α, with α denoting different strings.

Further simple examples are the following [452, 619, 650, 651]:

A *circular loop* given by $\mathbf{y}(\sigma) = (a\cos\sigma, a\sin\sigma, 0)$, with $0 \le \sigma \le 2\pi$. Then

$$G^{01}(x) = \delta\left[x_2 - \sqrt{a^2 - x_1^2}\right]\delta(x_3),\tag{8.18a}$$

$$G^{02}(x) = \delta\left[x_1 - \sqrt{a^2 - x_2^2}\right]\delta(x_3),\tag{8.18b}$$

$$G^{03}(x) = G^{ij}(x) = 0.\tag{8.18c}$$

A *straight line* along the third axis moving in the x_1 direction with velocity v is given by $\mathbf{y}(\sigma, \tau) = (v\tau, 0, \sigma)$, $y_0(\sigma, \tau) = \tau$, from which

$$G^{03}(x) = v\,\delta(x_1 - vt)\,\delta(x_2),\tag{8.19a}$$

$$G^{13}(x) = v\,\delta(x_1 - vt)\,\delta(x_2).\tag{8.19b}$$

Vortex solutions in $3 + 1$-dimensional Abelian Higgs model (Nielsen–Olesen vortex solutions) and cosmic strings are discussed in Sections 10.4.1 and 10.5.2, respectively.

8.3 Topological solitons in gauge theories

By following closely the presentation of [440], we now consider some examples of gauge theories, also including the effect of temperature on the extended objects generated by boson transformation.

We have already considered temperature effects on boson condensates in Chapter 5. In this Section we focus on the case of the vortex in scalar electrodynamics, the monopole in $SO(3)$ non-Abelian gauge theory and the sphaleron solution in the $SU(2)$ version of this theory. We will see that a critical temperature T_c exists at which symmetry restoration is obtained. The Higgs field condensation, which we will show participates to the formation of the topological soliton together with the NG bosons, is temperature-dependent and vanishes at the phase transition point T_c. The vortex shape depends on the temperature through the Higgs field condensation and the vortex core grows as temperature rises. Both monopoles and sphaleron have temperature-dependent "magnetic" charges. The monopole magnetic charge decreases with increasing temperature, whereas the sphaleron magnetic charge increases. At $T = 0$ it is zero.

In the framework of the covariant operator quantization approach to gauge theories [392, 488], we consider the following Lagrangian density for a complex scalar field $\phi(x)$ coupled to a gauge field:

$$\mathcal{L}_c = -\frac{1}{4}F^a_{\mu\nu}F^{a\mu\nu} + (D_\mu\phi)^\dagger_i(D^\mu\phi)_i - U(\phi)\,, \tag{8.20}$$

where

$$F^a_{\mu\nu} = \partial_\mu A^a_\nu - \partial_\nu A^a_\mu + gf_{abc}A^b_\mu A^c_\nu\,, \tag{8.21a}$$

$$(D_\mu\phi)_i = \partial_\mu\phi_i - ig(A_\mu)_{ij}\phi_j\,, \tag{8.21b}$$

$$(A_\mu)_{ij} = A^a_\mu T^a_{ij} \in g\,. \tag{8.21c}$$

$(A_\mu)_{ij}$ is an element of the gauge group Lie algebra g. G will denote the gauge group. The potential $U(\phi)$ is

$$U(\phi) = \lambda\left(\phi^\dagger_i\phi_i - \frac{1}{2}v^2\right)^2\,. \tag{8.22}$$

The classical Euler–Lagrange equations are:

$$-\partial_\mu F^{\mu\nu a} = gf_{abc}A^b_\mu F^{c\mu\nu} + ig\left(\phi^\dagger T^a D^\nu\phi - (D^\nu\phi)^\dagger T^a\phi\right)\,, \tag{8.23}$$

$$(D_\mu D^\mu\phi)_i = -2\lambda\left(\phi^\dagger_j\phi_j - \frac{1}{2}v^2\right)\phi_i\,. \tag{8.24}$$

Such equations have (classical) soliton solutions. In order to proceed with quantization, it is necessary to add to $\mathcal{L}_c$ a gauge fixing and a ghost term:

$$\mathcal{L} = \mathcal{L}_c + \mathcal{L}_{gf} + \mathcal{L}_{gh}\,. \tag{8.25}$$

We represent $\phi_i(x)$ in terms of $\chi(x) = \chi^a(x)T^a$ and of the Higgs field $\bar\rho$:

$$\phi_i(x) = \frac{1}{\sqrt{2}}\left(e^{ig\chi(x)}\right)_{ij} n_j(\bar\rho(x) + \sigma)\,, \tag{8.26}$$

where n_j is an arbitrary unit vector ($\mathbf{n} \cdot \mathbf{n} = 1$). In Eqs. (8.24) and (8.26) v and σ are c-number quantities. The gauge fixing term is:

$$\mathcal{L}_{gf} = -\partial_\mu B^a A^{a\mu} + \frac{1}{2}\xi B^a B^a + E^a B^a, \qquad (8.27)$$

with ξ being an arbitrary non-vanishing constant and B^a is the auxiliary field. The ghost field is denoted by $c(x)$ and the ghost term is [440]:

$$\mathcal{L}_{gh} = -\partial_\mu \bar{c}^a D^{\mu ab} c^b + ig^2\sigma^2 g_{abc} c^a c^b. \qquad (8.28)$$

In these equations

$$E^a = g^2 \xi \sigma^2 g^{ab} \chi^b, \qquad (8.29a)$$

$$g_{ab} = \frac{1}{2}\left(n\left\{T^a, T^b\right\} n\right). \qquad (8.29b)$$

$\mathcal{L}_c$ is invariant under the following (gauge) transformation:

$$\bar{\rho} \to \bar{\rho}' = \rho, \qquad (8.30a)$$

$$\chi^a \to \chi'^a = \chi^a + \alpha^a, \qquad (8.30b)$$

$$A_\mu^a \to A_\mu'^a = A_\mu^a + D_\mu^{ab}\alpha^b + \dots, \qquad (8.30c)$$

with $D_\mu^{ab} = \delta_{ab}\partial_\mu - gf_{abc}A_\mu^c$. From the gauge fixing term $\mathcal{L}_{gf}$, Eq. (8.27), we obtain the condition

$$B^a = -\frac{1}{\xi}\left(\partial_\mu A^{a\mu} + E^a\right), \qquad (8.31)$$

which in general breaks the gauge invariance. However, the total Lagrangian Eq. (8.25) is invariant under the BRST transformation [343]:

$$A_\mu^a \to A_\mu'^a = A_\mu^a + \theta D_\mu^{ab} c^b, \qquad (8.32a)$$

$$B^a \to B'^a = B^a, \qquad (8.32b)$$

$$\bar{\rho} \to \bar{\rho}' = \rho, \qquad (8.32c)$$

$$\chi^a \to \chi'^a = \chi^a + \theta c^a, \qquad (8.32d)$$

$$c^a \to c'^a = c^a - \frac{1}{2}\theta g f_{abc} c^b c^c, \qquad (8.32e)$$

$$\bar{c}^a \to \bar{c}'^a = \bar{c}^a + i\theta B^a, \qquad (8.32f)$$

where θ is a Grassmann number ($\theta^2 = 0$). Note that the BRST symmetry can be thought as a generalization of the local gauge symmetry on the Grassmannian algebra (with $\alpha^a(x) = \theta c^a(x)$). The associated (asymptotic) BRST charge is

$$Q_{BRST} = \int d^3x \, B^a \overset{\leftrightarrow}{\partial}_0 c^a, \qquad (8.33)$$

with $Q^2_{BRST} = 0$. The subspace of physical states is then defined as:

$$\mathcal{H}_{phys} \equiv \{|\alpha\rangle, \; Q_{BRST}|\alpha\rangle = 0\} \,. \tag{8.34}$$

Such a condition means that the scalar photons, the ghosts c^a and $\bar{c}^a$, and the Goldstone bosons (for $\sigma \neq 0$), do not belong to $\mathcal{H}_{phys}$. We remark that $\mathcal{H}_{phys}$ possess a zero norm subspace. However, this subspace does not have any role in computing the average of the observables (which commute with Q_{BRST}).

We now consider a trial Lagrangian for the effective description of the system in terms of the asymptotic free particle excitation fields $\bar{\rho}$, $\bar{\chi}^a$, $\bar{A}^a_\mu$:

$$\begin{aligned}
\mathcal{L}_0 = {} & -\frac{1}{4}\mathcal{G}^a_{\mu\nu}\mathcal{G}^{a\mu\nu} - \partial_\mu B^a \bar{A}^{a\mu} + \frac{1}{2}B^a B^a + \frac{1}{2}M^2_{ab}(x)\bar{A}^a_\mu \bar{A}^{b\mu} \\
& + \frac{1}{2}\partial_\mu\bar{\rho}\partial^\mu\bar{\rho} - \frac{1}{2}m^2(x)\bar{\rho}^2 + \frac{1}{2}g^2\sigma^2\,\partial_\mu\bar{\chi}^a\partial^\mu\bar{\chi}^a \\
& - \frac{1}{2}g^2\sigma^2 M^2_{ab}(x)\xi\bar{\chi}^a\bar{\chi}^b - i\partial_\mu\bar{c}^a\partial^\mu c^a + iM^2_{ab}(x)\xi\bar{c}^a c^b , \tag{8.35}
\end{aligned}$$

where $\mathcal{G}^a_{\mu\nu} \equiv \partial_\mu\bar{A}^a_\nu - \partial_\nu\bar{A}^a_\mu$. We note that $\mathcal{L}_0$ is not BRST invariant. However, it possesses another supersymmetry with the same asymptotic BRST charge (8.33) [439]. In this sense $\mathcal{L}_0$ does not change the physical subspace $\mathcal{H}_{phys}$. Also note that the spacetime dependence of the mass $m^2(x)$ implies that it behaves rather as a potential term [440].

Next, we assume the following relation between the Heisenberg fields (ρ, χ^a, A^a_μ) and the asymptotic fields $(\bar{\rho}, \bar{\chi}^a, \bar{A}^a_\mu)$:

$$\rho = \bar{\rho} + \sigma \,, \tag{8.36a}$$

$$\chi^a = \bar{\chi}^a + \kappa^a \,, \tag{8.36b}$$

$$A^a_\mu = \bar{A}^a_\mu + \alpha^a_\mu \,. \tag{8.36c}$$

These can be thought as boson transformations and they imply, at the level of states, a change in the vacuum state (boson condensation):

$$|0\rangle \;\; \rightarrow \;\; |0(\sigma, \kappa^a, \alpha^a_\mu)\rangle = U(\sigma)U(\kappa^a)U(\alpha^a_\mu)|0\rangle \,, \tag{8.37}$$

where

$$U(\sigma) = \exp\sum_n(\sigma_n\rho^\dagger_n - \sigma^*_n\rho_n) \,. \tag{8.38}$$

Here, ρ_n and σ_n are the Fourier components of $\bar{\rho}$ and σ. $U(\sigma)$ induces the shift in the $\bar{\rho}$ field: $\bar{\rho} \to U^+(\sigma)\bar{\rho}U(\sigma) = \bar{\rho} + \sigma$. $U(\kappa^a)$ and $U(\alpha^a_\mu)$ are defined and operate in a similar way. The classical fields σ, κ^a and α^a_μ thus appear as the result of boson condensation.

Gauge transformations correspond to a special case of the above boson transformation (8.36). For example, in electrodynamics, we have a gauge transformation for

$$\sigma = 0, \quad \kappa = \alpha(x), \quad \alpha_\mu(x) = \partial_\mu \alpha(x). \tag{8.39}$$

In general, the boson transformations can give rise to non-trivial topological configurations. The asymptotic condition for the gauge field at infinity is then:

$$\alpha^a_{\mu,\infty} = -\frac{i}{g}(\partial_\mu U)U^\dagger \quad \text{with} \quad U = e^{ig\kappa(x)}, \tag{8.40}$$

where the index ∞ denotes the fields at spatial infinity. Moreover, if $\sigma \to \sigma_\infty = \sigma_0$, then

$$D_\mu \phi_{i,\infty} = 0 \quad \text{with} \quad \phi_{i,\infty} = \frac{1}{\sqrt{2}} U_{ij} n_j \sigma_0, \tag{8.41}$$

and the solution at infinity is determined by the potential $U(\phi)$.

In the following, in considering thermal effects, we will use the analysis similar to the one done in Sections 5.7 and 7.3.1. Together with Eq. (5.179), we also have for the gauge field

$$\bar{A}^a_\mu \bar{A}^b_\nu = \; : \bar{A}^a_\mu \bar{A}^b_\nu :_\beta + \langle : \bar{A}^a_\mu \bar{A}^b_\nu : \rangle_0, \tag{8.42a}$$

$$\bar{A}^a_\mu \bar{A}^b_\nu \bar{A}^c_\rho = \; : \bar{A}^a_\mu \bar{A}^b_\nu \bar{A}^c_\rho :_\beta + \bar{A}^a_\mu \langle : \bar{A}^b_\nu \bar{A}^c_\rho : \rangle_0$$
$$+ \; \bar{A}^b_\nu \langle : \bar{A}^a_\mu \bar{A}^c_\rho : \rangle_0 + \bar{A}^c_\rho \langle : \bar{A}^a_\mu \bar{A}^b_\nu : \rangle_0. \tag{8.42b}$$

8.3.1 *Homogeneous boson condensation*

Let us first consider homogeneous condensation of Higgs bosons, with

$$\kappa^a(x) = 0 \quad \Rightarrow \quad U = \mathbb{1}, \tag{8.43a}$$

$$\sigma = \sigma_0 = \text{const}, \tag{8.43b}$$

$$\alpha^a_\mu = 0. \tag{8.43c}$$

σ_0 is assumed to be non-vanishing below a certain temperature T_c. For $T > T_c$ it is assumed to be zero. We obtain

$$\phi_i = \frac{1}{\sqrt{2}} n_i(\bar{\rho} + \sigma), \tag{8.44a}$$

$$A^a_\mu = \bar{A}^a_\mu. \tag{8.44b}$$

Substitute Eqs. (8.44) into the Euler–Lagrange equation (8.23), and take thermal averages with respect to the effective system $\mathcal{L}_0$. We thus obtain

$$\Box \bar{A}^{a\nu} = M^{2a\nu}_{d\mu} \bar{A}^{d\mu} + \dots, \tag{8.45}$$

$$M^{2a\nu}_{d\mu} = g^2 f_{abc} f_{cbd} \left[\langle : \bar{A}^b_\mu \bar{A}^{b\nu} : \rangle_0 - \langle : \bar{A}^b_\rho \bar{A}^{b\rho} : \rangle_0 \delta^\nu_\mu \right]$$
$$+ g^2 g_{ad} \delta^\mu_\nu \left[\sigma_0^2 + \langle : \bar{\rho}^2 : \rangle_0 \right]. \tag{8.46}$$

We consider the case when the gauge field squared mass $M_{d\mu}^{2a\nu}$ is a scalar with respect to spacetime indices, i.e.,

$$\langle : \bar{A}_\mu^b \bar{A}^{b\nu} : \rangle_0 = \frac{1}{4} \delta_\mu^\nu \langle : \bar{A}_\rho^b \bar{A}^{b\rho} : \rangle_0 . \tag{8.47}$$

In this approximation, we obtain:

$$M_{d\mu}^{2a\nu} = \delta_\mu^\nu \left[-\frac{3}{4} g^2 f_{abc} f_{cbd} \langle : \bar{A}_\rho^b \bar{A}^{b\rho} : \rangle_0 + g^2 g_{ad} \left(\sigma_0^2 + \langle : \bar{\rho}^2 : \rangle_0 \right) \right] . \tag{8.48}$$

From the second Euler–Lagrange equation (8.24), we get:

$$\Box \bar{\rho} = \sigma_0 \left[-g^2 g_{ab} \langle : \bar{A}_\mu^a \bar{A}^{b\mu} : \rangle_0 + 3\lambda \langle : \bar{\rho}^2 : \rangle_0 + \lambda(\sigma_0^2 - v^2) \right]$$
$$+ \bar{\rho} \left[-g^2 g_{ab} \langle : \bar{A}_\mu^a \bar{A}^{b\mu} : \rangle_0 + 3\lambda \langle : \bar{\rho}^2 : \rangle_0 + 3\lambda \sigma_0^2 - \lambda v^2 \right] , \tag{8.49}$$

or, equivalently, the Klein–Gordon equation

$$\Box \bar{\rho} = m^2 \bar{\rho} + \dots , \tag{8.50}$$

and the equation

$$\sigma_0 \left[-g^2 g_{ab} \langle : \bar{A}_\mu^a \bar{A}^{b\mu} : \rangle_0 + 3\lambda \langle : \bar{\rho}^2 : \rangle_0 + \lambda(\sigma_0^2 - v^2) \right] = 0 . \tag{8.51}$$

If $\sigma_0 \neq 0$, then

$$m^2 = 2\lambda \sigma_0^2 , \tag{8.52}$$

$$\sigma_0^2 = v^2 - 3\langle : \bar{\rho}^2 : \rangle_0 + \frac{g^2}{\lambda} g_{ab} \langle : \bar{A}_\mu^a \bar{A}^{b\mu} : \rangle_0 . \tag{8.53}$$

We remark that the factor $\frac{1}{\lambda}$ in this equation has a meaning as far as λ is non-vanishing, which expresses the non-perturbative character of the gauge field contribution. This non-perturbative feature is common to all soliton solutions, as already mentioned in previous Chapters. These solutions indeed could not be found if one would obtain them in a perturbative approach by using expansions around the vanishing λ point.

Compare Eqs. (8.52) and (8.53) with Eqs. (7.62) and notice the contribution of the $: \bar{A}_\mu^a \bar{A}^{b\mu} :$ term. Eq. (8.53) allows us to calculate the transition temperature T_c for which $\sigma_{\beta_c} = 0$ (disordered phase). It has to be such that

$$v^2 = 3\langle : \bar{\rho}^2 : \rangle_{\beta_c} - \frac{g^2}{\lambda} g_{ab} \langle : \bar{A}_\mu^a \bar{A}^{b\mu} : \rangle_{\beta_c} . \tag{8.54}$$

For $T > T_c$, it is assumed $\sigma_0 = 0$. The equation for the thermal mass is

$$m^2 = -g^2 g_{ab} \langle : \bar{A}_\mu^a \bar{A}^{b\mu} : \rangle_0 + 3\lambda \langle : \bar{\rho}^2 : \rangle_0 - \lambda v^2 . \tag{8.55}$$

Eqs. (8.54) and (8.55) are both self-consistent equations, because of the dependence of $\langle : \bar{\rho}^2 : \rangle_0$ on m^2. Notice that for $T = 0$, $\sigma_0 = v$, thus recovering the zero temperature symmetry breaking condition.

Finally we note that the above treatment is valid only for second order (continuous) phase transitions. For discontinuous transitions, one must consider the free energy, as shown in [440].

8.3.2 *The vortex of scalar electrodynamics*

We now consider non-homogeneous condensation with space-dependent shift functions. The simplest case of a gauge theory of the form (8.25) is the scalar electrodynamics. In this case, the gauge group is Abelian, $G = U(1)$, and $f_{abc} = 0$, $n = 1$, $g_{ab} = \mathbb{1}$, $g = e$.

Let us introduce cylindrical coordinates, so that:

$$x^1 = r\cos\phi, \quad x^2 = r\sin\phi, \quad x^3 = z, \tag{8.56}$$

and

$$g_{\mu\nu} = \begin{pmatrix} +1 & & & \\ & -1 & & \\ & & -r^2 & \\ & & & -1 \end{pmatrix}. \tag{8.57}$$

Then, the asymptotic gauge field configuration defined in (8.40) gives:

$$U(x) = e^{in\phi}, \quad \alpha_\infty^i = -\frac{n}{er}\mathbf{e}_\phi^i, \tag{8.58}$$

with

$$\mathbf{e}_\phi^i = \frac{1}{r^2}\epsilon_{i31}x^1. \tag{8.59}$$

Here n is the winding number. In the case of the homogeneous condensation U is trivial ($U = 1$) and no topological charge arise. The shift function giving rise to the vortex follows from Eqs. (8.39) and (8.40) as:

$$\kappa(x) = \frac{n}{e}\phi; \quad r \neq 0, \tag{8.60a}$$

$$\kappa(x) = 0; \quad r = 0. \tag{8.60b}$$

The factor $\frac{1}{e}$ in Eqs. (8.56) and (8.60) expresses the non-perturbative character of the solution. Boson condensation is typically a non-perturbative phenomenon, as stressed many times in this book.

From the above, we see that only the condensation of Goldstone bosons is responsible for the appearance of vortices. Although the Goldstone bosons are not present in the spectrum of the physical excitations, nonetheless their (localized) quantum condensation is observable as a vortex.

Let us consider in more detail the vortex solution by making the Ansatz:

$$\alpha^i(x) = -\frac{n}{er}(1 - K(r))\mathbf{e}_\phi^i = -\frac{n}{er^3}\epsilon_{i31}x^1(1 - K(r)), \tag{8.61a}$$

$$\sigma(x) = \sigma_0 f(x), \tag{8.61b}$$

with σ_0 being the shift for the Higgs field for the case of the homogeneous condensation as described by the equations:

$$m^2 = 2\lambda\sigma_0^2 , \tag{8.62a}$$

$$\sigma_0^2 = v^2 - 3\langle : \bar\rho^2 :\rangle_0 + \frac{e^2}{\lambda}\langle : \bar A_\mu \bar A^\mu :\rangle_0 , \tag{8.62b}$$

$$M^2 = e^2(\sigma_0^2 + \langle : \bar\rho^2 :\rangle_0) . \tag{8.62c}$$

Then one obtains [440] the temperature-dependent vortex equations

$$\frac{d}{dr}\left(\frac{1}{r}\frac{dK}{dr}\right) = e^2 K(r)\left[\langle : \bar\rho^2 :\rangle_0 + f^2\sigma_0^2\right] , \tag{8.63a}$$

$$\frac{1}{r}\frac{d}{dr}\left(r\frac{df}{dr}\right) = n^2 K^2(r)f - \lambda\sigma_0^2 f(1 - f^2)r^2 . \tag{8.63b}$$

Introducing $K(r) = 1 - \frac{r}{n}A(r)$, with $A(r) \to n/r$ for $r \to \infty$ (and $\alpha^i(x) \to \alpha^i_\infty(x)$), the above equations assume the form

$$\frac{d}{dr}\left(\frac{1}{r}\frac{d}{dr}(rA)\right) = e^2\left(A - \frac{n}{r}\right)(\langle : \bar\rho^2 :\rangle_0 + f^2\sigma_0^2) , \tag{8.64a}$$

$$\frac{1}{r}\frac{d}{dr}\left(r\frac{df}{dr}\right) = \left(A - \frac{n}{r}\right)^2 f - \lambda\sigma_0^2 f(1 - f^2) , \tag{8.64b}$$

which, for $T \to 0$, reduce to the well-known vortex equations.

We note (see also Section 8.2) that the vortices arise as the result of two types of boson condensates: the Higgs boson condensate controlled by $\sigma(x) \neq 0$, which gives the classical vortex envelope, and the Goldstone boson condensate induced by $K(x) \neq 0$, which is responsible for the (non-zero) topological charge. The vortex shape depends on the temperature through σ_0, which vanishes at $T = T_c$; in a first approximation, the vortex size goes as m^{-2} which means that the vortex core grows as temperature rises [440]. In fact, in the case of non-homogeneous condensation, the "masses" $m(x)$ of the Higgs field and $M(x)$ of the gauge field play the role of potentials:

$$M^2(x) = e^2\left[\langle : \bar\rho^2 :\rangle_0 + \sigma_0^2 f^2(x)\right] ; \tag{8.65a}$$

$$m^2(x) = 2\lambda\sigma_0^2 f^2(x) , \tag{8.65b}$$

and only in the limit $r \to \infty$ (when $f(x) \to 1$), they can be interpreted as masses. Thus we have the asymptotic (at spatial infinity) behavior:

$$K(r) \simeq e^{-Mr} = e^{-\frac{r}{R_0}} , \tag{8.66a}$$

$$f(r) \simeq 1 - f_0 e^{-mr} = 1 - f_0 e^{-\frac{r}{r_0}} . \tag{8.66b}$$

Here, $R_0 = 1/M$ is the size of the gauge field core and $r_0 = 1/m$ the Higgs field core. As $T \to T_c$ the Higgs field core increases but the gauge field core diminishes. At $T = T_c$ (i.e., $\sigma_0 = 0$) we have a pure gauge field core. Above T_c the gauge symmetry is restored, and the gauge field has only two physical polarizations. Notice, however, that as $T \to T_c$ (and $\sigma_0 \to 0$), since $\langle : \bar{\rho}^2 : \rangle_0 \to \langle : \bar{\rho}^2 : \rangle_{\beta_c} \neq 0$ we have an effect which is a remnant of symmetry breaking, namely the solution does not reduce to the one for $v = 0$. In a similar way, above the critical temperature T_c, where $\sigma_0 = 0$ is assumed, since the gauge field mass gets contributions from thermal averages (indeed in general $\langle : \bar{\rho}^2 : \rangle_{\beta > \beta_c} \neq 0$), the symmetry is "globally" restored (i.e., $m^2 = 0$), but not necessarily "locally" restored (i.e., $M^2 \neq 0$). In other words we could have local (unstable) domains where symmetry is broken even above T_c. We thus can have (transient) domain (bubble) structures for the vacuum state: non-vanishing gauge field mass signals a hysteresis phenomenon similar to the one observed in ferromagnets. In some sense, low temperature phase bubbles remain embedded (trapped) in the high temperature phase. Considering the vortex solution, this means that the magnetic field inside the vortex core may not be vanishing even above the phase transition point: this is a memory mechanism since the persistence of the string-like structure is reminiscent of the symmetry broken phase. The implications of such a phenomenon may be important in the problem of galaxies formation and cosmic strings scenarios [371].

We close this subsection by noticing that the singularity of $K(x)$ on the z axis is crucial in obtaining the vortex solution. The NG boson condensate $K(x)$ for $r \neq 0$ defines the homotopic mapping π of S^1 surrounding the $r = 0$ singularity to the group manifold of $U(1)$. This mapping is topologically characterized by the "winding number" $n \in \mathbb{Z} \in \pi_1(S^1)$ (see Chapter 10). The identity map ($n = 0$) corresponds to the perturbative vacuum (the homogeneous case). As a result we have the flux quantization

$$\Phi = \int d^3 x B_3(x, y) = \oint \mathbf{A} \cdot dl = -\frac{2\pi n}{e}, \qquad (8.67)$$

where

$$B_3 = \frac{1}{er} \frac{d}{dr}(rA) = -\frac{n}{er} \frac{dK}{dr}. \qquad (8.68)$$

8.3.3   *The 't Hooft–Polyakov monopole*

Though problematic and doubtful, the monopole existence seems to be relevant to an understanding of the evolution of the early Universe [306, 376, 425]. Together with the sphaleron solution considered in the following subsection, the presence of monopoles in a gauge theory leads to barion number violation [39]. Moreover, some defects in liquid crystals appear to be describable in terms of monopole-like structures [128, 395]. As a simple example of non-Abelian gauge field theory with monopole solution, we now consider 't Hooft–Polyakov monopole [524, 587], i.e., $G = SO(3)$.

The group structure constant is $f_{abc} = \epsilon_{abc}$ and the scalar matter field is in the adjoint representation $(T^a)_{ij} = -i\epsilon_{iaj}$. By choosing the unit vector $\mathbf{n}$ as $\mathbf{n} = (0, 0, 1)^T$, we get the g_{ab} matrix of the form:

$$g_{ab} = \begin{pmatrix} 1 & 0 & 0 \\ 0 & 1 & 1 \\ 0 & 0 & 0 \end{pmatrix}. \tag{8.69}$$

The mass for the asymptotic gauge field $\bar{A}_\mu^a$ is:

$$M_d^{2a} = -\frac{3}{4}g^2 \sum_{b \neq a} \langle: \bar{A}_\mu^b \bar{A}^{b\mu} :\rangle_0 \delta_d^a + g^2 g_{ad}(\sigma^2 + \langle: \bar{\rho}^2 :\rangle_0). \tag{8.70}$$

For $T \to 0$, we have:

$$M^2 = g^2 \sigma^2 \quad \text{for} \quad a = 1, 2, \tag{8.71a}$$

$$M^2 = 0 \quad \text{for} \quad a = 3. \tag{8.71b}$$

With our choice of $\mathbf{n}$, this implies that $\bar{A}_\mu^3 = n^a \bar{A}_\mu^a$ is massless, while the transversal components $\bar{A}_\mu^a(a = 1, 2)$ are massive. The same is true for arbitrary $\mathbf{n}$ and in particular for the case of a monopole configuration: $n^a = x^a/r$.

The mass of the Higgs field is (see Eqs. (8.52)–(8.53)):

$$m^2 = 2\lambda \sigma_0^2, \tag{8.72a}$$

$$\sigma_0^2 = v^2 - 3\langle: \bar{\rho}^2 :\rangle_0 + \frac{g^2}{\lambda} \sum_a \frac{2}{3} \langle: \bar{A}_\mu^a \bar{A}^{a\mu} :\rangle_0. \tag{8.72b}$$

The matrix U responsible for the boson condensation (see Eq. (8.37)) is now a rotation matrix:

$$U = e^{ig \sum_a K^a T^a} = U_3(\phi) U_1(\vartheta) U_3(\psi) = U(\eta) U_3(\psi), \tag{8.73}$$

$$U(\eta) = e^{(\eta T_+ - T_- \eta^*)}; \quad T_\pm = T_1 \pm iT_2; \quad \eta = \vartheta e^{i\phi}. \tag{8.74}$$

Let us denote by $H = SO(2) = \{U_3(\psi), U_3(\psi)\mathbf{n} = \mathbf{n}\}$ the little group with respect to the vector $\mathbf{n}$. On the other hand, $U\mathbf{n}$ belongs to the coset $G/H = SO(3)/SO(2) = \mathcal{S}^2$. The angles ϑ, ϕ may then be interpreted as the Goldstone boson condensates. We have:

$$(Un)_i = (U_3(\phi)U_1(\vartheta)n)_i = n_i(\vartheta, \phi) = \frac{x^i}{r}, \tag{8.75}$$

where x^i are expressed in spherical coordinates:

$$(x^1, x^2, x^3) = (r\sin\vartheta\cos\phi, r\sin\vartheta\sin\phi, r\cos\vartheta). \tag{8.76}$$

The interpretation of the Goldstone fields ϕ, ϑ as spatial coordinates leads to a mapping of the sphere $\mathcal{S}^2$, surrounding the singularity at $r = 0$, to $G/H = \mathcal{S}^2$. Thus a non-zero topological charge arises, associated to $\pi_2(\mathcal{S}^2) = \mathbb{Z}$ (see Chapter 10).

The asymptotic gauge field configuration defined in Eq. (8.41) can be obtained by means of the matrix U of Eq. (8.73), with the result:

$$\alpha^a_{i,\infty} = -\epsilon_{aik}\frac{x^k}{gr^2}. \tag{8.77}$$

Thus the Euler–Lagrange equations (8.23), (8.24) can be studied by means of the 't Hooft–Polyakov Ansatz [524, 587]:

$$\sigma = \frac{1}{gr}H(r), \tag{8.78a}$$

$$\alpha^a_i = -\epsilon_{aik}\frac{x^k}{gr^2}[1 - K(r)]. \tag{8.78b}$$

The field α^a_i is a classical gauge field that can be interpreted as a result of the NG boson condensation. Notice that, since $\alpha^a_i n^a = 0$ (i.e., the classical gauge field is always tangent to the sphere $\mathcal{S}^2$), such a condensation should involve only the massive gauge field, while the massless gauge field $\bar{A}_\mu = \bar{A}^a_\mu n^a$ does not condensate.

The finite temperature field equations are then:

$$r^2 K'' = K(K^2 - 1) + r^2 M^2(x)K + \frac{1}{2}g^2 r^2 \sum_b \langle: \bar{A}^b_\mu \bar{A}^{b\mu} :\rangle_0, \tag{8.79a}$$

$$r^2 H'' = 2K^2 H + \frac{\lambda}{g^2}\left(H^2 - g^2 r^2 \sigma_0^2\right)H, \tag{8.79b}$$

where

$$m^2(x) = 2\lambda\sigma^2 = 2\lambda\frac{H^2(x)}{g^2 r^2}, \tag{8.80a}$$

$$M^2(x) = g^2\left(\langle: \bar{\rho}^2 :\rangle_0 + \frac{H^2(x)}{g^2 r^2}\right), \tag{8.80b}$$

are the Higgs field and gauge field masses, respectively. In the zero-temperature limit, the above equations reproduce the well-known 't Hooft–Polyakov monopole equations [524, 587].

Similarly to the vortex case studied above, the Higgs boson envelope giving the monopole vanishes at the critical temperature T_c.

The behavior of the functions K and H at spatial infinity is

$$H(x) \to g r \sigma_0 , \tag{8.81a}$$

$$K(r) \to K_\infty = -\frac{3}{4} \frac{g^2}{M^2} \sum_b \frac{2}{3} \langle : \bar{A}_\mu^b \bar{A}^{b\mu} : \rangle_0 < 1 . \tag{8.81b}$$

Considering the classical asymptotic gauge field at finite temperature, we find:

$$\alpha_{i,\infty}^a (T \neq 0) = -\epsilon_{aik} \frac{x^k}{g r^2} (1 - K_\infty) , \tag{8.82}$$

with the corresponding magnetic field given by $B_k^a = \frac{1}{2} \epsilon_{aik} f^{ij,a}$, which has a radial magnetic field

$$B^k = B_a^k n^a = (1 - K_\infty^2) \frac{x^k}{g r^3} . \tag{8.83}$$

We have already observed that the massless radial gauge field $\bar{A}_\mu$ does not condensate: however, Eq. (8.82) shows that an asymptotic (radial) classical gauge field is generated by the condensation of the massive tangent gauge fields $\bar{A}_\mu^a$. Thus, for sufficiently large M^2 (or, equivalently, at sufficiently low energies), one would observe only the massless $U(1)$ gauge field $\bar{A}_\mu$ and the classical condensate α_i^a, which is reminiscent of the tangent gauge fields $\bar{A}_\mu^a$.

Finally, let us note that the above derived (radial) magnetic field (8.83) is of the same form as that of a Dirac monopole:

$$B_k = \frac{g_m}{r^3} x^k , \tag{8.84}$$

with g_m being the magnetic charge. Comparing the two expressions, we find the relation $g g_m = 1 - K_\infty^2$, which shows that in the model here considered the magnetic charge g_m decreases with increasing temperature.

8.3.4 *The sphaleron*

As a final example, we now consider the case of sphalerons [39,175,385,441]. They are static solutions of classical equations of motion that maximizes the potential energy. This corresponds to unstable (in conventional time) field configurations of the classical field equations. Sphalerons determine the hight V_{max} of the potential barrier for a given system and hence they may serve as a tool in obtaining the finite-temperature decay probability that is proportional to the Boltzmann factor $e^{-\beta V_{max}}$. Sphaleron transitions between false and true vacua are useful in describing bubble formation in first-order phase transitions. For instance, in the Standard Model of particle physics, sphalerons are directly involved in processes that violate baryon and lepton number, see, e.g., [318,572]. Here we shall discuss $SU(2)$ sphalerons. In this case, as well as in the monopole case, one has $f_{abc} = \epsilon_{abc}$. The only difference being that the Higgs ϕ field is in the adjoint representation $(T^a)_{ij} = -i\epsilon_{iaj}$ for the monopole case, while in the sphaleron case ϕ is in the fundamental representation. Thus we now have $T^a = \frac{1}{2}\sigma^a$, with the unit vector and the matrix g_{ab} given by:

$$\mathbf{n} = \begin{pmatrix} 0 \\ 1 \end{pmatrix} ; \qquad g_{ab} = \frac{1}{4} \begin{pmatrix} 1 & -i & 0 \\ i & 1 & 0 \\ 0 & 0 & 1 \end{pmatrix} . \tag{8.85}$$

Once diagonalized, such a matrix leaves one with the three massive gauge fields of the Standard Model:

$$W_\mu^\pm = \frac{1}{\sqrt{2}} \left(A_\mu^1 \pm i A_\mu^2 \right) , \tag{8.86}$$

$$Z_\mu = A_\mu^3 , \tag{8.87}$$

with the Weinberg angle equal to zero: $\vartheta_W = 0$. The gauge field masses, including the thermal corrections, are now:

$$M_N^2 = M_Z^2 = \frac{1}{4} g^2 \left(\sigma^2 + \langle : \bar{\rho}^2 : \rangle_0 \right) - \frac{3}{4} g^2 \sum_b \frac{2}{3} \langle : \bar{A}_\mu^b \bar{A}^{b\mu} : \rangle_0 . \tag{8.88}$$

In the case when $\vartheta_W \neq 0$ a massless gauge field can appear due to the presence of an additional $U(1)$ field. However, we consider for simplicity only the $\vartheta_W = 0$ case [440].

We can proceed now in a way similar to the one followed in the monopole case and decompose the $SU(2)$ group element as

$$U = e^{i\frac{\mu}{2}\sigma} e^{i\frac{\xi}{2}\sigma} e^{i\frac{\nu}{2}\sigma} = \begin{pmatrix} e^{i(\mu+\nu)/2} \cos\frac{\xi}{2} & e^{i(\mu-\nu)/2} \sin\frac{\xi}{2} \\ -e^{i(-\mu+\nu)/2} \sin\frac{\xi}{2} & e^{i(\mu+\nu)/2} \cos\frac{\xi}{2} \end{pmatrix} . \tag{8.89}$$

The unit vector (8.85) defines the $SU(2)/U(1) = CP^1$ coset space (Riemann sphere), also denoted as the one-dimensional complex projective space $P_1(\mathbb{C})$, which can be parametrized by setting:

$$\nu = \phi + \pi; \quad \mu = -\phi; \quad \xi = -2\vartheta. \tag{8.90}$$

We thus obtain:

$$U = in^a \sigma^a = i \begin{pmatrix} \cos\vartheta & e^{-i\phi}\sin\vartheta \\ e^{i\phi}\sin\vartheta & -\cos\vartheta \end{pmatrix} \in SU(2), \tag{8.91}$$

$$n^a = \frac{x^a}{r}. \tag{8.92}$$

By use of the matrix (8.91), we define the asymptotic gauge field configuration

$$\alpha_{\mu,\infty} = \alpha^a_{\mu,\infty} T^a = -\frac{i}{g}(\partial_\mu U)U^\dagger, \tag{8.93}$$

with

$$\alpha^a_{i,\infty} = -\frac{2}{gr^2}\epsilon_{iak}x^k. \tag{8.94}$$

The homotopy class is now $\pi_2(SU(2)/U(1)) = \mathbb{Z}$ (the same as for the monopole case, see also Section 10.5.2), describing the mapping π of the sphere $\mathcal{S}^2$, surrounding the spatial singularity at $r = 0$, to $G/H = SU(2)/U(1)$.

For the sphaleron, at $T = 0$, we have the asymptotic gauge field with $K_\infty = -1$ rather than $K_\infty = 0$ as in the monopole case. Thus one can make an Ansatz similar to the one for the monopole:

$$\alpha^a_i = -\epsilon_{aik}\frac{x^k}{gr^2}(1 - K(r)), \tag{8.95a}$$

$$\sigma = \sigma_0 f(r), \tag{8.95b}$$

$$\sigma_0^2 = v^2 - 3\langle : \vec{p}^2 : \rangle_0 + \frac{1}{4}\frac{g^2}{\lambda}\langle : \bar{A}^a_\mu \bar{A}^{a\mu} : \rangle_0. \tag{8.95c}$$

As in the monopole case, we have $\alpha^a_i n^a = 0$. This implies that the field $\bar{A}_\mu = \bar{A}^a_\mu n^a$ (in the homogeneous case such a field reduces to $\bar{A}^3_\mu = Z_\mu$) does not condensate. On the other hand, the condensation of the two gauge bosons $W^\pm_\mu$ produces the classical fields α^a_μ.

In the present case the field $\bar{A}_\mu$ is massive. For small energies, only the classical sphaleron will be observed. The fact that we have only $W^\pm_\mu$ gauge field condensation does not disturb the mixing between Z_μ and the additional $U(1)$ gauge field B_μ. As a consequence the electrical charge is conserved.

Applying the TFD procedure to the Euler–Lagrange equations (8.23), (8.24) for the sphaleron case we have

$$r^2 K'' = K(K^2 - 1) + M_g^2(K - 1)r^2 + M_s^2(K + 1)r^2, \quad (8.96a)$$

$$\frac{d}{dr}\left(r^2 \frac{df}{dr}\right) = \frac{1}{2}(1 + K^2)f + \lambda r^2 \sigma_0^2(f^2 - 1)f, \quad (8.96b)$$

$$M^2 = M_g^2 + M_s^2, \quad (8.96c)$$

$$M_g^2 = -\frac{3}{4}g^2 \sum_b \frac{2}{3}\langle: \bar{A}_\mu^b \bar{A}^{b\mu} :\rangle_0, \quad (8.96d)$$

$$M_s^2 = \frac{1}{4}g^2 \left(\sigma^2 + \langle: \bar{\rho}^2 :\rangle_0\right). \quad (8.96e)$$

As before, the Higgs boson envelope vanishes at T_c. At spatial infinity $(r \to \infty)$

$$K(r) \to K_\infty = -\frac{M_s^2 - M_g^2}{M_s^2 + M_g^2}, \quad (8.97)$$

with $K_\infty \to -1$ for $T \to 0$.

As done for the monopole, we define the "magnetic" field

$$B^k = B_a^k n^a = (1 - K_\infty^2)\frac{x^k}{gr^3}, \quad (8.98)$$

related to the massive Z_μ field (instead of to a massless field as in the case of the monopole). The sphaleron then exhibits a "magnetic charge" g_m connected to the neutral gauge field interaction, such that $gg_m = 1 - K_\infty$. Notice that for $T = 0$ such a charge vanishes. However, it increases with increasing temperature, contrary to what was found for the monopole case.

After the $U(1)$ group addition, the sphaleron has both a neutral "magnetic charge" ($\propto \cos\vartheta_W$) connected to the Z_μ field and a genuine magnetic charge ($\propto \sin\vartheta_W$) connected to the electrodynamics gauge field $\bar{A}_\mu$.

8.4   The $SU(2)$ instanton

Instantons [588], also called pseudoparticles [66, 525], are interpreted as quantum tunneling events between topologically distinct vacua [131, 347] and it has been proposed [525] that the Euclidean functional integral should be expanded around pseudoparticles, i.e., around stationary points of the classical Euclidean action (see Section 10.9), in order to obtain the theory ground state. The problem of determining the small oscillation modes (quantum fluctuations) about pseudoparticle solutions and their eigenfrequencies is therefore relevant and poses the further problem of the quantization "about" a Yang–Mills classical solution [66,229,507]. In this Section we will see how to approach such a problem in the frame of the QFT boson condensation formalism [184,602,603].

We consider the $SU(2)$ Yang–Mills Lagrangian in Euclidean spacetime (cf. also Section 10.9):

$$\mathcal{L} = \frac{1}{4}F^a_{\mu\nu}F^a_{\mu\nu}, \qquad a = 1,2,3; \quad \mu,\nu = 1,\ldots,4, \qquad (8.99a)$$

$$F^a_{\mu\nu} = \partial_\mu A^a_\nu - \partial_\nu A^a_\mu + e\,\epsilon_{abc}A^b_\mu A^c_\nu. \qquad (8.99b)$$

The equations of motion are:

$$\partial_\mu F_{\mu\nu} + [A_\mu, F_{\mu\nu}] = 0, \qquad (8.100)$$

where

$$A_\mu \equiv \frac{e}{2i}A^a_\mu \sigma^a; \quad F_{\mu\nu} \equiv \frac{e}{2i}F^a_{\mu\nu}\sigma^a = \partial_\mu A_\nu - \partial_\nu A_\mu + [A_\mu, A_\nu], \qquad (8.101)$$

and $\sigma^a = \sigma_a$ are the Pauli matrices. The dual tensor is defined as

$$^*F_{\mu\nu} \equiv \frac{1}{2}\epsilon_{\mu\nu\rho\sigma}F_{\rho\sigma}. \qquad (8.102)$$

The Pontryagin index Q is given by

$$Q = -\frac{1}{16\pi^2}\int d^4x\,\mathrm{Tr}\,{}^*F_{\mu\nu}F_{\mu\nu}. \qquad (8.103)$$

The index Q is an integer if A_μ leads to finite Euclidean action $\mathcal{S}_E$:

$$\mathcal{S}_E = -\frac{1}{2e^2}\int d^4x\,\mathrm{Tr}\,F_{\mu\nu}F_{\mu\nu}. \qquad (8.104)$$

The solutions of the self-dual (anti-self-dual) equation

$$^*F_{\mu\nu} = \overset{(-)}{+}\,F_{\mu\nu} \qquad (8.105)$$

are solutions of the field equations (8.102) [346,348]. Self-dual (anti-self-dual) solutions are called instantons (anti-instantons) and have $Q > 0$ ($Q < 0$).

It is possible [348,348] to study the properties of the theory under combined transformations of conformal and gauge group by projecting the 4-D Euclidean space onto the 4-D unit hypersphere in a 5-D Euclidean space, and extending the $SU(2) \times SU(2) \sim O(4)$ gauge group to an $O(4)$ gauge group. In this way, the most general non-trivial solution with $Q = 1$, which here we denote by $\hat{\varphi}_\mu(x)$, is found to be given by [66] (cf. also Section 10.9)

$$\hat{\varphi}_\mu(x) = -\frac{2i}{1+x^2}\Sigma_{\mu\nu}x_\nu \equiv \begin{pmatrix} \varphi_\mu(x) & 0 \\ 0 & \bar{\varphi}_\mu(x) \end{pmatrix}, \qquad (8.106)$$

where

$$x_5 = \frac{1-x^2}{1+x^2}, \qquad \Sigma_{\mu\nu} = \begin{pmatrix} \sigma_{\mu\nu} & 0 \\ 0 & \bar{\sigma}_{\mu\nu} \end{pmatrix}; \qquad \Sigma_{\mu5} = \frac{1}{2}\alpha_\mu, \qquad (8.107a)$$

$$\sigma_{ij} = \frac{1}{4i}[\sigma_i,\sigma_j]; \qquad \sigma_{i4} = \frac{1}{2}\sigma_i, \qquad \bar{\sigma}_{ij} \equiv \sigma_{ij}; \qquad \bar{\sigma}_{i4} = -\bar{\sigma}_{i4}, \qquad (8.107b)$$

$$\alpha_i = \begin{pmatrix} 0 & \sigma_i \\ \sigma_i & 0 \end{pmatrix}; \qquad \alpha_4 = i\begin{pmatrix} 0 & -\mathbb{1} \\ \mathbb{1} & 0 \end{pmatrix}. \qquad (8.107c)$$

We consider now the problem of the number of the degrees of freedom of the instanton solution and we will follow the discussion of [184].

The 1-instanton solution $\varphi_\mu(x)$ is invariant under the $O(5)$ group. This group has 10 generators[1]: the $M_{\mu\nu}$ generators for the $O(4)$ rotations and $R_\mu = \frac{1}{2}(P_\mu + K_\mu)$ with P_μ and K_μ being respectively the spacetime translation and the conformal transformation generators. The basic $O(5,1)$ conformal symmetry is broken, while the gauge symmetry is preserved. The 1-instanton depends on 5 parameters (recall that it lives in the 5-D Euclidean space), which are related to the five zero-frequency modes found among the small oscillations (quantum fluctuations) $a_\mu(x)$ around the classical instanton solution [66,507] $\varphi_\mu(x)$:

$$\varphi'_\mu = \varphi_\mu(x) + a_\mu(x). \qquad (8.108)$$

In the $Q = 1$ case as well as in the $Q > 1$ case, the search of solutions A_μ of the non-linear Eq. (8.100) of the form (8.108) leads to a system of linear equations for $a_\mu(x)$, namely of linear equations for fluctuations around a classical solution: this means that one considers solutions of the Yang–Mills equations under the condition of spontaneous breakdown of $O(5,1)$ conformal symmetry:

$$\langle\varphi'_\mu(x)\rangle = \varphi_\mu(x). \qquad (8.109)$$

Let us suppose that the linearization procedure can be fully carried out. Let χ_μ be the linearized fields solutions of the system of linear equations

$$K(\partial)\chi_\mu(x) = 0, \qquad (8.110)$$

---

[1] We recall that the number of generators of $SU(N)$, $O(N)$ and $E(N)$ is given by N^2-1, $\frac{N}{2}(N-1)$ and $\frac{N}{2}(N+1)$, respectively.

where $K(\partial)$ is some convenient linear differential operator. The linearization procedure is equivalent to finding the dynamical map for φ'_μ

$$\langle a|\varphi'_\mu(x)|b\rangle = \langle a|F_\mu[\chi_\mu(x)]|b\rangle\,, \tag{8.111}$$

where $F_\mu[\chi_\mu(x)]$ is a functional of $\chi_\mu(x)$ and $|a\rangle$ and $|b\rangle$ are vectors in the Hilbert space for the linearized fields $\chi_\mu(x)$.

Suppose now that $A_\mu(x)$ undergoes the conformal transformation $A_\mu(x) \to gA_\mu(x)$, with $g \in O(5,1)$. In view of Eq. (8.111) we expect that $\varphi'_\mu(x) \to g\varphi'_\mu(x)$ is induced by $\chi_\mu(x) \to h\chi_\mu(x)$ with h belonging to some group G:

$$\langle a|g\varphi'_\mu(x)|b\rangle = \langle a|F_\mu[h\chi_\mu(x)]|b\rangle\,. \tag{8.112}$$

We know from Chapters 3 and 4 that, due to the spontaneous symmetry breaking, the group G is not the same as the basic invariance group $O(5,1)$. As a matter of fact, G turns out to be the group contraction of $O(5,1)$, namely the group $E(5)$, and it is the group under which the linear equations (8.110) are invariant. We recall that the dynamical rearrangement of symmetry $O(5,1) \to E(5)$ occurs since in the dynamical mapping (8.111) infrared effects from the zero-frequency modes are missing (cf. Chapter 4). We also recall that the group contraction, leading to the $E(5)$ group in the present case, preserves the number of parameters (15 in total) of the original invariance group, which here is $O(5,1)$, and fully accounts for the invariance of the linearized equations: $E(5)$ is indeed the Euclidean group in five dimensions which is spanned by the $O(5) = (M_{\mu\nu}, R_\mu)$ subgroup (which is the "unbroken" part of $O(5,1)$) plus five "translations" generated by $S_\mu = \frac{1}{2}(P_\mu - K_\mu)$ and D, the dilation generator. These last "degrees of freedom" correspond to the five zero modes found in [507] and [66], namely the five parameters (position and size) on which $\varphi_\mu(x)$ depends [348]. Besides such degrees of freedom one may further add the ones (three) describing orientation in the $SU(2)$ space. Spontaneous breakdown of conformal symmetry has been also studied in [181, 269].

The case $Q > 1$ can be considered by noticing that the Pontryagin number is by definition the degree of the homotopic mapping and that a mapping of degree Q is obtained as a sum of Q mappings of degree one. One can see this by considering the $\mathcal{S}^4$ hypersphere of coordinates x, y, z, t, u in the five Euclidian space. On $\mathcal{S}^4$ we consider the circles (the "parallels") which are intersections of the hyperplanes $u = \frac{j}{Q}, j = 1, 2, \ldots, Q-1$ with the hypersphere surface. A mapping of $\mathcal{S}^4$ to Q hyperspheres $\mathcal{S}_i^4$, $i = 1, 2, \ldots, Q$ is then obtained by shrinking each circle to a point. The overlapping of these Q hyperspheres brings back to $\mathcal{S}^4$. The mapping of $\mathcal{S}_i^4$ to $\mathcal{S}^4$ is of degree Q, hence the mapping so induced from $\mathcal{S}^4$ to $\mathcal{S}^4$ is the mapping of Pontryagin index Q. We can therefore extend our group contraction argument to the case $Q > 1$ by considering Q 1-pseudoparticle systems and the associated invariance group

$$[O(5,1) \times O(5,1) \times \cdots \times O(5,1)]_Q \times [O(3) \times O(3) \times \cdots \times O(3)]_{Q-1}, \quad (8.113)$$

where the indices Q and $Q-1$ denote the number of factors entering the respective brackets. Here the $SU(2) \approx O(3)$ factors account for the relative orientations of the pseudoparticles in the gauge space. Notice that these ·factors account, indeed, for the degrees of freedom for the unbroken $SU(2)$ gauge invariance of the theory. We are disregarding the fictitious degrees of freedom due to the hyperspherical construction [348, 507]. Note that generators belonging to different factors commute with each other.

The linearization procedure outlined for the $Q = 1$ case can be applied again and we finally get the contracted group

$$[E(5) \times E(5) \times \cdots \times E(5)]_Q \times [O(3) \times O(3) \times \cdots \times O(3)]_{Q-1}. \quad (8.114)$$

From this we derive the number of degrees of freedom (zero-frequency modes) on which the Q-pseudoparticle solution depends: five (field "translation") parameters associated with each of the Q $E(5)$ factors plus three parameters for each of the $Q-1$ $O(3)$ factors, i.e., $8Q-3$ parameters, which is the result obtained by celebrated algebraic geometry methods [43, 556]. To these parameters one might add three more parameters accounting for the overall orientation in the $SU(2)$ space. See also Section 10.9.1 for a further discussion on the number of degrees of freedom of instanton solutions. An analysis in terms of differential geometry has been presented in [602, 603].

Chapter 9

Dissipation and quantization

9.1 Introduction

In Chapter 5 we briefly discussed non-unitary time evolution (the arrow of time) and trajectories in the space of representations [632, 636]. Actually, we have learned that any microscopic system cannot be considered to be an isolated closed system. It is in fact always in interaction with the quantum fluctuations of the vacuum. In this sense it is therefore imperative to study dissipative open systems in QFT.

A microscopic theory for a dissipative system must include the details of the processes responsible for dissipation: thus the total Hamiltonian must describe the system, the bath and the system-bath interaction. It turns out that the canonical commutation relations (CCR) are not preserved by time evolution due to damping terms. By including the bath one "closes" the system in order to recover the canonical formalism and one realizes that the role of fluctuating forces is in fact the one of preserving the canonical structure of the CCR [311, 659]. The description of the original dissipative system is recovered by means of the reduced density matrix, obtained by integrating out the bath variables which originate the damping and the fluctuations. However, it is not always possible to carry out such a computational program since the knowledge of the details of the processes inducing the dissipation may not always be achievable. These details may not be explicitly known and the dissipation mechanisms are sometimes globally described by such parameters as friction, resistance, viscosity etc. A possible strategy is then to double the degrees of freedom for the system under consideration, in order to close it. One has thus a mirror image of the system, which behaves effectively as a "reservoir" [153]. In this Chapter we consider the description of dissipative systems in the frame of the

373

quantum Brownian motion as described by Schwinger [560] and by Feynman and Vernon [239]. As we will see, such approach leads naturally to the doubling of the degrees of freedom intrinsic to QFT [153]. We show how the quantization of the damped harmonic oscillator can be achieved by taking advantage of inequivalent representations, the transition among representations being controlled by the free energy operator.

We then show that dissipative features arise also in neutrino mixing and that the process of flavor oscillations can be seen in terms of an interaction with an external gauge field, which act like a "reservoir".

As an interesting development of the above-mentioned formalism for dissipative quantum systems, we report on the dissipative quantum model of brain, which provides a conceptually consistent framework for the description of many important brain activities, from memory to consciousness.

Finally we consider the proposal by G. 't Hooft [589–593] that the information loss (dissipation) in a regime of deterministic dynamics at high energies, might be responsible for the quantum-like behavior of the world at lower energies.

In conclusion, in this Chapter several different lines of research are presented, which, however, have a common denominator, namely dissipation and the doubling of degrees of freedom. For the sake of brevity we do not consider other related applications such as the study of Chern–Simons-like dynamics of Bloch electrons in solids [91], unstable states [185] and expanding geometry model in inflationary cosmology [14,19] or the quantization of matter field in a curved background [14, 19, 338, 339, 444]. Moreover, we do not consider the relation between the dissipative systems and the Nelson stochastic quantization scheme [301, 496], which is beyond the task of this book.

9.2   The exact action for damped motion

In this Section we derive the exact action for a particle of mass M damped by a mechanical resistance R in a potential U [576]. We first focus on the special case of an isolated particle ($R = 0$). The Hamiltonian is

$$H = -\frac{\hbar^2}{2M}\left(\frac{\partial}{\partial x}\right)^2 + U(x). \qquad (9.1)$$

We consider the Wigner function [237,311] (cf. Eq. (5.1)),

$$W(p, x, t) = \frac{1}{2\pi\hbar}\int dy\, \psi^*\left(x - \frac{1}{2}y, t\right)\psi\left(x + \frac{1}{2}y, t\right)e^{-i\frac{py}{\hbar}}, \qquad (9.2)$$

and the related density matrix

$$W(x, y, t) = \langle x + \frac{1}{2}y|\rho(t)|x - \frac{1}{2}y\rangle = \psi^*\left(x - \frac{1}{2}y, t\right)\psi\left(x + \frac{1}{2}y, t\right). \quad (9.3)$$

For an isolated particle, the density matrix equation of motion is

$$i\hbar\frac{d\rho}{dt} = [H, \rho]. \quad (9.4)$$

By introducing

$$x_\pm = x \pm \frac{1}{2}y, \quad (9.5)$$

Eq. (9.4) reads, in coordinate representation,

$$i\hbar\frac{\partial}{\partial t}\langle x_+|\rho(t)|x_-\rangle = \left\{-\frac{\hbar^2}{2M}\left[\left(\frac{\partial}{\partial x_+}\right)^2 - \left(\frac{\partial}{\partial x_-}\right)^2\right]\right.$$
$$\left. + [U(x_+) - U(x_-)]\right\}\langle x_+|\rho(t)|x_-\rangle, \quad (9.6)$$

i.e.,

$$i\hbar\frac{\partial}{\partial t}W(x, y, t) = \mathcal{H}_0 W(x, y, t), \quad (9.7)$$

$$\mathcal{H}_0 = \frac{1}{M}p_x p_y + U(x_+) - U(x_-), \quad (9.8)$$

$$p_x = -i\hbar\frac{\partial}{\partial x}, \quad p_y = -i\hbar\frac{\partial}{\partial y}. \quad (9.9)$$

The Lagrangian, from which the Hamiltonian (9.8) is obtained, is

$$\mathcal{L}_0 = M\dot{x}\dot{y} - U(x_+) + U(x_-). \quad (9.10)$$

Suppose now that the particle interacts with a thermal bath at temperature T with interaction Hamiltonian

$$H_{int} = -fx, \quad (9.11)$$

where f is the random force on the particle at the position x due to the bath. In the Feynman–Vernon formalism, the effective action has the form

$$\mathcal{A}[x, y] = \int_{t_i}^{t_f} dt\, \mathcal{L}_0(\dot{x}, \dot{y}, x, y) + \mathcal{I}[x, y], \quad (9.12)$$

and

$$e^{\frac{i}{\hbar}\mathcal{I}[x,y]} = \left\langle\left(e^{-\frac{i}{\hbar}\int_{t_i}^{t_f} f(t)x_-(t)dt}\right)_-\left(e^{\frac{i}{\hbar}\int_{t_i}^{t_f} f(t)x_+(t)dt}\right)_+\right\rangle. \quad (9.13)$$

Here the average is with respect to the thermal bath; "$(\dots)_+$" and "$(\dots)_-$" denote time ordering and anti-time ordering, respectively. If the interaction H_{int} between the bath and the coordinate x were turned off, then f would develop in time according to $f(t) = e^{iH_R t/\hbar} f e^{-iH_R t/\hbar}$, where H_R is the Hamiltonian of the isolated bath (decoupled from the coordinate x). $f(t)$ is the force operator of the bath to be used in Eq. (9.13).

The reduced density matrix in Eq. (9.3) for the particle which first makes contact with the bath at the initial time t_i is, at a final time t_f,

$$W(x_f, y_f, t_f) = \int_{-\infty}^{\infty} dx_i \int_{-\infty}^{\infty} dy_i \, K(x_f, y_f, t_f; x_i, y_i, t_i) W(x_i, y_i, t_i) \,,$$
(9.14)

with the path integral representation for the evolution kernel

$$K(x_f, y_f, t_f; x_i, y_i, t_i) = \int_{x(t_i)=x_i}^{x(t_f)=x_f} \mathcal{D}x \int_{y(t_i)=y_i}^{y(t_f)=y_f} \mathcal{D}y \, e^{\frac{i}{\hbar} A[x,y]} \,.$$
(9.15)

The Green's functions for the evaluation of $\mathcal{I}[x, y]$ for a linear damping have been discussed by Schwinger [560]. Here we only mention that the fundamental correlation function for the random force on the particle due to the thermal bath is given by (see [576])

$$G(t - s) = \frac{i}{\hbar} \langle f(t) f(s) \rangle \,.$$
(9.16)

The retarded and advanced Green's functions are defined by

$$G_{ret}(t - s) = \theta(t - s)[G(t - s) - G(s - t)] \,,$$
(9.17a)

$$G_{adv}(t - s) = \theta(s - t)[G(s - t) - G(t - s)] \,.$$
(9.17b)

The mechanical impedance $Z(\zeta)$ (analytic in the upper half complex frequency plane $\mathcal{I}m \, \zeta > 0$) is given by

$$-i\zeta Z(\zeta) = \int_0^{\infty} dt \, G_{ret}(t) e^{i\zeta t} \,.$$
(9.18)

The quantum noise in the fluctuating random force is given by

$$N(t - s) = \frac{1}{2} \langle f(t) f(s) + f(s) f(t) \rangle \,,$$
(9.19)

distributed in the frequency domain according to the Nyquist theorem

$$N(t - s) = \int_0^{\infty} d\omega \, S_f(\omega) \cos[\omega(t - s)] \,,$$
(9.20)

$$S_f(\omega) = \frac{\hbar\omega}{\pi} \coth \frac{\hbar\omega}{2k_B T} \mathcal{R}e Z(\omega + i0^+) \,.$$
(9.21)

The mechanical resistance is defined by $R = \lim_{\omega \to 0} \mathcal{R}e Z(\omega + i0^+)$. Eq. (9.13) may now be evaluated [576] following Feynman and Vernon as,

$$\mathcal{I}[x, y] = \frac{1}{2} \int_{t_i}^{t_f} \int_{t_i}^{t_f} dt ds \, [G_{ret}(t-s) + G_{adv}(t-s)][x(t)y(s) + x(s)y(t)]$$

$$+ \frac{i}{2\hbar} \int_{t_i}^{t_f} \int_{t_i}^{t_f} dt ds N(t-s)y(t)y(s) \,. \tag{9.22}$$

By defining the retarded force on y and the advanced force on x as

$$F_y^{ret}(t) = \int_{t_i}^{t_f} ds \, G_{ret}(t-s)y(s) \,, \tag{9.23a}$$

$$F_x^{adv}(t) = \int_{t_i}^{t_f} ds \, G_{adv}(t-s)x(s) \,, \tag{9.23b}$$

respectively, the interaction between the bath and the particle is then

$$\mathcal{I}[x, y] = \frac{1}{2} \int_{t_i}^{t_f} dt \, \left[x(t)F_y^{ret}(t) + y(t)F_x^{adv}(t) \right]$$

$$+ \frac{i}{2\hbar} \int_{t_i}^{t_f} \int_{t_i}^{t_f} dt ds \, N(t-s)y(t)y(s) \,. \tag{9.24}$$

Thus the real and the imaginary parts of the action are

$$\mathcal{R}e\mathcal{A}[x, y] = \int_{t_i}^{t_f} dt \, \mathcal{L} \,, \tag{9.25}$$

$$\mathcal{L} = M\dot{x}\dot{y} - [U(x_+) - U(x_-)] + \frac{1}{2} \left[xF_y^{ret} + yF_x^{adv} \right] \,, \tag{9.26}$$

and

$$\mathcal{I}m\mathcal{A}[x, y] = \frac{1}{2\hbar} \int_{t_i}^{t_f} \int_{t_i}^{t_f} dt ds \, N(t-s)y(t)y(s) \,, \tag{9.27}$$

respectively. Eqs. (9.25), (9.26), (9.27) are *rigorously exact* for linear damping due to the bath when the path integral Eq. (9.15) is employed.

When the choice $F_y^{ret} = R\dot{y}$ and $F_x^{adv} = -R\dot{x}$ is made in Eq. (9.26), we obtain

$$\mathcal{L}(\dot{x}, \dot{y}, x, y) = M\dot{x}\dot{y} - U(x_+) + U(x_-) + \frac{R}{2}(x\dot{y} - y\dot{x}) \,. \tag{9.28}$$

9.2.1 Quantum Brownian motion

By following Schwinger [560], the description of a Brownian particle of mass M moving in a potential $U(x)$ with a damping resistance R interacting with a thermal bath at temperature T is provided by [107,576]

$$\mathcal{H}_{Brownian} = \mathcal{H} - \frac{ik_B TR}{\hbar}(x_+ - x_-)^2. \tag{9.29}$$

Here $\mathcal{H}$ is given by

$$\mathcal{H} = \frac{1}{2M}\left(p_+ - \frac{R}{2}x_-\right)^2 - \frac{1}{2M}\left(p_- + \frac{R}{2}x_+\right)^2 + U(x_+) - U(x_-), \tag{9.30}$$

where $p_\pm = -i\hbar\frac{\partial}{\partial x_\pm}$ and the evolution equation for the density matrix is

$$i\hbar\frac{\partial\langle x_+|\rho(t)|x_-\rangle}{\partial t} = \mathcal{H}\langle x_+|\rho(t)|x_-\rangle - \langle x_+|N[\rho]|x_-\rangle, \tag{9.31}$$

where $N[\rho]$, taken to be $(ik_B TR/\hbar)[x,[x,\rho]]$, describes the effects of the reservoir random thermal noise [107,576]. In general the density operator in the above expression describes a mixed statistical state. The thermal bath contribution to the r.h.s. of Eq. (9.29), proportional to the fluid temperature T, can be shown [107] to be equivalent to a white noise fluctuation source coupling the forward and backward motions according to

$$\langle y(t)y(t')\rangle_{noise} = \frac{\hbar^2}{2Rk_B T}\delta(t-t'), \tag{9.32}$$

so that thermal fluctuations are always occurring in the difference $y = x_+ - x_-$ between forward in time and backward in time coordinates.

Eq. (9.16) gives the correlation function for the random force f on the particle due to the bath. The retarded and advanced Green's functions are the ones studied above (see also [576]). The interaction between the bath and the particle is evaluated by following Feynman and Vernon and Eqs. (9.25) and (9.27) are found [576] for the real and the imaginary part of the action, respectively.

In the discussion above, we have considered the low temperature limit: $T \ll T_\gamma$ where $T_\gamma = \frac{\hbar\gamma}{k_B} = \frac{\hbar R}{2Mk_B}$. At high temperature, $T \gg T_\gamma$, the thermal bath motion suppresses the probability for $x_+ \neq x_-$ due to the thermal term $(k_B TR/\hbar)(x_+ - x_-)^2$ in Eq. (9.29) (cf. also Eq. (9.32)). By writing the diffusion coefficient $D = \frac{k_B T}{R}$ as

$$D = \frac{T}{T_\gamma}\left(\frac{\hbar}{2M}\right), \tag{9.33}$$

the condition for classical Brownian motion for high mass particles is $D \gg (\hbar/2M)$, and the condition for quantum interference with low mass particles is $D \ll (\hbar/2M)$. In colloidal systems, for example, classical Brownian motion for large particles would appear to be dominant. In a fluid at room temperature it is typically $D \sim (\hbar/2M)$ for a single atom, or, equivalently, $T \sim T_\gamma$, so that the role played by quantum mechanics, although perhaps not dominant, may be an important one in the Brownian motion.

We stress that the meaning of our result Eqs. (9.25)–(9.27) is that non-zero y yields an "unlikely process" in the classical limit "$\hbar \to 0$", in view of the large imaginary part of the action implicit in Eq. (9.27). On the contrary, at quantum level non-zero y may allow quantum noise effects arising from the imaginary part of the action [576]. We will come back to this point in the next Section, where we discuss the canonical quantization of the damped simple harmonic oscillator.

9.3 Quantum dissipation and unitarily inequivalent representations in QFT

We consider the classical damped harmonic oscillator (DHO) as a simple prototype for dissipative systems

$$M\ddot{x} + R\dot{x} + \kappa x = 0 \,. \tag{9.34}$$

It is a non-Hamiltonian system and therefore the canonical formalism, needed for its quantization, cannot be set up [62]. However, the problem can be faced by proceeding in the following way (cf. [107, 153, 233, 576]).

The equation of motion for the density matrix is given by Eq. (9.4), where the Hamiltonian in the (x_+, x_-) plane is given by Eq. (9.30) [107, 233, 576]. The real and imaginary parts of the action are given by Eqs. (9.25)–(9.27) for the linear passive damping resulting in the mechanical resistance R. By making the choice $U(x_\pm) = \frac{1}{2}\kappa x_\pm^2$, in terms of the doubled coordinates (x, y), $y = x_+ - x_-$, the Hamiltonian (9.30) can be derived from the Lagrangian [62, 153, 192, 482] (cf. (9.28))

$$L = M\dot{x}\dot{y} + \frac{1}{2}R(x\dot{y} - \dot{x}y) - \kappa xy \,, \tag{9.35}$$

which gives the DHO equation Eq. (9.34) and its complementary equation for the y coordinate

$$M\ddot{y} - R\dot{y} + \kappa y = 0 \,. \tag{9.36}$$

The y-oscillator is the time-reversed image $(R \to -R)$ of the x-oscillator. If from the manifold of solutions to Eqs. (9.34) and (9.36) we choose those for which the y coordinate is constrained to be zero, then Eqs. (9.34) and (9.36) simplify to

$$M\ddot{x} + R\dot{x} + \kappa x = 0, \qquad y = 0. \qquad (9.37)$$

Thus we obtain a classical damped equation of motion from a Lagrangian theory at the expense of introducing an "extra" coordinate y, later constrained to vanish. Note that $y(t) = 0$ is a true solution to Eqs. (9.34) and (9.36) so that the constraint is *not* in violation of the equations of motion. However, as already stressed in the previous Section and in Chapter 5, *the role of the "doubled" y coordinate is absolutely crucial in the quantum regime.* There it accounts for the quantum noise.

The system described by (9.35) is sometimes called the Bateman dual system of oscillators [62, 97, 482]. We observe that the doubling of the degrees of freedom, implied by the density matrix and the Wigner function formalism, here finds its physical justification in the fact that the canonical quantization scheme can only deal with an isolated system. In the present case our system has been assumed to be coupled with a reservoir and it is then necessary to *close* the system by including the reservoir.[1] This is achieved by doubling the phase-space dimensions [153, 233]. Eq. (9.35) is indeed the closed system Lagrangian: y may be thought of as describing an effective degree of freedom for the reservoir to which the system (9.34) is coupled. The canonical momenta are given by $p_x \equiv \frac{\partial L}{\partial \dot{x}} = M\dot{y} - \frac{1}{2}Ry$ and $p_y \equiv \frac{\partial L}{\partial \dot{y}} = M\dot{x} + \frac{1}{2}Rx$. For a discussion of Hamiltonian systems of this kind see also [50, 52, 366, 483, 502, 554, 601, 610].

Canonical quantization can now be performed by introducing the commutators

$$[x, p_x] = i\hbar = [y, p_y], \qquad [x, y] = 0 = [p_x, p_y], \qquad (9.38)$$

and the corresponding sets of annihilation and creation operators

$$\alpha \equiv \frac{1}{\sqrt{2\hbar\Omega M}}(p_x - iM\Omega x), \quad \alpha^\dagger \equiv \frac{1}{\sqrt{2\hbar\Omega M}}(p_x + iM\Omega x), \quad (9.39a)$$

$$\beta \equiv \frac{1}{\sqrt{2\hbar\Omega M}}(p_y - iM\Omega y), \quad \beta^\dagger \equiv \frac{1}{\sqrt{2\hbar\Omega M}}(p_y + iM\Omega y), \quad (9.39b)$$

$$[\alpha, \alpha^\dagger] = \mathbb{1} = [\beta, \beta^\dagger], \qquad [\alpha, \beta] = 0 = [\alpha, \beta^\dagger]. \qquad (9.39c)$$

---

[1]A different approach to the quantization of DHO is the one which makes use a time-dependent Lagrangian [130, 363].

We have introduced $\Omega \equiv \sqrt{\frac{\kappa}{M} - \frac{R^2}{4M^2}}$, the common frequency of the two oscillators Eq. (9.34) and Eq. (9.36), assuming Ω to be real, hence $\kappa > \frac{R^2}{4M}$ (no overdamping).

In Section 5.4 the modes α and β have been shown to be the ones involved in the coproduct operator of the underlying q-deformed Hopf algebra structure, the q-deformation parameter being a function of R, M and t.

By using $\Gamma \equiv \frac{R}{2M}$ and the canonical linear transformations $A \equiv \frac{1}{\sqrt{2}}(\alpha + \beta)$, $B \equiv \frac{1}{\sqrt{2}}(\alpha - \beta)$, the Hamiltonian H is obtained [153, 233] as

$$H = H_0 + H_I, \tag{9.40a}$$

$$H_0 = \hbar\Omega(A^\dagger A - B^\dagger B), \qquad H_I = i\hbar\Gamma(A^\dagger B^\dagger - AB), \tag{9.40b}$$

The dynamical group structure associated with the system of coupled quantum oscillators is that of $SU(1,1)$. The two mode realization of the algebra $su(1,1)$ is indeed generated by

$$J_+ = A^\dagger B^\dagger, \quad J_- = J_+^\dagger = AB, \tag{9.41a}$$

$$J_3 = \frac{1}{2}(A^\dagger A + B^\dagger B + 1), \tag{9.41b}$$

$$[J_+, J_-] = -2J_3, \quad [J_3, J_\pm] = \pm J_\pm. \tag{9.41c}$$

The Casimir operator $\mathcal{C}$ is given by

$$\mathcal{C}^2 \equiv \frac{1}{4} + J_3^2 - \frac{1}{2}(J_+ J_- + J_- J_+) = \frac{1}{4}(A^\dagger A - B^\dagger B)^2. \tag{9.42}$$

We also observe that $[H_0, H_I] = 0$. The time evolution of the vacuum $|0\rangle \equiv |n_A = 0, n_B = 0\rangle = |0\rangle \otimes |0\rangle$, $(A \otimes 1)|0\rangle \otimes |0\rangle \equiv A|0\rangle = 0$; $(1 \otimes B)|0\rangle \otimes |0\rangle \equiv B|0\rangle = 0$, is controlled by H_I

$$|0(t)\rangle = e^{-it\frac{H}{\hbar}}|0\rangle = e^{-it\frac{H_I}{\hbar}}|0\rangle = \frac{1}{\cosh(\Gamma t)}e^{\tanh(\Gamma t)A^\dagger B^\dagger}|0\rangle, \tag{9.43}$$

$$\langle 0(t)|0(t)\rangle = 1, \quad \forall t, \tag{9.44a}$$

$$\lim_{t\to\infty} \langle 0(t)|0\rangle \propto \lim_{t\to\infty} \exp(-t\Gamma) = 0. \tag{9.44b}$$

Once one sets the initial condition of positiveness for the eigenvalues of H_0, such a condition is preserved by the time evolution since H_0 is proportional to the Casimir operator (it commutes with H_I). Thus, there is no danger of transitions to negative energy states, i.e., of dealing with energy spectrum unbounded from below. Time evolution for creation and annihilation operators is given by

$$A \to A(t) = e^{-i\frac{t}{\hbar}H_I} A \, e^{i\frac{t}{\hbar}H_I} = A \cosh\left(\Gamma t\right) - B^\dagger \sinh\left(\Gamma t\right), \quad (9.45a)$$

$$B \to B(t) = e^{-i\frac{t}{\hbar}H_I} B \, e^{i\frac{t}{\hbar}H_I} = B \cosh\left(\Gamma t\right) - A^\dagger \sinh\left(\Gamma t\right), \quad (9.45b)$$

and h.c. One can show [153] that the creation of B modes is equivalent, up to a statistical factor, to the destruction of A modes, and that the states generated by $B^\dagger$ represent the sink where the energy dissipated by the quantum damped oscillator flows: the B-oscillator represents the reservoir or heat bath coupled to the A-oscillator.

Eqs. (9.45) are time-dependent Bogoliubov transformations: they are canonical transformations preserving the CCR. Eq. (9.44) expresses the instability (decay) of the vacuum under the evolution operator $\exp\left(-it\frac{H_I}{\hbar}\right)$. In other words, time evolution leads out of the Hilbert space of the states. *This means that the QM framework is not suitable for the canonical quantization of the damped harmonic oscillator.* A way out from such a difficulty is provided by QFT [153]: the proper way to perform the canonical quantization of the DHO turns out to be working in the framework of QFT. In fact, for many degrees of freedom the time evolution operator $U(t)$ and the vacuum are formally (at finite volume) given by

$$U(t) = \prod_{\mathbf{k}} \exp\left(\Gamma_{\mathbf{k}} t(A_{\mathbf{k}}^\dagger B_{\mathbf{k}}^\dagger - A_{\mathbf{k}} B_{\mathbf{k}})\right), \quad (9.46)$$

$$|0(t)\rangle = \prod_{\mathbf{k}} \frac{1}{\cosh\left(\Gamma_{\mathbf{k}} t\right)} \exp\left(\tanh\left(\Gamma_{\mathbf{k}} t\right) A_{\mathbf{k}}^\dagger B_{\mathbf{k}}^\dagger\right) |0\rangle, \quad (9.47)$$

respectively, with $\langle 0(t)|0(t)\rangle = 1$, $\forall t$. $\mathbf{k}$ is the momentum index. Using the continuous limit relation $\sum_{\mathbf{k}} \to \frac{V}{(2\pi)^3} \int d^3\kappa$, in the infinite volume limit we have (for $\int d^3\kappa \, \Gamma_{\mathbf{k}}$ finite and positive)

$$\langle 0(t)|0\rangle \to 0 \text{ as } V \to \infty \, \forall \, t, \quad (9.48)$$

and in general, $\langle 0(t)|0(t')\rangle \to 0$ as $V \to \infty \, \forall \, t$ and t', $t' \neq t$. At each time t a representation $\{|0(t)\rangle\}$ of the CCR is defined and turns out to be unitarily inequivalent to any other representation $\{|0(t')\rangle$, $\forall t' \neq t\}$ in the infinite volume limit. In such a way, the quantum DHO evolves in time through unitarily inequivalent representations of CCR (*tunneling;* trajectories in the representation space, cf. Section 5.8 [632, 636]). We remark that $|0(t)\rangle$ is a two-mode time-dependent generalized coherent state [380, 519] where the modes A and B are entangled (cf. Section 5.5.1). We have

$$\mathcal{N}_{A_\mathbf{k}}(t) = \langle 0(t)|A_\mathbf{k}^\dagger A_\mathbf{k}|0(t)\rangle = \sinh^2 \Gamma_\mathbf{k} t \ . \tag{9.49}$$

The Bogoliubov transformations, Eqs. (9.45), can be implemented for every $\mathbf{k}$ as inner automorphism for the algebra $su(1,1)$. At each time t one has a copy $\{A_\mathbf{k}(t), A_\mathbf{k}^\dagger(t), B_\mathbf{k}(t), B_\mathbf{k}^\dagger(t) \, ; \, |0(t)\rangle \, | \, \forall \, \mathbf{k}\}$ of the original algebra induced by the time evolution operator which can thus be thought of as a generator of the group of automorphisms of $\oplus_\mathbf{k} su(1,1)_\mathbf{k}$ parametrized by time t (we have a realization of the operator algebra at each time t, which can be implemented by a Gel'fand–Naimark–Segal construction in the C*-algebra formalism [123, 505]). The various copies become unitarily inequivalent in the infinite volume limit, as shown by Eqs. (9.48): the space of the states splits into unitarily inequivalent representations of the CCR, each one labeled by time parameter t. As usual, one works at finite volume, and only at the end of the computations the limit $V \to \infty$ is performed.

We remark that the "negative" kinematic term in the Hamiltonian (9.40) (or (9.30)) also appears in two-dimensional gravity models where, in general, two different strategies are adopted in the quantization procedure [137]: the Schrödinger representation approach, where no negative norm appears, and the string/conformal field theory approach, where negative norm states arise similarly as in Gupta–Bleurer electrodynamics.

Finally, we do not discuss here the structure of time-dependent states of Bateman dual system. These aspects and other properties, including geometric phases, can be found in [97, 153].

9.3.1 The arrow of time and squeezed coherent states

It is now possible to introduce the free energy functional (cf. Eq. (5.116)) and, by extremizing it under quasi-equilibrium conditions, the state $\{|0(t)\rangle\}$ is recognized to be a representation of the CCR at finite temperature (equivalent to the TFD representation $\{|0(\beta)\rangle\}$ [600]). Consistently with the fact that damping (or, more generally, dissipation) implies the choice of a privileged time-direction (*the arrow of time*), with the consequent breaking of time-reversal invariance, time evolution is recognized to be controlled by the entropy variations [153, 185] (cf. Eq. (5.118)). The change in time $d\mathcal{N}_A$ of particles condensed in the vacuum turns into heat dissipation $dQ_a = \frac{1}{\beta} d\mathcal{S}_a$ (cf. Eq. (5.119)).

The time evolution operator $U(t)$ written in terms of α and β modes (cf. Eqs. (9.39)) is given by

$$U(t) \equiv e^{-it\frac{H_I}{\hbar}} = \prod_{\mathbf{k}} e^{-\frac{\theta_{\mathbf{k}}}{2}\left(\alpha_{\mathbf{k}}^2 - \alpha_{\mathbf{k}}^{\dagger 2}\right)} e^{\frac{\theta_{\mathbf{k}}}{2}\left(\beta_{\mathbf{k}}^2 - \beta_{\mathbf{k}}^{\dagger 2}\right)} \equiv \prod_{\mathbf{k}} \hat{S}_\alpha(\theta_{\mathbf{k}}) \hat{S}_\beta(-\theta_{\mathbf{k}}) ,$$

(9.50)

with $\hat{S}_\alpha(\theta_{\mathbf{k}}) \equiv \exp\left(-\frac{\theta_{\mathbf{k}}}{2}\left(\alpha_{\mathbf{k}}^2 - \alpha_{\mathbf{k}}^{\dagger 2}\right)\right)$ and similar expression for $\hat{S}_\beta(-\theta_{\mathbf{k}})$ with β and $\beta^\dagger$ replacing α and $\alpha^\dagger$, respectively. The operators $\hat{S}_\alpha(\theta_{\mathbf{k}})$ and $\hat{S}_\beta(-\theta_{\mathbf{k}})$ are the squeezing operators for the $\alpha_{\mathbf{k}}$ and the $\beta_{\mathbf{k}}$ modes, respectively, well known in quantum optics [664]. The set $\theta \equiv \{\theta_{\mathbf{k}} \equiv \Gamma_{\mathbf{k}} t\}$ as well as each $\theta_{\mathbf{k}}$ for all κ is called the squeezing parameter. The state $|0(t)\rangle$ is thus a two-mode squeezed coherent states at each time t.

To illustrate the effect of the squeezing, we focus our attention only on the $\alpha_{\mathbf{k}}$ modes for sake of definiteness. For the β modes one can proceed in a similar way. As usual, for given $\mathbf{k}$ we express the α mode in terms of conjugate variables of the corresponding oscillator. By using dimensionless quantities we thus write $\alpha = X + iY$, with $[X, Y] = \frac{i}{2}$. The uncertainty relation is $\Delta X \Delta Y = \frac{1}{4}$, with $\Delta X^2 = \Delta Y^2 = \frac{1}{4}$ for (minimum uncertainty) coherent states. The squeezing occurs when $\Delta X^2 < \frac{1}{4}$ and $\Delta Y^2 > \frac{1}{4}$ (or $\Delta X^2 > \frac{1}{4}$ and $\Delta Y^2 < \frac{1}{4}$) in such a way that the uncertainty relation remains unchanged. Under the action of $U(t)$ the variances ΔX and ΔY are indeed squeezed as

$$\Delta X^2(\theta) = \Delta X^2 \exp(2\theta) , \quad \Delta Y^2(\theta) = \Delta Y^2 \exp(-2\theta). \quad (9.51)$$

For the tilde-mode similar relations are obtained for the corresponding variances, say $\tilde{X}$ and $\tilde{Y}$:

$$\Delta \tilde{X}^2(\theta) = \Delta \tilde{X}^2 \exp(-2\theta) , \quad \Delta \tilde{Y}^2(\theta) = \Delta \tilde{Y}^2 \exp(2\theta). \quad (9.52)$$

For positive θ, squeezing then reduces the variances of the Y and $\tilde{X}$ variables, while the variances of the X and $\tilde{Y}$ variables grow by the same amount so as to keep the uncertainty relations unchanged. This reflects, in terms of the A and B modes, the constancy of the difference $\mathcal{N}_{A_{\mathbf{k}}} - \mathcal{N}_{B_{\mathbf{k}}}$ against separate, but equal, changes of $\mathcal{N}_{A_{\mathbf{k}}}$ and $\mathcal{N}_{B_{\mathbf{k}}}$ (degeneracy of the states $|0(t)\rangle$ *labeled* by different $\mathcal{N}_{A_{\mathbf{k}}}$, or different $\mathcal{N}_{B_{\mathbf{k}}}$, cf. Eq. (9.49)).

In conclusion, the θ-set $\{\theta_{\mathbf{k}}(\mathcal{N}_{\mathbf{k}})\}$ is nothing but the squeezing parameter classifying the squeezed coherent states in the hyperplane $(X, \tilde{X}; Y, \tilde{Y})$. Note that to different squeezed states (different θ-sets) are associated with unitarily inequivalent representations of the CCRs in the infinite volume limit. Also note that in the limit $t \to \infty$ the variances of the variables Y and $\tilde{X}$ become infinite, making them completely spread out.

Further details on the squeezing states and their relation with deformed algebraic structures in QFT can be found in [142, 143, 151, 152, 341, 342].

From our discussion it appears that the role of the doubled y coordinate (the quantum β, or B mode) is absolutely crucial in the quantum regime. In the cases of the quantum Brownian motion and the damped oscillator, it accounts for the quantum noise in the fluctuating random force in the system-environment coupling [576]. In the two-slit experiment considered in Section 5.2.1, quantum effects are obtained provided $y \neq 0$, i.e., for $x_+ \neq x_-$, namely when there is "information loss" on "which one" is the slit through which the electron passes. Quantum effects are washed out as soon as $y = 0$, i.e., $x_+ = x_-$, i.e., when one of the slits is covered and thus one "knows" from where the electron passes.

9.4 Dissipative non-commutative plane

Systems whose non-commutative geometry has been so far studied in detail in literature are the harmonic oscillator on the non-commutative plane, the motion of a particle in an external magnetic field and the Landau problem on the non-commutative sphere. Non-commutative geometries are also of interest in Chern–Simons gauge theories, in the usual gauge theories and string theories and in gravity theory [40, 51, 170, 208, 340, 566]. In this Section we show that quantum dissipation induces non-commutative geometry in the (x_+, x_-) plane. We follow [574] in our presentation.

By using $\mathcal{H}$ given by Eq. (9.30), the components in the (x_+, x_-) plane of forward and backward in time velocity $v_\pm = \dot{x}_\pm$ are obtained as

$$v_\pm = \frac{\partial \mathcal{H}}{\partial p_\pm} = \pm \frac{1}{M} \left(p_\pm \mp \frac{R}{2} x_\mp \right). \qquad (9.53)$$

They do not commute

$$[v_+, v_-] = i\hbar \frac{R}{M^2}, \qquad (9.54)$$

and it is thus impossible to fix these velocities v_+ and v_- as being identical. Eq. (9.54) is similar to the usual commutation relations for the quantum velocities $\mathbf{v} = (\mathbf{p} - (e\mathbf{A}/c))/M$ of a charged particle moving in a magnetic field $\mathbf{B}$; i.e., $[v_1, v_2] = (i\hbar e B_3/M^2 c)$. Just as the magnetic field $\mathbf{B}$ induces the Bohm–Aharonov phase interference for the charged particle, the (Brownian motion) friction coefficient R induces an analogous phase interference between forward and backward motion, as we will now discuss in connection with non-commutative geometry induced by quantum dissipation.

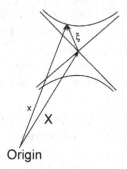

Origin

Fig. 9.1   The hyperbolic path of a particle moving in the $x = (x_+, x_-)$ plane. The non-commuting coordinate pairs $X = (X_+, X_-)$, which points from the origin to the hyperbolic center, and $\xi = (\xi_+, \xi_-)$, which points from the center of the orbit to the position on the hyperbola, are shown. $x = X + \xi$.

Similarly to Eq. (9.53), we also have

$$\dot{p}_\pm = -\frac{\partial \mathcal{H}}{\partial x_\pm} = \mp U'(x_\pm) \mp \frac{R v_\mp}{2} .\qquad(9.55)$$

From Eqs. (9.53) and (9.55) it follows that

$$M\dot{v}_\pm + R v_\mp + U'(x_\pm) = 0 .\qquad(9.56)$$

When the choice $U(x_\pm) = \frac{1}{2}\kappa x_\pm^2$ is made, these are equivalent to Eqs. (9.34) and (9.36). The classical equation of motion including dissipation thereby holds true if $x_+(t) = x_-(t) = x(t)$:

$$M\dot{v} + R v + U'(x) = 0 .\qquad(9.57)$$

If one defines $M v_\pm = \hbar K_\pm$, then Eq. (9.54) gives

$$[K_+, K_-] = \frac{iR}{\hbar} \equiv \frac{i}{L^2} ,\qquad(9.58)$$

and a canonical set of conjugate position coordinates (ξ_+, ξ_-) may be defined by

$$\xi_\pm = \mp L^2 K_\mp , \qquad [\xi_+, \xi_-] = iL^2 .\qquad(9.59)$$

Another independent canonical set of conjugate position coordinates (X_+, X_-) is defined by

$$x_\pm = X_\pm + \xi_\pm , \qquad [X_+, X_-] = -iL^2 .\qquad(9.60)$$

Note that $[X_a, \xi_b] = 0$, where $a = \pm$ and $b = \pm$.

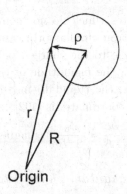

Fig. 9.2 A charge e moving in a circular cyclotron orbit. Non-commuting coordinate pairs are $\mathbf{R} = (X, Y)$, which points from the origin to the orbit center, and $\boldsymbol{\rho} = (\rho_x, \rho_y)$, which points from the center of the orbit to the charge position $\mathbf{r} = \mathbf{R} + \boldsymbol{\rho}$.

The commutation relations Eqs. (9.59) and (9.60) characterize the non-commutative geometry in the plane (x_+, x_-).

It is interesting to consider the case of pure friction in which the potential $U = 0$. Eqs. (9.30) and (9.59) then imply

$$\mathcal{H}_{friction} = \frac{\hbar^2}{2M}(K_+^2 - K_-^2) = -\frac{\hbar^2}{2ML^4}(\xi_+^2 - \xi_-^2). \tag{9.61}$$

The equations of motion are

$$\dot{\xi}_\pm = \frac{i}{\hbar}[\mathcal{H}_{friction}, \xi_\pm] = -\frac{\hbar}{ML^2}\xi_\mp = -\frac{R}{M}\xi_\mp = -\Gamma\xi_\mp, \tag{9.62}$$

with the solution

$$\begin{pmatrix} \xi_+(t) \\ \xi_-(t) \end{pmatrix} = \begin{pmatrix} \cosh(\Gamma t) & -\sinh(\Gamma t) \\ -\sinh(\Gamma t) & \cosh(\Gamma t) \end{pmatrix} \begin{pmatrix} \xi_+ \\ \xi_- \end{pmatrix}. \tag{9.63}$$

Eq. (9.63) describes the hyperbolic orbit

$$\xi_-(t)^2 - \xi_+(t)^2 = \frac{2L^2}{\hbar\Gamma}\mathcal{H}_{friction}. \tag{9.64}$$

The hyperbolae are defined by $(x - X)^2 - c^2(t - T)^2 = \Lambda^2$, where $\Lambda^2 = (\frac{mc}{\hbar}L^2)^2$, the hyperbolic center is at (X, cT) and one branch of the hyperbolae is a particle moving forward in time while the other branch is the same particle moving backward in time as an anti-particle (Fig. 9.1).

A physical realization of the mathematical non-commutative plane is present in every laboratory wherein a charged particle moves in a plane with a normal uniform magnetic field $\mathbf{B}$. For this case, there are two canonical pairs of position coordinates which do not commute: (i) the position $\mathbf{R}$ of the center of the cyclotron circular orbit and (ii) the radius vector $\boldsymbol{\rho}$ from the center of the circle to the charged particle position $\mathbf{r} = \mathbf{R} + \boldsymbol{\rho}$ (Fig. 9.2). Denoting by ϕ_0 the magnitude of the magnetic flux quantum associated with a charge e, the (Landau) magnetic length scale L of the non-commuting geometric coordinates is [402],

$$L^2 = \frac{\hbar c}{eB} = \frac{\phi_0}{2\pi B} \quad \text{(magnetic)}. \tag{9.65}$$

9.4.1 *The dissipative quantum phase interference*

We now observe that a quantum phase interference of the Aharanov–Bohm-type can always be associated with the non-commutative plane where

$$[X, Y] = iL^2, \tag{9.66}$$

with L denoting the geometric length scale in the plane. Suppose that a particle can move from an initial point in the plane to a final point in the plane via one of two paths, say $\mathcal{P}_1$ or $\mathcal{P}_2$. Since the paths start and finish at the same point, if one transverses the first path in a forward direction and the second path in a backward direction, then the resulting closed path encloses an area $\mathcal{A}$. The phase interference ϑ is determined by the difference between the actions for these two paths $\hbar\vartheta = \mathcal{S}(\mathcal{P}_1) - \mathcal{S}(\mathcal{P}_2)$, and, as shown below, it may be written as

$$\vartheta = \frac{\mathcal{A}}{L^2}. \tag{9.67}$$

For motion at fixed energy one may associate with each path $\mathcal{P}$ (in phase space) a phase space action integral

$$\mathcal{S}(\mathcal{P}) = \int_{\mathcal{P}} p_i dq^i. \tag{9.68}$$

As said previously, the phase interference ϑ between the two paths $\mathcal{P}_1$ and $\mathcal{P}_2$ is determined by the action difference

$$\hbar\vartheta = \int_{\mathcal{P}_1} p_i dq^i - \int_{\mathcal{P}_2} p_i dq^i = \oint_{\mathcal{P}=\partial\Omega} p_i dq^i, \tag{9.69}$$

wherein $\mathcal{P}$ is the closed path which goes from the initial point to the final point via path $\mathcal{P}_1$ and returns back to the initial point via $\mathcal{P}_2$. The closed $\mathcal{P}$ path may be regarded as the boundary of a two-dimensional surface Ω; i.e., $\mathcal{P} = \partial\Omega$. Stokes theorem yields

$$\vartheta = \frac{1}{\hbar} \oint_{\mathcal{P}=\partial\Omega} p_i dq^i = \frac{1}{\hbar} \int_{\Omega} (dp_i \wedge dq^i). \tag{9.70}$$

The phase interference ϑ between two alternative paths is thereby proportional to an "area" $\mathcal{A}$ of a surface Ω in phase space $(p_1, \ldots, p_f; q^1, \ldots, q^f)$ as described by the r.h.s. of Eq. (9.70).

If one reverts to the operator formalism and writes the commutation Eq. (9.66) in the non-commutative plane as

$$[X, P_X] = i\hbar \quad \text{where} \quad P_X = \left(\frac{\hbar Y}{L^2} \right), \tag{9.71}$$

then back in the path integral formalism, Eq. (9.70) reads

$$\vartheta = \frac{1}{\hbar} \int_\Omega (dP_X \wedge dX) = \frac{1}{L^2} \int_\Omega (dY \wedge dX) \tag{9.72}$$

and Eq. (9.67) is proven, i.e., the quantum phase interference between two alternative paths in the plane is determined by the non-commutative length scale L and the enclosed area $\mathcal{A}$.

We also remark that the existence of a phase interference is intimately connected to the zero point fluctuations in the coordinates; e.g., Eq. (9.66) implies a zero point uncertainty relation $\Delta X \Delta Y \geq (L^2/2)$.

Resorting back to Eq. (9.58) for the quantum dissipative case, i.e.,

$$L^2 = \frac{\hbar}{R} \quad \text{(dissipative)}, \tag{9.73}$$

one then concludes that, provided $x_+ \neq x_-$, the quantum dissipative phase interference $\vartheta = \frac{\mathcal{A}}{L^2} = \frac{\mathcal{A}R}{\hbar}$ is associated with two paths in the non-commutative plane, starting at the same point $\mathcal{P}_1$ and ending at the same point $\mathcal{P}_2$ so as to enclose the surface of area $\mathcal{A}$.

By comparing the non-commutative dissipative plane with the non-commutative Landau magnetic plane, the circular orbit in Fig. 9.2 for the magnetic problem is seen to be replaced by the hyperbolic orbit and the magnetic field is replaced by the electric field. The hyperbolic orbit in Fig. 9.1 is reflected in the classical orbit for a charged particle moving along the x-axis in a uniform electric field (see [574]).

Finally, we recall that the Lagrangian for the system of Eqs. (9.56) has been found [91] to be the same as the Lagrangian for the three-dimensional topological massive Chern–Simons gauge theory in the infrared limit. It is also the same as for a Bloch electron in a solid which propagates along a lattice plane with a hyperbolic energy surface [91]. In the Chern–Simons case, one has $\theta_{CS} = R/M = (\hbar/ML^2)$, with θ_{CS} the "topological mass parameter". In the Bloch electron case, it is $(eB/\hbar c) = (1/L^2)$, with B denoting the z-component of the applied external magnetic field. In [91] the symplectic structure for the system of Eqs. (9.56) in the case of strong damping $R \gg M$ (the so-called reduced case) has been considered in the Dirac constraint formalism as well as in the Faddeev and Jackiw formalism [231].

9.5   Gauge structure and thermal features in particle mixing

In this Section, we consider again flavor mixing, which has already been treated in Chapter 2. Here we discuss some aspects of flavor mixing which suggest that flavor oscillations may be viewed as a dissipative process, much in the spirit of the other topics treated in this Chapter.

The QFT treatment of flavor states presented in Section 2.7 leads to a vacuum state for the mixed fields, the flavor vacuum, which is orthogonal to the vacuum state for the fields with definite masses. The use of the flavor vacuum allows us to define correctly flavor states as eigenstates of the flavor charges. However, the Lorentz invariance is broken, since the flavor vacuum is explicitly time-dependent. As a consequence, flavor states cannot be interpreted in terms of irreducible representations of the Poincaré group. A possible way to recover Lorentz invariance for mixed fields has been explored in [104] where non-standard dispersion relations for the mixed particles have been related to non-linear realizations of the Poincaré group [434, 435].

The relation of neutrino masses and mixing with a possible violation of the Lorentz and CPT symmetries has been the subject of many efforts, see, e.g., [389]. A related line of research concerns the use of neutrino mixing and oscillations as a sensitive probe for quantum gravity effects, as quantum gravity-induced decoherence is expected to affect neutrino oscillations [21]. Such effects have also been connected [464] to the non-trivial structure of the flavor vacuum. For a review on CPT and Lorentz invariance in neutrino physics, see [611].

In [89] it has been shown that a non-Abelian gauge structure appears naturally in connection with flavor mixing. In this framework, it is then possible to account for the above-mentioned violation of Lorentz invariance due to the flavor vacuum having, at the same time, standard dispersion relations for flavor neutrino states.

To see how this is possible, let us start out by noting that Lagrangian (2.128) can be rewritten as describing a doublet of Dirac fields in interaction with an external Yang–Mills field:

$$\mathcal{L} = \bar{\nu}_f (i\gamma^\mu D_\mu - M_d)\nu_f, \tag{9.74}$$

where $\nu_f = (\nu_e, \nu_\mu)^T$ is the flavor doublet and $M_d = \text{diag}(m_e, m_\mu)$ is a diagonal mass matrix. Note that the mixing term, proportional to $m_{e\mu}$, is taken into account by the (non-Abelian) covariant derivative:

$$D_0 \equiv \partial_0 + i\, m_{e\mu}\, \beta\, \sigma_1, \tag{9.75}$$

where $m_{e\mu} = \frac{1}{2} \tan 2\theta\, \delta m$, and $\delta m \equiv m_\mu - m_e$.

We thus see that flavor mixing can be seen as an interaction of the flavor fields with an $SU(2)$ constant gauge field having the following structure:

$$A_\mu \equiv \frac{1}{2} A_\mu^a \sigma_a = n_\mu \delta m \frac{\sigma_1}{2} \in su(2), \qquad n^\mu \equiv (1,0,0,0)^T, \tag{9.76}$$

that is, having only the temporal component in spacetime and only the first component in $su(2)$ space. In terms of this connection, the covariant derivative can be written in the form:

$$D_\mu = \partial_\mu + i\,g\,\beta\,A_\mu, \tag{9.77}$$

where we have defined $g \equiv \tan 2\theta$ as the coupling constant for the mixing interaction. Note that in the case of maximal mixing ($\theta = \pi/4$), the coupling constant grows to infinity while δm goes to zero. We further note that, since the gauge connection is a constant, with just one non-zero component in group space, its field strength vanishes identically:

$$F_{\mu\nu}^a = \epsilon^{abc} A_\mu^b A_\nu^c = 0, \tag{9.78}$$

with $a, b, c = 1, 2, 3$. The fact that the gauge field has physical effects (despite $F_{\mu\nu}$ vanishes identically), leads to an analogy with the Aharonov–Bohm effect [11].

Here α_i, $i = 1, 2, 3$ and β are the usual Dirac matrices in a given representation. For definiteness, we choose the following representation:

$$\alpha_i = \begin{pmatrix} 0 & \sigma_i \\ \sigma_i & 0 \end{pmatrix}, \qquad \beta = \begin{pmatrix} \mathbb{1} & 0 \\ 0 & -\mathbb{1} \end{pmatrix}, \tag{9.79}$$

where σ_i are the Pauli matrices and $\mathbb{1}$ is the 2×2 identity matrix.

We now consider the energy momentum tensor associated with the flavor neutrino fields in interaction with the external gauge field. This is easily obtained by means of the standard procedure [421]:

$$\widetilde{T}_{\rho\sigma} = \bar{\nu}_f i\gamma_\rho D_\sigma \nu_f - \eta_{\rho\sigma} \bar{\nu}_f (i\gamma^\lambda D_\lambda - M_d)\nu_f. \tag{9.80}$$

This is to be compared with the canonical energy momentum tensor associated with the Lagrangian (2.128):

$$T_{\rho\sigma} = \bar{\nu}_f i\gamma_\rho D_\sigma \nu_f - \eta_{\rho\sigma} \bar{\nu}_f (i\gamma^\lambda D_\lambda - M_d)\nu_f + \eta_{\rho\sigma} m_{e\mu} \bar{\nu}_f \sigma_1 \nu_f. \tag{9.81}$$

We then define a 4-momentum operator as $\widetilde{P}^\mu \equiv \int d^3\mathbf{x}\, \widetilde{T}^{0\mu}$ and obtain a conserved 3-momentum operator:

$$\widetilde{P}^i = i \int d^3\mathbf{x}\, \nu_f^\dagger \partial^i \nu_f$$

$$= i \int d^3\mathbf{x}\, \nu_e^\dagger \partial^i \nu_e + i \int d^3\mathbf{x}\, \nu_\mu^\dagger \partial^i \nu_\mu$$

$$\equiv \widetilde{P}_e^i(x_0) + \widetilde{P}_\mu^i(x_0), \qquad i = 1,2,3 \qquad (9.82)$$

and a non-conserved Hamiltonian operator:

$$\widetilde{P}^0(x_0) \equiv \widetilde{H}(x_0) = \int d^3\mathbf{x}\, \bar\nu_f \left(i\gamma_0 D_0 - i\gamma^\mu D_\mu + M_d \right) \nu_f$$

$$= \int d^3\mathbf{x}\, \nu_e^\dagger \left(-i\boldsymbol{\alpha} \cdot \boldsymbol{\nabla} + \beta m_e \right) \nu_e + \int d^3\mathbf{x}\, \nu_\mu^\dagger \left(-i\boldsymbol{\alpha} \cdot \boldsymbol{\nabla} + \beta m_\mu \right) \nu_\mu$$

$$\equiv \widetilde{H}_e(x_0) + \widetilde{H}_\mu(x_0). \qquad (9.83)$$

Note that both the Hamiltonian and the momentum operators split in a contribution involving only the electron neutrino field and in another where only the muon neutrino field appears.

We remark that the tilde Hamiltonian is *not* the generator of time translations. This role competes to the complete Hamiltonian $H = \int d^3\mathbf{x}\, T^{00}$, obtained from the energy-momentum tensor Eq. (9.81).

We now show that it is possible to define flavor neutrino states which are simultaneous eigenstates of the 4-momentum operators above constructed and of the flavor charges. Such a non-trivial request requires a redefinition of the flavor vacuum. To this end, let us expand the flavor neutrino field operators in a different mass basis with respect to Eq. (2.105), as follows:

$$\nu_\sigma(x) = \& \int \tfrac{d^3 k}{(2\pi)^{3/2}} \sum_r \left[u_{\mathbf{k},\sigma}^r(x_0) \widetilde{\alpha}_{\mathbf{k},\sigma}^r(x_0) + v_{-\mathbf{k},\sigma}^r(x_0) \widetilde{\beta}_{-\mathbf{k},\sigma}^{r\dagger}(x_0) \right] e^{i\mathbf{k}\cdot\mathbf{x}}, \quad (9.84)$$

with $u_{\mathbf{k},\sigma}^r(x_0) = u_{\mathbf{k},\sigma}^r e^{-i\omega_{\mathbf{k},\sigma}x_0}$, $v_{-\mathbf{k},\sigma}^r(x_0) = v_{-\mathbf{k},\sigma}^r e^{i\omega_{\mathbf{k},\sigma}x_0}$. The new spinors are defined as the solutions of the equations:

$$(-\boldsymbol{\alpha} \cdot \mathbf{k} + m_\sigma \beta) u_{\mathbf{k},\sigma}^r = \omega_{\mathbf{k},\sigma} u_{\mathbf{k},\sigma}^r, \qquad (9.85a)$$

$$(-\boldsymbol{\alpha} \cdot \mathbf{k} + m_\sigma \beta) v_{-\mathbf{k},\sigma}^r = -\omega_{\mathbf{k},\sigma} v_{-\mathbf{k},\sigma}^r, \qquad (9.85b)$$

where $\omega_{\mathbf{k},\sigma} = \sqrt{\mathbf{k}^2 + m_\sigma^2}$ and $\sigma = e, \mu$.

The tilde flavor operators are connected to the previous ones by a Bogoliubov transformation:

$$\begin{pmatrix} \widetilde{\alpha}_{\mathbf{k},\sigma}^r(x_0) \\ \widetilde{\beta}_{-\mathbf{k},\sigma}^{r\dagger}(x_0) \end{pmatrix} = J^{-1}(x_0) \begin{pmatrix} \alpha_{\mathbf{k},\sigma}^r(x_0) \\ \beta_{-\mathbf{k},\sigma}^{r\dagger}(x_0) \end{pmatrix} J(x_0), \qquad (9.86)$$

with generator [270]:

$$J(x_0) = \prod_{\mathbf{k},r} \exp\left\{ i \sum_{(\sigma,j)} \xi_{\sigma,j}^{\mathbf{k}} \left[\alpha_{\mathbf{k},\sigma}^{r\dagger}(x_0)\beta_{-\mathbf{k},\sigma}^{r\dagger}(x_0) + \beta_{-\mathbf{k},\sigma}^r(x_0)\alpha_{\mathbf{k},\sigma}^r(x_0) \right] \right\},$$

(9.87)

with $(\sigma,j) = (e,1), (\mu,2)$, and $\xi_{\sigma,j}^{\mathbf{k}} = (\chi_\sigma - \chi_j)/2$ and $\chi_\sigma = \arctan(m_\sigma/|\mathbf{k}|)$, $\chi_j = \arctan(m_j/|\mathbf{k}|)$. The new flavor vacuum is given by

$$|\widetilde{0}(x_0)\rangle_{e\mu} = J^{-1}(x_0)|0(x_0)\rangle_{e\mu}.$$

(9.88)

Notice that the flavor charges are invariant under the above Bogoliubov transformations [109], i.e., $\widetilde{Q}_\sigma = Q_\sigma$, with:

$$\widetilde{Q}_\sigma(x_0) = \sum_r \int d^3\mathbf{k} \left(\widetilde{\alpha}_{\mathbf{k}\sigma}^{r\dagger}(x_0)\widetilde{\alpha}_{\mathbf{k}\sigma}^r(x_0) - \widetilde{\beta}_{-\mathbf{k}\sigma}^{r\dagger}(x_0)\widetilde{\beta}_{-\mathbf{k}\sigma}^r(x_0) \right). \quad (9.89)$$

In terms of the tilde flavor ladder operators, the Hamiltonian and momentum operators Eqs. (9.82) and (9.83) read:

$$\widetilde{\mathbf{P}}_\sigma(x_0) = \sum_r \int d^3\mathbf{k}\,\mathbf{k} \left(\widetilde{\alpha}_{\mathbf{k},\sigma}^{r\dagger}(x_0)\widetilde{\alpha}_{\mathbf{k},\sigma}^r(x_0) + \widetilde{\beta}_{\mathbf{k},\sigma}^{r\dagger}(x_0)\widetilde{\beta}_{\mathbf{k},\sigma}^r(x_0) \right), \quad (9.90a)$$

$$\widetilde{H}_\sigma(x_0) = \sum_r \int d^3\mathbf{k}\,\omega_{\mathbf{k},\sigma} \left(\widetilde{\alpha}_{\mathbf{k},\sigma}^{r\dagger}(x_0)\,\widetilde{\alpha}_{\mathbf{k},\sigma}^r(x_0) - \widetilde{\beta}_{\mathbf{k},\sigma}^r(x_0)\,\widetilde{\beta}_{\mathbf{k},\sigma}^{r\dagger}(x_0) \right). \quad (9.90b)$$

Since all the above operators are diagonal, we can define common eigenstates as follows:

$$|\widetilde{\nu}_{\mathbf{k},\sigma}^r(x_0)\rangle = \widetilde{\alpha}_{\mathbf{k},\sigma}^{r\dagger}(x_0)|\widetilde{0}(x_0)\rangle_{e\mu}$$

(9.91)

and similar ones for the antiparticles. In particular, these single particle states are eigenstates of both the Hamiltonian and the momentum operator:

$$\begin{pmatrix} \widetilde{H}_\sigma(x_0) \\ \widetilde{\mathbf{P}}_\sigma(x_0) \end{pmatrix} |\widetilde{\nu}_{\mathbf{k},\sigma}^r(x_0)\rangle = \begin{pmatrix} \omega_{\mathbf{k},\sigma} \\ \mathbf{k} \end{pmatrix} |\widetilde{\nu}_{\mathbf{k},\sigma}^r(x_0)\rangle,$$

(9.92)

making explicit the 4-vector structure.

Note that the above construction and the consequent Poincaré invariance holds at a given time x_0. Thus, for each different time, we have a different Poincaré structure. Flavor neutrino fields behave (locally in time) as ordinary on-shell fields with definite masses m_e and m_μ, rather than those of the mass eigenstates of the standard approach, m_1 and m_2. Flavor oscillations then arise as a consequence of the interaction with the gauge field, which acts as a sort of refractive medium – *neutrino aether*. This leads to an interesting analogy with some scenarios in which, for the case of photons, the vacuum has been thought to act as a refractive medium in consequence of quantum gravity fluctuations [216].

Let us now consider the interpretation of the Hamiltonian operator $\widetilde{H}$ which, as already remarked, does not take into account the interaction energy, i.e., the energy associated with mixing. One can view $\widetilde{H}$ as the sum of the kinetic energies of the flavor neutrinos, or equivalently as the energy which can be extracted from flavor neutrinos by scattering processes, the mixing energy being "frozen" (there's no way to turn off the mixing!). This suggests the interpretation of such a quantity as a "free" energy $F \equiv \widetilde{H}$, so that we can write:

$$H - F = TS. \tag{9.93}$$

This quantity defines an entropy associated with flavor mixing. It is natural to identify the "temperature" T with the coupling constant $g = \tan 2\theta$, thus leading to:

$$S = \int d^3\mathbf{x}\, \bar{\nu}_f A_0 \nu_f = \frac{1}{2} \delta m \int d^3\mathbf{x} \left(\bar{\nu}_e \nu_\mu + \bar{\nu}_\mu \nu_e \right). \tag{9.94}$$

The appearance of an entropy should not be surprising, since each of the two flavor neutrinos can be considered as an open system which presents some kind of (cyclic) dissipation. In Appendix Q, in a simplified QM context, it is shown that at a given time, the difference of the expectation values of the muon and electron free energies is less than the total initial energy of the flavor neutrino state. The missing part is proportional to the expectation value of the entropy.

Finally we point out that the thermodynamical considerations developed in this Section, fit well with the interpretation of the gauge field as a reservoir, as discussed in Section 5.6.

Phenomenological consequences

The above analysis leads us to the view that the flavor fields ν_e and ν_μ should be regarded as fundamental. This fact has some interesting consequences at phenomenological level. Indeed, if we consider a charged current process in which for example an electron neutrino is created, we see that the hypothesis that mixing is due to interaction with an external field, implies that what is created in the vertex is really $|\nu_e\rangle$, rather than $|\nu_1\rangle$ or $|\nu_2\rangle$. As remarked above, such an interpretation is made possible because we can regard, at any given time, flavor fields as on-shell fields, associated with masses m_e and m_μ.

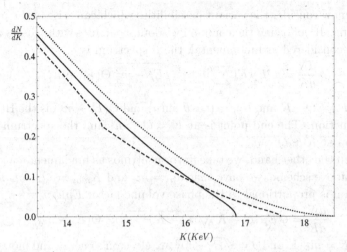

Fig. 9.3 The tail of the tritium β spectrum for: - a massless neutrino (dotted line); - fundamental flavor states (continuous line); - superposed prediction for 2 mass states (short-dashed line): notice the inflexion in the spectrum where the most massive state switches off. We used $m_e = 1.75$ KeV, $m_1 = 1$ KeV, $m_2 = 4$ KeV, $\theta = \pi/6$.

We consider the case of a beta decay process, say for definiteness tritium decay, which allows for a direct investigation of neutrino mass. In the following we compare the various possible outcomes of this experiment predicted by the different theoretical possibilities for the nature of mixed neutrinos. As we shall see, the scenario described above presents significant phenomenological differences with respect to the standard theory.

Let us then consider the decay:

$$A \to B + e^- + \bar{\nu}_e,$$

where A and B are two nuclei (e.g., ^{3}H and ^{3}He).

The electron spectrum is proportional to phase volume factor $EpE_e p_e$:

$$\frac{dN}{dK} = CEp(Q - K)\sqrt{(Q - K)^2 - m_\nu^2}, \tag{9.95}$$

where $E = m + K$ and $p = \sqrt{E^2 - m^2}$ are electron's energy and momentum. The endpoint of β decay is the maximal kinetic energy K_{max} the electron can take (constrained by the available energy $Q = E_A - E_B - m \approx m_A - m_B - m$). In the case of tritium decay, $Q = 18.6$ KeV. Q is shared between the (unmeasured) neutrino energy and the (measured) electron kinetic energy K. It is clear that if the neutrinos were massless, then $m_\nu = 0$ and $K_{max} = Q$. On the other hand, if the neutrinos were a mass eigenstate with $m_\nu = m_1$, then $K_{max} = Q - m_1$.

We now consider the various possibilities which can arise in the presence of mixing. If, following the common wisdom, neutrinos with masses m_1 and m_2 are considered as fundamental, the β spectrum is:

$$\frac{dN}{dK} = CEp\,E_e \sum_j |U_{ej}|^2 \sqrt{E_e^2 - m_j^2}\, \Theta(E_e - m_j)\,, \qquad (9.96)$$

where $E_e = Q - K$ and $U_{ej} = (\cos\theta, \sin\theta)$ and $\Theta(E_e - m_j)$ is the Heaviside step function. The end point is at $K = Q - m_1$ and the spectrum has an inflexion at $K \simeq Q - m_2$.

If, on the other hand, we take flavor neutrinos as fundamental according to the above scheme, we have that $m_\nu = m_e$ and $K_{max} = Q - m_e$ and the spectrum is proportional to the phase volume factor $EpE_e p_e$:

$$\frac{dN}{dK} = CEp\,(Q - K)\sqrt{(Q - K)^2 - m_e^2}\, \Theta(E_e - m_e)\,, \qquad (9.97)$$

where $E = m + K$ and $p = \sqrt{E^2 - m^2}$ are electron's energy and momentum.

The above discussed possibilities are plotted in Fig. 9.3, together with the spectrum for a massless neutrino, for comparison.

We remark also that in the neutrino detection process, it would be possible to discriminate among the various scenarios considered above. In such a case, our scheme would imply that in each detection vertex, either an electron neutrino or a muon neutrino would take part in the process. Again, this is in contrast with the standard view, which assumes that either ν_1 or ν_2 are entering into the elementary processes.

9.6 Dissipation and the many-body model of the brain

One application of quantum dissipation formalism has been made in the study of neural dynamics in the frame of the many-body model of the brain proposed by Umezawa and Ricciardi in 1967 [534, 582, 583]. In this Section we summarize very briefly the dissipative many-body model of the brain [641, 643, 645] and give an account of how it fits some experimental data [260, 261].

The mesoscopic neural activity of the neocortex appears in laboratory observations to consist of the dynamical formation of spatially extended domains in which widespread cooperation supports brief epochs of patterned oscillations. Imaging of scalp potentials (electroencephalograms, EEGs) and cortical surface potentials (electrocorticograms, ECoGs) of animals and humans has demonstrated the formation of large-scale patterns of synchronized oscillations in the neocortex in the $12 - 80\ Hz$ range (denoted as the β or γ range). These patterns are also observed by magnetoencephalografic (MEG) imaging in the resting state and in motor task-related states of the human brain [61]. Observations show that large-scale neuronal assemblies in β and γ ranges re-synchronize in frames at frame rates in the $3 - 12\ Hz$ range (θ and α ranges) [244, 246–250, 260, 261] and appear to extend over spatial domains covering much of the hemisphere in rabbits and cats [249, 250, 260], and over domains of linear size of $\approx 19\ cm$ in the human cortex with near zero phase dispersion [256, 259]. Karl Lashley, by operating on trained rats, was led, in the first half of the 20th century, to the hypothesis of "mass action" in the storage and retrieval of memories in the brain: "Here is the dilemma. Nerve impulses are transmitted ... from cell to cell through definite intercellular connections. Yet, all behavior seems to be determined by masses of excitation ... within general fields of activity, without regard to particular nerve cells ... What sort of nervous organization might be capable of responding to a pattern of excitation without limited specialized path of conduction? The problem is almost universal in the activity of the nervous system" (pp. 302–306 of [409]). The presence of repeated phase transitions in collective cortical dynamics has been subsequently confirmed in neurophysiological experiments. In the sixties, Karl Pribram, on the basis of his own laboratory observations confirming Lashley's hypothesis of mass action, described the fields of neural activity in brain by use of the hologram conceptual framework [528, 529]. In 1967 Umezawa and Ricciardi [534, 582, 583] proposed the description of the collective neural activity, which manifests in the formation of spatially extended domains, by using the mechanism of SSB in QFT. By resorting to preceding studies on the physics of living matter [193–195] and to the QFT formalism for dissipative systems [153], the extension to the dissipative dynamics of the Umezawa and Ricciardi many-body model has been worked out [641, 645] and the comparison of the predictions of the dissipative quantum model of brain with the laboratory observations has been pursued [260–262]. It appears that the dissipative model predicts the experimentally observed coexistence of physically distinct amplitude modulated (AM) and phase modulated (PM) patterns, correlated with categories of conditioned stimuli, and the remarkably rapid onset of AM patterns into irreversible sequences that resemble cinematographic frames [260,261]. The dissipative model also accounts for the formation of phase cones (Fig. 9.4) and vortices in the transition from one AM pattern to another one [251,261]. In the following we follow closely the discussion presented in [645].

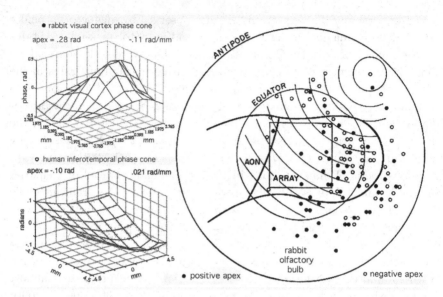

Fig. 9.4   The silhouette shows the outline of the olfactory bulb, on which is superimposed a 4 × 4 mm rectangle giving the position of the surgically placed 8 × 8 electrode array. The two circles specify the flattened surface of the spherical bulb. The solid and open dots show the locations of conic apices. The arcs show the isophase contours of one phase cone at 0.1 rad intervals. The insets at left give examples of phase cones respectively from rabbit and human. [Adapted from [255]]

The functional stability and efficiency of living matter

The collective behavior of a large ensemble of elementary components is the object of study of Statistical Mechanics. In the case of neural components, Hopfield [332] asked whether stability of memory and other macroscopic properties of neural nets are also derivable as collective phenomena and emergent properties. Classical Statistical Mechanics provides very powerful tools in answering Hopfield's question [22, 473]. However, at a classical level analysis, the electric field of the extracellular dendritic current and the magnetic fields inside the dendritic shafts appear to be far too weak and the chemical diffusion appears to be far too slow to be able to fully account for the cortical collective activity observed in the laboratory [250,261]. Molecular (neuro-)biology provides crucial tools in the discovery of many mechanisms and chemical functions in the brain and living matter in general. The question is then how to put together all this data so as to derive the observed complex behavior of the whole system. In this connection, Schrödinger has introduced the important distinction between the "two ways of producing orderliness" (p. 80 of [553]): ordering generated by the "statistical mechanisms" and ordering generated by "dynamical" interactions among the atoms and the molecules, which, as is well known, are *necessarily* quantum interactions.

The functional stability in living systems is characterized by the time ordering of pathways of biochemical reactions sequentially interlocked. Common laboratory experience is that even the simplest chemical reaction pathway, once embedded in a random chemical environment, soon collapses. Chemical efficiency and functional stability to the degree observed in living matter, i.e., not as "regularity only in the average" [553], appears to be out of reach of any probabilistic approach *solely* based on microscopic random kinematics. Thus, it is still a *matter of belief* that out of purely random kinematics there may arise with high probability a unique, time ordered sequence of chemical reactions like the one required by the macroscopic history of the system. It is a fact that there is no available computation or abstract proof which shows how to obtain the characteristic chemical efficiency and stability of living matter by resorting uniquely to statistical concepts. Even in the (seemingly) simpler case of the generation of spatially ordered domains and tissues in living systems, the evident failure of any model solely based on random chemical kinematics, or even on short range forces assembling cells one-by-one, is the real (unfortunately dramatic) obstacle in biology and medicine preventing the understanding of how and why cells are assembled in healthy tissues, and how and why a healthy tissue might evolve into a cancer. In Schrödinger's words: "it needs no poetical imagination but only clear and sober scientific reflection to recognize that we are here obviously faced with events whose regular and lawful unfolding is guided by a mechanism entirely different from the "probability mechanism of physics" (p. 79 of [553]). Classical Statistical Mechanics and short range forces of molecular biology, *although necessary*, do not seem to be completely adequate tools. Therefore, it is necessary to supplement them with a further step so as to include underlying quantum *dynamical* features. Moreover, one more motivation to follow the research path pioneered by Schrödinger is in the fact that there is no conventional neural network offered by neuroscientists or neuroengineers that can adequately model or simulate the neurophysiological data on brain activity which are available today [262, 263].

The many-body model of the brain

In such a conceptual frame and aware of the state of the experimental observations, Umezawa and Ricciardi [534, 616] observed that large-scale neuron assemblies might be described in terms of SSB long-range correlations induced as a response to the external stimuli. They proposed their QFT model for the brain guided by the observation that any modeling of its functioning cannot rely on the knowledge of the behavior of any single neuron. They remarked [534] that it is in fact pure optimism to hope to determine the numerical values for the coupling coefficients and the thresholds of all neurons by means of anatomical or physiological methods. On the other hand, the behavior of any single neuron should not be significant for the functioning of the whole brain, otherwise a higher and higher degree of malfunctioning should be observed as some of the neurons die. This clearly excludes that the high stability of brain functions, e.g., of memory, over a long period of time could be explained solely in terms of specific, localized arrangements of biomolecules. Observations [252–254, 297, 298, 528, 529] show, on the contrary, that long-range correlations appear in the brain as a response to external stimuli.

In the many-body model, patterns of correlated elements are described by the mechanism of SSB. The order parameter, related to the density of coherently condensed NG bosons in the vacuum, is the macroscopic observable specifying the degree of ordering of the system vacuum, the *vacuum code* classifying it among many possible degenerate vacua. In the model, the memory content is specified by such a code. The external informational input triggers the breakdown of symmetry out of which the NG bosons and their condensation are generated. The recall of the recorded information occurs under the input of a stimulus "similar" to the one responsible for the memory recording [534] (see also [573]).

We stress that the external stimulus only acts as a trigger inducing SSB: the ordered pattern is generated by the "internal" brain dynamics, which, except for the breakdown of the symmetry, is not conditioned by the external stimulus (*spontaneous* breakdown of the symmetry). This model feature is perfectly consistent with laboratory observations; it accounts for the observed lack of invariance of AM patterns with invariant stimuli [244, 245, 260] (Fig. 9.5).

A different regime is, in contrast, obtained when symmetry is *explicitly* broken, as, for example, under the effect of an electric shock or a highly stressing stimulus by which the cortex dynamics is enslaved,

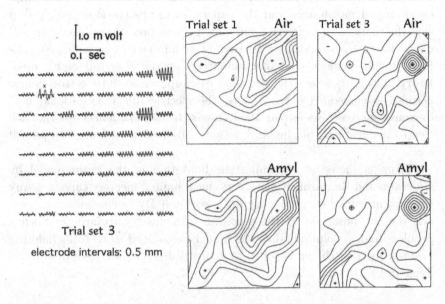

Fig. 9.5 Left: Example of an AM carrier wave in one frame. Center: contour plots of AM patterns from 4×4 mm frames showing differences correlated with stimuli. Right: Demonstration of lack of invariance of AM patterns with respect to stimuli, showing that AM patterns are not representations of stimuli; they are operators containing activated memory about stimuli (center frames), which is subject to continual updating with new experience (right frames). The vertical differences reflect short-term memory changes upon the formation of an assembly. The horizontal differences reflect long-term memory changes in consolidation. From [252]

with consequent response by the so-called evoked or *event-related potential* (ERP). The explicit breakdown in cortical dynamics is observed by resort to stimulus-locked averaging across multiple presentations in order to remove or attenuate the background activity, so as to demonstrate that the location, intensity and detailed configuration of the ERP is predominantly determined by the stimulus.

It has been recognized that the symmetry which gets broken is the rotational symmetry of the electrical dipoles of the water molecules [356–358] and the NG modes are the vibrational dipole wave quanta (DWQ) [193–195]. The whole brain dynamics is indeed embedded in a matrix of molecules carrying the quantum degree of freedom associated with the electric dipole oscillatory motion. The well-known electrical properties of cell membranes and the experimental observations [310, 522] of slow fluctuations in neuronal membrane polarization (the so-called up and down states), corresponding to that of spontaneous fluctuations in the fMRI signal, confirm that one cannot ignore the electrical dipole oscillatory matrix in which the neuronal electrophysical and electrochemical activity is embedded [531]. Thus, the electric dipole vibrational field dynamics at basic molecular level cannot be without effect on the rich electrochemical and biochemical activity observed at the neural classical level.

There is, however, an important limitation in the brain model by Umezawa and Ricciardi: it has a very much limited memory capacity. Any subsequent stimulus, producing the associated DWQ condensation, cancels the one produced by the preceding stimulus (overprinting). Such a problem finds a solution when the model is modified so as to include the brain property of being an open system ruled, therefore, by a dissipative dynamics [641].

The dissipative many-body model of the brain

The brain is an *open system* continuously coupled with the environment. Brain dynamics is thus intrinsically dissipative. Once dissipation is considered, the coexistence of physically distinct amplitude modulated and phase modulated patterns, correlated with categories of conditioned stimuli, and their extremely rapid onset into irreversible sequences are predicted by the model. These are indeed two main features of neurophysiological data [260–263] fitted by the dissipative model of brain.

In the previous Sections, we have seen that the procedure of the canonical quantization of a dissipative system requires the "doubling" of the degrees of freedom of the system [153] in order to ensure that the flow of the energy exchanged between the system and the environment is balanced. When considering the brain system, let us denote by A_k and $\tilde{A}_k$ the annihilation operators for the DWQ mode and its "doubled mode", respectively. Here k denotes the momentum $\mathbf{k}$ and other specifications of the A operators ($A_k^\dagger$ and $\tilde{A}_k^\dagger$ denote the creation operators).

Let $\mathcal{N}$ be the *code* imprinted in the vacuum at the initial time $t_0 = 0$ by the external input and representing the *memory record* of the input. The code $\mathcal{N}$ is the set of the numbers $\mathcal{N}_{A_k}$ of modes A_k, for any k, condensed in the vacuum state denoted by $|0\rangle_{\mathcal{N}}$, which thus represents the memory state at $t_0 = 0$ [20, 641]. $\mathcal{N}_{A_k}(t)$ is given, at each t, by (cf. Eq. (9.49)):

$$\mathcal{N}_{A_k}(t) \equiv {}_{\mathcal{N}}\langle 0(t)|A_k^\dagger A_k|0(t)\rangle_{\mathcal{N}} = \sinh^2\left(\Gamma_k t - \theta_k\right), \qquad (9.98)$$

and similarly for the modes $\tilde{A}_k$. The state $|0(t)\rangle_{\mathcal{N}} \equiv |0(\theta, t)\rangle$ is the time-evolved of the state $|0\rangle_{\mathcal{N}}$. It is a generalized $SU(1,1)$ squeezed coherent state, where the A and B modes are entangled. Γ is the damping constant (related to the memory life-time) and θ_k fixes the code value at $t_0 = 0$. $|0\rangle_{\mathcal{N}}$ and $|0(t)\rangle_{\mathcal{N}}$ are normalized to 1. In the infinite volume limit it is

$$_{\mathcal{N}}\langle 0(t)|0\rangle_{\mathcal{N}'} \xrightarrow[V\to\infty]{} 0 \quad \forall t \neq t_0, \ \forall \mathcal{N}, \mathcal{N}', \qquad (9.99a)$$

$$_{\mathcal{N}}\langle 0(t)|0(t')\rangle_{\mathcal{N}'} \xrightarrow[V\to\infty]{} 0, \quad \forall t, t' \ \text{with} \ t \neq t', \ \forall \mathcal{N}, \mathcal{N}', \qquad (9.99b)$$

with $|0(t)\rangle_{\mathcal{N}'} \equiv |0(\theta', t)\rangle$ given by Eq. (9.47) where the $B^\dagger$ operator is substituted by $\tilde{A}^\dagger$ and $\Gamma_k t$ is substituted by $\Gamma_k t - \theta_k$. Eqs. (9.99) also hold for $\mathcal{N} \neq \mathcal{N}'$, $t = t_0$ and $t = t'$, respectively. They show that in the infinite volume limit, the vacua of the same code $\mathcal{N}$ at different times t and t', for any t and t', and, similarly, at equal times, but different $\mathcal{N}$s, are orthogonal states. The corresponding Hilbert spaces are unitarily inequivalent spaces. The number $\left(\mathcal{N}_{A_k} - \mathcal{N}_{\tilde{A}_k}\right)$ is a constant of motion for any k and θ. The physical meaning of the $\tilde{A}$ system is the one of the sink where the energy dissipated by the A system flows. The $\tilde{A}$ modes describe the thermal bath or the environment modes (cf. Section 9.3).

The balance of energy flow between the system and the environment is ensured by the requirement $\mathcal{N}_{A_k} - \mathcal{N}_{\tilde{A}_k} = 0$, for any k. Such a requirement, however, does not uniquely fix the code $\mathcal{N} \equiv \{\mathcal{N}_{A_k}, \text{ for any } k\}$. In addition, $|0\rangle_{\mathcal{N}'}$ with $\mathcal{N}' \equiv \{\mathcal{N}'_{A_k}; \mathcal{N}'_{A_k} - \mathcal{N}'_{\tilde{A}_k} = 0, \text{ for any } k\}$ ensures the energy flow balance. Thus, also $|0\rangle_{\mathcal{N}'}$ is an available memory state: it corresponds, however, to a different code number (i.e., $\mathcal{N}'$) and therefore to an information different from the one of code $\mathcal{N}$. In the infinite volume limit, $\{|0\rangle_{\mathcal{N}}\}$ and $\{|0\rangle_{\mathcal{N}'}\}$ are representations of the canonical commutation relations each other unitarily inequivalent for different codes $\mathcal{N} \neq \mathcal{N}'$. Thus, an infinite number of memory (vacuum) states may exist, each one of them corresponding to a different code $\mathcal{N}$. A huge number of sequentially recorded inputs may *coexist* without destructive interference since infinitely many vacua $|0\rangle_{\mathcal{N}}$, for all $\mathcal{N}$, are *independently* accessible in the sequential recording process.

In conclusion, the "brain (ground) state" is represented as the collection (or the superposition) of the full set of states $|0\rangle_{\mathcal{N}}$, for all $\mathcal{N}$. The brain is thus described as a complex system with a huge number of macroscopic states (the space of the memory states).

The degree of the coupling of the system A with the system $\tilde{A}$ can be parametrized by an index, say n, in such a way that in the limit of $n \to \infty$ the possibilities of the system A to couple to $\tilde{A}$ (the environment) are "saturated": the system A then gets *fully* coupled to $\tilde{A}$. As a matter of fact, as a result of the dissipative dynamics, the brain is entangled with its environment. A higher or lower *degree of openness* (measured by n) to the external world may produce a better or worse ability in setting up neuronal correlates, respectively (different under different circumstances, and so on, e.g., during sleep or awake states, childhood or older ages) [20]. The functional or effective connectivity (here we do not consider the structural or anatomical one) is highly dynamic in the dissipative model. Once these functional connections are formed, they are not necessarily fixed. On the contrary, they may quickly change and new configurations of connections may be formed extending over a domain including a larger or smaller number of neurons. The finite size of the correlated domain implies a non-zero effective mass of the DWQs. These propagate through the domain with a greater inertia than in the case of large (infinite) volume where they are (quasi-)massless. The domain correlations are then established with a certain time-delay, which contributes to the delay observed in the recruitment of neurons in a correlated assembly under the action of an external stimulus.

As seen in Section 5.5, the minimization of the free energy, $d\mathcal{F}_A = dE_A - \frac{1}{\beta}d\mathcal{S}_A = 0$, is ensured and the change in time $d\mathcal{N}_A$ of particles condensed in the vacuum turns into heat dissipation $dQ_A = \frac{1}{\beta}d\mathcal{S}_A$ (cf. Eq. (5.119)). See Appendix R for a discussion of further aspects of the dissipative many-body model of brain.

We close this Section by stressing that *neurons and other brain cells are by no means considered quantum objects in the Umezawa and Ricciardi many-body model and in the dissipative many-body model.* These QFT models differ in a substantial way from brain models formulated in the Quantum Mechanics frame, such as those discussed in [312, 517, 577] (see also [463]). In contrast with such models, the neurons and other brain cells are classical objects in the dissipative many-body model of brain.

9.7   Quantization and dissipation

G. 't Hooft has proposed [589–593] that the quantum nature of our world may emerge from an underlying deterministic dynamics acting at an energy scale much higher than the one of our observations. More specifically,

Quantum Mechanics would result from a more fundamental deterministic theory as a consequence of a process of information loss. The long-standing puzzle of the contrast between a probabilistic view of natural laws and a deterministic one may find a solution in 't Hooft's scenario, together with the problem of quantizing gravity. It is important to remark that this scenario is quite different from previous attempts relying on the existence of hidden variables (for a recent review see [276]). It is also interesting that such a conjecture has renewed the debate about more speculative implications including the free will problem [594].

't Hooft considers a class of deterministic Hamiltonian systems which can be described by means Hilbert space techniques. The quantum systems are obtained when constraints implementing the information loss are imposed on the original Hilbert space. The Hamiltonian for such systems is of the form

$$H = \sum_i p_i \, f_i(q) \,, \tag{9.100}$$

where $f_i(q)$ are non-singular functions of the coordinates q_i. The equations for the q's (i.e., $\dot{q}_i = \{q_i, H\} = f_i(q)$) are decoupled from the conjugate momenta p_i and this implies [589, 590] that the system can be described deterministically even when expressed in terms of operators acting on the Hilbert space. The condition for the deterministic description is the existence of a complete set of observables commuting at all times, called *beables* [68]. For the systems of Eq. (9.100), such a set is given by the $q_i(t)$'s [589, 590].

Hamiltonians of the type (9.100) are not bounded from below. This may be cured by splitting H in Eq. (9.100) as [589, 590]:

$$H = H_I - H_{II} \,, \tag{9.101a}$$

$$H_I = \frac{1}{4\rho} \left(\rho + H\right)^2 \,, \qquad H_{II} = \frac{1}{4\rho} \left(\rho - H\right)^2 \,, \tag{9.101b}$$

where ρ is a time-independent, positive function of q_i. H_I and H_{II} are then positively (semi)definite and $\{H_I, H_{II}\} = \{\rho, H\} = 0$. The constraint

$$H_{II}|\psi\rangle = 0 \,, \tag{9.102}$$

projecting out the states responsible for the negative part of the spectrum, then ensures that the Hamiltonian is bounded from below. In other words, one thus gets rid of the unstable trajectories [589, 590].

In [103] and [85] it has been shown that the Bateman system of classical damped-antidamped oscillators discussed in Section 9.3 does provide an explicit realization of 't Hooft's mechanism. Moreover, a connection exists between the zero point energy of the quantum harmonic oscillator and the geometric phase of the damped-antidamped oscillator (deterministic) system. Indeed, the Hamiltonian Eq. (9.40) is of the type (9.100) with $i = 1, 2$ and $f_1(q) = 2\Omega$, $f_2(q) = -2\Gamma$, provided a set of canonical transformations is used, which for brevity we do not report here (see [103]). With the choice $\rho = 2\Omega\mathcal{C}$ in Eq. (9.101), and by using $J_2 = -\frac{i}{2}(J_+ - J_-)$ and $\mathcal{C} = \frac{1}{2}(A^\dagger A - B^\dagger B)$, Eq. (9.40) becomes

$$H = H_I - H_{II} , \tag{9.103a}$$

$$H_I = \frac{1}{2\Omega\mathcal{C}}(2\Omega\mathcal{C} - \Gamma J_2)^2 , \quad H_{II} = \frac{\Gamma^2}{2\Omega\mathcal{C}}J_2^2 . \tag{9.103b}$$

Since $\mathcal{C}$ is the Casimir operator, it is a constant of motion and, as already observed, this ensures that once it has been chosen to be positive it will remain such at all times. The physical states $|\psi\rangle$ are defined by imposing the constraint (9.102), which now reads

$$J_2|\psi\rangle = 0 . \tag{9.104}$$

It is convenient to introduce

$$x_1 = \frac{x + y}{\sqrt{2}} = r \cosh u , \quad x_2 = \frac{x - y}{\sqrt{2}} = r \sinh u , \tag{9.105}$$

in terms of which [91]

$$\mathcal{C} = \frac{1}{4\Omega m}\left[p_r^2 - \frac{1}{r^2}p_u^2 + m^2\Omega^2 r^2\right] , \quad J_2 = \frac{1}{2}p_u . \tag{9.106}$$

Of course, only non-zero r^2 should be taken into account in order for $\mathcal{C}$ to be invertible. If one does not use the operatorial formalism, then the constraint $p_u = 0$ implies $u = -\frac{\gamma}{2m}t$. Eq. (9.104) implies

$$H|\psi\rangle = H_I|\psi\rangle = 2\Omega\mathcal{C}|\psi\rangle = \left(\frac{1}{2m}p_r^2 + \frac{K}{2}r^2\right)|\psi\rangle , \tag{9.107}$$

where $K \equiv m\Omega^2$. H_I thus reduces to the Hamiltonian for the linear harmonic oscillator $\ddot{r} + \Omega^2 r = 0$. The physical states are even with respect to time-reversal ($|\psi(t)\rangle = |\psi(-t)\rangle$) and periodical with period $\tau = \frac{2\pi}{\Omega}$.

The states $|\psi(t)\rangle_H$ and $|\psi(t)\rangle_{H_I}$ are introduced, satisfying the equations:

$$i\hbar\frac{d}{dt}|\psi(t)\rangle_H = H|\psi(t)\rangle_H , \tag{9.108}$$

$$i\hbar\frac{d}{dt}|\psi(t)\rangle_{H_I} = 2\Omega\mathcal{C}|\psi(t)\rangle_{H_I} . \tag{9.109}$$

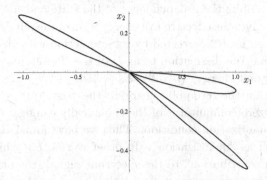

Fig. 9.6 Trajectories for $r_0 = 0$ and $v_0 = \Omega$, after three half-periods for $\kappa = 20$, $\gamma = 1.2$ and $m = 5$. The ratio $\int_0^{\tau/2}(\dot{x}_1 x_2 - \dot{x}_2 x_1)dt/\mathcal{E} = \pi\frac{\Gamma}{m\Omega^3}$ is preserved. $\mathcal{E}$ is the initial energy: $\mathcal{E} = \frac{1}{2}mv_0^2 + \frac{1}{2}m\Omega^2 r_0^2$.

Eq. (9.109) describes the two-dimensional "isotropic" (or "radial") harmonic oscillator. $H_I = 2\Omega\mathcal{C}$ has the spectrum $\mathcal{H}_I^n = \hbar\Omega n$, $n = 0, \pm 1, \pm 2, \ldots$. According to the choice for $\mathcal{C}$ to be positive, only positive values of n will be considered. The generic state $|\psi(t)\rangle_H$ can be written as

$$|\psi(t)\rangle_H = \hat{T}\left[\exp\left(\frac{i}{\hbar}\int_{t_0}^t 2\Gamma J_2 dt'\right)\right]|\psi(t)\rangle_{H_I}, \qquad (9.110)$$

where $\hat{T}$ denotes time-ordering and $\hbar$ is introduced on dimensional grounds. Its actual value cannot be fixed by the present analysis. One has [103]:

$$_H\langle\psi(\tau)|\psi(0)\rangle_H = {}_{H_I}\langle\psi(0)|\exp\left(i\int_{C_{0\tau}}A(t')dt'\right)|\psi(0)\rangle_{H_I} \equiv e^{i\phi}. \quad (9.111)$$

The contour $C_{0\tau}$ goes from $t' = 0$ to $t' = \tau$ and back, and $A(t) \equiv \frac{\Gamma m}{\hbar}(\dot{x}_1 x_2 - \dot{x}_2 x_1)$. Note that $(\dot{x}_1 x_2 - \dot{x}_2 x_1)dt$ is the area element in the (x_1, x_2) plane enclosed by the trajectories (see Fig. 9.6). The evolution (or dynamical) part of the phase does not enter in ϕ, as the integral in Eq. (9.111) picks up a purely geometric contribution [23].

We consider the periodic physical states $|\psi\rangle$ and write [23]

$$|\psi(\tau)\rangle = e^{i\phi - \frac{i}{\hbar}\int_0^\tau\langle\psi(t)|H|\psi(t)\rangle dt}|\psi(0)\rangle = e^{-i2\pi n}|\psi(0)\rangle, \qquad (9.112)$$

i.e., $\frac{\langle\psi(\tau)|H|\psi(\tau)\rangle}{\hbar}\tau - \phi = 2\pi n$, $n = 0, 1, 2, \ldots$, and, using $\tau = \frac{2\pi}{\Omega}$ and

$\phi = \alpha\pi$,

$$\mathcal{H}_{I,\text{eff}}^n \equiv \langle\psi_n(\tau)|H|\psi_n(\tau)\rangle = \hbar\Omega\left(n + \frac{\alpha}{2}\right). \qquad (9.113)$$

The index n exhibits the n dependence of the state and the corresponding energy. $\mathcal{H}^n_{I,eff}$ gives the effective nth energy level of the physical system, i.e., the energy given by $\mathcal{H}^n_I$ corrected by its interaction with the environment. We thus see that the dissipation term J_2 of the Hamiltonian is responsible for the "zero point energy" ($n = 0$): $E_0 = \frac{\hbar}{2}\Omega\alpha$.

We recall that the zero point energy is the "signature" of quantization, since the non-zero commutator of the canonically conjugate q and p operators *is* the quantization condition. Thus we have found that dissipation manifests itself as "quantization". In other words, E_0, which appears as the "quantum contribution" to the spectrum, signals the underlying dissipative dynamics. If one wants to match the Quantum Mechanics zero point energy, one has to fix $\alpha = 1$, which gives [103] $\Omega = \frac{\gamma}{m}$.

The thermodynamical features of the dynamical role of J_2 (cf. Section 9.3 and Chapter 5) can be revealed by rewriting Eq. (9.110) as

$$|\psi(t)\rangle_H = \hat{T}\left[\exp\left(i\frac{1}{\hbar}\int_{u(t_0)}^{u(t)} 2J_2 du'\right)\right]|\psi(t)\rangle_{H_I}, \qquad (9.114)$$

where $u(t) = -\Gamma t$ has been used. Thus,

$$-i\hbar\frac{\partial}{\partial u}|\psi(t)\rangle_H = 2J_2|\psi(t)\rangle_H. \qquad (9.115)$$

$2J_2$ appears then to be responsible for shifts (translations) in the u variable, as it has to be expected since $2J_2 = p_u$ (cf. Eq. (9.106)). Indeed, one can write: $p_u = -i\hbar\frac{\partial}{\partial u}$. Then, in full generality, Eq. (9.104) defines families of physical states, representing stable, periodic trajectories (cf. Eq. (9.107)). $2J_2$ implements transition from family to family, according to Eq. (9.115). Eq. (9.108) can then be rewritten as

$$i\hbar\frac{d}{dt}|\psi(t)\rangle_H = i\hbar\frac{\partial}{\partial t}|\psi(t)\rangle_H + i\hbar\frac{du}{dt}\frac{\partial}{\partial u}|\psi(t)\rangle_H. \qquad (9.116)$$

The first term on the r.h.s. denotes the derivative with respect to the explicit time dependence of the state. The dissipation contribution to the energy is thus described by the "translations" in the u variable. Now we consider the relation

$$\frac{\partial S}{\partial U} = \frac{1}{T}. \qquad (9.117)$$

From Eq. (9.103), by using $S \equiv \frac{2J_2}{\hbar}$ and $U \equiv 2\Omega\mathcal{C}$, one obtains $T = \hbar\Gamma$. Eq. (9.117) is the defining relation for temperature in thermodynamics (with $k_B = 1$), so that one could formally regard $\hbar\Gamma$ (which is an energy dimensionally) as the temperature, provided the dimensionless quantity

S is identified with the entropy. In such a case, the "full Hamiltonian" Eq. (9.103) plays the role of the free energy F: $H = 2\Omega\mathcal{C} - (\hbar\Gamma)\frac{2J_2}{\hbar} = U - TS = F$. Thus $2\Gamma J_2$ represents the heat contribution in H. Of course, consistently, $\frac{\partial F}{\partial T}\big|_{\Omega} = -\frac{2J_2}{\hbar}$. In conclusion $\frac{2J_2}{\hbar}$ behaves as the entropy, which is not surprising since it controls the dissipative (thus irreversible) part of the dynamics (cf. Sections 5.5 and 9.3). It is also suggestive that the temperature $\hbar\Gamma$ is actually given by the background zero point energy: $\hbar\Gamma = \frac{\hbar\Omega}{2}$. Finally, we observe that

$$\frac{\partial F}{\partial \Omega}\bigg|_T = \frac{\partial U}{\partial \Omega}\bigg|_T = mr^2\Omega\,, \qquad (9.118)$$

which is the angular momentum, as expected since it is the conjugate variable of the angular velocity Ω.

Note that Eq. (9.104) can be interpreted as a condition for an adiabatic physical system. $\frac{2J_2}{\hbar}$ might be viewed as an analogue of the Kolmogorov–Sinai entropy for chaotic dynamical systems. Further developments along this line, which we do not discuss here, are reported in [85, 97–99, 217–219, 594, 595].

Composite systems

The above scheme can also be extended to composite systems. We review an explicit example of this studied in [100]. The Hamiltonians of two Bateman oscillators, denoted by subscripts A and B, are

$$
\begin{aligned}
H_i &= \frac{1}{m_i}p_{x_i}p_{y_i} + \frac{\gamma_i}{2m_i}(y_ip_{y_i} - x_ip_{x_i}) + \left(\kappa_i - \frac{\gamma_i^2}{4m_i}\right)x_iy_i \\
&= 2\left(\Omega_i\mathcal{C}_i - \Gamma_iJ_{2i}\right),
\end{aligned}
\qquad (9.119)
$$

where $i = A, B$. We now consider the composite system with Hamiltonian

$$H_T = H_A + H_B = 2(\Omega_A\mathcal{C}_A + \Omega_B\mathcal{C}_B) - 2(\Gamma_AJ_{2A} + \Gamma_BJ_{2B})\,, \qquad (9.120)$$

and implement the quantization procedure above outlined. The Casimir operators $\mathcal{C}_i$ of the respective $su(1,1)$ algebras and J_{2i} are constants of motion. Once $\mathcal{C}_i$ are chosen to be positive (as we do from now on), they remain such at all times. Then, we can define new integrals of motion:

$$\mathcal{C} \equiv \frac{\Omega_A\mathcal{C}_A + \Omega_B\mathcal{C}_B}{\Omega}\,, \quad J \equiv \frac{\Gamma_AJ_{2A} + \Gamma_BJ_{2B}}{\Gamma}\,, \qquad (9.121)$$

where Ω and Γ are numbers to be defined shortly. Using the fact that $\Omega_i > 0$ and assuming that $\Omega > 0$, we may conclude that $\mathcal{C}_A, \mathcal{C}_B > 0 \Rightarrow \mathcal{C} > 0$.

The total Hamiltonian will be

$$H_T = 2\Omega\mathcal{C} - 2\Gamma J.$$ (9.122)

Note that the Hamiltonian (9.122) for the total system reproduces exactly the Hamiltonian of each of the two subsystems (cf. Eq. (9.119)). The system is auto-similar. With the choice $\rho = 2\Omega\mathcal{C}$ (cf. Eq. (9.101)), H_T can be split as

$$H_+ = \frac{(H_T + 2\Omega\mathcal{C})^2}{8\Omega\mathcal{C}} = \frac{1}{2\Omega\mathcal{C}}(2\Omega\mathcal{C} - \Gamma J)^2,$$ (9.123a)

$$H_- = \frac{(H_T - 2\Omega\mathcal{C})^2}{8\Omega\mathcal{C}} = \frac{1}{2\Omega\mathcal{C}}\Gamma^2 J^2.$$ (9.123b)

We note that $\mathcal{C}, J$ are again beables because they are functions of beables.

Let us now impose the constraint on the Hilbert space as

$$H_-|\psi\rangle_{phys} = J|\psi\rangle_{phys} = 0.$$ (9.124)

This implies

$$H_T \approx H_+ \approx 2\Omega\mathcal{C}$$ (9.125)

($\approx$ indicates that operators are equal only on the physical states). Since $J = (\Gamma_A J_{2A} + \Gamma_B J_{2B})/\Gamma$, the condition $J \approx 0$ implies a relation between J_{2A} and J_{2B}. Solving with respect to J_{2B}, Eq. (9.124) gives

$$J_{2B} \approx -\frac{\Gamma_A}{\Gamma_B} J_{2A},$$ (9.126)

which, when substituted into H_T, yields

$$H_T \approx H_+ \approx \left(\frac{p_{r_A}^2}{2m_A} - \frac{2J_{2A}^2}{m_A\, r_A^2} + \frac{1}{2}m_A\,\Omega_A^2 r_A^2 \right)$$
$$+ \left(\frac{p_{r_B}^2}{2m_B} + \frac{1}{2}m_B\,\Omega_B^2 r_B^2 \right) - \frac{2}{m_B}\frac{\Gamma_A^2}{\Gamma_B^2}\frac{1}{r_B^2}\,J_{2A}^2.$$ (9.127)

The emergent Hamiltonian H_T, Eq. (9.127), represents (on the physical states) a good quantum system, since, by its very construction, it is bounded from below. The term inside the first parenthesis is $2\,\Omega_A\mathcal{C}_A$, which is constant, because $\mathcal{C}_A$ is an integral of motion. The second term represents a QM oscillator, and the third one corresponds to a centripetal barrier. The potential $1/r^2$ is analogous to the centrifugal contribution in polar coordinates and one may thus expect an exact solvability. The only difference here is that $r \in \mathbb{R}$ and not merely $\mathbb{R}^+$. The system with the Hamiltonian

$$H = \frac{N^2}{2}\,p_{r_B}^2 + \frac{Q^2}{2}\,r_B^2 + \frac{R^2 - N^2/4}{2r_B^2},\qquad r_B \in \mathbb{R},\ N, Q, R \in \mathbb{R}^+,$$ (9.128)

is known as the *isotonic* oscillator [402]. Its spectrum can be exactly solved by purely algebraic means since the Hamiltonian admits a shape-invariant factorization [205]. The energy eigenvalues read [205, 299]

$$E_{n,\mp} = QN\left(2n \mp \frac{R}{N} + 1\right), \quad n \in \mathbb{N}. \tag{9.129}$$

If $R/N \leq 1/2$, the potential is attractive in the origin, and both the negative and positive sign must be taken into account [299] (this is indicated by the notation $\mp$). Whenever $R/N > 1/2$, then the positive sign in front of R/N has to be taken, and the inverse square potential is repulsive at the origin, so the motion takes place only in the domain $r_B > 0$. Since in our case

$$N^2 = \frac{1}{m_B}, \quad Q^2 = m_B\Omega_B^2, \quad R^2/N^2 = \frac{1}{4} - \left(\frac{2\Gamma_A}{\Gamma_B}\mu_A\right)^2, \tag{9.130}$$

with μ_A being the eigenvalue of J_{2A}, the actual spectrum of (9.127) is [100]

$$E_{n,\mp}(c, \mu_A) = \Omega_B\left(2n \mp \sqrt{\frac{1}{4} - \left(\frac{2\Gamma_A}{\Gamma_B}\mu_A\right)^2} + 1\right) + c, \quad n \in \mathbb{N}, \tag{9.131}$$

where c is a shift constant term due to the presence of $2\Omega_A\mathcal{C}_A$. Note that when Γ_A is small (in particular $2\Gamma_A\mu_A/\Gamma_B \ll 1/2$), the inverse square potential in (9.127) can be neglected and the system reduces to that of a QM linear oscillator with a shift term c. This follows also directly from the spectrum (9.131) provided we set $\Gamma_A = 0$ and consider both signs.

In conclusion, we have considered the possibility of regarding, according to 't Hooft's proposal, the quantization as a low energy manifestation of loss of information in deterministic systems occurring at high energy (Planck scale). Our presentation shows how these subjects are related with the general mathematical structure of QFT described in this book.

Appendix Q

Entropy and geometrical phases in neutrino mixing

In order to clarify some aspects of neutrino mixing and oscillations discussed in Section 9.5, we consider here the simplified situation of neutrino oscillations in a QM context.

Let us define flavor (fermionic) annihilation operators as [89]:

$$\alpha_e(t) = \cos\theta\,\alpha_1(t) + \sin\theta\,\alpha_2(t)\,, \tag{Q.1}$$

$$\alpha_\mu(t) = -\sin\theta\,\alpha_1(t) + \cos\theta\,\alpha_2(t), \tag{Q.2}$$

where $\alpha_i(t) = e^{i\omega_i t}\alpha_i$, $i = 1, 2$. The Pontecorvo flavor states are given by:

$$|\nu_\sigma(t)\rangle = \alpha_\sigma^\dagger(t)|0\rangle_m, \qquad \sigma = e, \mu, \tag{Q.3}$$

where $|0\rangle_m = |0\rangle_1 \otimes |0\rangle_2$ is the vacuum for the mass eigenstates. We use the notation $|\nu_\sigma\rangle = |\nu_\sigma(t = 0)\rangle$. The Hamiltonian of the system is:

$$
\begin{aligned}
H &= \omega_e\alpha_e^\dagger(t)\alpha_e(t) + \omega_\mu\alpha_\mu^\dagger(t)\alpha_\mu(t) + \omega_{e\mu}\left[\alpha_e^\dagger(t)\alpha_\mu(t) + \alpha_\mu^\dagger(t)\alpha_e(t)\right] \\
&= \omega_1\alpha_1^\dagger\alpha_1 + \omega_2\alpha_2^\dagger\alpha_2, \tag{Q.4}
\end{aligned}
$$

where $\omega_e = \omega_1 \cos^2\theta + \omega_2 \sin^2\theta$, $\omega_\mu = \omega_1 \sin^2\theta + \omega_2 \cos^2\theta$, $\omega_{e\mu} = (\omega_2 - \omega_1)\sin\theta\cos\theta$.

In analogy with the QFT case Eq. (9.75), we now define a covariant derivative:

$$D_t = \frac{d}{dt} + igA = \frac{d}{dt} + i\omega_{e\mu}\sigma_1, \tag{Q.5}$$

where $\omega_{e\mu} = \frac{1}{2}\tan 2\theta\delta\omega$, $\delta\omega = \omega_\mu - \omega_e$ and $A \equiv \delta\omega\frac{\sigma_1}{2}$. Then the equations of motion read:

$$D_t\,\alpha_f = -i\omega_d\,\alpha_f, \tag{Q.6}$$

where $\alpha_f = (\alpha_e, \alpha_\mu)^T$ and $\omega_d = \text{diag}(\omega_e, \omega_\mu)$. The Hamiltonian becomes

$$H = \alpha_f^\dagger\omega_d\alpha_f + g\alpha_f^\dagger A\alpha_f. \tag{Q.7}$$

	H	F_e	F_μ	$TS_e = TS_\mu$
$|\nu_e(0)\rangle$	ω_e	$\omega_e(1 - P(t))$	$\omega_\mu P(t)$	$\frac{1}{2}\delta\omega P(t)$
$|\nu_\mu(0)\rangle$	ω_μ	$\omega_\mu P(t)$	$\omega_e(1 - P(t))$	$-\frac{1}{2}\delta\omega P(t)$

Fig. Q.1   Energetic balance for flavor neutrino states. $P(t)$ denotes the transition probability $P_{\nu_e \to \nu_\mu}(t)$.

The diagonal part of the above expression can be readily split into separate contributions for each flavor

$$\widetilde{H}(t) = \alpha_f^\dagger \omega_d \alpha_f = \omega_e \alpha_e^\dagger(t)\alpha_e(t) + \omega_\mu \alpha_\mu^\dagger(t)\alpha_\mu(t) = \widetilde{H}_e(t) + \widetilde{H}_\mu(t). \quad \text{(Q.8)}$$

The expectation values of the flavor number operators on the flavor neutrino states (Q.3) at time zero give the oscillation probabilities (cf. Eqs. (H.6)):

$$\langle \nu_e(0)|N_e(t)|\nu_e(0)\rangle = P_{\nu_e \to \nu_e}(t) = 1 - \sin^2 2\theta \, \sin^2\left(\frac{\omega_2 - \omega_1}{2}t\right), \quad \text{(Q.9)}$$

$$\langle \nu_e(0)|N_\mu(t)|\nu_e(0)\rangle = P_{\nu_e \to \nu_\mu}(t) = \sin^2 2\theta \, \sin^2\left(\frac{\omega_2 - \omega_1}{2}t\right). \quad \text{(Q.10)}$$

Thus we have:

$$\langle \nu_e(0)|\widetilde{H}_e(t)|\nu_e(0)\rangle = \omega_e P_{\nu_e \to \nu_e}(t), \quad \text{(Q.11)}$$

$$\langle \nu_e(0)|\widetilde{H}_\mu(t)|\nu_e(0)\rangle = \omega_\mu P_{\nu_e \to \nu_\mu}(t). \quad \text{(Q.12)}$$

In analogy with the field theoretical case, we regard these "free" Hamiltonians as free energies, and we write:

$$H = \sum_{\sigma = e,\mu} (F_\sigma(t) + TS_\sigma(t)), \quad \text{(Q.13)}$$

where we set $g \equiv T$ and

$$S_\sigma(t) = \frac{1}{4} \, \delta\omega \left[\alpha_e^\dagger(t)\alpha_\mu(t) + \alpha_\mu^\dagger(t)\alpha_e(t)\right]. \quad \text{(Q.14)}$$

We have:

$$\langle \nu_e(0)|S_e(t)|\nu_e(0)\rangle = -\frac{1}{4} \, \delta\omega \sin 4\theta \sin^2\left[\frac{1}{2}(\omega_2 - \omega_1)t\right], \quad \text{(Q.15)}$$

with the same result for $S_\mu(t)$.

All the expectation values obtained are summarized in Fig. Q.1, from which we see how the energetic balance is recovered. The situation for an electron neutrino state is represented in Fig. Q.2 for sample values of the parameters.

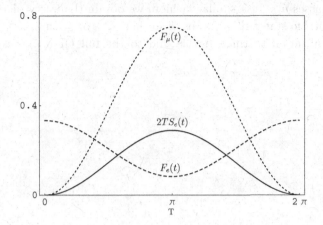

Fig. Q.2 Plot of expectation values on $|\nu_e(0)\rangle$ of $F_e(t)$ (long-dashed line), $F_\mu(t)$ (short-dashed line) and $2TS_e(t)$ (solid line). We used rescaled dimensionless time $T = (\omega_2 - \omega_1)t$ and $\theta = \pi/6$. The scale on the vertical axis is normalized to ω_μ.

Note finally that the integral of the entropy expectation value over an oscillation cycle, is only dependent on the mixing angle:

$$\int_0^T \langle \nu_e(0)|S_e(t)|\nu_e(0)\rangle \, dt \; = \; \pi \cos^2 2\theta \, \sin 2\theta. \qquad (Q.16)$$

where the period $T = \frac{2\pi}{\omega_2 - \omega_1}$.

It is interesting to compare this result with other geometric invariants appearing in neutrino oscillations [94]. To see this, let us write the electron neutrino state at time t as

$$|\nu_e(t)\rangle = e^{-i\omega_1 t}\left(\cos\theta \, |\nu_1\rangle \; + \; e^{-i(\omega_2 - \omega_1)t} \, \sin\theta \, |\nu_2\rangle \right). \qquad (Q.17)$$

The state $|\nu_e(t)\rangle$, apart from a phase factor, reproduces the initial state $|\nu_e(0)\rangle$ after a period $T = \frac{2\pi}{\omega_2 - \omega_1}$:

$$|\nu_e(T)\rangle = e^{i\phi}|\nu_e(0)\rangle, \qquad\qquad \phi = -\frac{2\pi\omega_1}{\omega_2 - \omega_1}. \qquad (Q.18)$$

Such a time evolution does contain a purely geometric part. Geometric and dynamical phases are separated following the standard procedure [23]:

$$\beta_e \; = \; \phi + \int_0^T \langle\nu_e(t)| \, i\partial_t \, |\nu_e(t)\rangle \, dt \; = \; 2\pi \sin^2\theta.$$

We thus see that there is indeed a non-zero geometrical phase β_e, related to the mixing angle θ, and that it is independent of the neutrino energies

ω_i and masses m_i. In a similar fashion, we obtain the Berry phase for the muon neutrino state: $\beta_\mu = 2\pi \cos^2 \theta$ so that $\beta_e + \beta_\mu = 2\pi$.

Generalization to three flavors and to the full QFT case is given in [81, 654].

Appendix R

Trajectories in the memory space

In the infinite volume limit, Eqs. (9.99) also hold true for $\mathcal{N} = \mathcal{N}'$. Time evolution of the state $|0\rangle_{\mathcal{N}}$ is thus represented as the (continual) transition through the representations $\{|0(t)\rangle_{\mathcal{N}}, \forall \mathcal{N}, \forall t\}$, namely by the "trajectory" through the "points" $\{|0(t)\rangle_{\mathcal{N}}, \forall \mathcal{N}, \forall t\}$ in the space of the representations (on each trajectory the free energy functional is minimized). The trajectory initial condition at $t_0 = 0$ is specified by the $\mathcal{N}$-set. We have seen in Section 5.8 that: *a)* these trajectories are classical trajectories and *b)* they are chaotic trajectories [520, 521, 632]. In other words, the trajectories are bounded and each trajectory does not intersect itself; there are no intersections between trajectories specified by different initial conditions and trajectories of different initial conditions are diverging trajectories. The meaning of these properties has been commented upon in Section 5.8. In the case of brain dynamics, the property that trajectories specified by different initial conditions ($\mathcal{N} \neq \mathcal{N}'$) never cross each other implies that no *confusion* (interference) arises among the codes of different neuronal correlates, even as time evolves. In realistic situations of finite volume, states with different codes may have non-zero overlap (the inner products, Eqs. (9.99), are not zero). In such a case, at a "crossing" point between two, or more than two, trajectories, there can be "ambiguities" in the sense that one can switch from one of these trajectories to another one which crosses. This may indeed be felt as an *association* of memories.

In Section 5.8 we have seen that for a very small difference $\delta\theta_k \equiv \theta_k - \theta'_k$ in the initial conditions of two initial states, the modulus of the difference $\Delta\mathcal{N}_{A_k}(t)$ and its time derivative diverge, for large enough t, as $\exp(2\Gamma_k t)$, for all ks. This may account for the high perceptive resolution in the recognition of the perceptual inputs. Moreover, the difference between k-*components* of the codes $\mathcal{N}$ and $\mathcal{N}'$ may become zero at a given time

$t_k = \frac{\theta_k}{\Gamma_k}$. However, the difference between the codes $\mathcal{N}$ and $\mathcal{N}'$ does not necessarily become zero. The codes are different even if a finite number of their components are equal since they are made up of a large number of $\mathcal{N}_{A_k}(\theta, t)$ components (infinite in the continuum limit). On the other hand, suppose that, for very small $\delta\theta_k \equiv \theta_k - \theta'_k$, the time interval $\Delta t = \tau_{max} - \tau_{min}$, with τ_{min} and τ_{max} the minimum and the maximum, respectively, of $t_k = \frac{\theta_k}{\Gamma_k}$, for *all* k's, are very small. Then the codes are recognized to be *almost* equal in such a Δt, which then expresses the recognition (or recall) process time. This shows how it is possible that "slightly different" $\mathcal{N}_{A_k}$-patterns (or codes) are recognized to be the *same code* even if corresponding to slightly different inputs.

We also remark that the vacua $\{|0(t)\rangle_{\mathcal{N}}, \forall\mathcal{N}, \forall t\}$, through which trajectories go, behave as non-linear dynamical "attractors" since they are the system ground states (for each $\mathcal{N}$ at each t). Their manifold, which constitutes the space of the brain states, may thus be thought of as the attractor landscape for the (non-linear) brain dynamics. It is possible to show that the trajectories in such a landscape are classical ones [261, 262].

The brain engagement with its environment and its Double

In the engagement (entanglement, see the comment after Eq. (9.98)) of the brain with the environment in the action-perception cycle, the existing attractor landscape provides the "accumulated experience", or the "knowledge", or the "context" in which the new perception happens to be situated. Under the influence of the new perception the brain reaches a new "balance" in a new vacuum, which enters the attractor landscape producing its global rearrangement. The process of the formation of the *meaning* of the newly acquired information and of its "contextualization" consists in such a global rearrangement of the attractor landscape, which, in turn, defines the conditions to the subsequent action.

The whole process involving the brain engagement with its environment has been referred to as the "dialog" between the brain and its *Double* (the environment is formally described by the "doubled" degrees of freedom, expressing at the same time the environment and the brain image, see Section 9.3). Consciousness has been proposed to be rooted in such a permanent dialog between the subject and its Double [641, 645].

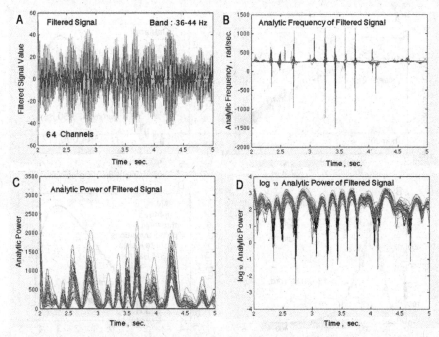

Fig. R.1 Null spikes are observed by band pass filtering the EEG (A), applying the Hilbert transform to get the analytic power (B), and taking the logarithm (C). On each channel the downward spikes coincide with spikes in analytic frequency (D) reflecting increased analytic phase variance. The flat segment between spikes reflects the stability of the carrier frequency of AM patterns. The spikes form clusters in time but are not precisely synchronized. One or more of these null spikes coincides with phase transitions leading to emergence of AM patterns. The modal repetition rate of the null spikes in Hz is predicted to be 0.641 times the pass band width in Hz (Eqs. (3.8)–(3.15) in [535]).

Time dependence of the frequency of the dipole wave quanta implies that higher momentum components of the $\mathcal{N}$-set possess longer lifetimes. Momentum is proportional to the inverse distance over which the mode propagates, thus modes with a shorter range of propagation (more "localized" modes) survive longer, modes with a longer range of propagation decay sooner.

The dissipative model of brain thus predicts the existence of correlated domains of different finite sizes with different degrees of stability [20]. They are described by the condensation function $f(x)$, which acts as a "form factor" specific for the considered domain [15, 17, 617]. In the presence of

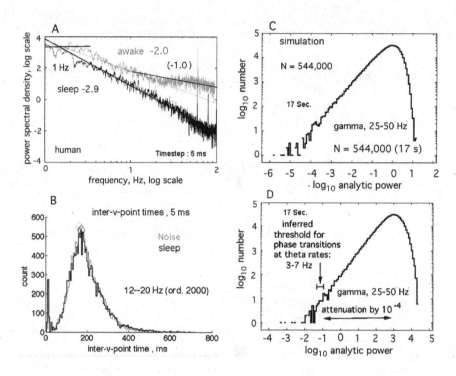

Fig. R.2 Evidence is summarized showing that the mesoscopic background activity conforms to scale-free, low-dimensional noise [257]. Engagement of the brain in perception and other goal-directed behaviors is accompanied by departures from randomness upon the emergence of order (A), as shown by comparing power spectral densities (PSD) in sleep, which conforms to black noise, vs. PSD in an aroused state showing excess power in the theta $(3 - 7$ Hz$)$ and gamma $(25 - 100$ Hz$)$ ranges. (B) The distributions of time intervals between null spikes of brown noise and sleep ECoG are superimposed. (C), (D) The distributions are compared of $\log_{10}$ analytic power from noise and ECoG. Hypothetically the threshold for triggering a phase transition is 10^{-4} down from modal analytic power. From [258,264].

the electromagnetic field, $f(x)$ has to carry some topological singularity in order for the condensation process to be physically detectable, and phase transitions driven by boson condensation are always associated with some singularity (indeterminacy) in the field phase at the phase transition point (cf. Section 4.3.2). These features hold in the dissipative model too, which thus accounts for a crucial mechanism observed in laboratory experiments: the event that initiates a perceptual phase transition is an abrupt decrease, named *null spike*, in the analytic power of the background activity to near zero (Fig. R.1). The model also accounts for the observation of vortices and propagating waves of phase gradients, called phase cones, in perceptual phase transitions. These features are described in detail in [262, 263].

In Section 1.8.1 we have seen that coherent states exhibit self-similarity. The dissipative model describes the brain activity in terms of coherent states and thus predicts self-similarity as a characterizing property of brain dynamics [634, 635]. This prediction is confirmed by laboratory observations which show that brain background activity is characterized by self-similarity. Indeed, measurements of the durations, recurrence intervals and diameters of neocortical EEG phase patterns have power-law distributions with no detectable minima. The power spectral densities in time and space of ECoGs from surface arrays conform to power-law distributions [121, 246, 247, 335, 426], suggesting that the activity patterns generated by neocortical neuropil might be scale-free [248, 655], with self-similarity in ECoGs patterns over distances ranging from hypercolumns to an entire cerebral hemisphere [260] (Fig. R.2). For more details on these topics see [634]. For a general review on scaling laws with particular reference to cognitive sciences see [368].

Chapter 10

Elements of soliton theory and related concepts

10.1 Introduction

In previous Chapters, we have seen that the macroscopic envelopes of localized condensates may manifest themselves as extended correlated domains, which can be identified, in the classical limit, with topological solitons, such as kinks, domain walls, vortices, etc. The topological properties of these "extended objects" are determined by those of the boson transformation function.

In this closing Chapter, which may be omitted in a first reading, we review some topics of soliton theory for the reader's convenience. More complete reviews of solitons may be found in [163, 232, 442, 471, 532, 623].

In Sections 10.2–10.4 we discuss some examples of soliton solutions of classical non-linear equations. We consider Korteweg–de Vries and sine-Gordon soliton solutions as well as the kink for the $\lambda\phi^4$ theory. In Appendix S we show how to generate multi-soliton solutions in the sine-Gordon model by means of Bäcklund transformation. As examples of solitons in gauge theories, we study the Nielsen–Olesen and the 't Hooft–Polyakov solutions.

Another topic of our discussion is the connection between topological defects and homotopy groups (Section 10.5). Such a connection gives a criterion for the existence of topologically stable defects formed during the spontaneous breakdown of symmetry [371, 372, 471, 609, 647]. We also briefly discuss the Kibble–Zurek mechanism for defect formation [371, 675]. Basic notions of homotopy theory are presented in Appendix T.

Section 10.6 deals with the issue of stability of soliton solutions. A necessary condition under which topology can stabilize soliton solutions is given by the Derrick–Hobart theorem [201, 329], which limits the number of spatial dimensions in which static solitons can exist. Some applications of this theorem are also presented.

While the homotopy considerations together with the Derrick–Hobart theorem can help to predict the existence of topological defects in a given theory, it may be difficult to find out their explicit form. In most cases only numerical solutions are available. There is, however, a method which allows us to derive a lower bound on the energy functional in terms of the topological charge [8, 114, 518], the so-called Bogomol'nyi bound [8, 116, 526]. It leads to a series of inequalities which are useful for solving soliton equations. This approach is discussed in Section 10.7.

Though the topological stability is a dominant stabilizing mechanism in the soliton dynamics, there exists also the possibility that solitons may be stabilized via Noether charge conservation. The example of a non-topological soliton, the Q-ball [166], is given in Section 10.8.

We then discuss instantons, which are solutions of classical field equations in Euclidean space with finite action. Instantons can be regarded as static solitons in higher-dimensional field theory in Minkowski spacetime. Like solitons, they are non-perturbative field configurations as their masses and actions are inversely proportional to coupling constant. Instanton effects appear in confinement, CP violation, tunneling and band structure in solids. In Section 10.9 we discuss the general properties of $SU(2)$ instantons as an example of Yang–Mills instantons. There we also show how instantons shape the vacuum structure of Yang–Mills theories. In particular, they describe the tunneling process between different vacua [131, 163, 348], labeled by different topological indices. The strong CP problem and the θ angle issue are also discussed [513–515, 612, 662].

Finally, in Section 10.9.1, we study the so-called collective coordinates on which instanton and soliton solutions depend [113, 132, 281, 348, 510]. Their connection with zero modes for the Dirac equation in the instanton background is also discussed.

10.2   The Korteweg–de Vries soliton

In shallow water, waves of small amplitude become slightly dispersive. In this case, a localized disturbance on the water surface is susceptible to spread during its propagation, according to Newton and Bernoulli's theories of hydrodynamics. However, if its amplitude is sufficiently large, the tendency to spread due to dispersion may, under special circumstances, be inhibited: the result is a localized hump of water, of symmetrical shape, which does not spread at all.[1]

The Korteweg–de Vries (KdV) equation for the propagation of waves in the x-direction on the surface of shallow water has the form

$$\frac{\partial \eta}{\partial t} + u_0 \frac{\partial \eta}{\partial x} + \alpha \eta \frac{\partial \eta}{\partial x} + \beta \frac{\partial^3 \eta}{\partial x^3} = 0, \tag{10.1}$$

where u_0, α and β are constants. α and β are positive. The first two terms describe wave evolution in the shallow water with speed u_0, i.e., they describe a linear non-dispersive limit. The third term is due to finite amplitude effects, and the last term describes dispersion.[2] It is convenient to pass to the frame S_{u_0} which moves with speed u_0. The Galileo transformations $x \to x - u_0 t$ and $t \to t$ lead then to the "canonical" KdV equation for the surface elevation:

$$\frac{\partial \eta}{\partial t} + \alpha \eta \frac{\partial \eta}{\partial x} + \beta \frac{\partial^3 \eta}{\partial x^3} = 0. \tag{10.2}$$

To find a soliton solution of Eq. (10.1), we look for a traveling wave solution that is at rest in some inertial frame S_u (in general, $u \neq u_0$), so that $\eta(t, x) = \eta(X)$ with $X = x - ut$ and with $\eta \to 0$ as $|x| \to \infty$. By plugging the Ansatz $\eta(t, x) = \eta(X)$ into Eq. (10.1), we obtain the ordinary differential equation

$$(u_0 - u)\eta' + \alpha \eta \eta' + \beta \eta''' = 0, \tag{10.3}$$

where $\eta' \equiv d\eta/dX$. If we now integrate Eq. (10.3) we have

$$(u_0 - u)\eta + \frac{\alpha}{2}\eta^2 + \beta \eta'' = A, \tag{10.4}$$

with A an integration constant. By multiplying Eq. (10.4) by $2\eta'$ and integrating again, we obtain

$$(u_0 - u)\eta^2 + \frac{\alpha}{3}\eta^3 + \beta(\eta')^2 = 2A\eta + B. \tag{10.5}$$

B is another integration constant. To obtain localized wave solutions, we require that $\eta \to 0$, $\eta' \to 0$ and $\eta'' \to 0$ as $|x| \to \infty$. We have

$$(\eta')^2 = a_0 \eta^2 (a - \eta), \tag{10.6}$$

---

[1] This phenomenon was first observed on a canal near Edinburgh in 1834 by J.S. Russell. In 1895 the first mathematical model explaining Russell's soliton was formulated by D.J. Korteweg and G. de Vries in the context of non-linear hydrodynamics.

[2] If we neglect the non-linear term and substitute the plane wave solution $\eta = \eta_0 \exp(ikx - i\omega t)$, we obtain the dispersion relation $\omega = ku_0(1 - \beta/u_0 k^2)$, or, in terms of a phase velocity $u = \omega/k$, we would have $u = u_0(1 - \beta/u_0 k^2)$. Waves with different k move with different velocity. Any wave packet solution must thus inevitably spread. This is the phenomenon of wave dispersion.

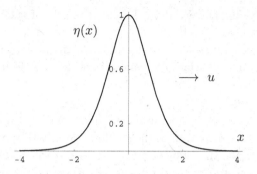

Fig. 10.1   Profile of the KdV soliton at $t = 0$. Soliton parameters are chosen to be: $a = 1$, $x_0 = 0$ and the soliton width is 1.

by setting $A = 0 = B$ and $a_0 = \frac{\alpha}{3\beta} > 0$, $a = \frac{3u_0}{\alpha}\left(\frac{u}{u_0} - 1\right)$. From Eq. (10.6) we see that $\eta \leq a$ and that $\eta' = 0$ only when $\eta = 0$ or $\eta = a$. These, together with the assumption $\eta \to 0$ at $|x| \to \infty$, imply that $\eta \geq 0$ and is bounded on the whole $\mathbb{R}$. So taking the square root, Eq. (10.6) can be integrated once again to yield:

$$\int_{\eta(X_0)}^{\eta(X)} \frac{d\eta}{\eta\sqrt{a_0(a - \eta)}} = X - X_0. \tag{10.7}$$

Inverting, we find

$$\text{arccosh}\left(\sqrt{\frac{a}{\eta}}\right) = \frac{\sqrt{a\,a_0}}{2}(X_0 - X) \quad \Rightarrow \quad \eta = a\,\text{sech}^2\left(\frac{\sqrt{a\,a_0}}{2}(X - X_0)\right).$$

This represents an exact KdV soliton. Here X_0 is such that $\eta(X_0) = a$, that is X_0 is the point where η reaches its maximum. By returning to the original frame we obtain

$$\eta(t, x) = a\,\text{sech}^2\left(\sqrt{\frac{a\alpha}{12\beta}}(x - x_0 - ut)\right), \tag{10.8}$$

where the soliton velocity is given by

$$u = u_0\left(1 + \frac{a\alpha}{3u_0}\right), \tag{10.9}$$

showing that the propagation velocity increases with the amplitude of the hump. Taking into account that the soliton width is $(a\alpha/12\beta)^{-1/2}$, we see that the smaller the width the larger the height and hence taller KdV solitons travel faster. The solitary wave profile (10.8) is depicted in Fig. 10.1.

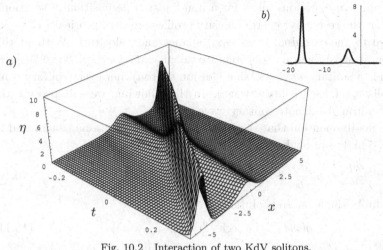

Fig. 10.2   Interaction of two KdV solitons.

An important feature of the KdV soliton (10.8) is that it travels without change of the shape. The localized behavior, as already mentioned, results from a balance between non-linear steepening and dispersive spreading and so it cannot be attained in linear equations because an appropriate amount of non-linearity is necessary.

When both A and B are non-zero, another type of solutions with unchanging profile (non-dispersive waves) is possible, namely periodic solutions. They are known as cnoidal waves, because they are expressed in terms of Jacobian elliptic functions $\mathrm{cn}(x)$. Here, however, we will not discuss periodic functions.

The KdV equation allows for a superposition of solitons and hence it can describe an interaction between solitons. For instance, numerical investigations of the interaction of two solitons reveal that if a high, narrow soliton is formed behind a low, broad one, it will catch up with the low one; they undergo a non-linear interaction — the high soliton passes through the low one — and both emerge with their shape unchanged; see Fig. 10.2.

The KdV solitons, as the historically first soliton solutions, have been taken up as a basis for (physicists') definition of soliton solutions also in other non-linear physical systems. In this view, solitons are solutions of the (field) equation of the motion, whose energy-density is non-zero only in a finite region of space (so the total energy is finite) and whose energy-density profile moves undistorted with constant velocity. It is also required that the dynamical equations allow for a non-linear superposition of solutions, such that the respective energy-density profiles are asymptotically $(t \to \infty)$ restored to their original $(t \to -\infty)$ shapes and velocities. Without this last condition the solitary solutions are called *solitary waves*. So while every soliton is a solitary wave (cf. sine-Gordon soliton), not every solitary wave is a soliton (cf. $\lambda\phi^4$ solitary wave). In the following, we will also use the term "soliton" to denote solitary waves.

We finally mention that the KdV equation (10.1) can be generalized to so-called higher-order KdV equations

$$\frac{\partial \eta}{\partial t} + u_0 \frac{\partial \eta}{\partial x} + (p+1)\eta^p \frac{\partial \eta}{\partial x} + \frac{\partial^3 \eta}{\partial \eta^3} = 0, \quad p \geq 2, \quad (10.10)$$

that admits single-soliton solutions

$$\eta(x,t) = a \operatorname{sech}^{2/p}(b(x - x_0 - ut)). \quad (10.11)$$

The same analysis we did for ordinary KdV solitons leads to

$$a = \left[\frac{p+2}{2} u_0 \left(\frac{u}{u_0} - 1 \right) \right]^{1/p} \quad \text{and} \quad b = p\sqrt{\frac{a^p}{2(p+2)}}. \quad (10.12)$$

This again indicates that taller and slimmer higher-order KdV solitons travel faster than smaller and thicker ones. Numerical simulations indicate that for $p > 2$ the higher-order KdV soliton equations do not allow for a non-linear superposition of solitons.

10.3   Topological solitons in $(1+1)$-d relativistic field theories

Systems which exhibit soliton solutions are $1+1$-dimensional relativistic field theories with non-derivative interactions. Consider the Lagrange density for a real scalar field ϕ,

$$\mathcal{L} = \frac{1}{2}(\dot{\phi})^2 - \frac{1}{2}(\phi')^2 - V(\phi), \quad (10.13)$$

and the corresponding Euler–Lagrange equation,

$$\ddot{\phi} - \phi'' = -\frac{dV(\phi)}{d\phi}. \quad (10.14)$$

The energy, obtained integrating the component T^{00} of the energy-momentum tensor, is

$$E[\phi] = \int dx \, T^{00}(\phi) = \int dx \left(\frac{1}{2}(\dot{\phi})^2 + \frac{1}{2}(\phi')^2 + V(\phi) \right). \quad (10.15)$$

A way of finding the solitons is by solving Eq. (10.14) in the soliton's rest frame and then boosting the solution with a Lorentz transformation. The rest-frame (i.e., static) soliton must obey the dynamical equation

$$\phi'' = \frac{dV(\phi)}{d\phi} \quad \Rightarrow \quad \frac{1}{2}((\phi')^2)' = \frac{dV(\phi)}{d\phi}\phi'$$

$$\Rightarrow \quad \frac{1}{2}(\phi')^2 = V(\phi) + A, \qquad (10.16)$$

where A is an integration constant. Since the soliton solution and its energy must be localized, it must be $\phi'(|x| \to \infty) = 0$ and $E[\phi(|x| \to \infty)] = 0$, which means (cf. Eq. (10.15)) that $V(\phi(|x| \to \infty)) = 0$. The boundary condition $V(\phi) = 0$ and $\phi' = 0$ at $|x| \to \infty$ imply that $A = 0$ in Eq. (10.16).

We now take the square root in Eq. (10.16) and integrate. This gives

$$\pm \int_{\phi(x_0)}^{\phi(x)} \frac{d\phi}{\sqrt{2V(\phi)}} = x - x_0, \qquad (10.17)$$

which finally yields the soliton solution $\phi = \phi(\phi(x_0), x - x_0)$. We now discuss two paradigmatic examples of $(1+1)$-d relativistic soliton solutions.

10.3.1 *The sine-Gordon soliton*

The so-called sine-Gordon system, also called $\sin\phi_2$ model [232], is described by the Lagrange density

$$\mathcal{L} = \frac{1}{2}(\partial_\mu \phi)(\partial^\mu \phi) + \frac{m^4}{\lambda}\left[\cos\left(\frac{\sqrt{\lambda}}{m}\phi\right) - 1\right], \qquad (10.18)$$

($\mu = 0, 1$). The Euler–Lagrange equation reads

$$\Box\phi(t, x) + \frac{m^3}{\sqrt{\lambda}}\sin\left(\frac{\sqrt{\lambda}}{m}\phi(t, x)\right) = 0. \qquad (10.19)$$

After substitution $x_\mu \to mx_\mu$ and $\phi \to \sqrt{\lambda}\phi/m$, Eq. (10.19) becomes

$$\Box\phi(t, x) + \sin\phi(t, x) = 0, \qquad (10.20)$$

which is analogue to the dynamical equation for the pendulum. By inserting the corresponding potential into Eq. (10.17) we obtain

$$\pm \int_{\phi(x_0)}^{\phi(x)} \frac{d\phi}{2\sin(\phi/2)} = x - x_0. \qquad (10.21)$$

By solving this equation with respect to ϕ we have

$$\phi(x) = \pm 4\arctan[\exp(x - x_0)], \qquad (10.22)$$

which, after boosting to the frame that moves with the velocity u, yields

$$\phi(t, x) = \pm 4 \arctan \left[\exp \left(\frac{x - x_0 - ut}{\sqrt{1 - u^2}} \right) \right]. \tag{10.23}$$

These soliton solutions are called *kinks* ($+$ sign) or *antikinks* ($-$ sign). In contrast to KdV solitons, their velocities do not depend on the wave amplitude. The energy density $T^{00}(t, x)$ for the soliton solution (10.23) is

$$T^{00}(t, x) = 16 \frac{\exp \left(\frac{2(x - x_0 - ut)}{\sqrt{1 - u^2}} \right)}{\left(1 + \exp \left(\frac{2(x - x_0 - ut)}{\sqrt{1 - u^2}} \right) \right)^2 (1 - u^2)}. \tag{10.24}$$

The resulting static kink/antikink energy density thus reads

$$\varepsilon(x) = \frac{1}{2}(\phi')^2 + V(\phi) = 2V(\phi) = 16 \frac{e^{2(x + x_0)}}{(e^{2x} + e^{2x_0})^2}, \tag{10.25}$$

or in the original (unscaled) variables it is

$$\varepsilon(x) = \frac{16 m^4}{\lambda} \frac{e^{2m(x + x_0)}}{(e^{2mx} + e^{2mx_0})^2}. \tag{10.26}$$

The total energy of the sine-Gordon kink/antikink is

$$E[\phi] = \int_{-\infty}^{\infty} dx \, T^{00}(t, x) = \int_{-\infty}^{\infty} dx \, T^{00}(0, x)$$

$$= \frac{1}{\sqrt{1 - u^2}} \frac{8 m^3}{\lambda} \equiv \frac{M}{\sqrt{1 - u^2}}, \tag{10.27}$$

where M is the so-called sine-Gordon kink/antikink mass. The plot of the energy density and the kink solution is given in Fig. 10.3.

Similarly as in the KdV case, one of the most important features of the sine-Gordon solitons is their stability, which can be conveniently characterized by a conserved quantity known as topological charge (or topological index). Eq. (10.23) implies that the kink profile has the asymptotes: $\phi(t, x \to \infty) = \pi/2$ and $\phi(t, x \to -\infty) = 0$ (similarly in unscaled variables) at any time t. In fact, it takes an infinite amount of energy to change the kink to one of its constant vacuum configurations $\phi = 2\pi N; \ N \in \mathbb{Z}$. Because all the solitary solutions for the sine-Gordon system should be, by definition, finite energy solutions, for $x \to \pm\infty$ they must tend towards one of the vacuum values, labeled by an integer N. The current J^μ defined as

$$J^\mu(t, x) = \frac{1}{2\pi} \epsilon^{\mu\nu} \partial_\nu \phi(t, x), \tag{10.28}$$

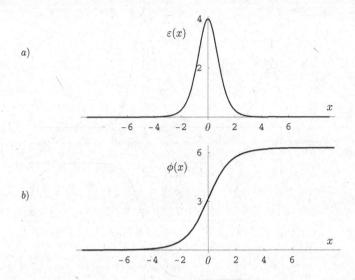

Fig. 10.3 *a)* Energy density and *b)* kink solution for the sine-Gordon system. Plots are made in the scaled variables (no dependence on constants). We choose $x_0 = 0$.

where $\epsilon^{\mu\nu}$ is the two-dimensional Levi–Civita tensor, satisfies the continuity equation $\partial_\mu J^\mu = 0$. The conserved (topological) charge Q is

$$Q = \int_{-\infty}^{\infty} dx \ J^0(t, x) \equiv \frac{1}{2\pi} \int_{-\infty}^{\infty} dx \ \phi'(t, x)$$

$$= \frac{1}{2\pi} [\phi(t, x \to \infty) - \phi(t, x \to -\infty)] = N_2 - N_1 \,. \qquad (10.29)$$

N_1 and N_2 are integers corresponding to the asymptotic values of the field. The factor $1/2\pi$ ensures that the charge is an integer number. Since Q is a constant topological charge, solitons with one value of Q cannot decay into solutions with a different value of Q. Note that Q cannot be derived from the Noether theorem since it is *not* related to any particular continuous symmetry of the Lagrangian, rather it is derived from the topological properties of regular (i.e., differentiable) finite-energy solutions of the sine-Gordon equation.

10.3.2 *The $\lambda\phi^4$ kink*

If we consider in the sine-Gordon equation only relatively small field elevations, then we can expand the sine function in Eq. (10.19) up to the third

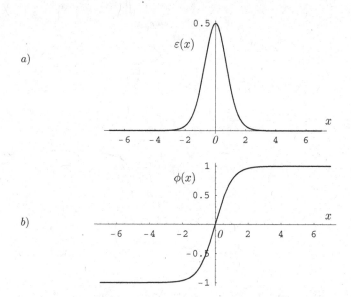

Fig. 10.4 a) Energy density and b) kink solution for the $\lambda\phi^4$ system. Units are chosen so that $m = \lambda = 1$. We choose the extremum of the energy density to be at $x_0 = 0$.

order.[3] After rescaling $\lambda/3!$ to λ the resulting dynamical equation acquires the form

$$\Box\phi(t,x) + m^2\phi(t,x) - \lambda\phi^3(t,x) = 0. \tag{10.30}$$

Up to an irrelevant constant the Lagrange density can be written as

$$\mathcal{L} = \frac{1}{2}(\partial_\mu\phi)(\partial^\mu\phi) - \frac{\lambda}{4}\left(\phi^2 - \frac{m^2}{\lambda}\right)^2. \tag{10.31}$$

Inserting the $V(\phi)$ into Eq. (10.17) we obtain the static *kink* equation

$$\pm\int_{\phi(x_0)}^{\phi(x)} \frac{d\phi}{\sqrt{\frac{\lambda}{2}\left(\phi^2 - \frac{m^2}{\lambda}\right)}} = x - x_0, \tag{10.32}$$

which gives

$$\phi(x) = \pm\frac{m}{\sqrt{\lambda}}\tanh\left[\frac{m}{\sqrt{2}}(x - x_0)\right]. \tag{10.33}$$

The boosted solution then reads

[3]Should we consider only the leading term expansion, $\sin\phi \approx \phi$, we would arrive at the Klein–Gordon equation that is dispersive and hence does not possess soliton solution.

$$\phi(t,x) = \pm \frac{m}{\sqrt{\lambda}} \tanh \left[\frac{m}{\sqrt{2}} \frac{(x - x_0 - ut)}{\sqrt{1 - u^2}} \right]. \tag{10.34}$$

Solutions with "+" sign are called *kinks*; the ones with "-" sign, *antikinks*. The resulting kink/antikink energy density T^{00} is

$$T^{00}(t,x) = \frac{1}{2}(\dot{\phi})^2 + \frac{1}{2}(\phi')^2 + V(\phi)$$

$$= \frac{m^4}{2\lambda(1 - u^2)} \operatorname{sech}^4 \left[\frac{m}{\sqrt{2}} \frac{(x - x_0 - ut)}{\sqrt{1 - u^2}} \right], \tag{10.35}$$

and hence the energy density of the static kink/antikink is

$$\varepsilon(x) = \frac{1}{2}(\phi'(x))^2 + V(\phi(x)) = \frac{m^4}{2\lambda} \operatorname{sech}^4 \left[\frac{m}{\sqrt{2}}(x - x_0) \right]. \tag{10.36}$$

Profiles of both the kink solution $\phi(x)$ and the energy density $\varepsilon(x)$ are depicted in Fig. 10.4. The kink/antikink energy is

$$E[\phi] = \int_{-\infty}^{\infty} dx\, T^{00}(t,x) = \int_{-\infty}^{\infty} dx\, T^{00}(0,x)$$

$$= \frac{1}{\sqrt{1 - u^2}} \frac{m^3 2\sqrt{2}}{3\lambda} \equiv \frac{\bar{M}}{\sqrt{1 - u^2}}, \tag{10.37}$$

where $\bar{M}$ is known as the ϕ^4 kink/antikink mass.

Eq. (10.34) implies that the kink profile has the asymptotes $\phi(t, x \to \infty) \equiv \phi_+ = m/\sqrt{\lambda}$ and $\phi(t, x \to -\infty) \equiv \phi_- = -m/\sqrt{\lambda}$ at any fixed time t. The topological current J^μ is defined as

$$J^\mu(t,x) = \frac{\sqrt{\lambda}}{2m} \epsilon^{\mu\nu} \partial_\nu \phi(t,x), \tag{10.38}$$

and the conserved charge Q is

$$Q = \int_{-\infty}^{\infty} dx J^0(t,x) \equiv \frac{\sqrt{\lambda}}{2m} \int_{-\infty}^{\infty} dx \phi'(t,x) = \frac{\sqrt{\lambda}}{2m} [\phi_+ - \phi_-]. \tag{10.39}$$

Possible values of Q are $\{-1, 0, 1\}$. $Q = 0$ means that the field lies in the same topological class as the vacuum solution $\phi = \pm m/\sqrt{\lambda}$, to which it may (and for energy reasons it will) be continuously deformed. $Q = \pm 1$ correspond to non-trivial topological configurations (kink $(Q = +1)$ and antikink $(Q = -1)$) that cannot be continuously deformed to the vacuum configuration nor to each other.

Note that $|N| > 1$ is not allowed for soliton solutions since the corresponding field configurations do not have boundary conditions compatible with finite energy. No multi-soliton solutions with $|N| > 1$ can exist. Nevertheless, field configurations that contain a finite mixture of kinks and antikinks alternating along the x-direction, so that $N = \{-1, 0, 1\}$, can be constructed [677], but there are no *static* solutions of this type.

10.4   Topological solitons in gauge theories

We now discuss two typical examples of soliton solutions in gauge theories.

10.4.1   *The Nielsen–Olesen vortex*

Vortex solutions were found by Abrikosov [4] in type II superconductors[4] and by Nielsen and Olesen [498] in the context of the $2 + 1$-dimensional $U(1)$ Higgs model. The corresponding soliton solution is also known as the Abrikosov–Nielsen–Olesen vortex (or string).

Let us consider the action for the $3+1$-dimensional Abelian Higgs model

$$ S = \int d^4x \left[-\frac{1}{4} F_{\mu\nu} F^{\mu\nu} + |D_\mu \phi|^2 - \frac{g}{2} \left(|\phi|^2 - v^2 \right)^2 \right], \qquad (10.40) $$

where A_μ is a $U(1)$ gauge field and ϕ a complex scalar Higgs field of charge e. The covariant derivative and the field strength are respectively

$$ D_\mu \phi = (\partial_\mu + i e A_\mu) \phi, \qquad F_{\mu\nu} = \partial_\mu A_\nu - \partial_\nu A_\mu. \qquad (10.41) $$

The action (10.40) is invariant under the $U(1)$ gauge transformations

$$ \phi(x) \rightarrow e^{i\alpha(x)} \phi(x), \qquad A_\mu(x) \rightarrow A_\mu(x) + \frac{1}{e} \partial_\mu \alpha(x). \qquad (10.42) $$

We will show that static string-like solitons (vortices) do exist in this model. By static solitons we mean a configuration in the Weyl (i.e., temporal) gauge $A_0 = 0$, with time-independent fields $A_i = A_i(\mathbf{x})$ and $\phi = \phi(\mathbf{x})$. The corresponding energy functional has the form

$$ E[F^{\mu\nu}, \phi] = \int d^3\mathbf{x} \left[\frac{1}{4} F_{ij} F^{ij} + |D_i \phi|^2 + \frac{g}{2} (|\phi|^2 - v^2)^2 \right]. \qquad (10.43) $$

Our considerations can be simplified by assuming that the soliton configurations are straight, static vortices along the z-direction. Their energy functional per unit length is

$$ E[F^{\mu\nu}, \phi] = \int d^2\mathbf{x} \left[\frac{1}{2} B^2 + |D_x \phi|^2 + |D_y \phi|^2 + \frac{g}{2} (|\phi|^2 - v^2)^2 \right]. \qquad (10.44) $$

Here $B = \partial_x A_y - \partial_y A_x$ represents the z-component of the magnetic field. A necessary condition for finite-energy configurations is that

$$ |\phi|^2|_{|\mathbf{x}|\to\infty} = v^2, \qquad (10.45) $$

---

[4]Type II superconductor is phenomenologically described by the Ginzburg–Landau model which is a non-relativistic analogue of the Abelian Higgs model. There the energy functional is replaced by the (Landau) free energy and the Higgs field is replaced by the order parameter. The corresponding Abrikosov vortices are better known as *flux lines*, *flux tubes* or *fluxons*.

along with $D_{x,y}\phi|_{|\mathbf{x}|\to\infty} = \mathcal{O}(1/r)$ and $B|_{|\mathbf{x}|\to\infty} = \mathcal{O}(1/r)$. (We interchangeably use $|\mathbf{x}| = r$.) Eq. (10.45) reveals that the (classical) vacuum manifold for the ϕ field is a circle $\mathcal{S}^1_{vac}$ parametrized by the phase angle ϑ ($\phi = |\phi|e^{i\vartheta}$). So for the field ϕ in the remote spatial distance we can generally write

$$\phi|_{|\mathbf{x}|\to\infty} = \phi(\mathbf{n}) = v\, e^{i\vartheta(\mathbf{n})}, \qquad (10.46)$$

where $\mathbf{n} = (x/r, y/r)$ is a unit radius vector in the xy plane. Thus, with each finite-energy static configuration ϕ is associated a map from the circle at infinity $\mathcal{S}^1_{\infty}$ to the vacuum manifold $\mathcal{S}^1_{vac}$. Because the field is single-valued, the angle $\vartheta(\mathbf{n})$ must have the property $\vartheta(2\pi) = \vartheta(0) + 2\pi n$, for some $n \in \mathbb{Z}$. The number n is known as the *winding number*, and it is the *topological charge/number/index* of the field configuration. In Appendix T we show that the map $\mathcal{S}^1_{\infty} \to \mathcal{S}^1_{vac}$ is topologically characterized by the fundamental homotopy group $\pi_1(\mathcal{S}^1) = \mathbb{Z}$. Thus, the static solutions to the equations of motion fall into disjoint homotopy classes, each characterized by an element of $\pi_1(\mathcal{S}^1)$, i.e., by its winding number. Topological charges are conserved quantities (cf. Appendix T); therefore decays between vortices belonging to different homotopy classes are impossible — the solutions are topologically stable.

We write $\phi(\mathbf{n})$, with the winding number n as

$$\phi(\mathbf{n}) = v\, e^{in\vartheta}, \qquad (10.47)$$

with ϑ now being the polar angle in the (physical) xy space. Because the covariant derivative $D_{x,y}\phi$ tends to zero at infinity, the gauge potential can be asymptotically written as

$$A_{x,y}|_{|\mathbf{x}|\to\infty} = -\frac{i}{e}\partial_{x,y}\log(\phi(\mathbf{n})). \qquad (10.48)$$

Hence $A_{x,y}|_{|\mathbf{x}|\to\infty}$ has the azimuthal component

$$A_{\vartheta}|_{|\mathbf{x}|\to\infty} = -\frac{i}{e}\frac{1}{r}\frac{d\log(\phi(\mathbf{n}))}{d\vartheta} = \frac{n}{er}. \qquad (10.49)$$

The asymptotic forms (10.48)–(10.49) suggest that the vortex solution in the whole physical space can be assumed in the form

$$\phi(r,\vartheta,z) = vf(r)e^{in\vartheta},$$

$$A_{\vartheta}(r,\vartheta,z) = \frac{n}{er}h(r) \quad \Rightarrow \quad B(r,\vartheta,z) = \frac{n}{er}h'(r). \qquad (10.50)$$

Here r,ϑ,z are cylindrical polar coordinates. The function f and g are assumed to be smooth with limiting values $f(0) = g(0) = 0, f(\infty) = g(\infty) =$

1. The fact that $f(0) = h(0) = 0$ corresponds to the requirement that the fields ϕ and A_ϑ are smooth (at least twice differentiable) at $r = 0$ and that the energy functional (10.44) is finite for small r.

The functions f and h are determined by minimizing the energy functional (10.44). By rewriting the energy functional (10.44) in the polar coordinates one can check that f and h satisfy the equations

$$f'' + \frac{1}{r}f' - \frac{n^2}{r^2}(1 - h)^2 f + gv^2(1 - f^2)f = 0, \quad (10.51a)$$

$$h'' - \frac{1}{r}h' + 2e^2v^2f^2(1 - h) = 0. \quad (10.51b)$$

Suitability of the Ansatz (10.50) is justified when Eqs. (10.51) yield a nontrivial solution. Although, no analytic solution is known so far, the consistency of the Ansatz (10.50) can be checked numerically. One can obtain some non-trivial insight into the structure of the solutions by confining to the asymptotic behavior

$$|\phi|_{|\mathbf{x}|\to\infty} = v. \quad (10.52)$$

In this case the second equation in (10.51) can be solved exactly. One finds that (c_1 and c_2 are constants of integration)

$$1 - h(r) = c_1 r J_1(i\sqrt{2}erv) + c_2 r Y_1(-i\sqrt{2}erv)$$

$$\stackrel{|\mathbf{x}|\to\infty}{\approx} c\sqrt{\frac{2}{\pi\sqrt{2}ev}}r^{1/2}e^{-\sqrt{2}evr}, \quad (10.53)$$

where J_1 and Y_1 are the Bessel functions of the first and second kind, respectively, and $c \equiv c_1 = ic_2$ is chosen so as to ensure the correct asymptotic behavior. The first equation in (10.51) can be linearized for large r by assuming that $f(r) = 1 - \delta f(r)$ with $\delta f_{|\mathbf{x}|\to\infty} = 0$. In such a case the differential equation for δf reads

$$(\delta f)'' + \frac{1}{r}(\delta f)' - 2gv^2\delta f = -c^2\frac{n^2}{r}\frac{2}{\pi\sqrt{2}ev}e^{-2\sqrt{2}evr}. \quad (10.54)$$

The homogenous solution can be easily obtained via direct integration yielding (a_1 and a_2 are constants of integration)

$$\delta f_{hom}(r) = a_1 J_0(-i\sqrt{2g}vr) + a_2 Y_0(-i\sqrt{2g}vr)$$

$$\stackrel{|\mathbf{x}|\to\infty}{\approx} \sqrt{\frac{1}{2\pi\sqrt{2g}vr}}\left[ie^{\sqrt{2g}vr}(a_1 - a_2) - e^{-\sqrt{2g}vr}(a_1 + a_2)\right]. \quad (10.55)$$

The particular solution can be obtained, for example, via variation of constants. In this case we have

$$
\delta f_{par}(r) = \frac{c^2 n^2}{\sqrt{2}ev} \left(J_0(i\sqrt{2g}vr) \int_1^r ds\, e^{-2\sqrt{2}evs} Y_0(-i\sqrt{2g}vs) \right.
$$

$$
\left. - Y_0(-i\sqrt{2g}vr) \int_1^r ds\, e^{-2\sqrt{2}evs} J_0(i\sqrt{2g}vs) \right)
$$

$$
\overset{|\mathbf{x}|\to\infty}{\approx} \frac{c^2 n^2}{\pi\sqrt{g}ev^2 r} \left[\frac{e^{-2\sqrt{2}evr}}{2\sqrt{2}ev + \sqrt{2g}v} - i\frac{e^{-(2\sqrt{2}ev+2\sqrt{2g}v)r}}{2\sqrt{2}ev + \sqrt{2g}v} \right]. \tag{10.56}
$$

The asymptotic behavior for f satisfying the condition $f_{|\mathbf{x}|\to\infty} = 1$ can therefore be written as

$$
1 - f(r) \overset{|\mathbf{x}|\to\infty}{\propto} \begin{cases} r^{-1}\, e^{-2\sqrt{2}evr} & \text{if } 2e < \sqrt{g} \\ r^{-1/2}\, e^{-\sqrt{2g}vr} & \text{if } 2e > \sqrt{g}. \end{cases} \tag{10.57}
$$

Note that there are two length scales governing the large r behavior of f and h, namely the inverse masses of the scalar and vector excitations (Higgs and gauge particles), i.e.,

$$
m_s^2 = 2gv^2, \qquad m_v^2 = 2e^2 v^2. \tag{10.58}
$$

From Eq. (10.57) we see that the asymptotic behavior depends on the ratio of these two

$$
\kappa = \frac{m_s^2}{m_v^2} = \frac{g}{e^2}. \tag{10.59}
$$

In particular, for $\kappa < 4$ the large r behavior of f is controlled by m_s while for $\kappa > 4$, m_v controls the asymptotic behavior of both f and h.

In superconductors, the two length scales are known as the correlation length $\xi = 1/m_s$ and the London penetration depth $\lambda = 1/m_v$. There the small and large values of κ (Ginzburg–Landau parameter) distinguish type II and type I superconductors. In type II superconductors, vortices with $|n| > 1$ are unstable; there is a repulsive force between parallel $n = 1$ vortices which can stabilize a lattice of vortices (Abrikosov vortex lattice). Because the correlation length ξ (i.e., mean free path of charge carriers), is smaller than the London penetration depth λ, type II superconductors can be penetrated by magnetic flux lines. Hence there is an intermediate range of magnetic field strength known as the mixed state or vortex state within which the field penetrates the superconductor but it stays confined to a lattice of flux tubes. In contrast, in type I superconductors the correlation length is larger than the London penetration depth, meaning that such superconductors cannot be penetrated by magnetic flux lines. This expulsion of flux lines is know as the Meissner–Ochsenfeld effect.

Let us finally observe that, by integrating around a large loop encircling the string, we find that the string carries a quantized magnetic flux. Indeed, by utilizing Stoke's theorem we obtain (see Eq. 10.49)

$$\Phi = \int d^2\mathbf{x}B = \lim_{r\to\infty} \oint_r dl_i A_i = \lim_{r\to\infty} \int_0^{2\pi} rA_\vartheta d\vartheta = \frac{2\pi n}{e}. \quad (10.60)$$

Relation (10.60) indicates that the magnetic flux is "quantized". Notice that (10.60) is similar to the flux quantization in type II superconductors. However, our result was derived in a purely classical context while flux quantization in superconductors has a quantum origin and it reads

$$\Phi = \frac{2\pi n\hbar}{2q}. \quad (10.61)$$

Here $2q$ corresponds to the charge of a Cooper pair. In particular, $\hbar$ never appears in our reasonings and e is just a coupling constant in the Lagrangian. At quantum level, the coupling constant e and the charge of the Cooper pair $2q$ are related through $2q = e\hbar$ and Eq. (10.60) represents the flux quantization condition with the magnetic flux quantum being $\pi\hbar/q$.

10.4.2 *The 't Hooft–Polyakov monopole*

In 1974 't Hooft [587] and Polyakov [524] independently discovered a topologically non-trivial finite energy solution in the $SO(3)$ Georgi–Glashow model [277]. This soliton solution is known as the 't Hooft–Polyakov monopole or non-Abelian magnetic monopole.

The Georgi–Glashow model is based on the $SO(3)$ gauge group and on the triplet of real Higgs scalar fields (isovector) ϕ that transform under the fundamental representation of $SO(3)$. The Lagrangian density is

$$\mathcal{L} = -\frac{1}{2}\mathrm{Tr}(\mathbf{F}_{\mu\nu}\mathbf{F}^{\mu\nu}) + \frac{1}{2}(D_\mu\phi)\cdot(D^\mu\phi) - \frac{g}{4}(\phi\cdot\phi - v^2)^2$$

$$= -\frac{1}{4}F_{\mu\nu}^a F^{\mu\nu a} + \frac{1}{2}(D_\mu\phi)^a(D^\mu\phi)^a - \frac{g}{4}(\phi^a\phi^a - v^2)^2, \quad (10.62)$$

with the Lorentz indices $\mu, \nu = 0, 1, 2, 3$. We have chosen the (Killing) normalization convention for T^a such that $\mathrm{Tr}(T^aT^b) = \frac{1}{2}\delta^{ab}$. The 3×3 matrices T^a are the $SO(3)$ generators and $\mathbf{F}_{\mu\nu} = F_{\mu\nu}^a T^a$. We observe that as a rule, gauge fields correspond to elements of a Lie algebra in its adjoint representation.

The covariant derivative $D_\mu\phi$ appearing in (10.62) is defined as

$$D_\mu\phi = \partial_\mu\phi - ieA_\mu^a T^a \phi, \qquad (10.63)$$

where e is the (real) gauge coupling constant. Under the gauge transformation

$$\phi(x) \to g(x)\phi(x), \qquad (10.64\text{a})$$

$$A_\mu^a(x)T^a = \mathbf{A}_\mu(x) \to g(x)\mathbf{A}_\mu(x)g^{-1}(x) + ie^{-1}(\partial_\mu g(x))g^{-1}(x), \quad (10.64\text{b})$$

the covariant derivative and the field strength $\mathbf{F}_{\mu\nu}$ transform covariantly, i.e.,

$$D_\mu\phi(x) \to g(x)D_\mu\phi(x), \qquad (10.65\text{a})$$

$$\mathbf{F}_{\mu\nu} = \partial_\mu\mathbf{A}_\nu - \partial_\nu\mathbf{A}_\mu + ie[\mathbf{A}_\mu, \mathbf{A}_\nu] = -ie^{-1}[D_\mu, D_\nu] \to g\mathbf{F}_{\mu\nu}g^{-1}. \quad (10.65\text{b})$$

$\mathcal{L}$ is invariant under the $SO(3)$ gauge group.

We will show that topological soliton configurations may exist in this model. We consider finite energy static configurations $A_i^a = A_i^a(\mathbf{x})$ and $\phi^a = \phi^a(\mathbf{x})$ in the Weyl gauge $A_0^a = 0$. Then the energy functional is

$$E[\mathbf{F}^{\mu\nu}, \phi] = \int d^3\mathbf{x} \left[\frac{1}{4}F_{ij}^a F_{ij}^a + \frac{1}{2}(D_i\phi)^a (D_i\phi)^a + \frac{g}{4}(\phi^a\phi^a - v^2)^2 \right]. \quad (10.66)$$

A necessary condition for the finiteness of the (Higgs field) energy is

$$(\phi^a\phi^a)|_{|\mathbf{x}|\to\infty} = v^2. \qquad (10.67)$$

Here the direction of the fields ϕ^a in the internal space may depend on the direction in the physical three-dimensional space because

$$\phi^a|_{|\mathbf{x}|\to\infty} = \phi^a(\mathbf{n}), \qquad (10.68)$$

with $\mathbf{n} = \mathbf{x}/r$ being a unite radius vector. Thus, each configuration of fields with finite energy is necessarily associated with a mapping of an infinitely remote sphere $\mathcal{S}_\infty^2$ to the sphere $\mathcal{S}_{vac}^2$ (i.e., vacuum manifold) in the space of the Higgs fields defined by the equation

$$\phi^a\phi^a = v^2. \qquad (10.69)$$

$\mathcal{S}_{vac}^2$ is called the classical vacuum of the model. In Appendix T we will see that the second homotopy group $\pi_2(\mathcal{S}^2) = \mathbb{Z}$, and thus the mapping $\mathcal{S}_\infty^2 \to \mathcal{S}_{vac}^2$ is classified by integer topological numbers $n = 0, \pm1, \pm2, \ldots$, which are conserved; they do not change under small deformations of the field configuration $\phi^a(\mathbf{x})$ for which the energy remains finite.

We write the field $\phi^a(\mathbf{n})$, corresponding to the mapping $S^2_\infty \to S^2_{vac}$ with a unit topological number, as

$$\phi^a|_{|\mathbf{x}|\to\infty} = n^a v. \tag{10.70}$$

The finiteness of the soliton energy requires that the covariant derivative Eq. (10.63) must decrease at the spatial infinity faster than $1/r$. The derivative $\partial_\mu \phi$ decreases only as $1/r$ because from (10.70) we have

$$\partial_i \phi^a|_{|\mathbf{x}|\to\infty} = \frac{1}{r}(\delta^{ai} - n^a n^i)v. \tag{10.71}$$

This behavior must be compensated by the second (i.e., gauge) term in (10.63). Thus the gauge field must have the asymptotic behavior:

$$A_i^a(\mathbf{x})|_{|\mathbf{x}|\to\infty} = \frac{1}{er}\epsilon^{aij}n^j. \tag{10.72}$$

Here we have used the fact that $(T_a)_{bc} = i\epsilon_{bac}$ and the identity

$$\epsilon^{ijk}\epsilon^{ilm} = \delta^{jl}\delta^{km} - \delta^{jm}\delta^{kl}. \tag{10.73}$$

Notice that (10.72) implies $F_{ij}^a F_{ij}^a \approx r^{-4}$ at large radius, and thus the $F_{ij}^a F_{ij}^a$ part of the gauge field sector has a finite contribution to the energy.

To obtain a smooth soliton solution in the whole $\mathbb{R}^3$ we write

$$\phi^a = n^a v f(r), \qquad A_i^a = \frac{1}{er}\epsilon^{aij}n^j(1 - h(r)), \tag{10.74}$$

where $f(r)$ and $h(r)$ are unknown smooth functions of r with asymptotes $f(\infty) = h(0) = 1$, $f(0) = h(\infty) = 0$. Then, the static energy functional (10.66) is

$$E[f, h] = 4\pi \int_0^\infty dr \left[\frac{1}{e^2}(h')^2 + \frac{r^2 v^2}{2}(f')^2 + \frac{1}{2e^2 r^2}(1 - h^2)^2 \right.$$

$$\left. + v^2 f^2 h^2 + \frac{gr^2 v^4}{4}(f^2 - 1)^2 \right]. \tag{10.75}$$

The ensuing static equations of motion are

$$f'' + \frac{2}{r}f' = \frac{2f}{r^2}h^2 + gv^2 f(f^2 - 1), \tag{10.76a}$$

$$h'' = \frac{h}{r^2}(h^2 - 1) + e^2 v^2 f^2 h. \tag{10.76b}$$

The exact analytic solutions of these equations are difficult to find. However, it is possible to verify the existence of non-trivial solutions numerically. The existence, though non-uniqueness, of the solution of Eqs. (10.76) for any $g \geq 0$ is proven, e.g., in [467]. Eqs. (10.74) represent the 't Hooft–Polyakov monopole, i.e., soliton solution to the $SO(3)$ Georgi–Glashow model with the topological number 1. Polyakov [524] has called "hedgehog" ϕ^a in Eq. (10.74) to stress that the Higgs isovector ϕ at a given point of space is directed along the radius vector. Note that the presence of gauge fields is essential in order to interpret the hedgehog as a solitary solution, i.e., an autonomous particle-like object. In fact, without gauge fields the Higgs field (10.74) would be linearly divergent at large distances, corresponding thus to the infinite rather than finite energy solution.

Let us see why the 't Hooft–Polyakov soliton is a *magnetic* monopole. For $g > 0$ we have SSB ($SO(3) \to SO(2)$). One may identify the unbroken $SO(2) \cong U(1)$ symmetry with the electromagnetic field. To do that 't Hooft proposed the gauge invariant definition of the electromagnetic field

$$\mathcal{F}_{\mu\nu} = \frac{\phi^a}{|v|} F_{\mu\nu}^a - \frac{1}{e|v|^3} \epsilon_{abc}\, \phi^a (D_\mu \phi)^b (D_\nu \phi)^c. \tag{10.77}$$

Choosing $\phi^a = \delta^{3a}|v|$ and using Eq. (10.63), $\mathcal{F}_{\mu\nu}$ is

$$\mathcal{F}_{\mu\nu} = \partial_\mu A_\nu^3 - \partial_\nu A_\mu^3. \tag{10.78}$$

Thus, the massless gauge boson corresponding to the unbroken $U(1)$ group is identified with the photon $A_\mu^3 = A_\mu$, provided $\phi^a = \delta^{3a}|v|$. $\mathcal{F}_{\mu\nu}$ is then a 4×4 matrix of magnetic and electric fields

$$\mathcal{F}_{ij} = \epsilon_{ijk} B^k, \qquad \mathcal{F}_{0i} = E_i = 0. \tag{10.79}$$

The static soliton configuration thus carries only magnetic field. $F_{\mu\nu}^a$ has at large distances the asymptotic behavior (cf. (10.65) and (10.72)):

$$F_{ij}^a\big|_{|\mathbf{x}|\to\infty} = \frac{n^k n^a v}{er^2}\, \epsilon_{ijk} = \frac{n^k}{er^2}\, \epsilon_{ijk}\, \phi^a\big|_{|\mathbf{x}|\to\infty}, \qquad F_{0i}^a = 0, \tag{10.80}$$

so that

$$B^k\big|_{|\mathbf{x}|\to\infty} = \frac{1}{2}\epsilon^{kij}\, \mathcal{F}_{ij}\big|_{|\mathbf{x}|\to\infty} = \frac{n^k}{er^2}. \tag{10.81}$$

Applying Gauss law, the total magnetic-monopole charge reads

$$Q = \lim_{r\to\infty} \oint_{S_r^2} \mathbf{B} \cdot d\mathbf{s} = \frac{4\pi}{e}, \tag{10.82}$$

with the Dirac-like charge quantization condition

$$Qe = 2\pi n, \qquad n \in \mathbb{N}_+, \tag{10.83}$$

for $n = 2$. The origin of the "quantization" condition is purely topological and is not related to the dynamics nor to quantum theory. In fact, true Dirac quantization condition is $Qq = 2\pi\hbar n$ (in the CGS units $Qq = \hbar n/2$), with $q = \hbar e$ being the electric charge.

10.5   Topological defect classification and the Kibble–Zurek mechanism for defect formation

The soliton solutions discussed above may appear in Nature as topological defects in ordered media. Due to their relevance in many contexts, ranging from condensed matter physics to cosmology, much attention has been devoted to their study. The classification of topological defects is based on homotopy theory; for a review see, for example, [471]. Basic notions of homotopy theory are given in Appendix T.

The appearance of topological defects is a common feature of symmetry breaking phase transitions [373]. Within this framework the necessary condition for the existence of topologically stable defects can be expressed in terms of the topology of the vacuum manifold, specifically its homotopy group. The connection between topological defects and homotopy groups $\pi_n(\mathcal{M})$ is established when the manifold $\mathcal{M}$ is identified with the vacuum manifold which is related to the order parameter space (see Appendices C and I). Then, non-trivial $\pi_n(\mathcal{M})$ can predict the possibility of the existence of defects of certain types in a given ordered medium and the corresponding topological charges.

As first proposed by Kibble [371], stable topological defects could have been formed during early-universe phase transitions. Indeed, the phase transition would occur independently in causally disconnected regions, so that the original (symmetric) phase could remain trapped in some regions at the end of the transition. A natural question is then how many defects would have been formed in the phase transition process [372]. Later, Zurek [675] argued that during a continuous phase transition defect formation is ruled by the non-equilibrium aspects of the process. The resulting scenario for defect formation is known as the *Kibble–Zurek mechanism*.

By following closely [374], we briefly review this mechanism. The early argument by Kibble was that in a real system going through the transition at a finite rate, the true correlation length ξ_ϕ cannot become infinite, since it cannot increase faster than the speed of light, or, in non-relativistic systems, some characteristic speed. The adiabatic approximation, that $\xi_\phi \approx \xi_{eq}(T)$ is not valid beyond the point where $\dot{\xi}_{eq} = c$, and thus one assumes that ξ_ϕ is more or less constant until the end the transition, at least to the point where it again becomes equal to the decreasing ξ_{eq}.

Similar conclusions have been reached by Zurek, comparing the quench rate and relaxation rate of the system. Near the transition, assuming that temperature varies linearly with time, we have

$$\epsilon \equiv \frac{T_c - T}{T_c} = \frac{t}{\tau_q}, \tag{10.84}$$

where τ_q is the quench time and we take $t = 0$ when $T = T_c$. The equilibrium correlation length near T_c has the form $\xi_{eq}(T) = \xi_0 |\epsilon|^{-\nu}$, where ν is a critical index. Also the relaxation time diverges at T_c as $\tau(T) = \tau_0 |\epsilon|^{-\mu}$, where μ is another critical index. This is known as *critical slowdown*.

The characteristic velocity is

$$c(T) = \frac{\xi_{eq}(T)}{\tau(T)} = \frac{\xi_0}{\tau_0} |\epsilon|^{\mu - \nu}, \tag{10.85}$$

vanishing at T_c. By using the causality bound, according to which the information about the phase of the order parameter cannot propagate faster than $c(T)$, the defect density in the case of strings or vortices is approximately given by

$$L(t_Z) = \frac{\kappa}{\xi_0^2} \left(\frac{\tau_0}{\tau_q} \right)^{\frac{2\nu}{1+\mu}}, \tag{10.86}$$

where κ is a constant and t_Z is the Zurek time, defined as the time at which the system goes out of equilibrium. One finds [374, 676]:

$$t_Z = \left[(1 + \mu - \nu) \tau_0 \tau_q^\mu \right]^{\frac{1}{1+\mu}}. \tag{10.87}$$

The Kibble–Zurek mechanism provides a very valuable tool for the understanding of the non-equilibrium dynamics in the process of phase transitions. Such aspects are indeed very difficult to describe in a QFT framework, where the methods introduced in Chapter 6 play a central role. Examples of this kind of analysis can be found in [12, 119, 364, 430, 539, 540].

The universal character of the Kibble–Zurek mechanism has led to the proposal of testing theories of cosmological phase transitions in condensed matter systems [675]. On this basis, the theory of defect formation has been tested in a many physical systems [375], such as liquid crystals [117, 159], liquid helium [63, 543] and Josephson junctions [479, 480].

As an illustration of the above scenario, let us consider the formation of line defects, i.e., strings, in a model with $U(1)$ broken symmetry. Examples include vortices in superfluid ^4He and magnetic flux lines (Abrikosov vortices) in type II superconductor. In these cases the expectation value of $\hat{\Phi}$ in the broken phase is a complex scalar which can be parametrized as $\Phi^0 = v\, e^{i\vartheta}$. The magnitude v comes from minimization of the Ginzburg–Landau free energy, while the phase ϑ is arbitrary. Below the critical temperature T_c, the value of ϑ is selected randomly among the set of values

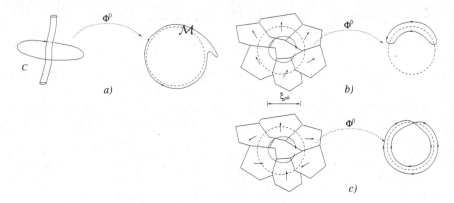

Fig. 10.5 a) A thin tube-like region of the disordered phase — string, is trapped in the ordered medium. A loop $C \cong \mathcal{S}^1$ is mapped to $\mathcal{M} \cong \mathcal{S}^1$ with a non-trivial winding number. b) The non-uniform order parameter (in the physical space) with $n = 0$. The resulting map $\Phi^0 : C \to \mathcal{M}$ can be shrunk to a point. c) The non-uniform order parameter with $n = 2$. Orientations of the arrows represent angle positions on $\mathcal{M}$.

corresponding to minima of the free energy. As observed above, in large systems there is no reason why the same choice should be made every-where. It is, further, reasonable to assume that short-range interactions cause the order parameter to be a continuous everywhere in the physical system except, perhaps, at a point or line.

In continuous phase transitions, as T decreases to T_c, the equilibrium correlation length $\xi_{eq}(T)$, which is the average distance over which the vacuum values of the order parameter are correlated, diverges. In practice, however, correlation lengths always remain finite. Thus, in systems that are much larger than a maximal *physical* correlation length ξ_{ph}, it is possible that the total change in ϑ, around some large loop C in physical space, is 2π or a n-multiple of it. One then has a mapping from a loop (i.e., from $\mathcal{S}^1$) in physical space into order parameter space $\mathcal{M} \cong U(1) \cong \mathcal{S}^1$. A smooth contraction of such a loop does not change the homotopy class, i.e., does not change n (the homotopy group here is the fundamental group $\pi_1(U(1) \cong \mathcal{S}^1)$). The winding number n describes the number of times one traverses the circle in $\mathcal{M} \cong \mathcal{S}^1$ while going from 0 to 2π around the loop in the physical system; see Fig. 10.5.

For non-zero n, there must be a point inside the loop where both the gradient $\nabla \Phi^0 \sim \mathbf{e}_\theta \, ive^{i\vartheta}/r$ diverges ($\mathbf{e}_\theta$ is the unit vector tangent to the loop and r is the radius of the loop) and the phase ϑ is undefined. This situation can be avoided only if $v = 0$ at the singular point. So, inside the loop must be trapped the original disordered phase, the core of the defect (string). Strings are either infinite (in finite-size media they terminate at boundaries) or closed in form of loops.

In realistic situations, the string cannot be considered as a one-dimensional (in space) object. For strings with non-zero thickness, one must consider v to be constant only far from the string core (cf., e.g., Section 7.7). The actual profile of $v(\mathbf{x})$ can be obtained by minimizing the Ginzburg–Landau free energy under conditions $v(0) = 0$ and $v(\mathbf{x})|_{|\mathbf{x}| \to \infty} = v$. The above homotopical considerations are true also for thick strings. The only difference is that the fundamental group $\pi_1(\mathcal{S}^1) = \mathbb{Z}$ now classifies the mappings $\mathcal{S}^1_\infty \to \mathcal{S}^1_{vac}$ ($\mathcal{S}^1_\infty$ corresponds to a loop very far from the string core and $\mathcal{S}^1_{vac}$ corresponds to the vacuum manifold). We have seen this type of argument already in Section 10.4 and we will repeatedly use it in the following.

10.5.1 *Exact homotopy sequences*

The above discussion for strings has been based on the homotopical properties of $\mathcal{M}$. Analogous considerations can be done for a large class of systems with SSB. As in the case of strings in $D = 3$, stable topological defects can exist only when the corresponding homotopy group $\pi_n(\mathcal{M})$ is non-trivial; see Fig. 10.6. It is normally quite difficult to calculate π_n for a general manifold $\mathcal{M}$. For instance, the higher homotopy groups of $\mathcal{S}^2$ are still unknown. If, however, the order parameter space has a group theoretic characterization (as in the case in SSB phase transitions), the homotopy groups can be calculated. The computational tool is known as *exact homotopy sequences* or *fiber homotopy sequence*.

An exact sequence (Fig. 10.7) is a (finite or infinite) sequence of sets A_i and maps such that the image of one map equals the kernel of the next:

$$\cdots \to A_{i-1} \xrightarrow{f_{i-1}} A_i \xrightarrow{f_i} A_{i+1} \to \cdots, \qquad \mathrm{Im} f_{i-1} = \mathrm{Ker} f_i. \qquad (10.88)$$

An exact sequence is called a *short exact sequence* if :

$$0 \to A_1 \xrightarrow{f_1} A_2 \xrightarrow{f_2} A_3 \to 0. \qquad (10.89)$$

A particularly important short exact sequence is a fiber bundle

$$\begin{array}{c} F \to E \\ \quad \downarrow \pi \\ B \ , \end{array} \qquad (10.90)$$

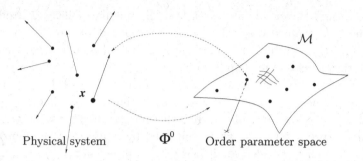

Physical system Φ^0 Order parameter space

Fig. 10.6   The order parameter field $\Phi^0(x)$ can be visualized in two different ways. On the one hand, one can think of a vector attached to each point in space. On the other hand, one can think of it as a mapping from a physical system into order parameter space $\mathcal{M}$. That is, Φ^0 is a function which takes different points in a physical system onto $\mathcal{M}$. This mapping is described homotopy theory.

---

where F represents the *fiber*, E is the *bundle space*, B is the *base space* and π is the projection from the bundle space to the base space. A discussion on fiber bundles can be found, e.g., in [388]. A theorem in algebraic topology ensures that the short exact sequence (10.90) induces an exact sequence of homomorphisms known as a *fiber homotopy sequence*:

$$\cdots \to \pi_n(F) \to \pi_n(E) \to \pi_n(B) \to \pi_{n-1}(F) \to \pi_{n-1}(E) \to \pi_{n-1}(B) \to \cdots .$$
(10.91)

From Eq. (10.88) we see that if $A_1 \to A_2 \overset{f}{\to} A_3 \to A_4$ is an exact homotopy sequence with A_1 and A_4 being unity, then the homomorphism f between A_2 and A_3 is an isomorphism. In particular, if $\pi_n(E) = \pi_{n-1}(E) = \{e\}$, then $\pi_n(B) \cong \pi_{n-1}(F)$.

The SSB scenario naturally induces the *principal* fiber bundle

$$
\begin{array}{c}
H \;\to\; G \\
\downarrow \pi \\
G/H \, ,
\end{array}
$$
(10.92)

which gives rise to the exact homotopy sequence

$$\cdots \to \pi_n(H) \to \pi_n(G) \to \pi_n(G/H) \to \pi_{n-1}(H) \to \cdots$$

$$\cdots \to \pi_0(H) \to \pi_0(G) \to \pi_0(G/H) \to \{e\}. \quad (10.93)$$

Simple illustration of the usefulness of the fiber homotopy sequence is provided by the Hopf bundle

$$
\begin{array}{c}
\mathcal{S}^1 \;\to\; \mathcal{S}^3 \\
\downarrow \pi \\
\mathcal{S}^2 \;\cong\; P_1(\mathbb{C}) \, .
\end{array}
$$
(10.94)

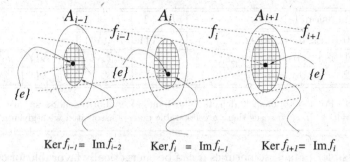

$$\text{Ker}\, f_{i-1} = \text{Im}\, f_{i-2} \qquad \text{Ker}\, f_i = \text{Im}\, f_{i-1} \qquad \text{Ker}\, f_{i+1} = \text{Im}\, f_i$$

Fig. 10.7 An exact sequence represents a sequence of maps $f_i : A_i \to A_{i+1}$ between a sequence of sets A_i.

Here $P_1(\mathbb{C})$ represents the one-dimensional complex projective space. The Hopf bundle is equivalent to the principal fiber bundle

$$
\begin{array}{c}
U(1) \;\to\; SU(2) \\
\downarrow \pi \\
SU(2)/U(1)
\end{array}
\qquad (10.95)
$$

This gives rise to the fiber homotopy sequence

$$\{e\} \;=\; \pi_2(SU(2)) \to \pi_2(SU(2)/U(1))$$

$$\to \pi_1(U(1)) \to \pi_1(SU(2)) \;=\; \{e\} . \qquad (10.96)$$

In (10.96) we have used the Bott periodicity theorem (T.17). The sequence (10.96) implies that $\pi_2(SU(2)/U(1)) \cong \pi_1(U(1)) \cong \mathbb{Z}$, or equivalently $\pi_2(\mathcal{S}^2) \cong \pi_1(\mathcal{S}^1) \cong \mathbb{Z}$ (cf. the Bott–Tu theorem (T.20)). From the above example we see that whenever in a principal fiber bundle the bundle space group G is simply connected and compact (i.e., $\pi_2(G) = \pi_1(G) = \{e\}$) then $\pi_2(G/H) \cong \pi_1(H)$.

10.5.2 Topological defects in theories with SSB

We now discuss some applications of homotopy theory (see Appendix T) in the classification of topological defects arising during SSB phase transitions.

In general, d-dimensional topological defects in D-dimensional *configuration* space are described by the homotopy group $\pi_n(\mathcal{M})$ (see Fig. 10.8) with $n = D - d - 1$.

Homotopy classification cannot answer the important question as to whether configurations with $|n| > 1$ are stable, it can only tell that they

d-dimensional defects	$D = 1$	$D = 2$	$D = 3$	$D = 4$
$d = 0$ (e.g., monopoles, ...)	$\pi_0(\mathcal{M})$	$\pi_1(\mathcal{M})$	$\pi_2(\mathcal{M})$	$\pi_3(\mathcal{M})$
$d = 1$ (e.g., cosmic strings, ...)	$\Diamond$	$\pi_0(\mathcal{M})$	$\pi_1(\mathcal{M})$	$\pi_2(\mathcal{M})$
$d = 2$ (e.g., domain walls, ...)	$\Diamond$	$\Diamond$	$\pi_0(\mathcal{M})$	$\pi_1(\mathcal{M})$

Fig. 10.8   Possible topological defects. The diamond symbol $\Diamond$ denotes situations where $\pi_n(\mathcal{M}) = \{e\}$. In such cases there are no stable defects associated with given d and D.

---

are possible. In fact, sometimes it can be energetically favorable for configurations with the winding number $n = 2$ to split into two separate configurations with $n = 1$. This is, for instance, the case in type-II superconductors. The detailed dynamics must be always examined in order to answer questions like the one above.

In the following, we adopt the usual description of symmetry breaking by which the residual symmetry after SSB is only the stability group. As discussed in Chapters 3 and 4, however, the complete picture of SSB involves also, in an essential way, the Abelian subgroups associated to translations of NG bosons in the group contraction mechanism.

Disclinations in uniaxial nematics

Nematic crystals (nematics) represent the simplest phase of liquid crystals. The building blocks of nematics are molecules that have a typically cigar-like shape (uniaxial nematics) or disc-like shape (biaxial nematics).

Nematic transitions in uniaxial liquid crystals can be identified with SSB transitions from the disordered (isotropic) phase described by the $SO(3)$ group to the ordered phase described by the dihedral group D_∞. The group D_∞ represents continuous rotations about the molecular axis and rotations by π about axes perpendicular to the molecular axis. According to our previous discussion, the order parameter space $\mathcal{M}$ can be identified with the coset space $SO(3)/D_\infty$. The corresponding order parameter is called a *director* and it can be naturally represented by a unit vector but without associated orientation — a headless arrow. This is because there is no sense of the direction, like, for instance, in ferromagnets. In the nematic phase the molecules are just preferentially aligned, but on average there are as many parallel as antiparallel oriented molecules. This implies that the order parameter space can be equivalently identified with the surface of the unit sphere $\mathcal{S}^2$ which has the antipodal points identified, i.e., with two-dimensional real *projective space* $P_2(\mathbb{R})$. Thus $\mathcal{M} \cong SO(3)/D_\infty \cong P_2(\mathbb{R})$.

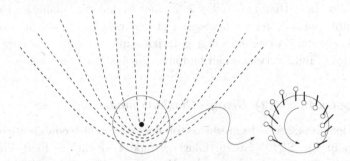

Fig. 10.9 Line-like defect in uniaxial nematics (left figure). The order parameter (head-less arrow) rotates by π when the defect is encircled (right figure). For visualization simplicity we have endowed the order parameter with a head, but one end should still be considered indistinguishable from the other.

Typical examples of uniaxial nematics are N-(p-methoxybenzylidene)-p-butylaniline which has the isotropic-to-nematic transition temperature of 315.5°C and p-azoxyanisole with the transition temperature of 135°C.

The line-like defects *do exist* in uniaxial nematics because $\pi_1(P_2(\mathbb{R}))$ is non-trivial. In fact, the bundle structure associated with the nematic transition can be written as

$$D_\infty \;\to\; SO(3)$$
$$\downarrow \pi \qquad\qquad (10.97)$$
$$SO(3)/D_\infty .$$

A relevant piece of the exact homotopy sequence reads

$$\{e\} \to \pi_1(D_\infty) \;\to\; \pi_1(SO(3))$$
$$\to \pi_1(SO(3)/D_\infty) \cong \pi_1(P_2(\mathbb{R})) \to \pi_0(D_\infty) \;\to\; \{e\}, \qquad (10.98)$$

which clearly indicates that

$$\pi_1(P_2(\mathbb{R})) \cong \pi_1(SO(3)/D_\infty) \cong \pi_1(SO(3)) \cong \mathbb{Z}_2 . \qquad (10.99)$$

The ensuing line-like defects are called *disclinations*. In addition, because $\pi_1(P_2(\mathbb{R}))$ is isomorphic to the cyclic group $\mathbb{Z}_2$, i.e., the additive group of integers modulo 2, there must be only one topologically distinct type of disclinations. The group structure of $\mathbb{Z}_2$ then dictates that when two such disclinations meet they must "annihilate", i.e., the resulting configuration is defect free. An illustration of a disclination line is given in Fig. 10.9.

The reader may notice that by traversing a closed loop around the singularity line once, the order parameter rotates by π. The latter explains why the nematic line-like defect is often called a π-disclination.

We also note that $\pi_2(P_2(\mathbb{R})) \cong \mathbb{Z}$, and so uniaxial nematics can also have stable point defects, *hedgehogs*. Both hedgehogs and disclinations can be experimentally observed via a polarizing microscope. A survey of the properties of liquid crystals can be found, for instance, in [186].

Monopoles in $SU(2)$ Georgi–Glashow model

In Section 10.4.2 we discussed a topologically non-trivial finite energy solution in the $SU(2)$ Georgi–Glashow model — the 't Hooft–Polyakov monopole. There we have seen that the ground state is not invariant under the *global $SU(2)$* symmetry (remaining after the gauge fixing). In fact, the classical vacuum manifold is the sphere $\mathcal{S}^2_{vac}$. In this case the stability group $H = U(1)$ is the subgroup of rotations about an arbitrarily chosen direction. The corresponding SSB is then described by the breaking scheme

$$SU(2) \rightarrow U(1), \tag{10.100}$$

where $U(1)$ is the stability group. The associated principal fiber bundle (Hopf bundle) is given by

$$\begin{array}{rl} U(1) \rightarrow & SU(2) \\ & \downarrow \pi \\ & SU(2)/U(1). \end{array} \tag{10.101}$$

The relevant part of the fiber homotopy sequence reads

$$\{e\} = \pi_2(SU(2)) \rightarrow \pi_2(SU(2)/U(1))$$
$$\rightarrow \pi_1(U(1)) \rightarrow \pi_1(SU(2)) = \{e\}. \tag{10.102}$$

This clearly shows that $\pi_2(\mathcal{M}) \cong \pi_2(SU(2)/U(1)) \cong \pi_1(U(1)) \cong \mathbb{Z}$, meaning that during the transition from the disordered $SU(2)$ phase to the ordered $U(1)$ phase, *monopoles* can be formed.

The 't Hooft–Polyakov monopole carries a magnetic charge (see Section 10.4.2). In contrast to the Dirac monopole, which is a point-like object, the 't Hooft–Polyakov monopole has a finite core and there is no need for the Dirac string.

The name *'t Hooft–Polyakov monopole* is used in a wider context than the original one. In fact, it is common to talk about a 't Hooft–Polyakov monopole whenever in a Yang–Mills–Higgs theory the homotopy group $\pi_2(\mathcal{M} \cong G/H)$ is non-trivial. One does not even need to require the existence of a scalar Higgs field. The only condition is that the global symmetry G should be broken to H. In practice this means that most symmetry breaking mechanisms (e.g., technicolor) would also give rise to a 't Hooft–Polyakov monopole.

Monopoles in Grand Unified Theories

Among the various models proposed as candidates for a Grand Unified Theory (GUT), we consider two examples: the Georgi–Glashow $SU(5)$ model [278] which has the symmetry breaking scheme

$$SU(5) \rightarrow SU(3)_C \times SU(2)_I \times U(1)_Y$$
$$\rightarrow SU(3)_C \times U(1)_Q, \tag{10.103}$$

and the Pati–Salam model [512] which is based on the breaking scheme

$$SO(10) \rightarrow SU(4) \times SU(2)_R \times SU(2)_L$$
$$\rightarrow SU(3)_C \times SU(2)_I \times U(1)_Y \rightarrow SU(3)_C \times U(1)_Q. \tag{10.104}$$

Subscripts I and Y denote respectively the weak isospin and weak hypercharge, C stands for the strong color charge, L (R) denote the left (right) parity symmetry and Q is the electric charge. Since the fermions are assumed to be in the fundamental spinor representation, the GUT symmetry is not really $SO(10)$, but its simply-connected[5] covering group Spin(10). The principal fiber bundle for the Pati–Salam SSB scenario is:

$$SU(3)_C \times U(1)_Q \rightarrow \text{Spin}(10)$$
$$\downarrow \pi \tag{10.105}$$
$$\text{Spin}(10)/[SU(3)_C \times U(1)_Q].$$

Using the relevant fiber homotopy sequence

$$\{e\} = \pi_2(\text{Spin}(10)) \rightarrow \pi_2(\text{Spin}(10)/[SU(3) \times U(1)])$$
$$\rightarrow \pi_1(SU(3) \times U(1)) \rightarrow \pi_1(\text{Spin}(10)) = \{e\}, \tag{10.106}$$

we obtain that

$$\pi_2(\text{Spin}(10)/[SU(3) \times U(1)]) \cong \pi_1(SU(3) \times U(1))$$
$$\cong \pi_1(SU(3)) \otimes \pi_1(U(1)) \cong \mathbb{Z}, \tag{10.107}$$

and so the standard model based on the $SO(10)$ grand-unifying group predicts monopoles. The same result holds also in the Georgi–Glashow $SU(5)$ SSB scenario because $\pi_2(SU(5)) = \pi_1(SU(5)) = \{e\}$. Whenever the GUT group G is compact and simply connected with unbroken $U(1)$ embedded in it, the theory predicts magnetic monopoles [527]. Because GUT fermions are in the fundamental spinor representation, the GUT group G is compact and simply connected. In addition, since the $U(1)$ is essential for describing electromagnetism, virtually all GUT theories predict magnetic monopoles.

---

[5]Spin group Spin(n) is a double cover of $SO(n)$.

In all GUTs the monopole mass m_M is $\sim m_B/\alpha$ where m_B is the mass of the superheavy vector boson which determines the unification scale, and α is the fine structure constant. For $SU(5)$ GUT, m_B is $\sim 10^{14}$ GeV which implies $m_M \sim 10^{16}$ GeV. In $SO(10)$ GUT, monopoles with masses $\sim 10^4$ GeV could exist. On the other hand, magnetic monopoles from GUTs with still larger groups, such as various supersymmetric theories, have higher monopole mass with $m_M \gtrsim 10^{19}$ GeV.

Cosmic strings

It has been proposed that during a succession of cosmological phase transitions, cosmic strings could have been formed [371, 372]. One GUT scenario of cosmological interest that can produce cosmic strings is described by the SSB breaking scheme

$$\text{Spin}(10) \;\rightarrow\; SU(5) \times \mathbb{Z}_2 \;\rightarrow\; \dots. \tag{10.108}$$

In this case the relevant piece of the homotopy sequence is

$$\{e\} = \pi_1(\text{Spin}(10)) \;\rightarrow\; \pi_1(\text{Spin}(10)/[SU(5) \times \mathbb{Z}_2])$$
$$\rightarrow\; \pi_0(SU(5) \times \mathbb{Z}_2) \;\rightarrow\; \pi_0(\text{Spin}(10)) = \{e\}. \tag{10.109}$$

Here we have used the fact that Spin(10) is both connected and single connected. The sequence (10.109) implies that

$$\pi_1(\text{Spin}(10)/[SU(5) \times \mathbb{Z}_2]) \;\cong\; \pi_0(SU(5) \times \mathbb{Z}_2)$$
$$\cong\; \pi_0(SU(5)) \otimes \pi_0(\mathbb{Z}_2) \;\cong\; \mathbb{Z}_2. \tag{10.110}$$

Hence there is only one class of stable line-like defects, the GUT $\mathbb{Z}_2$-string.

It has been thought that cosmic strings could provide the seeds for the structure formation in the Universe, as an alternative to the possibility that the seeds originated as quantum fluctuations during inflation. Recent measurements of the Cosmic Microwave Background (CMB) anisotropy [122, 241, 413, 497] seem to rule out cosmic strings as the main source of the large-scale structure formation. They can, however, have other cosmological implications, such as the generation of magnetic fields, high-energy cosmic rays or baryogenesis. Further details on cosmic strings can be found, e.g., in [631, 646].

10.6 Derrick theorem and defect stability

Derrick (or Derrick–Hobart) theorem provides a necessary condition under which topology can stabilize non-trivial stationary field configurations [201, 329]. The theorem is based on the analysis of the behavior of the energy functional under spatial rescaling of the fields representing candidate configuration. When there is no local minimum for the rescaled fields (i.e., there is no fundamental "soliton" scale) there are no stable and static field configurations. Though this type of a scaling argument can be applied in a broad field-theoretical context, we will consider, for simplicity, only real scalar fields in k spatial dimensions. Let $\tilde{\phi}(\mathbf{x}) = \{\tilde{\phi}_a(\mathbf{x})\}$ represent the time-independent, topologically non-trivial field configuration. The energy functional

$$E[\phi] = -\int d^k x \, \mathcal{L}(\phi_a, \nabla \phi_a), \qquad (10.111)$$

is stationary at $\phi = \tilde{\phi}$, i.e.,

$$\frac{\delta E[\phi]}{\delta \phi_a} = 0, \quad \text{for} \quad \phi = \tilde{\phi}. \qquad (10.112)$$

Let us further assume that $E[\phi]$ can be decomposed as

$$E[\phi] = E_1[\phi] + E_2[\phi]$$
$$E_1[\phi] = \frac{1}{2} \int d^k x \sum_a \nabla \phi_a \cdot \nabla \phi_a, \quad E_2[\phi] = \int d^k x \, V(\phi_a). \qquad (10.113)$$

We also require that $E_2[\phi]$ does not depend on derivatives of ϕ_a and that $E_2[\phi] \geq 0$. Now we define a one-parameter family of fields labeled by a parameter $\lambda \geq 0$: $\phi_\lambda(\mathbf{x}) = \tilde{\phi}(\lambda \mathbf{x})$. For $\lambda < 1$, the rescaled configuration $\phi_\lambda(\mathbf{x})$ is more spread out while for $\lambda > 1$ it is more concentrated. By changing variables in the integrals in (10.113) we obtain

$$E[\phi_\lambda] = \lambda^{2-k} E_1[\tilde{\phi}] + \lambda^{-k} E_2[\tilde{\phi}]. \qquad (10.114)$$

We require that

$$\left. \frac{dE[\phi_\lambda]}{d\lambda} \right|_{\lambda=1} = (2-k)E_1[\tilde{\phi}] - kE_2[\tilde{\phi}] = 0. \qquad (10.115)$$

For $k > 2$, we see from Eq. (10.114) that $E[\phi_\lambda]$ decreases for increasing $\lambda > 1$. The field $\tilde{\phi}(\mathbf{x})$ is then unstable against collapse. The only possible stable configuration is the one with $E_1[\tilde{\phi}] = E_2[\tilde{\phi}] = 0$, which means that $\tilde{\phi}(\mathbf{x})$ is a constant field taking a value in the minimum of the potential. Such a field configuration is trivial having zero energy density everywhere. Accordingly, there are no static, topologically non-trivial and space-dependent field configurations in space dimensions larger than two.

For $k = 2$, Eq. 10.115 implies that $E_2[\tilde{\phi}] = 0 \Rightarrow V(\tilde{\phi}_a) = 0$, i.e., the field $\tilde{\phi}(\mathbf{x})$ takes values in the minimum of V for all $\mathbf{x}$, but need not be constant. Such configurations are described by the Euler–Lagrange equation

$$\Delta\tilde{\phi}(\mathbf{x}) = \left(\frac{\partial^2}{\partial^2 x} + \frac{\partial^2}{\partial^2 y}\right)\tilde{\phi}(\mathbf{x}) = 0 \,, \tag{10.116}$$

with the boundary conditions determined by the finite-energy condition, i.e., $E[\tilde{\phi}] < \infty$. When the field $\tilde{\phi}(\mathbf{x})$ is defined in the entire $\mathbb{R}^2$ then the finite energy condition can be conveniently written in polar coordinates as

$$E[\tilde{\phi}] = \sum_a \int_0^\infty \int_0^{2\pi} dr d\theta \frac{r}{2}\left[\left(\frac{\partial\tilde{\phi}^a}{\partial r}\right)^2 + \left(\frac{1}{r}\frac{\partial\tilde{\phi}^a}{\partial\theta}\right)^2\right] < \infty \,. \tag{10.117}$$

In particular, when r is large the following asymptotic relations must hold

$$r\frac{\partial\tilde{\phi}^a}{\partial r} \sim \frac{1}{r^{\epsilon/2}} \,, \quad \frac{\partial\tilde{\phi}^a}{\partial\theta} \sim \frac{1}{r^{\epsilon/2}} \,, \quad \epsilon > 0 \,. \tag{10.118}$$

Now, because $\tilde{\phi}^a(\mathbf{x})$ is a harmonic function and because it is bounded at the spatial infinity, Liouville theorem [354] ensures that $\tilde{\phi}^a(\mathbf{x})$ must be a *constant* over the whole plane $\mathbb{R}^2$. In order to evade this trivial solution in $k = 2$ one must introduce either subsidiary constraints on the fields (to circumvent Liouville theorem), or add, e.g., gauge fields (i.e., invalidate the simple structure (10.113)).

In definitive, only for $k = 1$, we can have stable non-trivial stationary field configurations. We also note that the Derrick theorem does not prohibit the existence of non-static solitons.

In the following we present a couple of applications of the Derrick theorem.

Non-linear $O(3)$ σ model in $2 + 1$ dimensions

The $O(3)$ spin model in $2+1$ dimensions (known also as non-linear $O(3)$ σ model) has three real scalar fields ϕ_a that are constrained to the unit circle

$$\phi_1^2 + \phi_2^2 + \phi_3^2 = 1 \,, \tag{10.119}$$

and the Lagrangian

$$\mathcal{L} = \frac{1}{2}(\partial_\mu\boldsymbol{\phi}) \cdot (\partial^\mu\boldsymbol{\phi}) \,. \tag{10.120}$$

The dot product means that the standard Euclidean metric in $\mathbb{R}^3$ is used.[6] We saw above that static solitary solutions in $k = 2$ spatial dimensions are allowed provided appropriate constraints on the field manifold are imposed. The non-linear $O(3)$ σ model provides an example of this type.

The field equation is obtained from the extremization of the functional

$$S_\alpha[\phi] = \frac{1}{2}\int d^3x \left[(\partial_\mu\phi)\cdot(\partial^\mu\phi) + \alpha(x)(\phi\cdot\phi - 1)\right]. \qquad (10.121)$$

Here α is the Lagrange multiplier. The Euler–Lagrange equation reads

$$\partial_\mu\partial^\mu\phi - \alpha\phi = 0. \qquad (10.122)$$

One can find the Lagrange multiplier by taking scalar product of (10.122) with ϕ and using the constraint (10.119). This gives

$$\alpha = \phi\cdot\partial_\mu\partial^\mu\phi, \qquad (10.123)$$

which brings (10.122) to the form

$$\partial_\mu\partial^\mu\phi - (\phi\cdot\partial_\mu\partial^\mu\phi)\phi = 0. \qquad (10.124)$$

The static-field configurations must thus fulfill the equation

$$\nabla^2\phi - (\phi\cdot\nabla^2\phi)\phi = 0 \qquad (10.125)$$

and the energy functional has the form

$$E[\phi_a] = \frac{1}{2}\int d^2x \,(\partial_i\phi)\cdot(\partial_i\phi). \qquad (10.126)$$

The index i runs from 1 to 2. The lowest value of E — i.e., ground state, is clearly zero. This can be reached only with a constant field configuration.

Due to its finite energy, the static soliton configurations must satisfy $\phi_a|_{|\mathbf{x}|\to\infty} = c_a$ ($\mathbf{c}$ is a constant unit vector). Because of the global $O(3)$ symmetry, we can choose the ground state configuration as $\phi_a = \delta_{a3}$. The field $\phi(\mathbf{x})$ is a function that maps the two-dimensional space xy, into a vector ϕ in $O(3)$ isotopic space. Since at $|\mathbf{x}| \to \infty$ the field ϕ_a approaches δ_{a3}, the xy-space is described by $\mathcal{S}^2$ since the values of ϕ are at spatial infinity all identical (i.e., we replace $\mathbb{R}^2$ with a sphere, where infinity has been transformed into the north pole). In addition, since the isotopic space reduces also to a sphere $\mathcal{S}^2$ (since $\phi\cdot\phi = 1$), the mapping from xy-space $\mathcal{S}^2$ ($\cong \mathbb{R}^2 \cup \infty$) to isotopic space is classified by the homotopy group $\pi_2(\mathcal{S}^2)$. According to the Bott–Tu theorem (cf. Appendix T) the topological charge

---

[6]The space $\mathbb{R}^3$ of internal symmetry is also known as the *isotopic space*.

must be represented by elements of $\pi_2(\mathcal{S}^2) \cong \mathbb{Z}$. An explicit formula for the topological charge (degree of mapping) is in this case

$$Q = \frac{1}{8\pi} \int d^2x \; \epsilon_{ik} \, \boldsymbol{\phi} \cdot (\partial_i \boldsymbol{\phi} \wedge \partial_k \boldsymbol{\phi}) \,. \tag{10.127}$$

This form is justified by realizing that an infinitesimal surface element $d\sigma_a$ in the isovector space $\mathcal{S}^2$ is related to the surface element in the physical coordinate space $\mathcal{S}^2 \cong \mathbb{R}^2 \cup \infty$ via

$$d\sigma_a = d^2\zeta \left(\frac{1}{2} \, \epsilon_{ik} \, \epsilon_{abc} \frac{\partial \phi_b}{\partial \zeta_i} \frac{\partial \phi_c}{\partial \zeta_k} \right) \,. \tag{10.128}$$

Here $\{\zeta_1, \zeta_2\}$ are arbitrary local variables parametrizing xy-space.

The soliton solution can be found using a technique that is akin to the Bogomol'nyi trick (see Section 10.7). To this end we consider the inequality

$$\int d^2x [(\partial_i \boldsymbol{\phi} \pm \epsilon_{ij} \boldsymbol{\phi} \wedge \partial_j \boldsymbol{\phi}) \cdot (\partial_j \boldsymbol{\phi} \pm \epsilon_{ik} \boldsymbol{\phi} \wedge \partial_k \boldsymbol{\phi})] \;\geq 0 \,. \tag{10.129}$$

This is equivalent to

$$\int d^2x \; [(\partial_i \boldsymbol{\phi}) \cdot (\partial_i \boldsymbol{\phi}) + \epsilon_{ij} (\boldsymbol{\phi} \wedge \partial_j \boldsymbol{\phi}) \cdot \epsilon_{ik} (\boldsymbol{\phi} \wedge \partial_k \boldsymbol{\phi})]$$

$$\geq \pm 2 \int d^2x \, [\epsilon_{ik} \, \boldsymbol{\phi} \cdot (\partial_i \boldsymbol{\phi} \wedge \partial_k \boldsymbol{\phi})] \,. \tag{10.130}$$

Using the vector identity $(\mathbf{a} \wedge \mathbf{b}) \cdot (\mathbf{c} \wedge \mathbf{d}) = (\mathbf{a} \cdot \mathbf{c})(\mathbf{b} \cdot \mathbf{d}) - (\mathbf{a} \cdot \mathbf{d})(\mathbf{b} \cdot \mathbf{c})$, and the relation $\epsilon_{ij} \epsilon_{kl} = \delta_{jk} \delta_{jl} - \delta_{il} \delta_{jk}$, we get that the two terms in the first integral in (10.130) are identical, and so

$$E \geq 4\pi |Q| \,. \tag{10.131}$$

This inequality is saturated only for self-dual solutions (cf. Eq. (10.129))

$$\partial_i \boldsymbol{\phi} = \pm \, \epsilon_{ij} \boldsymbol{\phi} \wedge \partial_j \boldsymbol{\phi} \,. \tag{10.132}$$

The solution can be anticipated in the form of the Ansatz

$$\phi_3 = \cos f(r) \,, \qquad \phi_a = n_a \sin f(r) \,, \qquad a = 1, 2 \,, \tag{10.133}$$

where $f(r)|_{r \to \infty} = 0$ and $n_a = x_a / r$. With this (10.132) turns into

$$\frac{df}{dr} = \pm \frac{1}{r} \sin f \,. \tag{10.134}$$

The solutions that are compatible with the boundary condition $\phi_a|_{|\mathbf{x}| \to \infty} = \delta_{a3}$ exist only for "$-$" sign, in which case they read

$$f(r) = \pm 2 \arccos \frac{r}{\sqrt{r_0^2 + r^2}} = \pm 2 \arcsin \frac{r_0}{\sqrt{r_0^2 + r^2}} \,, \tag{10.135}$$

so that

$$\phi_3 = \frac{r^2 - r_0^2}{r^2 + r_0^2}, \qquad \phi_a = \pm 2\,\frac{x_a r_0}{r^2 + r_0^2}, \qquad a = 1, 2. \qquad (10.136)$$

The collective coordinate r_0 plays the role of soliton size. Other collective coordinates are related to constant spatial translation and rigid $O(3)$ rotation. In this respect, both configurations in (10.136) are equivalent since they are related by $O(3)$ rotation. By examining the value of the topological charge (10.127), we find that both configurations have the topological charge $Q = -1$. The $Q = 1$ soliton is obtained via the Ansatz

$$\phi_3 = -\cos f(r), \qquad \phi_a = n_a \sin f(r), \qquad a = 1, 2, \qquad (10.137)$$

with $f(r)|_{r \to \infty} = \pi$.

Since the above static solitons/antisolitons are finite energy configurations, they also represent finite Euclidean-action configurations in $1 + 1$ dimensions — i.e., instantons. For this reason the static solitonic/antisolitonc solution Eq. (10.136) is often referred to as Belavin–Polyakov instanton [65].

Pure Yang–Mills theory in $3 + 1$ dimensions

A further example is provided by pure Yang–Mills theory. We do not need to assume any particular dimension of spacetime. The Lagrange density reads

$$\mathcal{L} = -\frac{1}{2}\mathrm{Tr}(\mathbf{F}_{\mu\nu}\mathbf{F}^{\mu\nu}) = -\frac{1}{4}(F_{\mu\nu})^a (F^{\mu\nu})^a, \qquad (10.138)$$

where $\mathbf{F}_{\mu\nu} = T^a F_{\mu\nu}^a$, with T^a being the generator of the corresponding gauge group. In the Abelian case the internal index would be dropped. By setting the temporal gauge, i.e., $A_0^a = 0$, and considering only the static field configurations, we obtain the energy functional

$$E[A_i^a] = -\frac{1}{2}\int d^k x\,\mathrm{Tr}(\mathbf{F}_{ij}\mathbf{F}_{ij}). \qquad (10.139)$$

We assume that $\tilde{\mathbf{A}}(\mathbf{x})$ is a stationary classical solution and define Derrick one-parameter family of fields

$$\mathbf{A}_\lambda(\mathbf{x}) = \lambda\tilde{\mathbf{A}}(\lambda\mathbf{x}), \qquad (10.140)$$

where $\lambda \geq 0$. Note that, in comparison with scalar fields, there is an extra factor λ in front of $\tilde{\mathbf{A}}$. It ensures that the scaling behavior of the covariant derivative of a scalar field multiplet ϕ is the same as in the case of the ordinary derivative. Indeed,

$$D_\mu^A \phi(\mathbf{x}) = \partial_\mu \phi(\mathbf{x}) + ie A_\mu^a(\mathbf{x}) T^a \phi(\mathbf{x})$$

$$\rightarrow \quad D_\mu^{A_\lambda} \phi_\lambda(\mathbf{x}) = \partial_\mu \phi_\lambda(\mathbf{x}) + ie(A_\mu^a)_\lambda(\mathbf{x}) T^a \phi_\lambda(\mathbf{x})$$

$$= \lambda D_\mu^A \phi(\lambda \mathbf{x}). \tag{10.141}$$

A simple consequence of Derrick rescaling is that

$$\mathbf{F}_{ij} = -ie^{-1}[D_i^A, D_j^A] \quad \rightarrow \quad (\mathbf{F}_{ij})_\lambda = -ie^{-1}[D_i^{A_\lambda}, D_j^{A_\lambda}]$$

$$= \lambda^2 \mathbf{F}_{ij}, \tag{10.142}$$

or $(\mathbf{F}_{ij})_\lambda(\mathbf{x}) = \lambda^2 \tilde{\mathbf{F}}_{ij}(\lambda \mathbf{x})$. Inserting this into (10.139) we obtain

$$E[(A_i^a)_\lambda] = -\frac{1}{2} \int d^k x \ \text{Tr}[(\mathbf{F}_{ij})_\lambda(\mathbf{x})(\mathbf{F}_{ij})_\lambda(\mathbf{x})]$$

$$= -\lambda^{4-k} \frac{1}{2} \int d^k x \ \text{Tr}(\tilde{\mathbf{F}}_{ij}(\mathbf{x}) \tilde{\mathbf{F}}_{ij}(\mathbf{x})) = \lambda^{4-k} E[\tilde{A}_i^a]. \tag{10.143}$$

This directly implies that

$$\left. \frac{dE[(A_i^a)_\lambda]}{d\lambda} \right|_{\lambda=1} = (4-k) E[\tilde{A}_i^a]. \tag{10.144}$$

For the stationarity of $\tilde{A}_i^a$, the l.h.s. should be zero, which can happen only for $k = 4$. In the case when $k \neq 4$ one would need to have $E[\tilde{A}_i^a] = 0$ which can be fulfilled only by the constant vacuum solution $\tilde{\mathbf{F}}_{ij}(\mathbf{x}) = 0$.

Thus, in pure Yang–Mills theory in $3+1$ dimensions, Derrick theorem prohibits the existence of (classical) static solitons. However, if $k = 4$ then $E[(A_i^a)_\lambda]$ is scale-independent and non-trivial physical configurations may exist. Indeed, instantons of a pure Yang–Mills theory *do exist* in four Euclidean dimensions (see Section 10.9).

10.7   Bogomol'nyi bounds

Solitons can carry a topological charge Q coming from the non-trivial boundary conditions. It is often possible to derive a lower bound on the energy E of any field configuration in terms of topological charge. Such a bound depends on the field values at spatial infinity.

Let us consider a single scalar field in $1+1$ dimensions, for which, in the static case, the inequality holds (see Section 10.3):

$$\left(\frac{1}{\sqrt{2}} \phi' \pm \sqrt{V(\phi)} \right)^2 \geq 0. \tag{10.145}$$

By expanding the square and integrating over space we obtain

$$\int_{-\infty}^{\infty} dx \left(\frac{1}{2}(\phi')^2 + V(x) \right) \geq \mp \int_{-\infty}^{\infty} dx \sqrt{2V(\phi)}\, \phi'. \quad (10.146)$$

This implies that, for static field configurations, we have

$$E[\phi] \geq \left| \int_{-\infty}^{\infty} dx \sqrt{2V(\phi)}\, \phi' \right| = \left| \int_{\phi_-}^{\phi_+} d\phi \sqrt{2V(\phi)} \right|, \quad (10.147)$$

with $\phi_+ \equiv \phi(\infty)$ and $\phi_- \equiv \phi(-\infty)$. For time-dependent field configurations, we have

$$\frac{1}{2}(\dot{\phi})^2 + \left(\frac{1}{\sqrt{2}}\phi' \pm \sqrt{V(\phi)} \right)^2 \geq 0, \quad (10.148)$$

which leads to the same bound as in (10.147). By assuming that $V(\phi) \geq 0$ (this assumption is always legitimate) one can introduce the superpotential $W(\phi)$ as

$$V(\phi) = \frac{1}{2} \left(\frac{dW}{d\phi} \right)^2. \quad (10.149)$$

The bound (10.147) then becomes

$$E[\phi] \geq |W(\phi_+) - W(\phi_-)|$$
$$= \left| \frac{dW}{d\phi} \right|_{\phi \in [\phi_-, \phi_+]} |\phi_+ - \phi_-| \propto |Q|. \quad (10.150)$$

Q is the topological charge. The second line follows from (10.147) and the definition of topological charge. The result (10.150) is known as Bogomol'nyi bound [116]. In general, such terminology denotes conditions that provide lower energy bounds solely in terms of topological data.

The saturation of the Bogomol'nyi bound (10.147) is achieved (cf. Eq. (10.148)) by a static configuration $\dot{\phi} = 0$ fulfilling one of the first order differential equations

$$\phi'(x) = \pm\sqrt{2V(\phi(x))}. \quad (10.151)$$

Note that the solutions of the Bogomol'nyi equation (10.151) represent stationary points of the energy functional $E[\phi]$, which implies that they are static solutions of the Euler–Lagrange field equation (10.16). Indeed, by taking the spatial derivative of both sides of Eq. (10.151), we obtain

$$\phi'' = \pm \frac{1}{\sqrt{2V(\phi)}} \frac{dV(\phi)}{d\phi} \phi' = \frac{dV(\phi)}{d\phi}. \quad (10.152)$$

Due to its first-order nature, the Bogomol'nyi equation often provides a very efficient tool for obtaining soliton solutions. Field configurations saturating the Bogomol'nyi bound are often known as *BPS states*, after Bogomolny'nyi, Prasad and Sommerfield [526].

Bogomol'nyi bounds are often unaffected by quantum mechanical corrections. If a soliton attains the classical Bogomol'nyi bound it will do so also in the full quantum theory [8].

We now consider three applications of Bogomol'nyi bounds.

The sine-Gordon system

For the sine-Gordon system the Bogomol'nyi energy bound reads

$$E[\phi] \geq \left| \int_{2\pi N_1}^{2\pi N_2} d\phi \left| 2\sin\frac{\phi}{2} \right| \right| = \left| \left[-4\cos\frac{\phi}{2} \right]_0^{2\pi} \right| |N_2 - N_1| = 8|Q|. \quad (10.153)$$

$N_1, N_2 \in \mathbb{Z}$ and Q is the topological charge of Section 10.3.1. The inequality (10.153) is saturated when ϕ solves one of the Bogomol'nyi equations

$$\phi'(x) = \pm 2\sin\frac{\phi(x)}{2}. \quad (10.154)$$

These are the (scaled) kink and antikink solutions given in Section 10.3.1.

It should be noted that the general solution of the Bogomol'nyi equation corresponds to a kink/antikink with $|Q| = 1$. So, in particular, there are no multi-kink solutions of the Bogomol'nyi equation.

Note that while static multi-soliton solutions do not exist in the sine-Gordon system, time-dependent multi-soliton solutions do exist and they describe the scattering of two or more kinks/antikinks. Multi-kink solutions can be generated via Bäcklund transformations (see Appendix S).

The $\lambda\phi^4$ system

In the case of $\lambda\phi^4$ theory the Bogomol'nyi energy bound is given by

$$E[\phi] \geq \left| \int_{\phi_-}^{\phi_+} d\phi \sqrt{\frac{\lambda}{2}} \left(\phi^2 - \frac{m^2}{\lambda} \right) \right| = \left| \sqrt{\frac{\lambda}{2}} \left[\frac{\phi^3}{3} - \frac{m^2\phi}{\lambda} \right]_{\phi_-}^{\phi_+} \right|$$

$$= \sqrt{\frac{2}{\lambda}} \frac{m^2}{3} |\phi_+ - \phi_-| = \frac{4m^3}{3\lambda\sqrt{2}} |Q|, \quad (10.155)$$

where we have used the fact that $\phi_\pm = \{-m/\sqrt{\lambda}, m/\sqrt{\lambda}\}$ (cf. Section 10.3.2) and so $\phi_\pm^2 = m^2/\lambda$. Since both kink and antikink have $|Q| = 1$, their Bogomol'nyi bound is $E \geq \frac{4m^3}{3\lambda\sqrt{2}}$. The inequality (10.155) is saturated when one of the Bogomol'nyi equations

$$\phi'(x) = \pm\sqrt{\frac{\lambda}{2}}\left(\phi^2(x) - \frac{m^2}{\lambda}\right), \qquad (10.156)$$

is fulfilled. It is easy to see that the solution of Eq. (10.156) is indeed the kink/antikink solution presented in Section 10.3.2 and the energy of saturation (10.155) coincides with the energy (10.37).

Bogomol'nyi–Prasad–Sommerfield SO(3) monopoles

In Section 10.4.2 we obtained the "hedgehog" monopole solution in the form of the Ansatz without being able to solve the equations for f and h analytically. There is, however, one particular case in which the monopole solution of the $SO(3)$ Georgi–Glashow model can be found explicitly. This happens when the coupling constant g approaches 0. In this so-called Bogomol'nyi–Prasad–Sommerfield (BPS) limit, one may apply the Bogomol'nyi trick and cast the energy functional into the form (cf. Eq. (10.66))

$$E[\mathbf{F}^{\mu\nu}, \phi] = \int d^3x \left[\frac{1}{2}B_i^a B_i^a + \frac{1}{2}(D_i\phi)^a(D_i\phi)^a\right], \qquad (10.157)$$

where we have defined $B_i^a = \frac{1}{2}\epsilon_{ijk}F_{jk}^a$. The information about the original g and hence about the structure of the Higgs potential is still encoded in the boundary condition

$$(\phi \cdot \phi)|_{|\mathbf{x}|\to\infty} = v^2. \qquad (10.158)$$

Since

$$\int d^3x \, (B_i^a \pm (D_i\phi)^a)^2 \geq 0, \qquad (10.159)$$

we can write

$$E[\mathbf{F}_{\mu\nu}, \phi] \geq \left|\int d^3x \, B_i^a(D_i\phi)^a\right|, \qquad (10.160)$$

which saturates only when $B_i^a = \pm(D_i\phi)^a$. To cast Eq. (10.160) into Bogomol'nyi form, we should show that the r.h.s. is proportional to a topological charge. This can be seen by using the fact that[7]

$$\int d^3x B_i^a(D_i\phi)^a = \oint_{S_\infty^2} d^2\sigma_i \, B_i^a\phi^a - \int d^3x \, \phi^a(D_iB_i)^a$$

$$= \oint_{S_\infty^2} d^2\sigma_i \, B_i^a\phi^a. \qquad (10.161)$$

---

[7]The second term is zero due to the Bianchi identity $(D_\nu{}^*\mathbf{F}^{\mu\nu})^a = \frac{1}{2}\epsilon^{\mu\nu\alpha\beta}(D_\nu\mathbf{F}_{\alpha\beta})^a = 0$. By setting $\mu = 0$ we obtain $\frac{1}{2}\epsilon_{ijk}(D_i\mathbf{F}_{jk})^a = (D_iB_i)^a = 0$.

Using 't Hooft gauge invariant field strength $\mathcal{F}_{\mu\nu}$ (cf. Eq. (10.77)), we write

$$E[\mathbf{F}_{\mu\nu}, \phi] \geq \left| \oint_{S_\infty^2} d^2\sigma_i B_i^a \phi^a \right| = v \left| \frac{1}{2} \oint_{S_\infty^2} d^2\sigma_i \epsilon_{ijk} \mathcal{F}_{jk} \right| = \frac{4\pi v}{e} |n| . \quad (10.162)$$

Here $|n| \in \mathbb{N}_+$ represents the number of 't Hooft–Polyakov elementary magnetic charges. n is the topological charge (see Section 10.4.2).[8] The solution of the Bogomol'nyi equation

$$B_i^a = \pm(D_i\phi)^a , \quad (10.163)$$

is again assumed in the form of the Ansatz (10.74). In this case we have

$$(D_i\phi)^a = \frac{(\delta^{ia} - n^i n^a)}{r} vhf + n^i n^a v f' , \quad (10.164a)$$

$$B_i^a = \frac{1}{2}\epsilon_{ijk} F_{jk}^a = \frac{(\delta^{ia} - n^i n^a)}{er} h' - \frac{n^i n^a}{er^2}(1 - h^2) , \quad (10.164b)$$

and so the Bogomol'nyi equation (10.163) leads to two coupled equations

$$h' = \pm evfh , \qquad f' = \mp \frac{1}{evr^2}(1 - h^2) . \quad (10.165)$$

Performing the change of the variable $r = \zeta/ev$ and setting $f = F \pm 1/\zeta$, $h = \pm\zeta H$, we can cast the previous equations into a more convenient form

$$H'(\zeta) = \pm F(\zeta)H(\zeta) , \qquad F'(\zeta) = \pm H^2(\zeta) , \quad (10.166)$$

from which we have

$$(H^2)' = \pm 2FH^2 , \qquad (F^2)' = \pm 2FH^2$$

$$\Rightarrow \quad H^2 = -1 + F^2 \quad \Rightarrow \quad F' = \pm(F^2 - 1) ,$$

$$\Rightarrow \quad F(\zeta) = \mp\coth(\zeta) , \qquad H(\zeta) = \pm\mathrm{csch}(\zeta) . \quad (10.167)$$

However, only the lower-sign solutions fulfil the corresponding boundary conditions from Section 10.4.2. The *analytic* solution of (10.165) is thus

$$f(r) = \coth(\zeta) - \frac{1}{\zeta} , \qquad h(r) = \zeta\,\mathrm{csch}(\zeta) , \qquad \zeta = evr . \quad (10.168)$$

By inserting the above solutions into $E[\mathbf{F}_{\mu\nu}, \phi]$, one obtains the topological charge $n = 1$. Proceeding as in the 't Hooft–Polyakov case, one finds that the magnetic charge of the BPS $SO(3)$ monopole is $4\pi/e$. Note that from Eq. (10.66) the masses of the scalar and vector excitations are $m_s^2 = gv^2$ and $m_v^2 = e^2v^2$. So the BPS limit might be viewed as a case where $m_v \gg m_s$.

---

[8]In order to prevent confusion with magnetic charge, we will denote in this subsection the topological charge as n and not Q.

10.8 Non-topological solitons

Let us briefly discuss non-topological solitons [414, 415]. For these the stabilization mechanism is *not* provided via conservation of some topological charge, but instead via conservation of a Noether charge. Non-topological solitons approach the physical vacuum asymptotically. In contrast to topological solitons, non-topological solitons may exist in any space dimension [266, 267] and do not require SSB [414]. They are also known as Q-balls [166], to emphasize the stabilizing role played by the Noether charge Q.

· An example of non-topological soliton is provided by the Lee (also Friedberg–Lee) phenomenological model of strong interactions [266, 267], where hadrons appear as non-topological solitons. We consider a $1 + 1$-dimensional toy system that consists of one real scalar field ϕ and one complex field ψ, with Lagrangian density:

$$\mathcal{L} = \frac{1}{2}(\partial_\mu \phi)^2 - V(\phi) + (\partial_\mu \psi^*)(\partial^\mu \psi) - \lambda^2 \phi^2 |\psi|^2, \qquad (10.169)$$

where

$$V(\phi) = \frac{m_\phi^2}{2}(\phi - v)^2, \qquad (10.170)$$

and $m_\phi, v, \lambda > 0$.

The ground state configuration is determined by the energy functional:

$$E[\phi, \psi, \psi^*] = \int dx\, T^{00}(\phi, \psi, \psi^*) \qquad (10.171)$$

$$= \int dx \left(\frac{1}{2}(\partial_t \phi)^2 + \frac{1}{2}(\partial_x \phi)^2 + |\partial_t \psi|^2 + |\partial_x \psi|^2 + V(\phi) + \lambda^2 \phi^2 |\psi|^2 \right).$$

$E[\phi, \psi, \psi^*]$ acquires the unique (unbroken) ground state at the (t, x)-independent field configuration

$$\phi = v \quad \text{and} \quad \psi = 0. \qquad (10.172)$$

The Lagrangian (10.169) is invariant under the global $U(1)$ symmetry:

$$\phi(x) \to \phi(x), \quad \psi(x) \to e^{-i\theta} \psi(x), \quad \psi^*(x) \to e^{i\theta} \psi^*(x), \qquad (10.173)$$

and thus possesses the conserved Noether charge:

$$Q = i \int dx\, (\psi^* \partial_t \psi - \psi \partial_t \psi^*) = 2\Im m \int dx\, \psi \partial_t \psi^*. \qquad (10.174)$$

The ground state configuration (10.172) is also invariant under the $U(1)$ phase transformation (10.173) and so there is no SSB.

Consider the case of fields whose amplitudes are small in the large time limit and thus for $t \to \infty$, in the energy functional we retain only quadratic modes corresponding to linear perturbations around the vacuum configuration: $\eta(x) \equiv \phi(x) - v$. One therefore has

$$E[\eta, \psi, \psi^*] = \lim_{t \to \infty} \int dx \Big(\frac{1}{2}(\partial_t \eta)^2 + \frac{1}{2}(\partial_x \eta)^2 + |\partial_t \psi|^2 + |\partial_x \psi|^2$$
$$+ \frac{m_\eta^2}{2} \eta^2 + m_\psi^2 |\psi|^2 \Big), \qquad (10.175)$$

with $m_\psi = \lambda v$ and $m_\eta = m_\phi$. By applying the relation $(\Im m(z))^2 \le |z|^2$ together with Cauchy–Schwarz inequality we obtain

$$Q^2 \le 4 \left| \int dx \, \psi^* \partial_t \psi \right|^2 \le 4 \left(\int dx \, |\psi|^2 \right) \left(\int dx \, |\partial_t \psi|^2 \right), \qquad (10.176)$$

which obviously holds for all times (so also at $t \to \infty$). Because all terms in the integrand (10.175) are positive, the inequality (10.176) implies that field configurations satisfy

$$E^2 \ge \left[\int dx \, \left(|\partial_t \psi|^2 + m_\psi^2 |\psi|^2 \right) \right]^2 \ge 4 \left(\int dx \, |\partial_t \psi|^2 \right) \left(\int dx \, m_\psi^2 |\psi|^2 \right),$$

or equivalently

$$Q^2 \le E^2 / m_\psi^2 \quad \Rightarrow \quad E \ge m_\psi |Q| = \lambda v |Q|. \qquad (10.177)$$

Now we show that it is possible to construct non-singular field configurations that violate the inequality (10.177) and yet conserve the charge (10.174). When such solutions correspond to field configurations with minimal energy, they represent absolutely stable solutions — Q balls. To obtain these configurations we consider a trial static ϕ field with the spatial profile[9]

$$\phi(t, x) = \begin{cases} 0, & \text{for } x \in [-L/2, L/2] \\ v, & \text{otherwise} \end{cases}. \qquad (10.178)$$

Here L denotes the size of the region where the energy density of the ϕ field is localized. For simplicity, we assume that L is sufficiently large so that the energy from the transient region (wall) can be neglected. The energy associated with ϕ is then

$$E[\phi] = \int_{-L/2}^{L/2} dx \left(\frac{1}{2}(\partial_t \phi)^2 + V(\phi) \right) = LV(\phi(t,0)) = L \frac{m_\phi^2}{2} v^2. \qquad (10.179)$$

---

[9] In the context of Q-balls this profile is known as the thin-wall approximation [166].

We now wish to minimize the energy functional (10.171) with respect to the field ψ, under the condition that the charge Q is fixed and ϕ is given by Eq. (10.178). This can be achieved by minimizing the functional

$$E_\omega[\psi, \psi^*] = E[\psi, \psi^*] + \omega \left[Q - i \int dx \, (\psi^* \partial_t \psi - \psi \partial_t \psi^*) \right] \quad (10.180)$$

$$= \int dx \left[|\partial_t \psi + i\omega \psi|^2 + |\partial_x \psi|^2 + (\lambda^2 \phi^2 - \omega^2)|\psi|^2 \right] + \omega Q.$$

Here ω is a Lagrange multiplier. Since the first term in E_ω is positive, minimizing of this term implies $\psi(t, x) = e^{-i\omega t} X(x)$, while minimizing of the remaining part of E_ω yields the differential equation for X in the form

$$\frac{d^2 X}{dx^2} + (\omega^2 - \lambda^2 \phi^2) X = 0. \quad (10.181)$$

The function X fulfils the Noether charge normalization condition

$$Q = i \int dx \, (X(-i\omega)X - X(i\omega)X) = 2\omega \int dx \, X^2. \quad (10.182)$$

Finding ω is thus identical to solving the energy spectrum of a particle in a square potential well of size L. So we can immediately write $\omega = n\pi/L$, $n = 1, 2, \ldots$. The normalization condition (10.182) ensures that X decreases exponentially outside the potential well. The reader may notice that $Q > 0$ for the X configuration.

For this minimizing configuration the energy $E[\psi, \psi^*]$ reads

$$E[\psi, \psi^*] = \int dx \, \left(|\partial_t \psi|^2 + |\partial_x \psi|^2 + \lambda^2 \phi^2 |\psi|^2 \right)$$

$$= \int dx \, \left(\omega^2 X^2 - X(\partial_x^2 - \lambda^2 \phi^2)X \right) + [X \partial_x X]\big|_{x=-\infty}^{x=\infty}$$

$$= 2\omega^2 \int dx \, X^2 = \omega Q = \frac{n\pi}{L} Q, \quad (10.183)$$

which is smallest for the X configuration with $n = 1$. We thus see that at fixed L the minimal total energy for the considered field configurations is

$$E = E[\phi] + E[\psi, \psi^*] = L \frac{m_\phi^2}{2} v^2 + \frac{\pi Q}{L}. \quad (10.184)$$

So far we have not yet fixed L. We can now demand that L is chosen so as to minimize E. This happens at

$$L = \frac{\sqrt{2\pi Q}}{m_\phi v}. \quad (10.185)$$

The corresponding minimal energy is then

$$E = E\left(L = \frac{\sqrt{2\pi Q}}{m_\phi v}\right) = m_\phi v\sqrt{2\pi Q}. \qquad (10.186)$$

Comparing (10.186) with (10.177) we see that for sufficiently large Q the charge dominates over its square root and the inequality (10.177) is violated.

Note that the above configurations are indeed localized since the field $\phi(t, x)$ (cf. (10.178)) approaches rapidly its vacuum value at spatial infinity and $\psi(t, x)$ tends exponentially to its vacuum value (i.e., zero) for $|x| \to \infty$. Note also that the Q-ball is a non-static soliton — the $\psi(t, x)$ configuration oscillates with the frequency $\omega = \pi/L$.

Non-topological solitons can arise in many models where there is a Noether charge conservation and where the corresponding symmetry is *not* spontaneously broken. The higher-dimensional analogue of (10.178) has the spherically symmetric form [166]: $\phi(t, \mathbf{x}) = \phi(r) = \theta(r - L)v$. The spherical symmetry together with the conserved Q is the actual historical reason for the name "Q-ball".

As yet, it is not known whether Q-balls play any relevant role in physics. It has been theorized that dark matter might consist of Q-balls [268, 400] and that Q-balls might play a role in baryogenesis [203, 222]. It is also believed that Q-balls generically arise in supersymmetric field theories [399].

10.9   Instantons and their manifestations

One of the consequences of the Derrick–Hobart theorem is that a *pure* Yang–Mills theory does not allow for *static* solitons except in 4 spatial dimensions (cf. Section 10.6). However, the Yang–Mills theory in 4 Euclidean dimensions supports soliton solution. This is because static solitons in the $1 + 4$ Minkowskian Yang–Mills theory involve only the spatial coordinates, i.e., the four-dimensional Euclidean subspace. The only difference with $1 + 4$-dimensional static solitons is that, instead of the finite energy functional $E[\mathbf{A}_\mu]$, one requires that the Euclidean action itself should be finite. The corresponding soliton solutions are called *instantons*. They represent finite-action topological soliton solutions to the Euclidean equations of motion. The name *instanton* derives from the fact that these solitons are localized in space and at a specific instance in time. Their presence signals the possibility of tunneling between degenerate vacua and reveals a non-trivial vacuum structure of the non-Abelian Yang–Mills theories [131, 348].

Yang–Mills instantons

The instanton solution[10] was found in the context of the $SU(2)$ Yang–Mills theory [66]. We will discuss such a solution as an example of Yang–Mills instanton. We start by observing that in Euclidean theories, the real time t is substituted, via Wick rotation, by the imaginary time τ: $t \to -i\tau$. In gauge theories the substitution $\mathbf{A}_0 \to i\mathbf{A}_0$ ensures that the Euclidean action is real. In such a case, $S \to iS_E$ with $S_E(\mathbf{A}_\mu, \phi, \tau) = S(i\mathbf{A}_0, \mathbf{A}_i, \phi, -i\tau)$ and ϕ representing remaining non-gauge fields. In particular, for the pure Yang–Mills theory the Euclidean Yang–Mills action integral reads

$$S_E[\mathbf{A}_\mu] = \frac{1}{2} \int d^4x \, \mathrm{Tr}(\mathbf{F}_{\mu\nu}\mathbf{F}^{\mu\nu}) = \frac{1}{2} \int d^4x \, \mathrm{Tr}(\mathbf{F}_{\mu\nu}\mathbf{F}_{\mu\nu}), \quad (10.187)$$

where the Euclidean field strength is given by the standard expression,

$$\mathbf{F}_{\mu\nu} = \partial_\mu\mathbf{A}_\nu - \partial_\nu\mathbf{A}_\mu + ie[\mathbf{A}_\mu, \mathbf{A}_\nu]. \quad (10.188)$$

Note that the action (10.187) is non-negative — due to our normalization $\mathrm{Tr}(T_a, T_b) = \frac{1}{2}\delta_{ab}$ — and is equivalent to the energy functional for a static field. Let us now define the Hodge dual field strength $^*\mathbf{F}_{\mu\nu}$ as

$$^*\mathbf{F}_{\mu\nu} = \frac{1}{2} \epsilon_{\mu\nu\alpha\beta}\mathbf{F}^{\alpha\beta}, \quad (10.189)$$

with $\epsilon_{\mu\nu\alpha\beta}$ the totally skewsymmetric tensor and $\epsilon_{0123} = 1$. We will utilize the Bogomol'nyi trick of Section 10.7 and consider the inequality

$$\mathrm{Tr}\left(\mathbf{F}_{\mu\nu} \pm {}^*\mathbf{F}_{\mu\nu}\right)^2 \geq 0, \quad (10.190)$$

together with the identity

$$^*\mathbf{F}_{\mu\nu}{}^*\mathbf{F}_{\mu\nu} = \mathbf{F}_{\mu\nu}\mathbf{F}_{\mu\nu} = \mathbf{F}_{\mu\nu}\mathbf{F}^{\mu\nu} = {}^*\mathbf{F}_{\mu\nu}{}^*\mathbf{F}^{\mu\nu}. \quad (10.191)$$

The latter is a simple consequence of the fact that

$$\epsilon_{\mu\nu\sigma\varrho}\,\epsilon_{\mu\nu\alpha\beta} = 2\left(\delta_{\varrho\beta}\delta_{\sigma\alpha} - \delta_{\varrho\alpha}\delta_{\sigma\beta}\right). \quad (10.192)$$

From Eqs. (10.190)–(10.191) we have the inequality

$$\mathrm{Tr}\left(\mathbf{F}_{\mu\nu}\mathbf{F}^{\mu\nu}\right) \geq \pm\mathrm{Tr}\left(\mathbf{F}_{\mu\nu}{}^*\mathbf{F}^{\mu\nu}\right), \quad (10.193)$$

which saturates only when $\mathbf{F}_{\mu\nu} = \pm^*\mathbf{F}_{\mu\nu}$. Thus the Yang–Mills Euclidean action fulfils

$$S_E[\mathbf{A}_\mu] \geq \frac{1}{2} \left| \int d^4x \, \mathrm{Tr}(\mathbf{F}_{\mu\nu}{}^*\mathbf{F}^{\mu\nu}) \right|. \quad (10.194)$$

---

[10]In their 1975 paper, Belavin *et al.* called the instanton solution *pseudoparticle*. The name *instanton* was coined by 't Hooft.

Similarly as in the Minkowski case, the lower bound in the Bogomol'nyi inequality (10.194) is proportional to a topological charge. The latter is known in mathematics as the second Chern number or the Pontryagin number; physicists call it the instanton number or the topological number/index/charge. Note that $\mathrm{Tr}\,(\mathbf{F}_{\mu\nu}{}^{*}\mathbf{F}^{\mu\nu})$ is a total derivative. Indeed,

$$
\begin{aligned}
\mathrm{Tr}\,(\mathbf{F}_{\mu\nu}{}^{*}\mathbf{F}_{\mu\nu}) &= \frac{1}{2}\,\epsilon_{\mu\nu\alpha\beta}\,\mathrm{Tr}\,(\mathbf{F}_{\mu\nu}\mathbf{F}_{\alpha\beta}) \\
&= 2\epsilon_{\mu\nu\alpha\beta}\,\mathrm{Tr}\,((\partial_{\mu}\mathbf{A}_{\nu}+ie\mathbf{A}_{\mu}\mathbf{A}_{\nu})(\partial_{\alpha}\mathbf{A}_{\beta}+ie\mathbf{A}_{\alpha}\mathbf{A}_{\beta})) \\
&= 2\epsilon_{\mu\nu\alpha\beta}\,\mathrm{Tr}\,(\partial_{\mu}\mathbf{A}_{\nu}\partial_{\alpha}\mathbf{A}_{\beta}+ie2\,\mathbf{A}_{\mu}\mathbf{A}_{\nu}\partial_{\alpha}\mathbf{A}_{\beta}) \\
&= 2\partial_{\alpha}\epsilon_{\alpha\beta\mu\nu}\,\mathrm{Tr}\,\left(\mathbf{A}_{\beta}\partial_{\mu}\mathbf{A}_{\nu}+ie\frac{2}{3}\,\mathbf{A}_{\beta}\mathbf{A}_{\mu}\mathbf{A}_{\nu}\right) \\
&= 2\partial_{\alpha}K_{\alpha}\,,
\end{aligned}
\tag{10.195}
$$

where on the third line we have used the cyclic property of trace to eliminate the $\mathbf{A}_{\mu}\mathbf{A}_{\nu}\mathbf{A}_{\alpha}\mathbf{A}_{\beta}$ term. It is easy to see that K_{α} can also be written as

$$
K_{\alpha} = \frac{1}{2}\epsilon_{\alpha\beta\mu\nu}\,\mathrm{Tr}\,\left(\mathbf{F}_{\mu\nu}\mathbf{A}_{\beta}-ie\frac{2}{3}\,\mathbf{A}_{\beta}\mathbf{A}_{\mu}\mathbf{A}_{\nu}\right).
\tag{10.196}
$$

K_0 is also known as the (three-dimensional) Chern–Simons term. Using Eq. (10.195), the integral on the r.h.s. of the Bogomol'nyi inequality (10.194) can be rewritten in the form

$$
\int d^{4}x\,\mathrm{Tr}(\mathbf{F}_{\mu\nu}{}^{*}\mathbf{F}^{\mu\nu}) = 2\int d^{4}x\,\partial_{\alpha}K_{\alpha} = 2\oint_{\mathcal{S}_{\infty}^{3}} d^{3}\sigma_{\alpha}K_{\alpha}.
\tag{10.197}
$$

Since the action S_E is finite, the field strength $F_{\mu\nu}^{a}$ vanishes as $1/r^{2+\epsilon}$, $\epsilon > 0$ in the surface integral over the remote 3-sphere $\mathcal{S}_{\infty}^{3}$. This means that

$$
\mathbf{A}_{\mu}(x)\big|_{|x|\to\infty} = -ie^{-1}g(x)\partial_{\mu}g^{-1}(x)\,,
\tag{10.198}
$$

i.e., $\mathbf{A}_{\mu}(x)$ is a pure gauge potential on the boundary of Euclidean four-space. Here $g(x)$ belongs to the gauge group G. In this way, any gauge field configuration with finite Euclidean action defines a map $g : \mathcal{S}_{\infty}^{3} \to G$. Such maps are classified topologically by the homotopy group $\pi_3(G)$. Since for any simple compact Lie group one has $\pi_3(G) = \mathbb{Z}$ (cf. Appendix T), the field configurations can be divided into disjoint equivalence classes, enumerated by elements of $\pi_3(G)$, which, therefore, classify finite-action field configurations; in fact they are the only gauge invariant information determined by the asymptotic field behavior (10.198).

Homotopy index for the $SU(2)$ Y–M instantons

In the $SU(2)$ case, $\text{Tr}\,(\mathbf{F}_{\mu\nu}{}^*\mathbf{F}^{\mu\nu})$ is related to the homotopy index. Indeed, substitute (10.198) into Eq. (10.197) and use Eq. (10.196) with $\mathbf{F}_{\mu\nu} = 0$, then

$$\int d^4x \, \text{Tr}(\mathbf{F}_{\mu\nu}{}^*\mathbf{F}^{\mu\nu}) \; = \; \frac{2}{3e^2} \oint_{\mathcal{S}^3_\infty} d^3\sigma_\alpha \epsilon_{\alpha\beta\mu\nu} \text{Tr}\left(g\partial_\beta g^{-1} g\partial_\mu g^{-1} g\partial_\nu g^{-1}\right)$$

$$= \; \frac{16\pi^2}{e^2}\, Q \,, \tag{10.199}$$

where the quantity Q is defined as

$$Q \; = \; \frac{1}{24\pi^2} \oint_{\mathcal{S}^3_\infty} d^3\sigma_\alpha \epsilon_{\alpha\beta\mu\nu} \text{Tr}\left(g\partial_\beta g^{-1} g\partial_\mu g^{-1} g\partial_\nu g^{-1}\right). \tag{10.200}$$

Q is the sought homotopy index. Eq. (10.199) depends entirely on the group element $g(x)$ and not on the details of the field configuration at some finite x. In the $SU(2)$ case one can parametrize any $g : \mathcal{S}^3_\infty \to SU(2)$ as

$$g(\mathbf{n}) = y_0(\mathbf{n})\mathbb{1} + iy_i(\mathbf{n})\sigma_i \,, \qquad \mathbf{n} \in \mathcal{S}^3_\infty \,, \tag{10.201}$$

where σ_i are the Pauli matrices. Unitarity of g ensures that $y_\alpha \in \mathbb{R}$ and $\det(g) = 1$ implying $y_\alpha y_\alpha = 1$, so the group $SU(2)$ represents geometrically $\mathcal{S}^3$. The standard normalized group (Haar) measure on $SU(2)$ is

$$d\mu(g) \; = \; \frac{1}{24\pi^2} \, d^3\sigma_\alpha \epsilon_{\alpha\beta\mu\nu} \text{Tr}\left(g\partial_\beta g^{-1} g\partial_\mu g^{-1} g\partial_\nu g^{-1}\right). \tag{10.202}$$

Indeed, under the left/right multiplication by fixed elements $g_1, g_2 \in SU(2)$ we have $d\mu(g) = d\mu(g_1 g) = d\mu(g g_2) = d\mu(g_1 g g_2)$. In addition, because $SU(2) \times SU(2)/\mathbb{Z}_2 \cong SO(4)$, the identity $d\mu(g) = d\mu(g_1 g g_2)$ implies that $d\mu(g)$ is rotationally invariant. The normalization factor $1/24\pi^2$ is chosen so that when $\mathcal{S}^3_\infty$ wraps onto $SU(2)$ once then $Q = 1$. To check the normalization factor we choose $y_\alpha = n_\alpha = x_\alpha/|x|$. In such a case we map every point from $\mathcal{S}^3_\infty$ on the corresponding point at the same polar angle on $SU(2)$ and thus $\mathcal{S}^3_\infty$ wraps only once onto $SU(2)$. This allows us to write

$$1 \; = \; c \oint_{\mathcal{S}^3_\infty} d^3\sigma_\alpha \, \epsilon_{\alpha\beta\mu\nu} \text{Tr}\left(g\partial_\beta g^{-1} g\partial_\mu g^{-1} g\partial_\nu g^{-1}\right)$$

$$= \; c \oint_{\mathcal{S}^3} d^3\sigma_\alpha \, \frac{\epsilon_{\alpha\beta\mu\nu}}{|x|^3} \, n_\kappa \text{Tr}\left(s_\beta s_\mu^\dagger s_\nu s_\kappa^\dagger\right) = c \oint_{\mathcal{S}^3} d^3\sigma_\alpha \, \frac{12\, n_\alpha}{|x|^3}$$

$$= c \oint_{\mathcal{S}^3} d^3\sigma_\alpha \, 12 \, x_\alpha \; = \; c \int_V dV \, 12 \, \partial_\alpha x_\alpha$$

$$= \; -c2\pi^2 \int_0^1 dr \, 48r^3 \; = \; c\, 24\pi^2 \,, \tag{10.203}$$

which shows that the proportionality factor c is indeed $1/24\pi^2$. In Eq. (10.203) we have used $s_\alpha = (\mathbb{1}, i\sigma_1, i\sigma_2, i\sigma_3)$ and $dV = r^3 d\Omega dr$ (with $d\Omega$ representing the surface element of the unit sphere $\mathcal{S}^3$). In general, the sphere $\mathcal{S}^3_\infty$ can cover the gauge group $SU(2)$ more than once. In such a case, Q will be an integer counting how many times $\mathcal{S}^3$ wraps around $SU(2)$. This is precisely what $\pi_3(SU(2))$ homotopy index ought to do. Other simple gauge groups can be studied in a similar way. The general proof of Eq. (10.200) is, however, more complicated and we shall not pursue it here. The corresponding analysis can be found, for instance, in [532].

The importance of Q is that it is integer-valued when $S_E[\mathbf{A}_\mu]$ is finite. This happens when the field strength $\mathbf{F}_{\mu\nu}$ decreases to zero sufficiently rapidly as $|x| \to \infty$. Taking into account the results (10.194) and (10.199) we can finally cast the Bogomol'nyi inequality (10.194) into the form

$$S_E[\mathbf{A}_\mu] \geq \frac{8\pi^2}{e^2} |Q| \,. \tag{10.204}$$

As in the soliton case, the r.h.s. of the Bogomol'nyi inequality (10.204) is not merely a lover bound but it is a global minimum. This is a simple consequence of Eq. (10.190). From the inequality (10.204) we see that, in a given topological sector, the solution to the field equations that minimizes S_E has either a self-dual or anti-self-dual field strength, i.e.,

$$\mathbf{F}_{\mu\nu} = \pm {}^*\mathbf{F}_{\mu\nu} \,, \tag{10.205}$$

where "+" corresponds to $Q > 0$ (instanton configuration) while "−" corresponds to $Q < 0$ (anti-instanton configuration), cf. Eq. (10.199). Note that (anti-)self-duality is a gauge invariant property, even though $\mathbf{F}_{\mu\nu}$ itself is not. It is also worth stressing that a solution to the (anti-)self-duality equations must necessarily solve the Euler–Lagrange equations of motion since it minimizes the action in a given topological sector. In fact, since $D_\mu {}^*\mathbf{F}_{\mu\nu}$ vanishes identically due to the Bianchi identity, we obtain that $\mathbf{F}_{\mu\nu}$ fulfills the Euler–Lagrange equation $D_\mu \mathbf{F}_{\mu\nu} = 0$.

One-instanton/anti-instanton solution

We now consider the one-instanton solution of self-duality Eq. (10.205). The index $Q = 1$ is obtained whenever the mapping $g: \mathcal{S}^3_\infty \to SU(2)$ is such that $\mathcal{S}^3_\infty$ wraps once around $SU(2)$ and the Jacobian has a positive signature. We have seen that one of the simplest choices for such g is

$$g(\mathbf{n}) = n_\alpha s_\alpha \,, \quad n_\alpha = \frac{x_\alpha}{|x|} \,, \quad s_\alpha = (\mathbb{1}, i\sigma_1, i\sigma_2, i\sigma_3) \,. \tag{10.206}$$

This implies that the gauge potential has the asymptotic form

$$\mathbf{A}_\mu(x)|_{|x|\to\infty} = \frac{1}{ie}\, g(x)\partial_\mu g^{-1}(x) = \frac{s_\alpha s_\beta^\dagger}{ie}\frac{n_\beta n_\mu - \delta_{\beta\mu}}{|x|}n_\alpha. \quad (10.207)$$

Result (10.207) may be further simplified with the help of the 't Hooft symbol [596] (or η-symbol) which is defined as

$$\eta_{a\mu\nu} \equiv \epsilon_{a\mu\nu} + \delta_{a\mu}\delta_{\nu 0} - \delta_{a\nu}\delta_{\mu 0}. \quad (10.208)$$

Note that for the fixed a the 't Hooft tensor is self-dual, i.e.,

$$\eta_{a\mu\nu} = \tfrac{1}{2}\epsilon_{\mu\nu\alpha\beta}\,\eta_{a\alpha\beta}, \quad (10.209)$$

it is antisymmetric in μ, ν and it obeys the relations

$$s_\alpha s_\beta^\dagger = \delta_{\alpha\beta} + i\eta_{a\alpha\beta}\sigma^a, \quad (10.210)$$

$$\epsilon_{abc}\eta_{b\mu\nu}\eta_{c\alpha\beta} = \delta_{\mu\alpha}\eta_{a\nu\beta} + \delta_{\nu\beta}\eta_{a\mu\alpha} - \delta_{\mu\beta}\eta_{a\nu\alpha} - \delta_{\nu\alpha}\eta_{a\mu\beta}, \quad (10.211)$$

$$\epsilon_{\mu\nu\alpha\beta}\eta_{a\gamma\beta} = \delta_{\gamma\mu}\eta_{a\nu\alpha} + \delta_{\gamma\alpha}\eta_{a\mu\nu} - \delta_{\gamma\nu}\eta_{a\mu\alpha}. \quad (10.212)$$

Inserting (10.210) into (10.207), we find that

$$\mathbf{A}_\mu(x)|_{|x|\to\infty} = e^{-1}\eta_{a\alpha\mu}\frac{\sigma_a x_\alpha}{x^2}. \quad (10.213)$$

The strategy now is to make a suitable Ansatz for $\mathbf{A}_\mu(x)$, and then check that it solves the self-duality equation. The asymptotic behavior (10.213) suggests that the full one-instanton solution can be assumed as

$$\mathbf{A}_\mu(x) = e^{-1}\eta_{a\alpha\mu}\frac{\sigma_a x_\alpha}{x^2}\,\phi(|x|), \quad (10.214)$$

where $\phi(x)$ is some regular function with the asymptotic conditions $\phi_{|x|\to\infty} = 1$ and $\phi_{|x|\to 0} = O(|x|)$. Substituting (10.214) into $\mathbf{F}_{\mu\nu}$ yields

$$\mathbf{F}_{\mu\nu}(x) = e^{-1}2\sigma_a\eta_{a\mu\nu}\frac{\phi(1-\phi)}{x^2}$$
$$- e^{-1}\sigma_a\frac{x^\alpha}{x^2}\left[(\eta_{a\alpha\nu}x_\mu - \eta_{a\alpha\mu}x_\nu)\left(2\frac{\phi(1-\phi)}{x^2} - \frac{\phi'}{|x|}\right)\right]. \quad (10.215)$$

Here we have used the relation (10.211). In order to solve the self-duality equation $\mathbf{F}_{\mu\nu} = {}^*\mathbf{F}_{\mu\nu}$ we need ${}^*\mathbf{F}_{\mu\nu}$. This can be easily obtained with the help of (10.189) and (10.212). One then finds

$${}^*\mathbf{F}_{\mu\nu}(x) = e^{-1}\sigma_a\eta_{a\mu\nu}\frac{\phi'}{|x|}$$
$$- e^{-1}\sigma_a\frac{x^\alpha}{x^2}\left[(\eta_{a\alpha\nu}x_\mu - \eta_{a\alpha\mu}x_\nu)\left(2\frac{\phi(\phi-1)}{x^2} - \frac{\phi'}{|x|}\right)\right]. \quad (10.216)$$

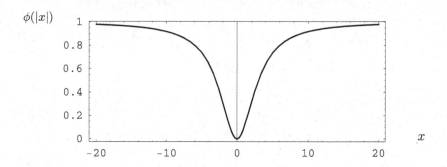

Fig. 10.10   Ansatz function $\phi(|x|)$ for $Q = 1$ $SU(2)$ Y–M instanton with $\rho = 3$.

Equating $\mathbf{F}_{\mu\nu}$ to $^*\mathbf{F}_{\mu\nu}$ we obtain the equation for ϕ, namely

$$\phi' = 2\frac{\phi(1-\phi)}{|x|}.$$ (10.217)

This can be readily integrated yielding (see also Fig. 10.10)

$$\phi(|x|) = \frac{x^2}{x^2 + \rho^2}.$$ (10.218)

So the one-instanton configuration reads

$$\mathbf{A}_\mu(x) = A_\mu^a(x)\left(\frac{\sigma_a}{2}\right) = e^{-1}\eta_{a\alpha\mu}\frac{\sigma_a x_\alpha}{x^2 + \rho^2}.$$ (10.219)

Here ρ is an arbitrary parameter called *collective coordinate* (cf. Section 10.9.1). Since ρ measures the region in which $\mathbf{A}_\mu(x)$ appreciably differs from its asymptotics, it can be viewed as the "size" of the instanton.

Taking into account Eq. (10.215), the field strength corresponding to the one-instanton gauge potential is

$$\mathbf{F}_{\mu\nu}(x) = F_{\mu\nu}^a(x)\left(\frac{\sigma_a}{2}\right) = 2e^{-1}\eta_{a\mu\nu}\sigma_a\frac{\rho^2}{(x^2 + \rho^2)^2}.$$ (10.220)

With this we can compute the action S_E by integrating the density

$$\mathrm{Tr}(\mathbf{F}_{\mu\nu}\mathbf{F}_{\mu\nu}) = 96e^{-2}\frac{\rho^4}{(x^2 + \rho^2)^4}.$$ (10.221)

One easily verifies that the ensuing S_E indeed corresponds to $Q = 1$.

The solution (10.219) represents the instanton with the position at the origin. Due to translational invariance of the self-duality equation, the one-instanton solution with the position at $x = x_0$ must have the form

$$\mathbf{A}_\mu(x) = e^{-1}\eta_{a\alpha\mu}\frac{\sigma_a(x-x_0)_\alpha}{(x-x_0)^2 + \rho^2},$$

$$\Rightarrow \mathbf{F}_{\mu\nu}(x) = 2e^{-1}\eta_{a\mu\nu}\sigma_a\frac{\rho^2}{((x-x_0)^2 + \rho^2)^2}. \tag{10.222}$$

This solution again corresponds to $Q = 1$. The four parameters x_0^μ are also called *collective coordinates*. In addition, one can obtain yet another solution by acting with a *global $SU(2)$* on the previous solution, namely

$$\mathbf{A}_\mu(x, x_0, \rho, \vartheta) = g(\vartheta)\mathbf{A}_\mu(x, x_0, \rho)g^{-1}(\vartheta). \tag{10.223}$$

This is because the theory is (even after the gauge fixing) still invariant under the global $SU(2)$ symmetry which introduces another three free parameters (e.g., Euler angles) ϑ_i. So in total, the one-instanton solution can be parametrized by eight collective coordinates, also called *moduli*.

The single anti-instanton solution can be obtained by introducing yet another 't Hooft symbol

$$\bar{\eta}_{a\mu\nu} = (-1)^{\delta_{0\mu}+\delta_{0\nu}}\eta_{a\mu\nu}. \tag{10.224}$$

It is easy to see that the latter is anti-self-dual in vector indices, i.e.,

$$\bar{\eta}_{a\mu\nu} = -\tfrac{1}{2}\epsilon_{\mu\nu\alpha\beta}\bar{\eta}_{a\alpha\beta}. \tag{10.225}$$

This straightforwardly implies that

$$\mathbf{F}_{\mu\nu}(x) = 2e^{-1}\bar{\eta}_{a\mu\nu}\sigma_a\frac{\rho^2}{((x-x_0)^2 + \rho^2)^2}, \tag{10.226}$$

is the one-instanton anti-self-dual solution for field strength (cf. (10.222)). Correspondingly

$$\mathbf{A}_\mu(x) = e^{-1}\bar{\eta}_{a\alpha\mu}\frac{\sigma_a(x-x_0)_\alpha}{(x-x_0)^2 + \rho^2}, \tag{10.227}$$

is the one-instanton anti-self-dual gauge potential solution.

We recall that there is a systematic algebraic procedure known as the Atiyah, Drinfeld, Hitchin and Manin (ADHM) construction [42] that allows us to find a complete set of the (anti-)self-dual gauge field configurations of arbitrary charge Q and for arbitrary $SU(N)$, $SO(N)$ and $Sp(N)$ gauge groups. The space of such solutions is known as the *Q-instanton moduli space.* It goes beyond the scope of this book to discuss the ADMH construction.

We note that finite-action soliton solutions in other equation systems are also frequently referred to as instantons. Typical examples of these are kinks and antikinks in the reversed double-well potential [532], instantons and anti-instantons in Coleman periodic-potential problem [163] and in CP_N model [292], or the Belavin–Polyakov instanton in the non-linear $SO(3)$ σ model (cf. Section 10.6). For instantons in $SU(N)$ Yang–Mills theories and in $\mathcal{N} = 1, 2, 4$ ($\mathcal{N}$ stands for the number of supersymmetries) $SU(N)$ supersymmetric Yang–Mills theories — supersymmetric instantons, the reader is referred, for instance, to [110, 532, 627]. Instantons can be found also in various gravitational models, see, e.g., [213].

Tunneling phenomena and vacuum structure

Instantonic solutions and their corresponding finite actions also arise in the study of problems concerning tunneling processes in quantum systems.

Instantons play an essential role in shaping the Yang–Mills vacuum. Let us consider the topological charge (10.200), i.e.,

$$Q = \frac{-ie}{24\pi^2} \oint_{\mathcal{S}^3_\infty} d^3\sigma_\alpha \, \epsilon_{\alpha\beta\mu\nu} \mathrm{Tr}\left(\mathbf{A}_\beta \mathbf{A}_\mu \mathbf{A}_\nu\right) . \tag{10.228}$$

Instead of the boundary $\mathcal{S}^3_\infty$ we can consider the equivalent (homeomorphic) boundary depicted in Fig.10.11. Consequently, we can write

$$Q = \frac{-ie}{24\pi^2} \left\{ \oint_{\mathrm{I+II}} d^3\sigma \, \epsilon_{0ijk} \mathrm{Tr}\left(\mathbf{A}_i \mathbf{A}_j \mathbf{A}_k\right) \right.$$

$$\left. + \int_{-\infty}^{\infty} dx_0 \oint_{\mathrm{III}} d^2\sigma_i \, \epsilon_{i\beta\mu\nu} \mathrm{Tr}\left(\mathbf{A}_\beta \mathbf{A}_\mu \mathbf{A}_\nu\right) \right\} . \tag{10.229}$$

We observe that although $\mathbf{A}_\mu$ and $\mathbf{F}_{\mu\nu}$ are not gauge invariant, the charge Q is. This allows us to calculate (10.229) in temporal gauge $\mathbf{A}_0 = \mathbf{0}$, which in turn provides a physical interpretation of Q.

In the temporal gauge, the integral over the hypersurface III vanishes and Q reduces to the difference between two integrals on the hypersurfaces $x_0 \equiv \tau \to \pm\infty$:

$$Q = \frac{-ie}{24\pi^2} \left\{ \oint_{\mathrm{II}} d^3\sigma \, \epsilon_{0ijk} \mathrm{Tr}\left(\mathbf{A}_i \mathbf{A}_j \mathbf{A}_k\right) - \oint_{\mathrm{I}} d^3\sigma \, \epsilon_{0ijk} \mathrm{Tr}\left(\mathbf{A}_i \mathbf{A}_j \mathbf{A}_k\right) \right\}$$

$$= n_{\mathrm{II}} - n_{\mathrm{I}} . \tag{10.230}$$

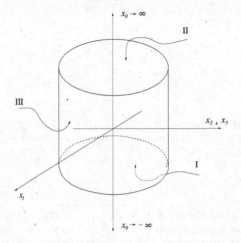

Fig. 10.11   Hypercylindrical boundary of the Euclidean spacetime. Region I corresponds to $x_0 \to -\infty$, while II corresponds to $x_0 \to \infty$.

Here n_{II} and n_{I} correspond to the topological vacuum indexes at $x_0 \to \infty$ and $x_0 \to -\infty$, respectively. Despite a formal similarity, the topological vacuum index is not the same as the topological charge of a Euclidean field configuration (cf. Eq. (10.200)). In the following we denote the *enumerable infinity* of degenerate vacuum states as $|n\rangle$ with the topological vacuum index $n = 0, \pm 1, \pm 2, \ldots$, as a labeling index.

Eq. (10.230) indicates that the instanton topological charge Q provides the difference between the topological sectors n_{II} and n_{I} of the pure gauges, between which the instanton interpolates. This is schematically depicted in Fig. 10.12. The situation with the Y–M vacuum is somehow similar to periodical potential known from QM (see, e.g., [163, 544]). A discrete set of distinct vacua $|n\rangle$, labeled by integer n, exists in Yang–Mills theory. By analogy with QM, the *true* vacuum can be thought to be a linear superposition of the pure gauge "would be" vacua $|n\rangle$. In ordinary QM, the degeneracy between the classical "would-be" vacua is lifted by quantum tunneling, leaving at the end only the true ground state. In QFT the situation is hindered by the existence of inequivalent representations: instead of a single vacuum state, we have a continuum of unitary inequivalent vacuum states. This means that Y–M theory has infinitely many inequivalent realizations. The particular realization relevant to the observations results either from particular initial conditions or from SSB mechanism.

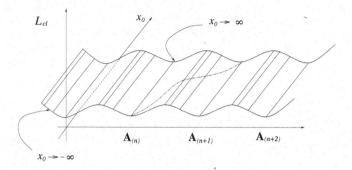

Fig. 10.12   The classical Euclidean Y–M Lagrangian. The degenerate absolute minima correspond to pure gauge configurations $\mathbf{A}_{(n),\mu}$ with topological vacuum index n.
Representative instanton trajectory that interpolates between two adjacent "would-be"
vacua (i.e., $Q = 1$ instanton) is depicted.

We observe that the "would be" vacua $|n\rangle$ are gauge-dependent, indeed

$$g_1|n\rangle = |n+1\rangle. \tag{10.231}$$

Here g_1 is the gauge transformation with the homotopy index $Q = 1$. Analogous results would hold for gauge transformations with higher homotopy
index. Such transformations are known as "large" gauge transformations.
The name "small" gauge transformation is reserved for the gauge transformations that preserve homotopy index. In particular, the Chern–Simons
term K_0 is invariant against small gauge transformations.

Since the true vacuum must be gauge invariant (both under large and
small gauge groups), it must have the form

$$|\theta\rangle = \sum_{n=-\infty}^{\infty} e^{in\theta}|n\rangle. \tag{10.232}$$

Here the angle $\theta \in [0, 2\pi]$ is an arbitrary parameter. The vacuum $|\theta\rangle$ is
known as the θ vacuum. Expansion (10.232) is a simple consequence of
Eq. (10.231) together with the gauge invariance of $\hat{H}$, i.e.,

$$[g_1, \hat{H}] = 0, \tag{10.233}$$

and unitarity of g_1 (last two conditions imply $g_1|\theta\rangle = e^{-i\theta}|\theta\rangle$). The reader
may notice that the result (10.232) is reminiscent of Bloch theorem (see,
e.g., [26,163]). Bloch theorem states that the energy eigenstates for a system
with a periodic potential must have the form

$$\psi_{m\mathbf{k}}(\mathbf{x}) = e^{i\mathbf{k}\cdot\mathbf{x}}u_{m\mathbf{k}}(\mathbf{x}), \tag{10.234}$$

Y–M vacuum structure	Bloch theorem
$[g_1, \hat{H}] = 0$	$[\hat{T}_\ell, \hat{H}] = 0$
$\|\theta\rangle = \sum_{n=-\infty}^{\infty} e^{in\theta} \|n\rangle$	$\psi_{mk}(x) = \sum_{n=-\infty}^{\infty} e^{ik\ell n} \psi_m(x - n\ell)$
$g_1\|\theta\rangle = e^{-i\theta}\|\theta\rangle$	$\hat{T}_\ell\|\psi_{mk}\rangle = e^{-ik\ell}\|\psi_{mk}\rangle$
$g_1\|n\rangle = \|n+1\rangle$	$\hat{T}_\ell\psi_m(x - n\ell) = \psi_m(x - (n-1)\ell)$
$g_Q = [g_1]^Q$	$\hat{T}_{n\ell} = [\hat{T}_\ell]^n$

Fig. 10.13   Comparison of mathematical structures of Y–M vacuum and Bloch state.

where $u_{mk}(\mathbf{x})$ has the same periodicity as the potential (e.g., periodicity of the crystal lattice), i.e., $u_{mk}(\mathbf{x}) = u_{mk}(\mathbf{x} + \ell)$, ℓ is the periodicity vector (the lattice displacement vector).

Eq. (10.234) implies that $\psi_{mk}(\mathbf{x} + \ell) = e^{ik \cdot \ell}\psi_{mk}(\mathbf{x})$, or equivalently $\hat{T}_\ell\psi_{mk}(\mathbf{x}) = e^{ik \cdot \ell}\psi_{mk}(\mathbf{x})$, where $\hat{T}_\ell = e^{\ell \cdot \nabla}$ is the (lattice) translation operator. The quantum number $\mathbf{k}$ thus determines how much the phase changes when one displaces the entire system by ℓ. Because of the single-valuedness of $\psi_{mk}(\mathbf{x})$, $\mathbf{k}$ is restricted to a finite region in $\mathbf{k}$-space, called the first Brillouin zone. The quantum number m, known as the *band index*, takes integer values which label the energy bands. It is often useful to use the tight-binding approximation, i.e., to assume that bound energy levels are well localized (e.g., near a single atom at the lattice point). Let $\psi_m(\mathbf{x})$ be the corresponding wave function located at site $\mathbf{x}$, then, in the periodic potential, $\psi_m(\mathbf{x} - \ell)$ represents the wave function located at $\mathbf{x} - \ell$. To preserve the Bloch description, one must find such a linear combination of these degenerate wave functions that satisfies Bloch theorem. The solution to this problem is[11]

$$\psi_{mk}(\mathbf{x}) = \sum_\ell e^{ik \cdot \ell}\psi_m(\mathbf{x} - \ell). \tag{10.235}$$

This may be seen directly from the fact that

$$\psi_{mk}(\mathbf{x} + \ell) = \sum_{\ell'} e^{ik \cdot \ell'}\psi_m(\mathbf{x} - \ell' + \ell) = e^{ik \cdot \ell} \sum_{\ell'} e^{ik \cdot (\ell' - \ell)}\psi_m(\mathbf{x} - (\ell' - \ell))$$

$$= \psi_{mk}(\mathbf{x}). \tag{10.236}$$

The analogy with Eq. (10.232) occurs when we consider a one-dimensional crystal (i.e., $\mathbf{k} = k$ is a number), band index $m = 1$, ℓ as the elementary period and identify $\hat{T}_\ell$ with g_1. In Fig. 10.13 we compare the major features

[11]It turns out that even without tight-binding any Bloch function ψ_{mk} can be written in the form (10.235). The function ψ_m is then known as Wannier function [41].

of the two theories. The comparison implies that the θ-angle can be formally identified with $k\ell$. Since ℓ is fixed, θ-angle plays the role of an index that parametrizes states within a lowest energy band ($m = 1$). This analogy fails in one important respect. In contrast with crystals there is here no continuum of states without a mass gap. Although one can construct stationary states for any value of θ, they are not excitations of the $\theta = 0$ vacuum because in Yang–Mills theory the value of θ cannot be changed. Theories with different values of θ are inequivalent: they correspond to different physical "worlds".

Since $|\theta\rangle$ describes energy ground states one can write the (Euclidean) vacuum-to-vacuum transition amplitude in the spectral form

$$\langle\theta|e^{-\tau\hat{H}}|\theta\rangle = \sum_{n,m} e^{i(n-m)\theta}\langle m|e^{-\tau\hat{H}}|n\rangle = \sum_{n}\left(\sum_{Q} e^{-iQ\theta}\langle n+Q|e^{-\tau\hat{H}}|n\rangle\right).$$
(10.237)

In the limit $\tau \to \infty$, internal transition amplitudes have the functional integral representation

$$\langle n+Q|e^{-\tau\hat{H}}|n\rangle = \int \mathcal{D}[A_\mu^a]_Q\, e^{-S_E}\,,$$
(10.238)

where the functional measure $\mathcal{D}[A_\mu^a]_Q$ indicates that the functional integration has to be taken over gauge fields in the Q sector only.

Thus, $\langle n+Q|e^{-\tau\hat{H}}|n\rangle$ does not depend on n, but only on Q, i.e., on the difference in the boundary values. We can thus write

$$\langle\theta|e^{-\tau\hat{H}}|\theta\rangle = \sum_{n}\left(\sum_{Q} e^{-iQ\theta}\int \mathcal{D}[A_\mu^a]_Q\, e^{-S_E}\right)$$

$$= 2\pi\delta(0)\left(\sum_{Q} e^{-iQ\theta}\int \mathcal{D}[A_\mu^a]_Q\, e^{-S_E}\right).$$
(10.239)

The δ function is a simple consequence of the identity

$$\sum_{n=-\infty}^{\infty} 1 = \sum_{n=-\infty}^{\infty} e^{in0} = 2\pi\delta(0)\,.$$
(10.240)

Since

$$Q = \frac{e^2}{16\pi^2}\int d^4x\, \mathrm{Tr}(\mathbf{F}_{\mu\nu}\,{}^*\mathbf{F}^{\mu\nu})\,,$$
(10.241)

we can absorb the phase factor $e^{-iQ\theta}$ together with $\sum_Q$ into the functional integral. We thus get

$$\langle\theta|e^{-\tau\hat{H}}|\theta\rangle = 2\pi\delta(0)\int\mathcal{D}[A^a_\mu]_{\text{all}Q}\,\exp\left(-S_E - \frac{ie^2\theta}{16\pi^2}\int d^4x\,\text{Tr}(\mathbf{F}_{\mu\nu}\,{}^*\mathbf{F}^{\mu\nu})\right),$$

$$(10.242)$$

where $\mathcal{D}[A^a_\mu]_{\text{all}Q}$ denotes the functional integral measure that is not restricted to any specific topological sector. One consequence of this is that the Minkowski continuation of Q, done via Wick rotation, does not pick up any extra factor i: one gets one i from d^4x since

$$\int d^4x\,\cdots \quad = \quad \int_{-\infty}^{\infty}dt\int_{\mathbb{R}^3}d^3\mathbf{x}\,\cdots = -i\int_{i\infty}^{-i\infty}d\tau\int_{\mathbb{R}^3}d^3\mathbf{x}\,\cdots$$

$$\stackrel{\text{Wick rot.}}{=}\quad -i\int_{-\infty}^{\infty}d\tau\int_{\mathbb{R}^3}d^3\mathbf{x}\,\cdots,\qquad(10.243)$$

and another i from $\mathbf{F}_{0i}$ since under Wick rotation $\mathbf{A}_0\to i\mathbf{A}_0$. Thus

$$d^4x\,\epsilon^{\mu\nu\alpha\beta}\text{Tr}(\mathbf{F}_{\mu\nu}\mathbf{F}_{\alpha\beta}) \quad = \quad 4\,dt d^3\mathbf{x}\,\epsilon^{ijk}\text{Tr}(\mathbf{F}_{0i}\mathbf{F}_{jk})$$

$$\rightarrow \quad 4\,d\tau d^3\mathbf{x}\,\epsilon^{ijk}\text{Tr}(\mathbf{F}_{0i}\mathbf{F}_{jk})\,.\qquad(10.244)$$

The net result is that an extra term appears in the Minkowski Lagrangian:

$$\triangle\mathcal{L}_\theta = -\frac{e^2\theta}{16\pi^2}\int d^4x\,\text{Tr}(\mathbf{F}_{\mu\nu}\,{}^*\mathbf{F}^{\mu\nu})\,.\qquad(10.245)$$

Eq. (10.245) and the fact that $S_E\to -iS$ under Wick rotation lead to the effective action vacuum-to-vacuum (Minkowski) transition amplitude:

$$\lim_{t\to\infty}\langle\theta|e^{-it\hat{H}}|\theta\rangle = 2\pi\delta(0)\int\mathcal{D}A^a_\mu\,\exp\left(iS_{\text{eff}}\right),\qquad(10.246)$$

$$S_{\text{eff}} = S - \frac{e^2\theta}{16\pi^2}\int d^4x\,\text{Tr}(\mathbf{F}_{\mu\nu}\,{}^*\mathbf{F}^{\mu\nu})\,.\qquad(10.247)$$

The net effect of the instantons tunneling is thus described by the extra $\triangle\mathcal{L}_\theta$ term in the effective action. The functional integral measure $\mathcal{D}A^a_\mu$ is then not restricted to any particular topological sector.

As seen above, $\triangle\mathcal{L}_\theta$ is a total derivative. In general, total derivatives do not contribute to surface integrations because fields vanish sufficiently rapidly at infinity. However, while $F^a_{\mu\nu}$ vanishes at infinity faster than r^{-2}, the A^a_μ fields tend asymptotically to pure gauge configurations with generally non-vanishing surface integral (10.199). Moreover, $\epsilon_{\mu\nu\alpha\beta}\text{Tr}(\mathbf{F}_{\mu\nu}\mathbf{F}_{\alpha\beta})$

is odd both under parity P and time-reversal T: the ϵ-symbol has only *one* *time* index and *three spatial* indices. At the same time,

$$\mathbf{F}_{0i} = -\mathbf{F}_{0i}^{P,T}, \mathbf{F}_{ij} = \mathbf{F}_{ij}^{PT} \quad \Rightarrow \quad {}^*\mathbf{F}_{0i} = {}^*\mathbf{F}_{0i}^{P,T}, {}^*\mathbf{F}_{ij} = -{}^*\mathbf{F}_{ij}^{PT} . \quad (10.248)$$

$\triangle \mathcal{L}_\theta$ thus leads to T, i.e., CP violation.

The above discussion on the $SU(2)$ gauge group also holds for all compact gauge groups G, since $\pi_3(G) = \mathbb{Z}$ (except for $SO(4)$ where $\pi_3(SO(4)) = \mathbb{Z} \otimes \mathbb{Z}$), including the $SU(3)$ color group in quantum chromodynamics. There $\theta \neq 0$ implies CP violation by strong interactions, with the consequence that the neutron should have a non-zero static electric dipole moment. Experimental measurements yield tight bounds, namely [513] $|\theta| < 10^{-9\pm1}$. We do not comment more on this subject, apart from mentioning that several strategies have been developed to avoid non-zero θ.[12] Among many, one of them introduces an extra $U(1)$ chiral symmetry (the Peccei–Quinn symmetry [514, 515]) and its related light pseudoscalar particle, the *axion* [612, 662].

Our analysis holds true not only in pure Yang–Mills theory, but can be extended also to models with matter fields, the Higgs mechanism, supersymmetry, etc. [67, 131, 596]. We mention that chiral symmetry is violated if fermions are massless, and then dependence on θ-angle disappears [131].

10.9.1   *Collective coordinates and fermionic zero modes*

A specific classical solution is often a member of a multi-parameter family of solutions with the same defining property. The defining property for solitons is the finite energy; for instantons the finite action, or equivalently, the fixed instanton number. The parameters labeling different degenerate solutions are called *collective coordinates* or *moduli* and the space of solutions of finite energy/action is called the *moduli space* of soliton/instanton solutions. For instantons it is the space of inequivalent solutions of the (anti-)self-dual equation (10.205). Here "inequivalent" means equivalent up to a *local* gauge transformation. Solutions differing by *global* gauge transformation are considered inequivalent, as already mentioned in Section 10.9.

Collective coordinates are widely used in many-body physics; an application to the polaron problem is due to Bogoliubov and Tyablikov

---

[12]The problem of the small θ is also known as the strong CP problem. It has nothing to do with the CP violation of purely electroweak origin, due to the δ-phase in the Cabbibo–Kobayashi–Maskawa matrix (see, e.g., [156]).

[113]; to meson physics to Pais [510]. For solitons, they were discussed in [132,232,281] and, for instantons, in [163,166,348]. Connection with the Atiyah–Singer index theorem for Dirac operators was studied in [70].

Here we consider the number of collective coordinates for instantons with homotopy index Q. This is strictly connected with the concept of *zero modes* (cf. Sections 7.2.1 and 8.4 where zero energy modes were considered in the QFT frame of boson condensation).

Let $\mathbf{A}_\mu$ be a given instanton solution. We study its small deformations: $\mathbf{A}_\mu + \delta \mathbf{A}_\mu$. In the linear approximation in $\delta \mathbf{A}_\mu$, the self-dual (anti-self-dual) equation reads

$$D_\mu \delta \mathbf{A}_\nu - D_\nu \delta \mathbf{A}_\mu = \pm {}^*(D_\mu \delta \mathbf{A}_\nu - D_\nu \delta \mathbf{A}_\mu) = \pm \epsilon_{\mu\nu\rho\sigma} D_\rho \delta \mathbf{A}_\sigma . \quad (10.249)$$

The sign "-" is for anti-self-dual equation. The corresponding covariant derivative depends on the background field $\mathbf{A}_\mu$ but not on $\delta \mathbf{A}_\mu$. To ensure that only *inequivalent* deformed solutions are counted, we require that the new solutions obtained from (10.249) are not related by a gauge transformation. This means that $\delta \mathbf{A}_\mu$ must be orthogonal to gauge transformations, with respect to a given inner product (Killing form with respect to internal indices, Euclidean inner product with respect to vector indices and an integral inner product for continuous real valued functions on $\mathbb{R}^4$)[13]

$$\langle \delta \mathbf{A}, \delta \mathbf{A}' \rangle = 2 \int d^4x \, \mathrm{Tr} \left(\delta \mathbf{A}_\mu(x) \delta \mathbf{A}'_\mu(x) \right) . \quad (10.250)$$

The factor 2 is due to the normalization $\mathrm{Tr}(T_a, T_b) = \frac{1}{2}\delta_{ab}$. Infinitesimal gauge transformations $g(\alpha(x)) \approx \mathbb{1} + i\alpha_a(x)T^a$ can be written as (cf. (10.64))

$$\mathbf{A}_\mu \rightarrow \mathbf{A}_\mu - e^{-1}\left(\partial_\mu \alpha + ie\,[\mathbf{A}_\mu, \alpha]\right) = \mathbf{A}_\mu - e^{-1}D_\mu\alpha, \quad (10.251)$$

which implies that for gauge transformed potentials $\delta \mathbf{A}_\mu = -e^{-1}D_\mu\alpha$. Here $\alpha \equiv \alpha_a T^a$. The requirement of orthogonality to gauge transformations for any $\alpha(x)$ in the gauge group algebra then yields

$$\int d^4x \, \mathrm{Tr} \left(\delta \mathbf{A}_\mu(x) D_\mu \alpha(x) \right) = 0 . \quad (10.252)$$

After partial integration, Eq. (10.252) gives the constraint on possible $\delta \mathbf{A}_\mu$:

$$D_\mu \delta \mathbf{A}_\mu = 0 . \quad (10.253)$$

Note that this gauge fixing condition does not remove the collective coordinates due to global gauge transformations, as required.

---

[13]We are thus implicitly discussing fluctuations $\delta \mathbf{A}_\mu$ which are square-integrable with respect to the inner product (10.250).

Fluctuations satisfying (10.249) together with (10.253) are known as zero modes, namely normalizable solutions of linearized field equations for quantum fluctuations.

Zero modes thus represent physical fluctuations in field space which *do not* change the value of the action (the energy, in the soliton case). Note that non-zero modes necessarily increase the action (energy) of the background instanton (soliton) solution.

Eqs. (10.249) and (10.253) are written in a compact form

$$s_\mu^\dagger s_\nu D_\mu \delta \mathbf{A}_\nu = 0, \qquad (10.254)$$

for the anti-self-dual case. The self-dual case would be analogously written as $s_\mu s_\nu^\dagger D_\mu \delta \mathbf{A}_\nu = 0$. Introducing the quaternionic expansions

$$\bar{\mathcal{D}} \equiv s_\mu^\dagger D_\mu, \qquad s_\mu \delta \mathbf{A}_\mu \equiv \mathbf{G}. \qquad (10.255)$$

Eq. (10.254) reduces to two spinor equations (adjoint-representations):

$$\bar{\mathcal{D}}\mathbf{G} = 0 \quad \Rightarrow \quad \bar{\mathcal{D}}\psi = 0, \quad \bar{\mathcal{D}}\bar{\psi} = 0; \quad \psi = \begin{pmatrix} a \\ b \end{pmatrix}, \quad \bar{\psi} = s_2^\dagger \psi^*, \qquad (10.256)$$

Since ψ and $\bar{\psi}$ solve the Dirac equation, one can construct [67] two independent $\mathbf{G}$; thus the number of solutions for $\mathbf{G}$ is twice the number of solutions for a single two-component adjoint spinor.

The number of zero modes of $\bar{\mathcal{D}}\mathbf{G}$ is identical to the Dirac index Ind $\bar{\mathcal{D}}$, see, e.g., [67, 206]. The result is

$$\text{Ind } \bar{\mathcal{D}} = \frac{R(N)}{8\pi^2} \int d^4x \, \text{Tr}(\mathbf{F}_{\mu\nu}{}^*\mathbf{F}^{\mu\nu}), \qquad (10.257)$$

where $R(N)$ depends on the number of group generators. This is the result of the Atiyah–Singer index theorem [44, 67, 206] for the Dirac operator in the anti-instanton background. An identical result applies to instantons. From Eq. (10.199) we know that Ind $\bar{\mathcal{D}} = 2N|Q|$, which proves that that Yang–Mills instantons have $4N|Q|$ collective coordinates (see also Section 8.4). For instance, the total number of collective coordinates is $4N|Q|$, $4(N-2)|Q|$ and $4(N+1)|Q|$ for the gauge groups $SU(N)$, $SO(N)$ and $Sp(N)$, respectively.

Appendix S

Bäcklund transformation for the sine-Gordon system

Multi-soliton solutions for the sine-Gordon system can be found by using some of the standard techniques known for integrable systems, such as the inverse scattering method [2] or the Lax Pair [665]. In this Appendix we present some foundations of the method known as Bäcklund transformation [45]. This technique belongs to the class of so-called *solution generating techniques*. In particular, from a given soliton solution $\phi_1(x)$ it allows us to generate a new solution $\phi_2(x)$.

To this end we write the sine-Gordon equation in terms of rotated (light-cone) coordinates $x_+ = (x + t)/2$ and $x_- = (x - t)/2$. The sine-Gordon equation (10.20) then gets the form

$$\frac{\partial^2 \phi}{\partial x_- \partial x_+} - \sin \phi = 0. \qquad (S.1)$$

The Bäcklund transformation is defined as:

$$\frac{1}{2} \frac{\partial}{\partial x_+} (\phi_2 - \phi_1) = a \sin\left(\frac{1}{2}(\phi_2 + \phi_1)\right), \qquad (S.2)$$

$$\frac{1}{2} \frac{\partial}{\partial x_-} (\phi_2 + \phi_1) = \frac{1}{a} \sin\left(\frac{1}{2}(\phi_2 - \phi_1)\right), \qquad (S.3)$$

where the constant $a \in \mathbb{R}/\{0\}$ is called the Bäcklund parameter. Next we differentiate Eq. (S.2) with respect to x_- and use Eq. (S.3) to arrive at

$$\frac{1}{2} \frac{\partial^2}{\partial x_+ \partial x_-} (\phi_2 - \phi_1) = \cos\left(\frac{1}{2}(\phi_2 + \phi_1)\right) \sin\left(\frac{1}{2}(\phi_2 - \phi_1)\right)$$

$$= \frac{1}{2} \sin \phi_2 - \frac{1}{2} \sin \phi_1. \qquad (S.4)$$

We thus see that ϕ_2 is a solution of the sine-Gordon equation provided ϕ_1 is, and vice-versa. The particular advantage of such a formulation is that one can now solve ϕ_2 in terms of ϕ_1, which in turn allows us to generate a new solution in terms of the old one. Note that the passage from Eqs. (S.2)–(S.3) to Eq. (S.4) works only under the implicit assumption that

$$\frac{\partial^2}{\partial x_- \partial x_+}\phi_2(x_+, x_-) = \frac{\partial^2}{\partial x_+ \partial x_-}\phi_2(x_+, x_-). \tag{S.5}$$

For ϕ_1 (as a true soliton solution) such a compatibility condition should hold automatically.

As an example we insert in Eqs. (S.2)–(S.3) the trivial no-soliton solution $\phi_1 = 0$, the corresponding equations for $\phi_2(x_+, x_-)$ then read:

$$\frac{1}{2}\frac{\partial}{\partial x_+}\phi_2 = \frac{1}{2}a^2\frac{\partial}{\partial x_-}\phi_2 = a\sin\frac{\phi_2}{2}. \tag{S.6}$$

These can be easily integrated back to the one-soliton solution, indeed the solution of Eq. (S.6) reads (cf. the Bogomol'nyi equation (10.154))

$$\phi_2(x_+, x_-) = \pm 4\arctg\left[\exp\left(ax_+ + \frac{1}{a}x_- + c\right)\right], \tag{S.7}$$

with c being the constant of integration. By choosing

$$a = \frac{\sqrt{1-u}}{\sqrt{1+u}}, \qquad c = -\frac{x_0}{\sqrt{1-u^2}},$$

we finally obtain

$$\phi_2(x) = \pm 4\arctg\left[\exp\left(\frac{x - x_0 - ut}{\sqrt{1-u^2}}\right)\right]. \tag{S.8}$$

The reader can recognize in Eq. (S.8) the Lorentz boosted one-kink (+ sign) or one-antikink (− sign) solution (10.23).

Although one may progress further by using the one-kink/antikink solution as a seed solution ϕ_1, etc., such a procedure becomes technically difficult for more complicated ϕ_1. Fortunately, there exists a purely *algebraic* way of constructing multi-kink solutions directly from the Bäcklund transformation. Consider two soliton solutions ϕ_2 and ϕ_3, obtained from Eqs. (S.2)–(S.3) by starting with the same seed solution ϕ_1, but choosing two different values of the Bäcklund parameter a, say a_2 and a_3. In particular, by manipulating the Bäcklund equations (S.2)–(S.3), it can be shown that a theorem of *permutability* holds, namely the solution ϕ_{23}, obtained by applying the Bäcklund transformation with parameter a_3 to the seed solution ϕ_2, is equal to the solution ϕ_{32}, obtained by applying the Bäcklund transformation with the parameter a_2 to the seed solution ϕ_3. The consistency condition $\phi_{23} = \phi_{32}$ yields the explicit relation

$$\phi_{23} = \phi_{32} = 4\arctan\left[\left(\frac{a_2 + a_3}{a_3 - a_2}\right)\tan\left(\frac{\phi_2 - \phi_3}{4}\right)\right] - \phi_1. \tag{S.9}$$

Thus, we have indeed obtained a new solution $\phi(x) \equiv \phi_{23}$ from the triplet of known solutions $\phi_1(x)$, $\phi_2(x)$ and $\phi_3(x)$.

To illustrate the efficiency of this approach we consider the triplet

$$\phi_1 = 0, \quad \phi_2 = 4 \arctan \left[\exp \left(a_2 x_+ + \frac{1}{a_2} x_- \right) \right],$$

$$\phi_3 = 4 \arctan \left[\exp \left(a_3 x_+ + \frac{1}{a_3} x_- \right) \right]. \tag{S.10}$$

In this case, we have the new multi-kink solution (cf. Eq. (S.9))

$$\phi(x_+, x_-) = 4 \arctan \left[\left(\frac{a_2 + a_3}{a_3 - a_2} \right) \frac{\sinh \left(\frac{y_3 - y_2}{2} \right)}{\cosh \left(\frac{y_3 + y_2}{2} \right)} \right], \tag{S.11}$$

where $y_j = a_j x_+ + x_-/a_j$. For simplicity's sake it is convenient to choose $a_2 = -1/a_3 \equiv a$. Then we get

$$\phi(x) = 4 \arctan \left[\frac{u \sinh \left(x/\sqrt{1 - u^2} \right)}{\cosh \left(ut/\sqrt{1 - u^2} \right)} \right]. \tag{S.12}$$

Since this solution goes from -2π to 2π as x changes from $-\infty$ to $+\infty$, it has the topological charge $Q = 2$. Thus Eq. (S.12) describes a time-dependent two-kink field configuration. This solution describes two kinks that are approaching the origin with the velocity u for negative t, and again separate at the positive t. Since they feel the repulsive kink-kink force, they smoothly bounce back off each other, rather than passing through each other. The time of the closest approach is at $t = 0$.

What is most remarkable about the solution (S.12) is that the shape of the individual kinks is preserved even after a collision. In fact, the following asymptotic behavior holds for large t:

$$\phi(x) \approx 4 \arctan \left[\exp \left(\frac{x - u(t - \delta/2)}{\sqrt{1 - u^2}} \right) \right]$$

$$+ 4 \arctan \left[\exp \left(\frac{x + u(t - \delta/2)}{\sqrt{1 - u^2}} \right) \right]. \tag{S.13}$$

Here the time delay is given by

$$\delta = \log u \left(\frac{1 - u^2}{u} \right). \tag{S.14}$$

It is also not difficult to find other multi-kink solutions. For example, the solution

$$\phi(x) = 4 \arctan \left[\frac{\sinh \left(ut/\sqrt{1 - u^2} \right)}{u \cosh \left(x/\sqrt{1 - u^2} \right)} \right], \tag{S.15}$$

represents the scattering of a kink off an antikink.

Appendix T

Elements of homotopy theory

As discussed in Chapters 4, 7, 8 and 10 the appearance of topological defects is a common feature of phase transitions. We have also seen in Chapter 4 that many typical phase transitions — namely continuous ones — are related to the concept of SSB. Within this framework the necessary (but not sufficient) condition for the existence of topologically stable defects can be expressed in terms of the topology of the vacuum manifold, and more specifically in terms of its *homotopy group*. Early applications of this idea can be found, e.g., in [371, 372, 471, 609, 647] (see also Section 10.5).

In this Appendix we present a brief survey of aspects from homotopy theory that are relevant to the study of defect formation in SSB phase transitions.

The fundamental group

We begin by defining the *fundamental group* (or the first homotopy group) $\pi_1(\mathcal{M})$ of a given manifold $\mathcal{M}$. The basic observation is that all loops on $\mathcal{M}$ that can be deformed continuously one into another form an equivalence class and the set of all those classes has a group structure. The corresponding group π_1 then represents global characteristics of topological spaces that is related to their *multiple-connectedness* (i.e., presence of holes or interconnections between various regions).

Let I denote the closed interval $[0, 1] \subset \mathbb{R}$. The continuous mapping $\omega : I \to \mathcal{M}$ with $\omega(0) = \omega(1) = x_0$ is said to form a loop with the base point x_0. Consider two loops $\omega_1 = \omega_1(t)$ and $\omega_2 = \omega_2(t)$ such that they have the same base point, that is,

$$\omega_1(0) = \omega_2(0) = x_0 = \omega_1(1) = \omega_2(1). \tag{T.1}$$

The loops $\omega_1(t)$ and $\omega_2(t)$ are said to be *homotopically* equivalent (we denote by the symbol $\sim$ such an equivalence) if there exists a continuous map

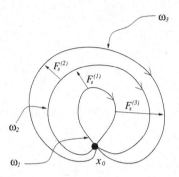

Fig. T.1   An illustration of transitivity of loops belonging to $[\omega]$.

$F : I \times I \to \mathcal{M}$ such that

$$F_0(t) = \omega_1(t), \quad F_1(t) = \omega_2(t), \quad \text{for } \forall t \in I$$
$$F_s(0) = F_s(1) = x_0. \tag{T.2}$$

Here $F(s,t) \equiv F_s(t)$ is called the *homotopy* or *deformation* (of ω_1 into ω_2). The parameter s thus plays the role of the deformation parameter that smoothly deforms ω_1 to ω_2 while changing from 0 to 1.

Homotopically equivalent loops form an equivalence class (or homotopy class) $[\omega]$. Indeed, the three conditions, i.e., reflectivity, symmetry and transitivity, are fulfilled for all elements in $[\omega]$. The *reflectivity*, i.e., $\omega \sim \omega$ for $\omega \in [\omega]$, results from the fact that one may choose $F_s(t) = \omega(t)$. The *symmetry* condition, i.e., when $\omega_1 \sim \omega_2$ then $\omega_2 \sim \omega_1$ for $\omega_1, \omega_2 \in [\omega]$, stems from the fact that if $F_s(t)$ is the homotopy from ω_1 to ω_2, then $F_{1-s}(t)$ is the homotopy from ω_2 to ω_1. Finally, the *transitivity*, i.e., when $\omega_1 \sim \omega_2$ and $\omega_2 \sim \omega_3$, then $\omega_1 \sim \omega_3$ for $\omega_1, \omega_2, \omega_3 \in [\omega]$, is a direct consequence of the fact that the homotopy $F_s^{(3)}(t)$ from ω_1 to ω_3 can be directly constructed in terms of the homotopy $F_s^{(1)}(t)$ from ω_1 to ω_2 and homotopy $F_s^{(2)}(t)$ from ω_2 to ω_3. Indeed, defining

$$F_s^{(3)}(t) = \begin{cases} F_{2s}^{(1)}(t) & \text{for } 0 \le s \le \frac{1}{2} \\ F_{2s-1}^{(2)}(t) & \text{for } \frac{1}{2} \le s \le 1 \end{cases}, \tag{T.3}$$

we may see that $F_s^{(3)}(t)$ represents, due to condition $F_1^{(1)}(t) = F_0^{(2)}(t) = \omega_2(t)$, a continuous mapping from ω_1 to ω_2. Hence the mapping defined in (T.3) is the required homotopy (see Fig. T.1). It is customary to denote the set of equivalence classes $[\omega]$ with the base point x_0 as $\pi_1(\mathcal{M}, x_0)$. Within

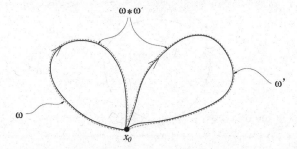

Fig. T.2   An illustration of the product law. The dashed loop corresponds to the product $\omega * \omega'$.

$\pi_1(\mathcal{M}, x_0)$ one may define a natural product law as follows: take two loops in $\mathcal{M}$ based at x_0, say $\omega(t)$ and $\omega'(t)$, and consider the loop.

$$\omega''(t) = \begin{cases} \omega(2t) & \text{for } 0 \leq t \leq \frac{1}{2} \\ \omega'(2t-1) & \text{for } \frac{1}{2} \leq t \leq 1 \end{cases}. \tag{T.4}$$

The situation is depicted in Fig. T.2. This loop is continuous ($\omega(0) = \omega'(1) = x_0$ while $\omega''(0) = \omega(0) = x_0$, and $\omega''(1) = \omega'(1) = x_0$) and it constitutes what is known as the product of two loops in $\mathcal{M}$. We shall write $\omega''(t) \equiv (\omega * \omega')(t)$. The product $\circ$ in $\pi_1(\mathcal{M}, x_0)$ is defined as

$$[\omega] \circ [\eta] = [\omega * \eta]. \tag{T.5}$$

For consistency reasons the product $\circ$ must be independent of the choice of representatives in $[\omega]$ and $[\eta]$. This is indeed the case, for if we choose $\omega' \in [\omega]$ and $\eta' \in [\eta]$ and ω is homotopic with ω' via $F_s^{(1)}$, and η is homotopic with η' via $F_s^{(2)}$, then

$$F_s^{(3)}(t) = \begin{cases} F_s^{(1)}(2t) & \text{for } 0 \leq t \leq \frac{1}{2} \\ F_s^{(2)}(2t-1) & \text{for } \frac{1}{2} \leq t \leq 1 \end{cases}, \tag{T.6}$$

is a homotopy from $\omega * \eta$ to $\omega' * \eta'$. Consequently $[\omega] \circ [\eta]$ is indeed representative independent.

The consequence of the previous reasoning is that $\pi_1(\mathcal{M}, x_0)$ equipped with the product (T.5) admits a group structure, the fundamental group. To see this we must prove the basic group conditions, that is; *associativity*, i.e., $([\omega] \circ [\eta]) \circ [\nu] = [\omega] \circ ([\eta] \circ [\nu])$, existence of an *identity*, i.e., existence of $[\omega_0]$ such that $[\omega_0] \circ [\omega] = [\omega] \circ [\omega_0] = [\omega]$, and existence of an *inverse element*, i.e., for given $[\omega]$ exists $[\omega']$ such that $[\omega] \circ [\omega'] = [\omega'] \circ [\omega] = [\omega_0]$.

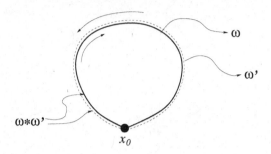

Fig. T.3   Loop ω' is an inverse loop to ω. So $\omega * \omega' = \omega_0$.

To prove associativity we use (T.4) and realize that for any $\omega \in [\omega]$, $\eta \in [\eta]$ and $\nu \in [\nu]$

$$((\omega * \eta) * \nu)(t) = \begin{cases} \omega(4t) & \text{for } 0 \leq t \leq \frac{1}{4} \\ \eta(4t - 1) & \text{for } \frac{1}{4} \leq t \leq \frac{1}{2} \\ \nu(2t - 1) & \text{for } \frac{1}{2} \leq t \leq 1 \end{cases}, \qquad (\text{T.7})$$

while

$$(\omega * (\eta * \nu))(t) = \begin{cases} \omega(2t) & \text{for } 0 \leq t \leq \frac{1}{2} \\ \eta(4t - 2) & \text{for } \frac{1}{2} \leq t \leq \frac{3}{4} \\ \nu(4t - 3) & \text{for } \frac{3}{4} \leq t \leq 1 \end{cases}. \qquad (\text{T.8})$$

The loop product (T.7) is just a reparametrization of the product (T.8) (i.e., loops pass over the same points in the same order, they are only traversed by t at different rates). Because reparametrized loops are mutually homotopic, the product in $\pi_1(\mathcal{M}, x_0)$ is associative.

As an identity element in $\pi_1(\mathcal{M}, x_0)$ we can choose loops homotopic with the loop $\omega_0(t) \equiv x_0$, in such case we have for any $\omega \in [\omega]$

$$(\omega_0 * \omega)(t) = \begin{cases} x_0 & \text{for } 0 \leq t \leq \frac{1}{2} \\ \omega(2t - 1) & \text{for } \frac{1}{2} \leq t \leq 1 \end{cases}, \qquad (\text{T.9})$$

which is just the reparametrized loop $\omega(t)$ (i.e., the constant loop). Hence $(\omega_0 * \omega)(t)$ is homotopic to $\omega(t)$. Similarly, $(\omega * \omega_0)(t)$ is homotopic to $\omega(t)$.

Finally the inverse element to $[\omega]$ can be constructed from loops that are homotopic to $\omega'(t) \equiv \omega(1 - t)$, indeed

$$(\omega * \omega')(t) = \begin{cases} \omega(2t) & \text{for } 0 \leq t \leq \frac{1}{2} \\ \omega(2 - 2t) & \text{for } \frac{1}{2} \leq t \leq 1 \end{cases}, \qquad (\text{T.10})$$

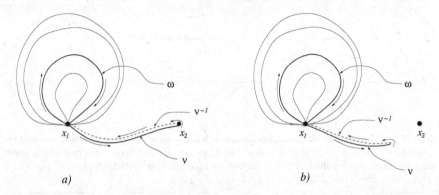

Fig. T.4 An illustration of the independence of $\pi_1(\mathcal{M})$ on the base point. The case $a)$ depicts $F_0(t)$, while the case $b)$ shows $F_s(t)$ with $s > 0$. When s reaches value 1 the loop $a)$ fully deforms to the loop ω.

and this is homotopic with $\omega_0(t)$ through the homotopy

$$F_s(t) = \begin{cases} \omega(2t(1-s)) & \text{for } 0 \leq t \leq \frac{1}{2} \\ \omega(-2t)(1-s)) & \text{for } \frac{1}{2} \leq t \leq 1 \end{cases} . \tag{T.11}$$

The same argument can be applied on $(\omega' * \omega)(t)$. Thus the class of loops that are homotopic to $\omega'(t) \equiv \omega(1-t)$ (i.e., the class of loops with the opposite orientation) represents the inverse element to $[\omega]$ in $\pi_1(\mathcal{M}, x_0)$; see Fig. T.3. To show that $\pi_1(\mathcal{M}, x_0)$ is independent of the choice of the base point x_0, we note that every loop in $\pi_1(\mathcal{M}, x_0)$ is homotopic to at least one loop in $\pi_1(\mathcal{M}, x_1)$ and, conversely, every loop in $\pi_1(\mathcal{M}, x_1)$ is homotopic to at leat one loop in $\pi_1(\mathcal{M}, x_0)$. Here x_0 and x_1 are to distinct points from $\mathcal{M}$. Indeed, take a loop $\omega \in \pi_1(\mathcal{M}, x_0)$ and assume that ν is a smooth curve connecting x_0 with x_1, such that $\nu(0) = x_0$ and $\nu(1) = x_1$. The homotopy

$$F_s(t) = \begin{cases} \omega(3t) & \text{for } 0 \leq t \leq \frac{1}{3} \\ \nu((3t-1)(1-s)) & \text{for } \frac{1}{3} \leq t \leq \frac{2}{3} , \\ \nu^{-1}((3t-2)(1-s)) & \text{for } \frac{2}{3} \leq t \leq 1 \end{cases} \tag{T.12}$$

then makes any $\omega \in \pi_1(\mathcal{M}, x_0)$ homotopic to a loop in $\pi_1(\mathcal{M}, x_1)$ (see Fig. T.4). The reverse statement could be obtained if the proof started with $\pi_1(\mathcal{M}, x_1)$ rather than $\pi_1(\mathcal{M}, x_0)$. Thus, provided any two points can be joined by a smooth curve (arcwise-connected topological spaces), the fundamental group does not refer to any particular base point and the fundamental group is denoted by $\pi_1(\mathcal{M})$.

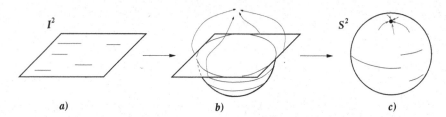

Fig. T.5   The 2-sphere $\mathcal{S}^2$ is homeomorphic to the cube I^n with the boundary ∂I^n identified. The square I^2 is shown in a). In b) the square has been deformed into a bowl and boundaries identified. In c) the 2-sphere emerges. Because all boundary points of I^2 are identified, the north pole in $\mathcal{S}^2$ is non-singular.

Higher homotopy groups

If, instead of the variable t, we used the scaled variable θ such that $\theta = 2\pi t$, then one could view the fundamental group $\pi_1(\mathcal{M})$ as a set of homotopic classes of continuous maps from a unit circle $\mathcal{S}^1$ to $\mathcal{M}$. Higher homotopy groups $\pi_n(\mathcal{M})$ $(n \geq 2)$ can then be defined in much the same way as $\pi_1(\mathcal{M})$. The only difference is that one considers homotopy classes of continuous maps from the n-spheres $\mathcal{S}^n$ to $\mathcal{M}$. In order to see that π_n is again a group, it is technically more convenient to utilize unit cubes $I^n = [0,1]^n$ instead of spheres $\mathcal{S}^n$. Indeed, the n-sphere $\mathcal{S}^n$ is topologically equivalent (homeomorphic) to the cube I^n with all boundary points identified (see Fig. T.5). The homotopy from ω_1 to ω_2 is again defined as a continuous map $F^{(n)} : I \times I^n \to \mathcal{M}$ such that

$$F_0^{(n)}(t) = \omega_1(t), \quad F_1^{(n)}(t) = \omega_2(t), \quad \forall t \in I^n,$$

$$F_s^{(n)}(\partial I^n) = x_0, \quad \forall s \in I. \tag{T.13}$$

The notation $F^{(n)}(s,t) \equiv F_s^{(n)}(t)$ has been used. The continuous maps $\omega : I^n \to \mathcal{M}$, $\partial I^n \to x_0$ are called spheroids ($n = 1$ spheroid is a loop). It is not difficult to see that all homotopically equivalent spheroids again form an equivalence class. The corresponding set of classes with the base point x_0 is denoted by $\pi_n(\mathcal{M}, x_0)$.

Consider two spheroids ω and ω' in $\mathcal{M}$ and the continuous maps $\omega, \omega' : I^n \to \mathcal{M}$, $\partial I^n \to x_0$; the product $\omega'' = \omega * \omega'$ is defined as

$$\omega''(t_1, \mathbf{t}) = \begin{cases} \omega(2t_1, \mathbf{t}) & \text{for } 0 \leq t_1 \leq \frac{1}{2} \\ \omega'(2t_1 - 1, \mathbf{t}) & \text{for } \frac{1}{2} \leq t_1 \leq 1 \end{cases}. \tag{T.14}$$

Here $\mathbf{t} = (t_2, \ldots, t_n)$. Since the points $(0, \mathbf{t})$ and $(1, \mathbf{t})$ belong to ∂I^n, $\omega(0, \mathbf{t}) = \omega''(0, \mathbf{t}) = x_0$ and $\omega'(1, \mathbf{t}) = \omega''(1, \mathbf{t}) = x_0$. The map ω'' defined

by (T.14) is continuous. It is customary to write $\omega''(t) \equiv (\omega*\omega')(t)$ $(t \in I^n)$.
As in the case of π_1 we can define a product in $\pi_n(\mathcal{M}, x_0)$ as

$$[\omega] \circ [\eta] = [\omega * \eta]. \tag{T.15}$$

The product $\circ$ is again independent of the choice of representatives in $[\omega]$
and $[\eta]$. With the product law (T.15) the homotopy classes of spheroids
admit a group structure. The group operations can be proven in a similar
way as for π_1. For instance, the identity element can be identified with the
homotopy class of the constant map $\omega_0 : I^n \to x_0$. Then, for any $\omega \in [\omega]$

$$(\omega_0 * \omega)(t_1, \mathbf{t}) = \begin{cases} x_0 & \text{for } 0 \le t_1 \le \frac{1}{2} \\ \omega(2t_1 - 1, \mathbf{t}) & \text{for } \frac{1}{2} \le t_1 \le 1 \end{cases}. \tag{T.16}$$

This is the reparametrized spheroid $\omega(t)$ and hence $\omega_0*\omega$ is homotopic to ω.
Spheroids in the class $[\omega_0]$, i.e., spheroids which can be smoothly deformed
to a point x_0, are called contractible spheroids (or contractible maps).

It is also not difficult to see that $\pi_n(\mathcal{M}, x_0)$ is independent of x_0, up to
an isomorphism, if $\mathcal{M}$ is an arcwise-connected topological space. This can
be proved in a similar way as for π_1.

In general, the fundamental group $\pi_1(\mathcal{M})$ is non-Abelian. A typical
example is π_1 on $\mathbb{R}^2$ with two holes. In this case we can choose a loop ω to
traverse around one hole and a loop η around the other. Fig. T.6 illustrates
that $\omega*\eta*\omega^{-1}$ cannot be continuously deformed into the loop η, so $\omega*\eta \not\sim$
$\eta*\omega$. The fundamental group of any Lie group is, however, Abelian. Higher
homotopy groups ($\pi_n(\mathcal{M})$, $n \ge 2$) are automatically Abelian since $\eta * \omega$ is
homotopic to $\omega*\eta$. The sphere $\mathcal{S}^0$ is defined by the prescription $t^2 = 1$, i.e.,
it consists of a pair of points $t = \pm 1$. Sometimes $\pi_0(\mathcal{M})$ is defined as a set
of equivalence classes of continuous maps from $\mathcal{S}^0$ to $\mathcal{M}$. Since one of the
points, usually $t = -1$, is mapped to the fixed base point $x_0 \in \mathcal{M}$, $\pi_0(\mathcal{M})$
represents the set of disconnected pieces of $\mathcal{M}$. As the set of disconnected
pieces of $\mathcal{M}$ is an invariant, π_0 cannot depend on the base point and we write
$\pi_0(\mathcal{M}) \equiv \pi_0(\mathcal{M}, x_0)$. In general, $\pi_0(\mathcal{M})$ does not possess group properties.
Only when $\mathcal{M}$ itself is a Lie group, π_0 inherits its group structure.

The product on Abelian π_n is denoted by the addition, and the identity
element $e \equiv [\omega_0]$ represented by 0. So, $[\omega] + [\omega^{-1}] = 0$, $[\omega] + 0 = [\omega]$ and
$([\omega] + [\eta]) + [\nu] = [\omega] + ([\eta] + [\nu])$. The notation $[\omega^{-1}] = [-\omega]$ is also used.

Calculation of homotopy groups $\pi_n(\mathcal{M})$ is a major task of the algebraic
topology and, in general, it is not an easy task [344]. Without going to
further mathematical details we state now some useful technical results:

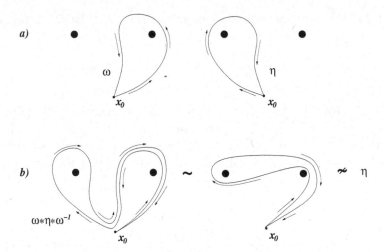

Fig. T.6   a) Loops ω and η. Black dots denote holes in $\mathbb{R}^2$. b) The loop $\omega * \eta * \omega^{-1}$ is not homotopic to η. This result is true for an arbitrary base point x_0.

The **Bott periodicity theorem:**

$$\pi_k(U(n)) \;\cong\; \pi_k(SU(n)) \;\cong\; \begin{cases} \{e\} \text{ if } k \text{ is even} \\ \mathbb{Z} \;\text{ if } k \text{ is odd} \end{cases}, \qquad (T.17)$$

for $n \geq (k+1)/2$. For example, $\pi_1(U(n)) \cong \mathbb{Z}$ and $\pi_2(U(n)) \cong \{e\}$.

The **J-isomorphism:**

$$\pi_k(SO(n)) \;\cong\; \pi_{n+k}(\mathcal{S}^n), \qquad (T.18)$$

implies, for instance, that $\pi_1(SO(2)) \cong \pi_3(\mathcal{S}^2) \cong \mathbb{Z}$.

The **stability property:**

$$\pi_{k+n}(\mathcal{S}^n) \;\cong\; \pi_{k+n+1}(\mathcal{S}^{n+1}), \quad \text{for} \quad n \geq k+2. \qquad (T.19)$$

In particular, for $k = 0$, $\pi_n(\mathcal{S}^n) \cong \mathbb{Z}$ for all $n > 2$. We also independently know that $\pi_1(\mathcal{S}^1) \cong \mathbb{Z}$, which can be obtained from the fundamental homotopy group relation for spheres, the Bott–Tu theorem.

The **Bott–Tu theorem:**

$$\pi_k(\mathcal{S}^n) \;\cong\; \begin{cases} \{e\} \text{ if } k < n \\ \mathbb{Z} \;\text{ if } k = n \end{cases}. \qquad (T.20)$$

The **Cross product theorem:**

$$\pi_n(X_1 \times X_2) \;\cong\; \pi_n(X_1) \otimes \pi_n(X_2), \qquad (T.21)$$

if X_1 and X_2 are arcwise connected. In words, if X is a topological product of X_1 and X_2 then $\pi_n(X)$ is the direct product of $\pi_n(X_1)$ and $\pi_n(X_2)$. For the torus $T^2 = \mathcal{S}^1 \times \mathcal{S}^1$, and so $\pi_1(T^2) = \pi_1(\mathcal{S}^1) \otimes \pi_1(\mathcal{S}^1) = \mathbb{Z} \oplus \mathbb{Z}$.

Homotopy groups of Lie groups:

$$\pi_2(G) \cong \{e\} \quad \text{for any compact Lie group},$$

$$\pi_n(G) = \pi_n(G/D), \quad n > 1 \quad D \text{ is a discrete normal subgroup of } G,$$

$$\pi_3(G) \cong \mathbb{Z} \quad \text{for any simple compact Lie group},$$

$$\pi_1(G/H) \cong \pi_0(H), \quad H \subset G,$$

$$\pi_2(G/H) \cong \pi_1(H), \quad H \subset G. \tag{T.22}$$

The last two relations are valid for any group G, both connected and simply connected, i.e., $\pi_0(G) \cong \pi_1(G) \cong \{e\}$. For not simply connected G the last relation is formulated in terms of universal covering group of G.

Bibliography

[1] Abdalla, E., Abdalla, M. C. B. and Rothe, K. D., (2001). *Nonperturbative Methods in Two-Dimensional Quantum Field Theory* (World Scientific, Singapore).

[2] Ablowitz, M. and Segur, H., (1981). *Solitons and the Inverse Scattering Transform* (SIAM, Philadelphia).

[3] Abramowitz, M. and Stegun, I. A., (1972). *Handbook of Mathematical Functions* (Dover Publications, New York).

[4] Abrikosov, A. A., (1957). *Sov. Phys. JETP* **5**, p. 1174.

[5] Abrikosov, A. A., Gor'kov, L. P. and Dzyaloshinski, I. E., (1959). *J. Exp. Theoret. Phys. (U.S.S.R.)* **36**, p. 900.

[6] Abud, M. and Sartori, G., (1983). *Annals Phys.* **150**, p. 307.

[7] Achiezer, N. I. and Glazman, I. M., (1993). *Theory of Linear Operators in Hilbert Space* (Dover Publications).

[8] Achucarro, A., (2003), p. 273 in *Patterns of Symmetry Breaking*, H. Arodz, J. Dziarmaga, and W. H. Zurek (Eds.) (Kluwer, London).

[9] Adler, S. L., (1965). *Phys. Rev.* **137**, p. B1022.

[10] Adler, S. L., (1965). *Phys. Rev.* **139**, p. B11638.

[11] Aharonov, Y. and Bohm, D., (1959). *Phys. Rev.* **115**, p. 485.

[12] Alamoudi, S. M., Boyanovsky, D. and Takakura, F. I., (1998). *Phys. Rev.* **D58**, p. 105003.

[13] Alfinito, E., Blasone, M., Iorio, A. and Vitiello, G., (1995). *Phys. Lett.* **B362**, p. 91.

[14] Alfinito, E., Manka, R. and Vitiello, G., (2000). *Class. Quant. Grav.* **17**, p. 93.

[15] Alfinito, E., Romei, O. and Vitiello, G., (2002). *Mod. Phys. Lett.* **B16**, p. 93.

[16] Alfinito, E., Viglione, R. G. and Vitiello, G., (2001). *Mod. Phys. Lett.* **B15**, p. 127.

[17] Alfinito, E. and Vitiello, G., (2002). *Phys. Rev.* **B65**, p. 054105.

[18] Alfinito, E. and Vitiello, G., (2003). *Mod. Phys Lett.* **B17**, p. 1207.

[19] Alfinito, E. and Vitiello, G., (1999). *Phys. Lett.* **A252**, p. 5.

[20] Alfinito, E. and Vitiello, G., (2000). *Int. J. Mod. Phys.* **B14**, p. 853.

[21] Amelino-Camelia, G., (2007). *Nature Phys.* **3**, p. 81.

[22] Amit, D. J., (1989). *Modeling Brain Function: the World of Attractor Neural Networks* (Cambridge University Press, Cambridge, U.K.).

[23] Anandan, J. and Aharonov, Y., (1990). *Phys. Rev. Lett.* **65**, p. 1697.

[24] Anderson, P. W., (1958). *Phys. Rev.* **110**, p. 827.

[25] Anderson, P. W., (1963). *Phys. Rev.* **130**, p. 439.

[26] Anderson, P. W., (1984). *Basic Notions of Condensed Matter Physics* (Benjamin, Menlo Park).

[27] Antunes, N. D., Bettencourt, L. M. A. and Zurek, W. H., (1999). *Phys. Rev. Lett.* **82**, p. 2824.

[28] Antunes, N. D., Lombardo, F. C. and Monteoliva, D., (2001). *Phys. Rev.* **E64**, p. 066118.

[29] Aoyama, S. and Kodama, Y., (1976). *Prog. Theor. Phys.* **56**, p. 1970.

[30] Araki, H., (1960). *J. Math. Phys.* **1**, p. 492.

[31] Araki, H. and Woods, E. J., (1963). *J. Math. Phys.* **4**, p. 637.

[32] Araki, H. and Wyss, W., (1964). *Helv. Phys. Acta* **37**, p. 136.

[33] Arecchi, F., Courtens, E., Gilmore, R. and Thomas, H., (1972). *Phys. Rev.* **A6**, p. 2211.

[34] Arimitsu, T., (1988). *Physica* **148A**, p. 427.

[35] Arimitsu, T., (1991). *Phys. Lett.* **A153**, p. 163.

[36] Arimitsu, T., Pradko, J. and Umezawa, H., (1986). *Physica* **135A**, p. 487.

[37] Arimitsu, T. and Umezawa, H., (1985). *Prog. Theor. Phys.* **74**, p. 429.

[38] Arimitsu, T. and Umezawa, H., (1987). *Prog. Theor. Phys.* **77**, p. 53.

[39] Arnold, P. and McLerran, L., (1987) Report Fermilab-Pub-87/84–7.

[40] Aschieri, P. *et al.*, (2005). *Class. Quant. Grav.* **22**, p. 3511.

[41] Ashcroft, N. W. and Mermin, N. D., (1976). *Solid State Physics* (Saunders College, Philadelphia).

[42] Atiyah, M., Drinfeld, V., Hitchin, N. and Manin, Y., (1978). *Phys. Lett.* **A65**, p. 185.

[43] Atiyah, M. F., Hitchin, N. J. and Singer, I. M., (1977). *Proc. Nat. Acad. Sci.* **74**, p. 2662.

[44] Atiyah, M. F. and Singer, I. M., (1968). *Ann. Math.* **87**, p. 484.

[45] Bäcklund, A. V., (1883). *Lunds Univ. Arsskr. Avd.* **2**, p. 19.

[46] Baier, R., Pire, B. and Schiff, D., (1990). *Phys. Lett.* **B238**, p. 367.

[47] Bak, P. and Creutz, M., (1995), p. 26 in *Fractals in Science*, A. Bunde and S. Havlin, (Eds.) (Springer-Verlag, Berlin).

[48] Balachandran, A. P., Jo, S. G. and Marmo, G., (2010). *Group Theory and Hopf Algebras* (World Scientific, Singapore).

[49] Ban, M. and Arimitsu, T., (1987). *Physica* **146A**, p. 89.

[50] Banerjee, R., (2002). *Mod. Phys. Lett.* **A17**, p. 631.

[51] Banerjee, R. and Ghosh, S., (2002). *Phys. Lett.* **B533**, p. 162.

[52] Banerjee, R. and Mukherjee, P., (2002). *J. Phys.* **A35**, p. 5591.

[53] Bang, O. and Christiansen, P. L., (1991). *Lecture Notes in Physics* (Springer, Berlin) **393**, p. 126.

[54] Bardeen, J., Cooper, L. N. and Schrieffer, J. R., (1957). *Phys. Rev.* **106**, p. 162.

[55] Bargmann, V., (1961). *Commun. Pure Appl. Math.* **14**, p. 187.

[56] Bargmann, V., Butera, P., Girardello, L. and Klauder, J. R., (1971). *Rep. Math. Phys.* **2**, p. 221.

[57] Barone, A., Esposito, F., Magee, C. and Scott, A., (1971). *Riv. Nuovo Cim.* **1**, p. 227.

[58] Barone, V., Penna, V. and Sodano, P., (1991). *Phys. Lett.* **A161**, p. 41.

[59] Bartnik, E. and Blinowska, K., (1989). *Phys. Lett.* **A134**, p. 448.

[60] Barut, A. O. and Girardello, L., (1971). *Commun. Math. Phys.* **21**, p. 41.

[61] Bassett, D. S. *et al.*, (2006). *PNAS* **103**, p. 19518.

[62] Bateman, H., (1931). *Phys. Rev.* **38**, p. 815.

[63] Bäuerle, C. *et al.*, (1996). *Nature* **382**, p. 332.

[64] Beige, A., Knight, P. L. and Vitiello, G., (2005). *New J. Phys.* **7**, p. 96.

[65] Belavin, A. A. and Polyakov, A. M., (1975). *JETP Lett.* **22**, p. 617.

[66] Belavin, A. A. *et al.*, (1975). *Phys. Lett.* **B59**, p. 85.

[67] Belitsky, V. S., Andrei V. and van Nieuwenhuizen, P., (2000). *Class. Quant. Grav.* **17**, p. 3521.

[68] Bell, J. S., (1987). *Speakable and Unspeakable in Quantum Mechanics. Collected Papers on Quantum Philosophy* (Cambridge University Press, Cambridge, U.K.).

[69] Bellman, R., (1961). *A Brief Introduction to Theta Functions* (Holt, Rinehart and Winston, New York).

[70] Bernard, C., Christ, N., Guth, A. and Weinberg, E., (1977). *Phys. Rev.* **D16**, p. 2967.

[71] Bernardini, A. E., Guzzo, M. M. and Nishi, C. C., (2010) arXiv: 1004.0734 [hep-ph].

[72] Berry, M. V., (1984). *Proc. Roy. Soc. Lond.* **A392**, p. 45.

[73] Bettencourt, L. M. A. *et al.*, (2000). *Phys. Rev.* **D62**, p. 065005.

[74] Beuthe, M., (2003). *Phys. Rept.* **375**, p. 105.

[75] Biedenharn, L. C., (1989). *J. Phys.* **22**, p. L873.

[76] Biedenharn, L. C. and Lohe, M. A., (1992). *Comm. Math. Phys.* **146**, p. 483.

[77] Bilenky, S. M. and Pontecorvo, B., (1978). *Phys. Rept.* **41**, p. 225.

[78] Birrell, N. D. and Davies, P. C., (1982). *Quantum Field in Curved Space* (Cambridge University Press, Cambridge, U.K.).

[79] Blaizot, J.-P. and Ripka, G., (1986). *Quantum Theory of Finite Systems* (MIT Press, Cambridge, MA).

[80] Blasone, M., Capolupo, A., Capozziello, S., Carloni, S. and Vitiello, G., (2004). *Phys. Lett.* **A323**, p. 182.

[81] Blasone, M., Capolupo, A., Celeghini, E. and Vitiello, G., (2009). *Phys. Lett.* **B674**, p. 73.

[82] Blasone, M., Capolupo, A., Ji, C.-R. and Vitiello, G., (2010). *Int. J. Mod. Phys.* **A25**, p. 4179.

[83] Blasone, M., Capolupo, A., Romei, O. and Vitiello, G., (2001). *Phys. Rev.* **D63**, p. 125015.

[84] Blasone, M., Capolupo, A. and Vitiello, G., (2002). *Phys. Rev.* **D66**, p. 025033.

[85] Blasone, M., Celeghini, E., Jizba, P. and Vitiello, G., (2003). *Phys. Lett.* **A310**, p. 393.

[86] Blasone, M., Dell'Anno, F., De Siena, S., Di Mauro, M. and Illuminati, F., (2008). *Phys. Rev.* **D77**, p. 096002.

[87] Blasone, M., Dell'Anno, F., De Siena, S. and Illuminati, F., (2009). *EPL* **85**, p. 50002.

[88] Blasone, M., Dell'Anno, F., De Siena, S. and Illuminati, F., (2010). *J. Phys. Conf. Ser.* **237**, p. 012007.

[89] Blasone, M., Di Mauro, M. and Vitiello, G., (2010) arXiv:1003.5812 [hep-ph].

[90] Blasone, M., Evans, T. S., Steer, D. A. and Vitiello, G., (1999). *J. Phys.* **A32**, p. 1185.

[91] Blasone, M., Graziano, E., Pashaev, O. K. and Vitiello, G., (1996). *Annals Phys.* **252**, p. 115.

[92] Blasone, M., Gui, Y.-X. and Vendrell, F., (1999) arXiv: gr-qc/9908024.

[93] Blasone, M., Henning, P. A. and Vitiello, G., (1999). *Phys. Lett.* **B451**, p. 140.

[94] Blasone, M., Henning, P. A. and Vitiello, G., (1999). *Phys. Lett.* **B466**, p. 262.

[95] Blasone, M. and Jizba, P., (2002). *Annals Phys.* **295**, p. 230.

[96] Blasone, M. and Jizba, P., (1999), p. 585 in *Basics and Highlights in Fundamental Physics*, A. Zichichi (Ed.) (World Scientific, Singapore).

[97] Blasone, M. and Jizba, P., (2004). *Annals Phys.* **312**, p. 354.

[98] Blasone, M., Jizba, P. and Kleinert, H., (2005). *Phys. Rev.* **A71**, p. 052507.

[99] Blasone, M., Jizba, P. and Kleinert, H., (2005). *Annals Phys.* **320**, p. 468.

[100] Blasone, M., Jizba, P., Scardigli, F. and Vitiello, G., (2009). *Phys. Lett.* **A373**, p. 4106.

[101] Blasone, M., Jizba, P. and Vitiello, G., (2001). *Phys. Lett.* **B517**, p. 471.

[102] Blasone, M., Jizba, P. and Vitiello, G., (2006), p. 221 in *Encyclopedia of Mathematical Physics*, J. P. Francoise, G. L. Naber and T. S. Tsun (Eds.) (Elsevier Acad. Press, Amsterdam).

[103] Blasone, M., Jizba, P. and Vitiello, G., (2001). *Phys. Lett.* **A287**, p. 205.

[104] Blasone, M., Magueijo, J. and Pires-Pacheco, P., (2005). *Europhys. Lett.* **70**, p. 600.

[105] Blasone, M. and Palmer, J., (2004). *Phys. Rev.* **D69**, p. 057301.

[106] Blasone, M., Pires Pacheco, P. and Tseung, H. W. C., (2003). *Phys. Rev.* **D67**, p. 073011.

[107] Blasone, M., Srivastava, Y. N., Vitiello, G. and Widom, A., (1998). *Annals Phys.* **267**, p. 61.

[108] Blasone, M. and Vitiello, G., (1995). *Annals Phys.* **244**, p. 283.

[109] Blasone, M. and Vitiello, G., (1999). *Phys. Rev.* **D60**, p. 111302.

[110] Blau, M. and Thompson, G., (1997). *Phys. Lett.* **B415**, p. 242.

[111] Blinov, L., (1983). *Russ. Chem. Rev.* **52**, p. 713.

[112] Bloch, F. and Nordsieck, A., (1937). *Phys. Rev.* **52**, p. 54.

[113] Bogoliubov, N. and Tyablikov, S., (1949). *Zh. Eksp. Teor. Fiz.* **19**, p. 256.

[114] Bogoliubov, N. N., Logunov, A. A. and Todorov, I. T., (1975). *Axiomatic*

Quantum Field Theory (Benjamin, New York).

[115] Bogoliubov, N. N. and Shirkov, D. V., (1959). *Introduction to the Theory of Quantized Fields* (Interscience, New York).

[116] Bogomolnyi, E. B., (1976). *Sov. J. Nucl. Phys.* **24**, p. 148.

[117] Bowick, M. J., Chandar, L., Schiff, E. A. and Srivastava, A. M., (1994). *Science* **263**, p. 943.

[118] Boyanovsky, D., (1984). *Phys. Rev.* **D29**, p. 743.

[119] Boyanovsky, D., de Vega, H. J., Holman, R., Prem Kumar, S. and Pisarski, R. D., (1998). *Phys. Rev.* **D57**, p. 3653.

[120] Braaten, E. and Pisarski, R., (1990). *Nucl. Phys.* **B337**, p. 569.

[121] Braitenberg, V. and Schüz, A., (1991). *Anatomy of the Cortex: Statistics and Geometry* (Springer-Verlag, Berlin).

[122] Brandenberger, R. H. and Martin, J., (2001). *Mod. Phys. Lett.* **A16**, p. 999.

[123] Bratteli, O. and Robinson, D. W., (1979). *Operator Algebras and Quantum Statistical Mechanics* (Springer-Verlag, Berlin).

[124] Buccella, F. and Ruegg, H., (1982). *Nuovo Cim.* **67A**, p. 61.

[125] Buccella, F., Ruegg, H. and Savoy, C. A., (1980). *Phys. Lett.* **B94**, p. 497.

[126] Bullough, R. K. and Caudrey, P. J. (Eds.)., (1980). *Solitons* (Springer-Verlag, Berlin).

[127] Bunde, A. and Havlin, S. (Eds.)., (1995). *Fractals in Science* (Springer-Verlag, Berlin).

[128] Cabrera, B., (1982). *Phys. Rev. Lett.* **48**, p. 1378.

[129] Cahill, K. E., (1974). *Phys. Lett.* **B53**, p. 174.

[130] Caldirola, P., (1941). *Nuovo Cim.* **3**, p. 393.

[131] Callan, C., Dashen, R. and Gross, D., (1976). *Phys. Lett.* **B63**, p. 334.

[132] Callan, C. and Gross, D., (1975). *Nucl. Phys.* **B93**, p. 29.

[133] Callan Jr., C. G., Coleman, S., Wess, J. and Zumino, B., (1969). *Phys. Rev.* **177**, p. 2247.

[134] Calvanese, R. and Vitiello, G., (1984). *Phys. Lett.* **A100**, p. 161.

[135] Campbell, D. K. and Bishop, A. R., (1981). *Phys. Rev.* **B24**, p. 4859.

[136] Campbell, D. K. and Bishop, A. R., (1982). *Nucl. Phys.* **B200**, p. 297.

[137] Cangemi, D., Jackiw, R. and Zwiebach, B., (1996). *Annals Phys.* **245**, p. 408.

[138] Capolupo, A., Capozziello, S. and Vitiello, G., (2007). *Phys. Lett.* **A363**, p. 53.

[139] Capolupo, A., Capozziello, S. and Vitiello, G., (2008). *Int. J. Mod. Phys.* **A23**, p. 4979.

[140] Capolupo, A., Capozziello, S. and Vitiello, G., (2009). *Phys. Lett.* **A373**, p. 601.

[141] Celeghini, E., De Martino, S., De Siena, S., Iorio, A., Rasetti, M. and Vitiello, G., (1998). *Phys. Lett.* **A244**, p. 455.

[142] Celeghini, E., De Martino, S., De Siena, S., Rasetti, M. and Vitiello, G., (1993). *Mod. Phys. Lett.* **B7**, p. 1321.

[143] Celeghini, E., De Martino, S., De Siena, S., Rasetti, M. and Vitiello, G., (1995). *Annals Phys.* **241**, p. 50.

[144] Celeghini, E., Giachetti, R., Sorace, E. and Tarlini, M., (1990). *J. Math. Phys.* **31**, p. 2548.

[145] Celeghini, E., Giachetti, R., Sorace, E. and Tarlini, M., (1991). *J. Math. Phys.* **32**, p. 1155.

[146] Celeghini, E., Graziano, E. and Vitiello, G., (1990). *Phys. Lett.* **A145**, p. 1.

[147] Celeghini, E., Graziano, E., Vitiello, G. and Nakamura, K., (1992). *Phys. Lett.* **B285**, p. 98.

[148] Celeghini, E., Graziano, E., Vitiello, G. and Nakamura, K., (1993). *Phys. Lett.* **B304**, p. 121.

[149] Celeghini, E., Magnollay, P., Tarlini, M. and Vitiello, G., (1985). *Phys. Lett.* **B162**, p. 133.

[150] Celeghini, E., Palev, T. D. and Tarlini, M., (1991). *Mod. Phys. Lett.* **B5**, p. 187.

[151] Celeghini, E., Rasetti, M., Tarlini, M. and Vitiello, G., (1989). *Mod. Phys. Lett.* **B3**, p. 1213.

[152] Celeghini, E., Rasetti, M. and Vitiello, G., (1991). *Phys. Rev. Lett.* **66**, p. 2056.

[153] Celeghini, E., Rasetti, M. and Vitiello, G., (1992). *Annals Phys.* **215**, p. 156.

[154] Celeghini, E., Tarlini, M. and Vitiello, G., (1984). *Nuovo Cim.* **84A**, p. 19.

[155] Chari, V. and Pressley, A., (1994). *A Guide to Quantum Groups* (Cambridge University Press, Cambridge, U.K.).

[156] Cheng, T. P. and Li, L. F., (1984). *Gauge Theory of Elementary Particle Physics* (Clarendon Press, Oxford, U.K.).

[157] Christiansen, P. L., Bang, O., Pagano, S. and Vitiello, G., (1992). *Nanobiol.* **1**, p. 229.

[158] Christiansen, P. L., Pagano, S. and Vitiello, G., (1991). *Phys. Lett.* **A154**, p. 381.

[159] Chuang, I., Yurke, B., Durrer, R. and Turok, N., (1991). *Science* **251**, p. 1336.

[160] Cini, M. and Serva, M., (1986). *J. Phys.* **A19**, p. 1163.

[161] Clark, T. D., Prance, H., Prance, R. J. and Spiller, T. P., (1991). *Macroscopic Quantum Phenomena* (World Scientific, Singapore).

[162] Coleman, S., (1965). *Phys. Lett.* **19**, p. 144.

[163] Coleman, S., (1985). *Aspects of Symmetry* (Cambridge University Press, Cambridge, U.K.).

[164] Coleman, S. R., (1973). *Commun. Math. Phys.* **31**, p. 259.

[165] Coleman, S. R., (1975). *Phys. Rev.* **D11**, p. 2088.

[166] Coleman, S. R., (1985). *Phys. Rev.* **B262**, p. 2929.

[167] Coleman, S. R. and Weinberg, E., (1973). *Phys. Rev.* **D7**, p. 1888.

[168] Coleman, S. R., Wess, J. and Zumino, B., (1969). *Phys. Rev.* **177**, p. 2239.

[169] Collins, J. C., (1984). *Renormalization* (Cambridge University Press, Cambridge, U.K.).

[170] Connes, A., Douglas, M. R. and Schwarz, A. S., (1998). *JHEP* **02**, p. 003.

[171] Craig, R. A., (1968). *J. Math. Phys.* **9**, p. 605.

[172] D'Ariano, G. and Rasetti, M., (1985). *Phys. Lett.* **A107**, p. 291.

[173] D'Ariano, G., Rasetti, M. and Vadacchino, M., (1985). *J. Phys.* **A18**, p. 1295.

[174] Das, A., (1997). *Finite Temperature Field Theory* (World Scientific, Singapore).

[175] Dashen, R. F., Hasslacher, B. and Neveu, A., (1974). *Phys. Rev.* **D10**, p. 4114.

[176] Dashen, R. F., Hasslacher, B. and Neveu, A., (1975). *Phys. Rev.* **D11**, p. 3424.

[177] Davydov, A. S., (1979). *Physica Scripta* **20**, p. 387.

[178] Davydov, A. S., (1982). *Biology and Quantum Mechanics* (Pergamon Press, Oxford, U.K.).

[179] Davydov, A. S., (1987). *Solitons in Molecular Systems* (Reidel, Dordrecht).

[180] Davydov, A. S. and Kislukha, N. I., (1976). *Sov. Phys. JEPT* **44**, p. 571.

[181] de Alfaro, V., Fubini, S. and Furlan, G., (1976). *Phys. Lett.* **B65**, p. 163.

[182] De Concini, C. and Vitiello, G., (1976). *Nucl. Phys.* **B116**, p. 141.

[183] De Concini, C. and Vitiello, G., (1977). *Phys. Lett.* **B70**, p. 355.

[184] De Concini, C. and Vitiello, G., (1979). *Nuovo Cim.* **A51**, p. 358.

[185] De Filippo, S. and Vitiello, G., (1977). *Nuovo Cim. Lett.* **19**, p. 92.

[186] de Gennes, P. G. and Prost, J., (1993). *The Physics of Liquid Crystals* (Clarendon Press, Oxford, U.K.).

[187] De Jongh, L. J. and Miedema, A. R., (1974). *Adv. Phys.* **23**, p. 1.

[188] De Martino, S., De Siena, S., Graziano, E. and Vitiello, G., (1993). *Phys. Lett.* **B305**, p. 119.

[189] De Martino, S., De Siena, S. and Vitiello, G., (1996). *Int. J. Mod. Phys.* **B10**, p. 1615.

[190] de Montigny, M. and Patera, J., (1991). *J. Phys.* **A24**, p. 525.

[191] de Vega, H. and Schaposnik, F., (1976). *Phys. Rev.* **D14**, p. 1100.

[192] Dekker, H., (1981). *Phys. Rept.* **80**, p. 1.

[193] Del Giudice, E., Doglia, S., Milani, M. and Vitiello, G., (1985). *Nucl. Phys.* **B251**, p. 375.

[194] Del Giudice, E., Doglia, S., Milani, M. and Vitiello, G., (1986). *Nucl. Phys.* **B275**, p. 185.

[195] Del Giudice, E., Mańka, R., Milani, M. and Vitiello, G., (1988). *Phys. Lett.* **B206**, p. 661.

[196] Del Giudice, E., Preparata, G. and Vitiello, G., (1988). *Phys. Rev. Lett.* **61**, p. 1085.

[197] Del Giudice, E. and Vitiello, G., (2006). *Phys. Rev.* **A74**, p. 022105.

[198] Delepine, D., Gonzalez Felipe, R. and Weyers, J., (1998). *Phys. Lett.* **B419**, p. 296.

[199] Dell'Antonio, G. and Umezawa, H., (1964) Unpublished, Naples.

[200] Dell'Antonio, G. F., Frishman, Y. and Zwanziger, D., (1972). *Phys. Rev.* **D6**, p. 988.

[201] Derrick, G. H., (1964). *J. Math. Phys.* **5**, p. 1252.

[202] Dirac, P., (1966). *Lectures on Quantum Field Theory* (Belfer Graduate School of Science, Monograph series, n.3, Yeshiva Univ., New York).

[203] Dodelson, S. and Widrow, L., (1990). *Phys. Rev. Lett.* **64**, p. 340.

[204] Dolan, L. and Jackiw, R., (1974). *Phys. Rev.* **D9**, p. 3320.

[205] Dongpei, Z., (1987). *J. Phys.* **A20**, p. 4331.

[206] Dorey, N., Hollowood, T. J., Khoze, V. V. and Mattis, M. P., (2002). *Phys. Rept.* **371**, p. 231.

[207] Drinfeld, V. G., (1986), p. 798 in *Proc. ICM Berkeley*, A. M. Gleason (Ed.), AMS, (Providence, R.I.).

[208] Dunne, G. V., Jackiw, R. and Trugenberger, C. A., (1990). *Phys. Rev.* **D41**, p. 661.

[209] Dür, W., Vidal, G. and Cirac, J. I., (2000). *Phys. Rev.* **A62**, p. 062314.

[210] Dyson, F. J., (1956). *Phys. Rev.* **102**, p. 1217.

[211] Dyson, F. J., (1949). *Phys. Rev.* **75**, p. 1736.

[212] Dzyaloshinkii, I. E., (1962). *Sov. Phys. JETP* **15**, p. 232.

[213] Eguchi, T., Gilkey, P. B. and Hanson, A. J., (1980). *Phys. Rept.* **66**, p. 213.

[214] Eilbeck, J., Lomdahl, P. and Scott, A., (1985). *Physica* **D 16**, p. 318.

[215] Einstein, A., Podolsky, B. and Rosen, N., (1935). *Phys. Rev.* **47**, p. 777.

[216] Ellis, J. R., Mavromatos, N. E. and Nanopoulos, D. V., (2008). *Phys. Lett.* **B665**, p. 412.

[217] Elze, H.-T., (2005). *Braz. J. Phys.* **352A**, p. 343.

[218] Elze, H.-T., (2006). *J. Phys. Conf. Ser.* **33**, pp. 399–404.

[219] Elze, H.-T. and Schipper, O., (2002). *Phys. Rev.* **D66**, p. 044020.

[220] Englert, F. and Brout, R., (1964). *Phys. Rev. Lett.* **13**, p. 321.

[221] Enomoto, H., Okumura, M. and Yamanaka, Y., (2006). *Annals Phys.* **321**, p. 1892.

[222] Enqvist, K. and McDonald, J., (1998). *Phys. Lett.* **B425**, p. 309.

[223] Evans, T., (1990). *Phys. Lett.* **B249**, p. 286.

[224] Evans, T. S., (1994), p. 77 in *Thermal Field Theories and their Applications* F.C. Khanna, R. Kobes, G. Kunstatter and H. Umezawa (Eds.) (World Scientific, Singapore).

[225] Evans, T. S. and Pearson, A. C., (1995). *Phys. Rev.* **D52**, p. 4652.

[226] Evans, T. S. and Steer, D. A., (1996). *Nucl. Phys.* **B474**, p. 481.

[227] Ezawa, H., Arimitsu, T. and Hashimoto, Y. (Eds.)., (1991). *Thermal Field Theories* (Elsevier Sc. Publ., Amsterdam).

[228] Ezawa, H., Tomozawa, Y. and Umezawa, H., (1957). *Nuovo Cim.* **5**, p. 810.

[229] Ezawa, Z. and Pottinger, D., (1977). *Phys. Lett.* **B66**, p. 445.

[230] Fabri, E. and Picasso, L. E., (1966). *Phys. Rev. Lett.* **16**, p. 408.

[231] Faddeev, L. D. and Jackiw, R., (1988). *Phys. Rev. Lett.* **60**, p. 1692.

[232] Faddeev, L. D. and Korepin, V. E., (1978). *Phys. Rep.* **42**, p. 1.

[233] Feshbach, H. and Tickochinski, Y., (1977). *Trans. NY Acad. Sci.* **38**, p. 44.

[234] Feynman, R. P., (1947). *Rev. Mod. Phys.* **20**, p. 367.

[235] Feynman, R. P., (1948). *Phys. Rev.* **74**, p. 1430.

[236] Feynman, R. P., (1950). *Phys. Rev.* **80**, p. 440.

[237] Feynman, R. P., (1972). *Statistical Mechanics* (W. A. Benjamin Publ. Co., Reading, MA).

[238] Feynman, R. P. and Hibbs, A. R., (1965). *Quantum Mechanics and Path Integrals* (McGraw-Hill, New York).

[239] Feynman, R. P. and Vernon, F. L., (1963). *Annals Phys.* **24**, p. 118.

[240] Fialowski, A., De Montigny, M., Novikov, S. and Schlichenmaier, M. (Eds.)., (2006). *Group Contraction in Quantum Field Theory* (Oberwolfach Reports, Rep. N.3/2006, European Math. Soc. Publ. House).

[241] Fixen, D. *et al.*, (1996). *Astrophys. J.* **437**, p. 576.

[242] Fock, V. A., (1928). *Z. Phys.* **49**, p. 339.

[243] Fradkin, E. and Stone, M., (1988). *Phys. Rev.* **B38**, p. 7215.

[244] Freeman, W., Burke, B. and Holmes, M., (2003). *Human Brain Mapping* **19**, p. 248.

[245] Freeman, W., Burke, B., Holmes, M. and Vanhatalo, S., (2003). *Clin. Neurophysiol.* **114**, p. 1055.

[246] Freeman, W. J., (2004). *Clin. Neurophysiol.* **115**, p. 2077.

[247] Freeman, W. J., (2004). *Clin. Neurophysiol.* **115**, p. 2089.

[248] Freeman, W. J., (2005). *J. Integrative Neuroscience* **4**, p. 407.

[249] Freeman, W. J., (2005). *Clin. Neurophysiol.* **116**, p. 1117.

[250] Freeman, W. J., (2006). *Clin. Neurophysiol.* **117**, p. 572.

[251] Freeman, W. J., (2010) Displays of movies of null spikes resembling tornados and vortices resembling hurricanes can be downloaded from: http://sulcus.berkeley.edu,.

[252] Freeman, W. J., (2001). *How Brains Make up their Minds* (Columbia University Press, New York).

[253] Freeman, W. J., (1975). *Mass Action in the Nervous System* (Academic Press, New York).

[254] Freeman, W. J., (2000). *Neurodynamics. An Exploration of Mesoscopic Brain Dynamics* (Springer-Verlag, London, U.K.).

[255] Freeman, W. J. and Baird, B., (1987). *Behavioral Neuroscience* **101**, p. 393.

[256] Freeman, W. J., Gaál, G. and Jornten, R., (2003). *Int. J. Bifurc. Chaos* **13**, p. 2845.

[257] Freeman, W. J., Kozma, R., Bollobás, B. and Riordan, O., (2008), Chap. 7 in *Handbook of Large-Scale Random Networks*. B. Bollobás, R. Kozma and D. Mikls (Eds.) (New York: Springer).

[258] Freeman, W. J., O'Nuillain, S. and Rodriguez, J., (2008). *J. Integr. Neurosci.* **7**, p. 337.

[259] Freeman, W. J. and Rogers, L., (2003). *Int. J. Bifurc. Chaos* **13**, p. 2867.

[260] Freeman, W. J. and Vitiello, G., (2006). *Phys. of Life Reviews* **3**, p. 93.

[261] Freeman, W. J. and Vitiello, G., (2008). *J. Phys.* **A41**, p. 304042.

[262] Freeman, W. J. and Vitiello, G., (2010). *Int. J. Mod. Phys.* **B24**, p. 3269.

[263] Freeman, W. J. and Vitiello, G., (2009). *J. Phys. Conf. Series* **174**, p. 012011.

[264] Freeman, W. J. and Zhai, J., (2008). *Cognitive Neurodynamics* **3**, p. 97.

[265] Frenkel, J. and Tylor, J., (1990). *Nucl. Phys.* **B334**, p. 199.

[266] Friedberg, R., Lee, T. D. and Sirlin, A., (1976). *Phys. Rev.* **D13**, p. 2739.

[267] Friedberg, R., Lee, T. D. and Sirlin, A., (1976). *Nucl. Phys.* **B115**, p. 32.

[268] Frieman, J., Gelmini, G., Gleiser, M. and Kolb, E., (1988). *Phys. Rev. Lett.* **60**, p. 2101.

[269] Fubini, S., (1976). *Nuovo Cim.* **34A**, p. 521.

[270] Fujii, K., Habe, C. and Yabuki, T., (1999). *Phys. Rev.* **D59**, p. 113003.
[271] Garbaczewski, P., (1985). *Classical and Quantum Field Theory of Exactly Soluble Nonlinear Systems* (World Scientific, Singapore).
[272] Gärding, L. and Wightman, A. S., (1954). *Proc. Nat. Acad. Sc.* **40**, p. 612.
[273] Gelfand, I. M. and Yaglom, A. M., (1960). *J. Math. Phys.* **1**, p. 48.
[274] Gell-Mann, M. and Lévy, M., (1960). *Nuovo Cim.* **16**, p. 705.
[275] Gell-Mann, M. and Low, F., (1951). *Phys. Rev.* **84**, p. 350.
[276] Genovese, M., (2005). *Phys. Rept.* **413**, p. 319.
[277] Georgi, H. and Glashow, S. L., (1972). *Phys. Rev. Lett.* **28**, p. 1494.
[278] Georgi, H. and Glashow, S. L., (1974). *Phys. Rev. Lett.* **32**, p. 438.
[279] Gerry, C. C. and Knight, P. L., (2005). *Introductory Quantum Optics* (Cambridge University Press, Cambridge, U.K.).
[280] Gerry, C. C. and Silverman, S., (1982). *J. Math. Phys.* **23**, p. 1995.
[281] Gervais, J.-L. and Sakita, B., (1975). *Phys. Rev.* **D11**, p. 2943.
[282] Ginzburg, V. and Landau, L., (1950). *Zh. Eksp. Teor. Fiz.* **20**, p. 64.
[283] Giulini, D. *et al.*, (1996). *Decoherence and the Appearance of a Classical World in Quantum Theory* (Springer-Verlag, Berlin).
[284] Giunti, C. and Kim, C. W., (2007). *Fundamentals of Neutrino Physics and Astrophysics* (Oxford University Press, Oxford, U.K.).
[285] Glauber, R. J., (1963). *Phys. Rev.* **131**, p. 2766.
[286] Glimm, J. and Jaffe, A. M., (1987). *Quantum Physics. A Functional Integral Point of View* (Springer-Verlag, New York).
[287] Globus, G., Pribram, K. H. and Vitiello, G. (Eds.)., (2004). *Brain and Being. At the Boundary between Science, Philosophy, Language and Arts.* (John Benjamins Publ. Co., Amsterdam).
[288] Goldberger, M. W. and Treimann, S. B., (1958). *Phys. Rev.* **110**, p. 1178.
[289] Goldstone, J., (1961). *Nuovo Cim.* **19**, p. 154.
[290] Goldstone, J. and Jackiw, R., (1975). *Phys. Rev.* **D11**, p. 1486.
[291] Goldstone, J., Salam, A. and Weinberg, S., (1962). *Phys. Rev.* **127**, p. 965.
[292] Golo, V. L. and Perelomov, A. M., (1978). *Phys. Lett.* **B79**, p. 112.
[293] Gomez Nicola, A. and Steer, D. A., (1999). *Nucl. Phys.* **B549**, p. 409.
[294] Gor'kov, L. P., (1959). *Zh. Eksp. Theor. Fiz.* **36**, p. 1918, [Sov. Phys. JEPT **9**, 1364 (1959)].
[295] Gor'kov, L. P., (1959). *Zh. Eksp. Theor. Fiz.* **37**, p. 1407, [Sov. Phys. JEPT **10**, 998 (1960)].
[296] Green, M. B. and Gross, D., (1985). *Proceedings of the Workshop on Unified String Theories, Santa Barbara 1985* (World Scientific, Singapore).
[297] Greenfield, S., (1997). *Communication and Cognition* **30**, p. 285.
[298] Greenfield, S., (1997). *The Brain: a Guided Tour* (Freeman, New York).
[299] Grosche, C., Pogosyan, G. S. and Sissakian, A. N., (1995). *Fortschr. Phys.* **43**, p. 6.
[300] Guerra, F., Rosen, L. and Simon, B., (1975). *Ann. Math.* **101**, p. 111.
[301] Guerra, F. and Ruggiero, P., (1973). *Phys. Rev. Lett.* **31**, p. 1022.
[302] Gui, Y.-X., (1990). *Phys. Rev.* **D42**, p. 1988.
[303] Gui, Y.-X., (1992). *Phys. Rev.* **D46**, p. 1869.
[304] Gui, Y.-X., (1992). *Phys. Rev.* **D45**, p. 697.

[305] Guralnik, G. S., Hagen, C. R. and Kibble, T. W. B., (1964). *Phys. Rev. Lett.* **13**, p. 585.

[306] Guth, A. and Weinberg, E., (1983). *Nucl. Phys.* **B 212**, p. 321.

[307] Haag, R., (1955). *Kgl. Danske Vidensk. Selsk. Mat.-Fys. Medd.* **29**, p. 12.

[308] Haag, R., (1992). *Local Quantum Physics: Fields, Particles, Algebras* (Springer-Verlag, Berlin).

[309] Haag, R., Hugenholtz, N. W. and Winnihk, M., (1977). *Comm. Math. Phys.* **105**, p. 215.

[310] Haider, B., Duque, A., Hasenstaub, A. R. and McCormick, D. A., (2006). *J. Neurosci.* **26**, p. 4535.

[311] Haken, H., (1984). *Laser Theory* (Springer-Verlag, Berlin).

[312] Hameroff, S. and Penrose, R., (1996). *J. Conscious. Stud.* **3**, p. 36.

[313] Hannabuss, K. C. and Latimer, D. C., (2000). *J. Phys.* **A33**, pp. 1369–1373.

[314] Hardman, I. and Umezawa, H., (1990). *Annals Phys.* **203**, p. 173.

[315] Hawking, S. W., (1975). *Commun. Math. Phys.* **43**, p. 199.

[316] Hawking, S. W. and Penrose, R., (1996). *Sci. Am.* **275**, p. 44.

[317] Heitler, W., (1954). *The Quantum Theory of Radiation* (Clarendon Press, Oxford, U.K.).

[318] Hellmund, M., Kripfganz, J. and Schmidt, M. G., (1994). *Phys. Rev.* **D50**, p. 7650.

[319] Henning, P. A., (1995). *Phys. Rept.* **253**, p. 235.

[320] Henning, P. A., Graf, M. and Matthaus, F., (1992). *Physica* **A182**, p. 489.

[321] Herman, R., (1966). *Lie Groups for Physicists* (Benjamin, New York).

[322] Herring, C., (1966) *Magnetism*, Vol. 4, G. T. Rado and H. Suhl (Eds.) (Academic Press, New York).

[323] Higgs, P. W., (1964). *Phys. Lett.* **12**, p. 132.

[324] Higgs, P. W., (1964). *Phys. Rev. Lett.* **13**, p. 508.

[325] Higgs, P. W., (1966). *Phys. Rev.* **145**, p. 1156.

[326] Hilborn, R., (1994). *Chaos and Nonlinear Dynamics* (Oxford University Press, Oxford, U.K.).

[327] Hindmarsh, M. and Rajantie, A., (2000). *Phys. Rev. Lett.* **85**, p. 4660.

[328] Hirota, R., (1972). *J. Phys. Soc. Japan* **33**, p. 1459.

[329] Hobart, R., (1963). *Proc. Phys. Soc. London* **82**, p. 201.

[330] Hohenberg, P. C., (1967). *Phys. Rev.* **158**, p. 383.

[331] Holstein, T. and Primakoff, H., (1940). *Phys. Rev.* **58**, p. 1098.

[332] Hopfield, J., (1982). *Proc. of Nat. Acad. Sc. USA* **79**, p. 2554.

[333] Hu, B.-L. B. and Calzetta, E. A., (2008). *Nonequilibrium Quantum Field Theory* (Cambridge University Press, Cambridge, U.K.).

[334] Huth, G. C., Gutmann, F. and Vitiello, G., (1989). *Phys. Lett.* **A140**, p. 339.

[335] Hwa, R. C. and Ferree, T., (2002). *Phys. Rev.* **E66**, p. 021901.

[336] Inönü, E. and Wigner, E. P., (1953). *Proc. Nat. Acad. Sci.* **39**, p. 510.

[337] Iorio, A., (2009). *Alternative Symmetries in QFT and Gravity. Habilitation Thesis* (Charles University, Prague, Cz. Rep.).

[338] Iorio, A., Lambiase, G. and Vitiello, G., (2001). *Annals Phys.* **294**, p. 234.

[339] Iorio, A., Lambiase, G. and Vitiello, G., (2004). *Annals Phys.* **309**, p. 151.

[340] Iorio, A. and Sykora, T., (2002). *Int. J. Mod. Phys.* **A17**, p. 2369.
[341] Iorio, A. and Vitiello, G., (1994). *Mod. Phys. Lett.* **B8**, p. 269.
[342] Iorio, A. and Vitiello, G., (1995). *Annals Phys.* **241**, p. 496.
[343] Itzykson, C. and Zuber, J. B., (1980). *Quantum Field Theory* (McGraw-Hill, New York).
[344] Iyanaga, S. and Kawada, Y. (Eds.)., (1980). *Encyclopedic Dictionary of Mathematics* (MIT Press, Cambridge, MA).
[345] Jackiw, R., (1986) , p. 81in *Current Algebra and Anomalies*, S. B. Treiman, E. Witten, R. Jackiw and B. Zumino (Eds.) (World Scientific, Singapore).
[346] Jackiw, R., (1977). *Rev. Mod. Phys.* **49**, p. 681.
[347] Jackiw, R. and Rebbi, C., (1976). *Phys. Rev.* **D13**, p. 3398.
[348] Jackiw, R. and Rebbi, C., (1976). *Phys. Rev.* **D14**, p. 517.
[349] Jackiw, R. and Rossi, P., (1981). *Nucl. Phys.* **B190**, p. 681.
[350] Jackiw, R. and Schrieffer, J. R., (1981). *Nucl. Phys.* **B190**, p. 253.
[351] Jauch, J., (1968). *Foundation of Quantum Mechanics* (Addison-Wesley Publ. Co., Reading, MA).
[352] Jauslin, H. and Schneider, T., (1982). *Phys. Rev.* **B26**, p. 5153.
[353] Jaynes, E. T. and Cumming, F. W., (1963). *Proc. IEEE* **51**, p. 89.
[354] Jeffreys, H. and Jeffreys, B. S., (1999). *Methods of Mathematical Phyisics* (Cambridge University Press, Cambridge, U.K.).
[355] Ji, C.-R. and Mishchenko, Y., (2002). *Phys. Rev.* **D65**, p. 096015.
[356] Jibu, M., Pribram, K. H. and Yasue, K., (1996). *Int. J. Mod. Phys.* **B10**, p. 1735.
[357] Jibu, M. and Yasue, K., (1992), p. 797 in *Cybernetics and System Research*, R. Trappl (Ed.) (World Scientific).
[358] Jibu, M. and Yasue, K., (1995). *Quantum Brain Dynamics and Consciousness. An Introduction* (John Benjamins Pub. Co., Amsterdam).
[359] Jimbo, M., (1989). *Int. J. Mod. Phys.* **A4**, p. 3759.
[360] Jona-Lasinio, G., (1964). *Nuovo Cim.* **34**, p. 1790.
[361] Joos, E. and Zeh, H. D., (1985). *Z. Phys.* **B59**, p. 223.
[362] Joos, H. and Weimar, E., (1975). *CERN preprint* **TH-2040**.
[363] Kanai, E., (1948). *Prog. Theor. Phys.* **3**, p. 440.
[364] Karra, G. and Rivers, R. J., (1998). *Phys. Rev. Lett.* **81**, p. 3707.
[365] Kastler, D., Pool, J. C. T. and Poulsen, E. T., (1969). *Commun. Math. Phys.* **12**, p. 175.
[366] Kaushal, R. S. and Korsch, H. J., (2000). *Phys. Lett.* **A276**, p. 47.
[367] Keldysh, L. V., (1964). *Sov. Phys. JEPT* **20**, p. 1018.
[368] Kello, C. T., Brown, G. D. and Ferrer-i Cancho, R. *et al.*, (2010). *Trends in Cognitive Sciences* **14**, p. 223.
[369] Khanna, F. C., Kobes, R., Kunstatter, G. and Umezawa, H. (Eds.), (1994). *Thermal Field Theories and Their Applications* (World Scientific, Singapore).
[370] Kibble, T. W. B., (1967). *Phys. Rev.* **155**, p. 1554.
[371] Kibble, T. W. B., (1976). *J. Phys.* **A9**, p. 1387.
[372] Kibble, T. W. B., (1980). *Phys. Rept.* **67**, p. 183.
[373] Kibble, T. W. B., (2000), p. 7 in *Topological Defects and the Nonequilibrium*

Dynamics of Symmetry Breaking Phase Transitions, Y. M. Bunkov and H. Godfrin (Eds.), (Kluwer, Dordrecht).

[374] Kibble, T. W. B., (2003), p. 3 in *Patterns of Symmetry Breaking*, H. Arodz, J. Dziarmaga and W. H. Zurek (Eds.) (Kluwer, London).

[375] Kibble, T. W. B., (2007). *Phys. Today* **60**, p. 47.

[376] Kirzhnts, D., (1972). *JEPT Lett.* **15**, p. 529.

[377] Klaiber, B., (1968) in *Lectures in Theoretical Physics*, Boulder, 1967, A. Barut and W. Brittin (Eds.) (Gordon and Breach, New York).

[378] Klauder, J. R., (1979). *Phys. Rev.* **D19**, p. 2349.

[379] Klauder, J. R. and Skagerstam, B., (1985). *Coherent States* (World Scientific, Singapore).

[380] Klauder, J. R. and Sudarshan, E. C., (1968). *Fundamentals of Quantum Optics* (Benjamin, New York).

[381] Kleinert, H., (1989). *Gauge Fields in Condensed Matter. Vol. 1: Superflow and Vortex Lines. Disorder Fields, Phase Transitions* (World Scientific, Singapore).

[382] Kleinert, H., (1989). *Gauge Fields in Condensed Matter. Vol. 2: Stresses and Defects. Differential Geometry, Crystal Melting* (World Scientific, Singapore).

[383] Kleinert, H., (2008). *Multivalued Fields. In Condensed Matter, Electromagnetism, and Gravitation* (World Scientific, Singapore).

[384] Kleinert, H., (2004). *Path Integrals in Quantum Mechanics, Statistics, Polymer Physics, and Financial Markets* (World Scientific, Singapore).

[385] Klinkhamer, F. and Manton, N., (1984). *Phys. Rev.* **D 30**, p. 2212.

[386] Klyachko, A. A., Öztop, B. and Shumovsky, A. S., (2007). *Phys. Rev.* **A75**, p. 032315.

[387] Knapp, A. W., (1986). *Representation Theory of Semisimple Groups* (Princeton University Press, Princeton).

[388] Kobayashi, S. and Nomizu, K., (1996). *Foundations of Differential Geometry, Vol. I & II.* (Willey, London).

[389] Kostelecky, V. A. and Mewes, M., (2004). *Phys. Rev.* **D69**.

[390] Kostyakov, I. V., Gromov, N. A. and Kuratov, V. V., (2001). hep-th/0102052.

[391] Kubo, R., (1957). *J. Phys. Soc. Japan* **12**, p. 570.

[392] Kugo, T. and Ojima, I., (1979). *Prog. Theor. Phys. Suppl.* **66**, p. 1.

[393] Kulish, P. P. and Reshetikhin, N. Y., (1989). *Lett. Math. Phys.* **18**, p. 143.

[394] Kundu, A., Makhankov, V. G. and Pashaev, O. K., (1982) Dubna preprint JINR E 17-82-677.

[395] Kunk, M. and Lavrentovich, O., (1982). *Mol. Cryst. Liq. Cryst. Lett.* **72**, p. 239.

[396] Kuratsuji, H. and Suzuki, T., (1980). *J. Math. Phys.* **21**, p. 472.

[397] Kurcz, A., Capolupo, A., Beige, A., Del Giudice, E. and Vitiello, G., (2010). *Phys. Rev.* **A81**, p. 063821.

[398] Kurcz, A., Capolupo, A., Beige, A., Del Giudice, E. and Vitiello, G., (2010). *Phys. Lett.* **A374**, p. 3726.

[399] Kusenko, A., (1997). *Phys. Lett.* **B405**, p. 108.

[400] Kusenko, A. and Shaposhnikov, M., (1998). *Phys. Lett.* **B418**, p. 46.
[401] Küster, F. W. and Thiel, A., (1988). *Tabelle per le Analisi Chimiche e Chimico-Fisiche* (Hoepli, Milano).
[402] Landau, L. D. and Lifshitz, E. M., (1977). *Quantum Mechanics* (Pergamon Press, Oxford, U.K.).
[403] Landau, L. D. and Lifshitz, E. M., (1996). *Statistical Mechanics, Part 1* (Pergamon Press, Oxford, U.K.).
[404] Landau, L. D. and Lifshitz, E. M., (1981). *Statistical Mechanics, Part 2* (Pergamon Press, Oxford, U.K.).
[405] Landsman, N., (1988). *Phys. Rev. Lett.* **60**, p. 1909.
[406] Landsman, N., (1989). *Phys. Lett.* **B232**, p. 240.
[407] Landsman, N. P., (1989). *Annals Phys.* **186**, p. 141.
[408] Landsman, N. P. and van Weert, C. G., (1987). *Phys. Rep.* **145**, p. 141.
[409] Lashley, K., (1948). *The Mechanism of Vision, XVIII, Effects of Destroying the Visual "Associative Areas" of the Monkey* (Journal Press, Provincetown, MA).
[410] Lawrie, I., (1989). *J. Phys.* **A21**, p. L823.
[411] Le Bellac, M., (1996). *Thermal Field Theory* (Cambridge University Press, Cambridge, U.K.).
[412] Le Bellac, M. and Mabilat, H., (1996). *Phys. Lett.* **B381**, p. 262.
[413] Lee, A. T. *et al.*, (2001). *Astrophys. J.* **561**, p. L1.
[414] Lee, T. D., (1976). *Phys. Rep.* **23**, p. 254.
[415] Lee, T. D. and Peng, Y., (1992). *Phys. Rep.* **221**, p. 251.
[416] Lehmann, H., (1954). *Nuovo Cim.* **11**, p. 342.
[417] Lehmann, H., (1958). *Nuovo Cim.* **10**, p. 579.
[418] Lehmann, H., Symanzik, K. and Zimmermann, W., (1955). *Nuovo Cim.* **1**, p. 1425.
[419] Lehmann, H., Symanzik, K. and Zimmermann, W., (1955). *Nuovo Cim.* **2**, p. 425.
[420] Lehmann, H., Symanzik, K. and Zimmermann, W., (1957). *Nuovo Cim.* **6**, p. 319.
[421] Leite Lopes, J., (1981). *Gauge Field Theories. An Introduction.* (Pergamon Press, Oxford, U.K.).
[422] Leplae, L., Mancini, F. and Umezawa, H., (1974). *Phys. Rep.* **10C**, p. 151.
[423] Leplae, L., Sen, R. N. and Umezawa, H., (1967). *Nuovo Cim.* **49**, p. 1.
[424] Leplae, L. and Umezawa, H., (1966). *Nuovo Cim.* **44**, p. 1.
[425] Linde, A. D., (1983). *Nucl. Phys.* **B216**, p. 421.
[426] Linkenkaer-Hansen, K., Nikouline, V. M., Palva, J. M. and Iimoniemi, R. J., (2001). *J Neurosci.* **15**, p. 1370.
[427] Lombardo, F. C. and Mazzitelli, F. D., (1996). *Phys. Rev.* **D53**, p. 2001.
[428] Lombardo, F. C., Mazzitelli, F. D. and Monteoliva, D., (2000). *Phys. Rev.* **D62**, p. 045016.
[429] Lombardo, F. C., Mazzitelli, F. D. and Rivers, R. J., (2001). *Phys. Lett.* **B523**, p. 317.
[430] Lombardo, F. C., Rivers, R. J. and Villar, P. I., (2007). *Phys. Lett.* **B648**, p. 64.

[431] Lomdahl, P. S., Olsen, O. H. and Christiansen, P. C., (1980). *Phys. Lett.* **A 78**, p. 125.

[432] Lurié, D., (1968). *Particles and Fields* (Interscience Publishers, New York).

[433] Macfarlane, A. J., (1989). *J. Phys.* **22**, p. 4581.

[434] Magueijo, J. and Smolin, L., (2002). *Phys. Rev. Lett.* **88**, p. 190403.

[435] Magueijo, J. and Smolin, L., (2003). *Phys. Rev.* **D67**, p. 044017.

[436] Mandelstam, S., (1975). *Phys. Rev.* **D11**, p. 3026.

[437] Manin, Y. I., (1988). *Quantum Groups and Non-Commutative Geometry* (Centre de Recherches Mathèmatiques, Montreal).

[438] Mańka, R., Kuczynski, J. and Vitiello, G., (1986). *Nucl. Phys.* **B276**, p. 533.

[439] Mańka, R. and Sladkowski, J., (1986). *Canad. J. Phys.* **B86**, p. 1026.

[440] Mańka, R. and Vitiello, G., (1990). *Annals Phys.* **199**, p. 61.

[441] Manton, N., (1983). *Phys. Rev.* **D 28**, p. 2019.

[442] Manton, N. S. and Sutcliffe, P., (2004). *Topological Solitons* (Cambridge University Press, Cambridge, U.K.).

[443] Marshak, R. E., (1993). *Conceptual Foundations of Modern Particle Physics* (World Scientific, Singapore).

[444] Martellini, M., Sodano, P. and Vitiello, G., (1978). *Nuovo Cim.* **A48**, p. 341.

[445] Martin, P. C. and Schwinger, J., (1959). *Phys. Rev.* **115**, p. 1342.

[446] Matsubara, T., (1955). *Progr. Theor. Phys.* **14**, p. 351.

[447] Matsumoto, H., Nakano, Y. and Umezawa, H., (1984). *Phys. Rev.* **D29**, p. 1116.

[448] Matsumoto, H., Ojima, I. and Umezawa, H., (1984). *Annals Phys.* **152**, p. 348.

[449] Matsumoto, H., Papastamatiou, N., Umezawa, H. and Vitiello, G., (1975). *Nucl. Phys.* **B97**, p. 61.

[450] Matsumoto, H., Papastamatiou, N. J. and Umezawa, H., (1974). *Nucl. Phys.* **B82**, p. 45.

[451] Matsumoto, H., Papastamatiou, N. J. and Umezawa, H., (1974). *Nucl. Phys.* **B68**, p. 236.

[452] Matsumoto, H., Papastamatiou, N. J. and Umezawa, H., (1975). *Nucl. Phys.* **B97**, p. 90.

[453] Matsumoto, H., Papastamatiou, N. J. and Umezawa, H., (1975). *Phys. Rev.* **D12**, p. 1836.

[454] Matsumoto, H., Papastamatiou, N. J. and Umezawa, H., (1976), p. 363 in *Functional and Probabilistic Methods in QFT.*, Vol.1, J. Lopuszanski and B. Jancewicz (Eds.), (Wrocław).

[455] Matsumoto, H., Papastamatiou, N. J. and Umezawa, H., (1976). *Phys. Rev.* **D13**, p. 1054.

[456] Matsumoto, H. and Sakamoto, S., (2002). *Prog. Theor. Phys.* **107**.

[457] Matsumoto, H., Sodano, P. and Umezawa, H., (1979). *Phys. Rev.* **D19**, p. 511.

[458] Matsumoto, H., Tachiki, M. and Umezawa, H., (1977). *Fortsch. Phys.* **25**, p. 273.

[459] Matsumoto, H. and Umezawa, H., (1976). *Fortsch. Phys.* **24**, p. 24.
[460] Matsumoto, H., Umezawa, H., Vitiello, G. and Wyly, J. K., (1974). *Phys. Rev.* **D9**, p. 2806.
[461] Matthews, P. T. and Salam, A., (1955). *Nuovo Cim.* **2**, p. 120.
[462] Mattis, D. C., (1965). *The Theory of Magnetism* (Harper and Row, New York).
[463] Mavromatos, N. E. and Nanopoulos, D. V., (1998). *Int. J. Mod. Phys.* **B12**, p. 517.
[464] Mavromatos, N. E. and Sarkar, S., (2008). *New J. Phys.* **10**, p. 073009.
[465] Mavromatos, N. E., Sarkar, S. and Tarantino, W., (2009). *Phys.Rev.* **D80**, p. 084046.
[466] Mazenko, G. F., (2002). *Fluctuations, Order and Defects* (Wiley-Interscience).
[467] McLeod, J. and Wang, C., (1999) Math-ph/9902002.
[468] Mekhfi, M., (1995). *Int. J. Theor. Phys.* **39**, p. 1163.
[469] Mercaldo, L., Rabuffo, I. and Vitiello, G., (1981). *Nucl. Phys.* **B188**, p. 193.
[470] Mermin, N. D., (1968). *Phys. Rev.* **176**, p. 250.
[471] Mermin, N. D., (1979). *Rev. Mod. Phys.* **51**, p. 591.
[472] Mermin, N. D. and Wagner, H., (1966). *Phys. Rev. Lett.* **17**, p. 1133.
[473] Mezard, M., Parisi, G. and Virasoro, M., (1987). *Spin Glass Theory and Beyond* (World Scientific, Singapore).
[474] Michel, L., (1980). *Rev. Mod. Phys.* **52**, p. 617.
[475] Mills, R., (1969). *Propagators for Many-Particle Systems* (Gordon and Breach Science Publisher, New York).
[476] Miransky, V. A., (1993). *Dynamical Symmetry Breaking in Quantum Field Theories* (World Scientific, Singapore).
[477] Moebius, D. and Kuhn, H., (1979). *Isr. J. Chem.* **18**, p. 382.
[478] Moebius, D. and Kuhn, H., (1988). *J. Appl. Phys.* **64**, p. 5138.
[479] Monaco, R., Mygind, J. and Rivers, R. J., (2002). *Phys. Rev. Lett.* **89**, p. 080603.
[480] Monaco, R., Mygind, J., Rivers, R. J. and Koshelets, V. P., (2009). *Phys. Rev.* **B80**, p. 180501(R).
[481] Moody, R. V. and Patera, J., (1991). *J. Phys.* **A24**, p. 2227.
[482] Morse, P. M. and Feshbach, H., (1953). *Methods of Theoretical Physics* (McGraw-Hill Book Co.).
[483] Mostafazadeh, A., (2001) . math-ph/0107001.
[484] Mumford, D., (1991). *Tata Lectures on Theta, III* (Birkhäuser, Boston).
[485] Nakagawa, K., Sen, R. N. and Umezawa, H., (1966). *Nuovo Cim.* **42**, p. 565.
[486] Nakanishi, N., (1973). *Prog. Theor. Phys.* **49**, p. 640.
[487] Nakanishi, N., (1973). *Prog. Theor. Phys.* **50**, p. 1388.
[488] Nakanishi, N. and Ojima, I., (1990). *Covariant Operator Formalism of Gauge Theories and Quantum Gravity*, Lect. Notes Phys., Vol. 27 (World Scientific, Singapore).
[489] Nambu, Y., (1960). *Phys. Rev. Lett.* **4**, p. 380.

[490] Nambu, Y., (1960). *Phys. Rev.* **117**, p. 648.

[491] Nambu, Y., (1979). *Prog. Theor. Phys.* **7**, p. 131.

[492] Nambu, Y. and Jona-Lasinio, G., (1961). *Phys. Rev.* **122**, p. 345.

[493] Nambu, Y. and Jona-Lasinio, G., (1961). *Phys. Rev.* **124**, p. 246.

[494] Naón, C. M., (1985). *Phys. Rev.* **D31**, p. 2035.

[495] Narnhofer, H., Requardt, M. and Thirring, W., (1983). *Commun. Math. Phys.* **92**, p. 247.

[496] Nelson, E., (1985). *Quantum Fluctuations* (Princeton University Press, Princeton).

[497] Netterfield, C. B. *et al.*, (2002). *Astrophys. J.* **571**, p. 604.

[498] Nielsen, H. B. and Olesen, P., (1973). *Nucl. Phys.* **B61**, p. 45.

[499] Nielsen, M. A. and Chuang, I. L., (2001). *Quantum Computation and Quantum Information* (Cambridge University Press, Cambridge, U.K.).

[500] Nieto, M. M. and Simmons Jr., L. M., (1978). *Phys. Rev. Lett.* **41**, p. 207.

[501] Nishi, C. C., (2008). *Phys. Rev.* **D78**, p. 113007.

[502] Nishiyama, S., Providencia, C., da Providencia, J., Tsue, Y. and Yamamura, M., (2005). *Prog. Theor. Phys.* **113**, p. 555.

[503] Oberlechner, G., Umezawa, M. and Zenses, C., (1978). *Lett. Nuovo Cim.* **23**, p. 641.

[504] Ohnuki, Y. and Kashiwa, T., (1978). *Prog. Theor. Phys.* **60**.

[505] Ojima, I., (1981). *Annals Phys.* **137**, p. 1.

[506] O'Raifeartaigh, L., (1979). *Rep. Prog. Phys.* **49**, p. 159.

[507] Ore, F., (1977). *Phys. Rev.* **D15**, p. 470.

[508] Osterwalder, K. and Schrader, R., (1973). *Comm. Math. Phys.* **31**, p. 83.

[509] Osterwalder, K. and Schrader, R., (1975). *Comm. Math. Phys.* **42**, p. 281.

[510] Papp, E., (1957). *Phys. Rev.* **105**, p. 1636.

[511] Parisi, G., (1988). *Statistical Field Theory (Frontiers in Physics, Vol. 66)* (Addison-Wesley, Redwood City).

[512] Pati, J. C. and Salam, A., (1974). *Phys. Rev.* **D10**, p. 275.

[513] Peccei, R. D., (1989) **38**, p. 503, in *CP Violation*, C. Jarlskog (Ed.) (World Scientific, Singapore).

[514] Peccei, R. D. and Quinn, H. R., (1977). *Phys. Rev.* **D16**, p. 1791.

[515] Peccei, R. D. and Quinn, H. R., (1977). *Phys. Rev. Lett.* **38**, p. 1440.

[516] Peitgen, H. O., Jürgens, H. and Saupe, D., (1986). *Chaos and Fractals. New Frontiers of Science* (Springer-Verlag, Berlin).

[517] Penrose, R., (1994). *Shadows of the Mind: A Search for the Missing Science of Consciousness* (Oxford University Press, Oxford, U.K.).

[518] Perelomov, A. M., (1972). *Commun. Math. Phys.* **26**, p. 222.

[519] Perelomov, A. M., (1986). *Generalized Coherent States and their Applications* (Springer-Verlag, Berlin).

[520] Pessa, E. and Vitiello, G., (2003). *Mind and Matter* **1**, p. 59.

[521] Pessa, E. and Vitiello, G., (2004). *Int. J. Mod. Phys.* **B18**, p. 841.

[522] Peterson, C. *et al.*, (2003). *Proc. Natl. Acad. Sci. U.S.A.* **100**, p. 13638.

[523] Pisarski, R., (1993). *Can. J. Phys.* **71**, p. 280.

[524] Polyakov, A. M., (1974). *JETP Lett.* **20**, p. 194.

[525] Polyakov, A. M., (1975). *Phys. Lett.* **B59**, p. 82.

[526] Prasad, M. K. and Sommerfield, C. M., (1975). *Phys. Rev. Lett.* **35**, p. 760.

[527] Preskill, J., (1984). *Annu. Rev. Nucl. Part. Sci.* **34**, p. 461.

[528] Pribram, K. H., (1991). *Brain and Perception* (Lawrence Erlbaum, Hillsdale, N. J.).

[529] Pribram, K. H., (1971). *Languages of the Brain* (Prentice-Hall, Engelwood Cliffs, N. J.).

[530] Pushkarov, D. I. and Pushkarov, K. I., (1977). *Phys. Stat. Sol.* **81b**, p. 703.

[531] Raichle, M., (2006). *Science* **314**, p. 1249.

[532] Rajaraman, R., (1982). *Solitons and Instantons. An Introduction to Solitons and Instantons in Quantum Field Theory* (North-Holland, Amsterdam).

[533] Rasetti, M., (1975). *Int. J. Theor. Phys.* **14**, p. 1.

[534] Ricciardi, L. M. and Umezawa, H., (1967). *Kibernetik* **4**, p. 44, Reprint in *Brain and Being*, G. G. Globus, K. H. Pribram and G. Vitiello (Eds.) (John Benjamins, Amsterdam, 2004) p. 255.

[535] Rice, S. O., (1950). *Mathematical Analysis of Random Noise – and Appendixes. Technical Publications Monograph B-1589* (Bell Telephone Labs Inc, New York).

[536] Rickayzen, G., (1965). *Theory of Superconductivity* (John Wiley and Sons, New York).

[537] Riesz, F. and Sz.-Nagy, B., (1955). *Functional Analysis* (F. Ungar Publ. Co., New York).

[538] Rindler, W., (1966). *Am. J. Phys* **34**, p. 1174.

[539] Rivers, R., (2001). *J. Low Temp. Phys.* **124**, p. 41.

[540] Rivers, R., Lombardo, F. and Mazzitelli, F., (2002). *Phys. Lett.* **B539**, p. 1.

[541] Rivers, R. J., (1987). *Path Integral Methods in Quantum Field Theory* (Cambridge University Press, Cambridge, U.K.).

[542] Rockafellar, R., (1972). *Convex Analysis* (Princeton University Press, Princeton).

[543] Ruutu, V. M. H. *et al.*, (1996). *Nature* **382**, p. 334.

[544] Ryder, L. H., (1985). *Quantum Field Theory* (Cambridge University Press, Cambridge, U.K.).

[545] Sachdev, S., (1999). *Quantum Phase Transitions* (Cambridge University Press, Cambridge, U.K.).

[546] Salam, A. and Strathdee, J., (1969). *Phys. Rev.* **184**, p. 1750.

[547] Saletan, E., (1961). *J. Math. Phys.* **2**, p. 1.

[548] Schmutz, M., (1978). *Z. Phys.* **B30**, p. 97.

[549] Schneider, T., (1981). *Phys. Rev.* **B24**, p. 5327.

[550] Schneider, T. and Stoll, E., (1982). *Phys. Rev.* **B25**, p. 4721.

[551] Schneider, T., Stoll, E. and Glaus, U., (1982). *Phys. Rev.* **B26**, p. 1312.

[552] Schrödinger, E., (1935). *Naturwiss.* **23**, pp. 807–812.

[553] Schrödinger, E., (1944). *What is Life?* (Cambridge University Press, Cambridge, U.K.).

[554] Schuch, D., (1997). *Phys. Rev.* **A55**, p. 935.

[555] Schulman, L. S., (1981). *Techniques and Applications of Path Integration* (Wiley, New York).

[556] Schwarz, A., (1977). *Phys. Lett.* **B 67**, p. 172.

[557] Schweber, S. and Wightman, A. S., (1955). *Phys. Rev.* **98**, p. 812.

[558] Schweber, S. S., (1961). *An Introduction to Relativistic Quantum Field Theory* (Harper & Row, New York).

[559] Schwinger, J. S., (1951). *Proc. Nat. Acad. Sci.* **37**, p. 455.

[560] Schwinger, J. S., (1961). *J. Math. Phys.* **2**, p. 407.

[561] Schwinger, J. S., (1951). *Phys. Rev.* **82**, p. 664.

[562] Schwinger, J. S., (1951). *Phys. Rev.* **82**, p. 914.

[563] Scott, A. C., Chu, F. Y. F. and McLaughlin, D. W., (1973). *Proc. IEEE* **61**, p. 1443.

[564] Segal, I. E., (1962). *Illinois J. Math.* **6**, p. 520.

[565] Segal, I. E., (1951). *Duke Math. J.* **18**, p. 221.

[566] Seiberg, N. and Witten, E., (1999). *JHEP* **09**, p. 032.

[567] Semenoff, G. W., Matsumoto, H. and Umezawa, H., (1982). *Phys. Rev.* **D25**, p. 1054.

[568] Sen, R. N. and Umezawa, H., (1967). *Nuovo Cim.* **50**, p. 53.

[569] Sewell, G. L., (1986). *Quantum Theory of Collective Phenomena* (Clarendon Press, Oxford, U.K.).

[570] Shah, M. N., Umezawa, H. and Vitiello, G., (1974). *Phys. Rev.* **B10**, p. 4724.

[571] Shah, M. N. and Vitiello, G., (1975). *Nuovo Cim.* **B30**, p. 21.

[572] Shaposhnikov, M. E., (1993). *Class. Quant. Grav.* **10**, p. S147.

[573] Sivakami, S. and Srinivasan, V., (1983). *J. Theor. Biol.* **102**, p. 287.

[574] Sivasubramanian, S., Srivastava, Y. N., Vitiello, G. and Widom, A., (2003). *Phys. Lett.* **A311**, p. 97.

[575] Spano, F. C., Kuklinski, J. R. and Mukamel, S., (1990). *Phys. Rev. Lett.* **65**, p. 211.

[576] Srivastava, Y. N., Vitiello, G. and Widom, A., (1995). *Annals Phys.* **238**, p. 200.

[577] Stapp, H. P., (2009). *Mind, Matter an Quantum Mechanics* (Springer-Verlag, Berlin).

[578] Steiner, M., Villain, J. and Windsor, C. G., (1976). *Adv. Phys.* **25**, p. 87.

[579] Sterman, G., (1993). *An Introduction to Quantum Field Theory* (Cambridge University Press, Cambridge, U.K.).

[580] Stone, M. H., (1930). *Proc. Nat. Acad. U.S.A.* **16**, p. 172.

[581] Streater, R. F. and Wightmann, A. S., (1964). *PCT, Spin and Statistics, and all that* (Benjamin, London, U.K.).

[582] Stuart, C. I. J., Takahashi, Y. and Umezawa, H., (1978). *J. Theor. Biol.* **71**, p. 605.

[583] Stuart, C. I. J., Takahashi, Y. and Umezawa, H., (1979). *Found. Phys.* **9**, p. 301.

[584] Su, W., Schriffer, J. R. and Heeger, A., (1979). *Phys. Rev.* **B22**, p. 1698.

[585] Sugi, M., (1985). *J. Molecular Electronics* **1**, p. 3.

[586] Symanzyk, K., (1954). *Zeit. Naturf.* **9**, p. 809.

[587] 't Hooft, G., (1974). *Nucl. Phys.* **B79**, p. 276.

[588] 't Hooft, G., (1976). *Phys. Rev. Lett.* **37**, p. 4.

[589] 't Hooft, G., (1999). *Class. Quant. Grav.* **16**, p. 3263.
[590] 't Hooft, G., (2001), p. 397 in *Basics and Highlights in Fundamental Physics*, Erice, A. Zichichi (Ed.) (World Scientific, Singapore).
[591] 't Hooft, G., (2001), p. 307 in *Quantum [Un]speakables, From Bell to Quantum Information*, R. A. Bertlmann and A. Zeilinger (Eds.) (Springer-Verlag).
[592] 't Hooft, G., (2001) in *PASCOS01*, P. Frampton and J. Ng (Eds.) (Princeton Unversity Press, Princeton, N.J.).
[593] 't Hooft, G., (2003). *Int. J. Theor. Phys.* **42**, p. 355.
[594] 't Hooft, G., (2007). *AIP Conf. Proc.* **957**, p. 154, arXiv:0707.4568 [hep-th].
[595] 't Hooft, G., (2007). *J. Phys.: Conf. Ser.* **67**, p. 012015, arXiv:quant-ph/0604008.
[596] 't Hooft, G., (1976). *Phys. Rev.* **D14**, pp. 3432–3450.
[597] Tajiri, M., (1984). *J. Phys. Soc. Jap.* **53**, p. 3759.
[598] Takahashi, Y., (1957). *Nuovo Cim.* **6**, p. 371.
[599] Takahashi, Y., (1969). *An Introduction to Field Quantization* (Pergamon Press, Oxford, U.K.).
[600] Takahashi, Y. and Umezawa, H., (1996). *Int. J. Mod. Phys.* **B10**, p. 1755, Originally in *Collective Phenomena* **2** (1975) 55.
[601] Tarasov, V. E., (2001). *Phys. Lett.* **A288**, p. 173.
[602] Tataru-Mihai, P. and Vitiello, G., (1982). *Physica* **A114**, p. 229.
[603] Tataru-Mihai, P. and Vitiello, G., (1982). *Lett. Math. Phys.* **6**, p. 277.
[604] Taylor, J. G., (1978). *Annals Phys.* **115**, p. 153.
[605] Terra Cunha, M., Dunningham, J. and Vedral, V., (2007). *Proc. Roy. Soc.* **A463**, p. 2277.
[606] Toda, M., (1967). *J. Phys. Soc. Japan* **22**, p. 431.
[607] Toda, M., (1967). *J. Phys. Soc. Japan* **23**, p. 501.
[608] Toda, M., (1970). *Progr. Theor. Phys. Suppl.* **45**, p. 174.
[609] Toulouse, G. and Kléman, M., (1976). *J. Phys. Lett.(Paris)* **137**, p. L149.
[610] Tsue, Y., Kuriyama, A. and Yamamura, M., (1994). *Prog. Theor. Phys.* **91**, p. 469.
[611] Tsukerman, I. S., (2010) arXiv:1006.4989 [hep-ph].
[612] Turner, M., (1990). *Phys. Rept.* **197**, p. 67.
[613] Umezawa, H., (1965). *Nuovo Cim.* **40**, p. 450.
[614] Umezawa, H., (1974) , p. 275 in *Renormalization and Invariance in Quantum Field Theory*, E. R. Caianiello (Ed.) (Plenum Press, New York).
[615] Umezawa, H., (1994), p. 109 in *Thermal Field Theories and their Applications* F. C. Khanna, R. Kobes, G. Kunstatter and H. Umezawa (Eds.) (World Scientific, Singapore).
[616] Umezawa, H., (1995). *Math. Japonica* **41**, p. 109.
[617] Umezawa, H., (1993). *Advanced Field Theory: Micro, Macro, and Thermal Physics* (AIP, New York).
[618] Umezawa, H., (1956). *Quantum Field Theory* (North-Holland, Amsterdam).
[619] Umezawa, H., Matsumoto, H. and Tachiki, M., (1982). *Thermo Field Dynamics and Condensed States* (North-Holland, Amsterdam).

[620] Umezawa, H. and Visconti, A., (1955). *Nuovo Cim.* **1**, p. 1079.
[621] Umezawa, H. and Vitiello, G., (1985). *Quantum Mechanics* (Bibliopolis, Napoli, Italy).
[622] Umezawa, H. and Yamanaka, Y., (1991). *Phys. Lett.* **A155**, p. 75.
[623] Vachaspati, T., (2006). *Kinks and Domain Walls: An Introduction to Classical and Quantum Solitons* (Cambridge University Press, Cambridge, U.K.).
[624] van Enk, S. J., (2005). *Phys. Rev.* **A72**, p. 064306.
[625] Van Hove, L., (1951). *Acad. Roy. de Belgique Bulletin* **37**, p. 1055.
[626] Van Hove, L., (1952). *Physica* **18**, p. 145.
[627] van Nieuwenhuizen, P. and Weldron, A., (1996). *Phys. Lett.* **B389**, p. 29.
[628] van Weert, C. G., (1994), p. 1 in *Thermal Field Theories and their Applications* F. C. Khanna, R. Kobes, G. Kunstatter and H. Umezawa (Eds.) (World Scientific, Singapore).
[629] Vilasi, G., (2001). *Hamiltonian Dynamics* (World Scientific, Singapore).
[630] Vilenkin, A., (1985). *Phys. Rept.* **121**, p. 263.
[631] Vilenkin, A. and Shellard, E., (1994). *Cosmic Strings and other Topological Defects* (Cambridge University Press, Cambridge, U.K.).
[632] Vitiello, G., (2004). *Int. J. Mod. Phys.* **B18**, p. 785.
[633] Vitiello, G., (2007), p. 165 in *Quantum Analogues: From Phase Transitions to Black Holes and Cosmology*, W. G. Unruh and R. Schuetzhold (Eds.), Lectures Notes in Physics 718 (Springer, Berlin).
[634] Vitiello, G., (2009). *New Mathematics and Natural Computation* **5**, p. 245.
[635] Vitiello, G., (2009), p. 6 in *Quantum Interaction*, P. Bruza and D. Sofge *et al* (Eds.), Lecture Notes in Artificial Intelligence, Edited by R. Goebel and J. Siekmann and W. Wahlster (Springer-Verlag, Berlin, Heidelberg).
[636] Vitiello, G., (2005). *Braz. J. Phys.* **35**, p. 351.
[637] Vitiello, G., (2000), p. 171 in *Topological Defects and the Non-Equilibrium Dynamics of Symmetry Breaking Phase Transitions*, Y. M. Bunkov and H. Godfrin (Eds.) , (Kluwer, Dordrecht).
[638] Vitiello, G., (2008). *Int. J. Theor. Phys.* **47**, p. 393.
[639] Vitiello, G., (1976), p. 405 in *Functional and Probabilistic Methods in QFT*, Vol. 1, J. Lopúszanski and B. Jancewicz (Eds.), (Wrocław).
[640] Vitiello, G., (1976). *Phys. Lett.* **A58**, p. 293.
[641] Vitiello, G., (1995). *Int. J. Mod. Phys.* **B9**, p. 973.
[642] Vitiello, G., (1982). *Czech. J. Phys.* **B32**, p. 575.
[643] Vitiello, G., (1998), p. 321 in *Toward a Science of Consciousness II. The Second Tucson Discussions and Debates*, S. R. Hameroff, A. W. Kaszniak and A. C. Scott (Eds.) (MIT Press, Cambridge, MA).
[644] Vitiello, G., (1975). *Dynamical Rearrangement of Symmetry*, Vol. B36 (Diss. Abstr. Int.).
[645] Vitiello, G., (2001). *My Double Unveiled* (John Benjamins, Amsterdam).
[646] Volovik, G. E., (2003). *The Universe in a Helium Droplet* (Clarendon Press, Oxford, U.K.).
[647] Volovik, G. E. and Mineev, V. P., (1977). *JETP* **45**, p. 1186.
[648] von Neumann, J., (1931). *Math. Ann.* **104**, p. 570.

[649] von Neumann, J., (1955). *Mathematical Foundations of Quantum Mechanics* (Princeton University Press, Princeton, N.J.).

[650] Wadati, M., Matsumoto, H. and Umezawa, H., (1978). *Phys. Lett.* **B73**, p. 448.

[651] Wadati, M., Matsumoto, H. and Umezawa, H., (1978). *Phys. Rev.* **D18**, p. 520.

[652] Wagner, D., (1972). *Introduction to the Theory of Magnetism* (Pergamon Press, Oxford, U.K.).

[653] Wagner, M., (1975). *Phys. Lett.* **A53**, p. 1.

[654] Wang, X.-B., Kwek, L. C., Liu, Y. and Oh, C. H., (2001). *Phys. Rev.* **D63**, p. 053003.

[655] Wang, X. F. and Chen, G. R., (2003). *IEEE Trans. Circuits Syst.* **31**, p. 6.

[656] Ward, J. C., (1950). *Phys. Rev.* **78**, p. 182.

[657] Weimar, E., (1966). *Phys. Rev. Lett.* **17**, p. 616.

[658] Weimar, E., (1973). *Nuovo Cim.* **15B**, p. 245.

[659] Weiss, U., (1999). *Quantum Dissipative Systems* (World Scientific, Singapore).

[660] Weyl, H., (1927). *Zeit. Phys.* **46**, p. 1.

[661] Weyl, H., (1931). *The Theory of Groups and Quantum Mechanics* (Dover, New York).

[662] Wilczek, F., (1978). *Phys. Rev. Lett.* **40**, p. 279.

[663] Witten, E., (1983). *Nucl. Phys.* **B223**, p. 422.

[664] Yuen, H. P., (1976). *Phys. Rev.* **A13**, p. 2226.

[665] Zahkharov, V. E. and Shabat, A. B., (1972). *Sov. Phys. JETP* **34**, p. 62.

[666] Zee, A., (2003). *Quantum Field Theory in a Nutshell* (Princeton University Press, Princeton, N.J.).

[667] Zeh, H. D., (1970). *Found. Phys.* **1**, p. 69.

[668] Zinn-Justin, J., (2007). *Phase Transitions and Renormalization Group* (Oxford University Press, Oxford, U.K.).

[669] Zralek, M., (1998). *Acta Phys. Pol.* **B29**, p. 3925.

[670] Zuo, J. and Gui, Y.-X., (1998) arXiv:hep-ph/9803235.

[671] Zuo, J. and Gui, Y.-X., (1995). *J. Phys.* **A28**, p. 4907.

[672] Zurek, W. H., (1996). *Phys. Rept.* **276**, p. 177.

[673] Zurek, W. H., (1981). *Phys. Rev.* **D24**, p. 1516.

[674] Zurek, W. H., (1982). *Phys. Rev.* **D26**, p. 1862.

[675] Zurek, W. H., (1985). *Nature* **317**, p. 505.

[676] Zurek, W. H., (1993). *Acta Phys. Pol.* **B7**, p. 1301.

[677] Zurek, W. H., Bettencourt, L. M. A., Dziarmaga, J. and Antunes, N. D., (2000). *Acta Phys. Pol.* **B31**, p. 2937.

Index

η-ξ spacetime, 261
  Euclidean section, 261, 264
  fields, 264
  Lorentzian section, 265, 268
    extended, 263
$\lambda\phi^4$ system, 460
$\lambda\phi^4$ kink, 304, 431
Q-ball, 463
S matrix, 238
σ model
  non-linear, 283, 454
θ vacuum, 476
't Hooft symbol, 471, 473
$U(1)$ Higgs model, 434
1PI thermal diagrams, 251

Abrikosov vortex, 340, 434
Adler theorem, 163
ADMH construction, 473
Algebra
  $e(2)$, 143, 181
  $e(3)$, 179
  $h(1)$, 207
  q-deformation, 21, 51, 194, 211, 381
  q-deformed $h(1)$, 207, 208
  q-deformed $su(2)$, 208
  q-deformed Hopf, 21, 26, 185, 206, 381
  q-Weyl–Heisenberg, 25, 49
  $su(1,1)$, 14, 46, 197, 381, 409
  $su(2)$, 76, 79, 80, 169, 180, 197, 208
  graded Hopf, 23

Hopf, 22, 26, 185, 381
  structure constants, 42, 362
  Virasoro, 179, 181
  Weyl, 11
  Weyl–Heisenberg, 9, 10, 15, 22, 143, 168
Angular momentum, 171, 207, 409
Antikink, 430
Asymptotic condition, vii, 357

Bäcklund
  parameter, 483
  quantum image, 324
  transformation, 324, 460, 483
Back-reaction potential, 334
Background
  curved, 374
  geometric, 260
  Minkowski, 261
  topologically non-trivial, 339
Baker–Hausdorff formula, 60
Band index, 477
Base space, 446
Basis, 4, 12, 19, 40, 266
Bianchi identity, 461
Bloch
  electrons, 374, 389
  function, 22, 26, 52, 477
  state, 477
  theorem, 476
Bogoliubov transformation, 13, 14, 60, 68, 194, 266, 382

Bogomol'nyi
  bound, 424, 458–460
  equation, 459–462
  trick, 456
Born approximation, 300, 314
Boson
  condensation, 20, 106, 138, 154,
      158, 165, 169, 312, 349, 357,
      368
  condensation function, 158, 159,
      310
  current, 152, 157
  non-homogeneous condensation,
      107, 126, 138, 154, 158, 221,
      227, 298
  transformation, 106, 153, 155, 299,
      349, 353
  transformation function, 152, 158,
      323, 349
  transformation theorem, 153, 300,
      307, 351
Bott periodicity theorem, 447, 494
Bott–Tu theorem, 455, 494
BPS $SO(3)$ monopoles, 461
BPS limit, 461
BPS states, 460
Brain dynamics
  dissipative many-body model of
      brain, 396, 402

Canonical
  commutation relations, 9, 18, 55,
      62, 98, 107, 159, 175, 189,
      202, 221, 226, 373
  transformation, 13, 16, 60, 63, 65
Casimir operator, 44, 46, 197, 207,
      381, 406
Cauchy sequence, 9, 12, 35
Cauchy–Schwarz inequality, 464
Chern number (second), 468
Chern–Simons term, 374, 389, 468,
  476
Classical limit, 277, 301, 317, 379
Closed time-path formalism, 235, 268
Coherent state, 37, 44, 49, 223, 326
  Fock–Bargmann, 37

  generalized, 23, 41, 51, 231, 275,
      279, 283, 382
    Bloch, 281
    spin, 281
    SU(1,1), 46, 195, 209, 231,
      282, 403
    SU(2), 44, 45, 76, 197, 281
  Glauber, xi, 19, 37, 41, 275
  in W_n, 278
  over-completeness (See also
      Over-complete set), 39
  squeezed, 13, 14, 25, 26, 30, 214,
      383, 384, 403
Collective
  coordinates, 304, 323, 424, 457,
      472, 480
  modes, 3, 168, 173, 175
Complete
  set, 3, 19, 39, 405
  space, 10, 36
Consciousness, 418
Continuous mass spectrum, 202
Convex envelope, 288
Cooper pair, 438
Correlation length, 228, 328, 437
  equilibrium, 444
  physical, 444
Coset space (see also Group), 42, 47,
  134, 280, 366, 448
Cosmic strings, 452
Countable
  basis, 7, 13, 35
  set, 4, 19, 35
Cross product theorem, 494
Cumulant, 250

Decoherence, 174, 177
Defect
  core, 445
  formation, 154, 158, 228, 298
  topological, 158
Dense set, 32, 35
Derrick
  Derrick–Hobart theorem, 423, 466
  scaling, 453
  theorem, 453

Deterministic systems, 374, 405
Dirac
  equation, 218, 334
  index, 482
Disclination, 448
  π, 449
Dissipation, 210, 373, 404
  dissipative noncommutative plane,
    385
  dissipative phase, 213, 389
  quantum, 213, 379, 385
Divergent singularities, 301
Double, 418
Doubling degrees of freedom, 186,
  188, 206, 235, 290, 374, 380
Dynamical map, 54, 60, 63, 69, 100,
  138, 144, 153, 154, 237, 299, 343,
  370
Dyson theorem, 163

Effective
  action, 235, 252
  potential, 291
Entanglement, 71, 91, 214, 215, 383,
  404, 418
Entire analytic function, 23, 29, 30,
  40, 49
Entropy, 212, 214, 234, 383, 409
Exact homotopy sequences, 445

Feynman–Matthews–Salam formula,
  246
Fiber, 446
Fiber bundle, 445, 446
  principal, 446
Field
  asymptotic, 33, 57, 97, 160, 164,
    299, 349, 356
  complex scalar, 247
  coset space, 282, 283
  free, 56, 200, 203, 237, 272, 317
  gauge, 106, 119, 129, 169, 215, 349,
    364, 434
  ghost, 125, 355
  Heisenberg, 33, 54, 116, 137, 147,
    237, 299, 350, 356

Higgs, 439
  interpolating, 57, 70, 114, 144
  physical, 32, 54, 56, 69, 100, 144,
    163
  translation, 56, 145, 154, 165
  Yang–Mills, 220
First Brillouin zone, 477
Flavor mixing, 71
Fock space, 7, 9, 16, 30–32, 53, 60,
  70, 107, 192, 330
Fock–Bargmann representation, 22,
  23, 26, 40, 49
Fractals, x, 26, 28, 421
  fractal dimension, 27
  Koch fractal, 28
  operator, 29
  self-similarity, 26
Fractional charge, 340
Free energy, 210, 212, 222, 225, 226,
  358, 383, 409
Fresnel integral, 243
Friedberg–Lee model, 463
Functional
  generating, 109, 265
    thermal, 241, 249
  integral, 19, 45, 48, 108, 110, 144,
    240, 243, 367
  integral for coherent states
    generalized, 279
    Glauber, 275
Functional methods
  at finite temperature, 236

Gauge (see also Transformation and
  Field)
  condition, 120
  linear, 202
  potential, 435, 473
  temporal, 434, 474
  Weyl, 434, 439
Gaussian decomposition, 73
Gell-Mann–Low formula, 236
Geometric phase, 406
Georgi-Glashow model, 438
  $SO(3)$, 438
  $SU(2)$, 450

$SU(5)$, 451
$SU(5)$ Grand Unified theory, 451
Ginzburg regime, 227
Ginzburg–Landau parameter κ, 437
Goldstone
    model, 130
    Nambu–Goldstone particle, 101,
        126, 137, 163, 196, 283, 299,
        302, 309, 349
    theorem, 97, 112, 126, 165, 175
Gradient expansion, 291
Grand Unified Theory, 451
Green's function, 111, 238, 299, 308,
    348, 351
    1PI, 256
    connected, 255
        thermal, 249
    in Mills' representation, 249
    two-point, 113, 126, 242, 348
Group
    $E(2)$, 142, 143, 166
    $E(3)$, 179
    $SO(2)$, 172
    $SO(3)$, 42, 179, 203, 354, 362, 438,
        448
    $SO(4)$, 179, 480
    $SU(1,1)$, 7, 14, 46, 195, 231, 282,
        381
    $SU(2)$, 1, 44, 97, 115, 118, 139,
        161, 166, 179, 231, 281, 365,
        368
    $SU(3)$, 164, 480
    $U(1)$, 44, 47, 52, 112, 119, 126, 139,
        143, 349
    center, 16
    contraction, 142, 159, 166, 179, 370
    coset, 42, 43, 134, 283, 363, 448
    cyclic $\mathbb{Z}_2$, 449
    dihedral, 448
    factor, 42
    generators, 42, 369
    Haar measure, 43
    homotopy, 159
    little, 164, 363
    maximal subgroup, 47, 164
    measure, 43

normal subgroup, 42
Poincaré, 202, 390
quotient, 42, 103
stability, 41, 103, 133
Weyl–Heisenberg, 16, 19

Haag
    expansion, 237, 299, 300, 312
    theorem, 54, 66, 236
Heisenberg equation, 68, 69, 130, 210,
    300, 305, 334
Higgs mechanism, 123, 149, 169, 349
Hilbert space, 3, 9, 35, 40, 41, 53, 54,
    59, 64, 69, 192, 196, 201, 210, 235,
    370, 382, 405
    separable, 4, 19, 35, 64
Holstein–Primakoff representation,
    168, 328
Homotopical equivalence, 488
Homotopy, 488
    classes, 493
    classification, 447
    group, 442
        fundamental π_1, 435, 487
        higher orders, 445, 492
        of Lie groups, 495
    index, 469
    theory, 487
Hopf bundle, 446

Imaginary-time formalism, 257–259,
    287
Index
    Keller–Maslov, 244
    Morse, 244
Inequality
    Bogoliubov, 293
    Jensen, 294
Inflation, 374
Infrared effect, 100, 159, 164
Inönü–Wigner group contraction, 102
Instanton (see also Soliton), 165, 424,
    467
    $SU(2)$, 470
    Belavin–Polyakov, 457
    number, 468

Yang–Mills, 467
Invariance, 97, 101, 117, 137, 144, 221
  chiral, 164
  gauge, 173, 184, 215, 371
  phase, 1, 112, 170
  rotational, 1, 170, 183
  time-reversal, 213, 383
  translational, 1, 20, 50, 56, 104, 246
Irreducible
  representation, 10, 13, 16, 46, 59,
       62, 67, 101, 162, 165
  set, 21, 58, 153

J-isomorphism, 494
Jaynes–Cummings Hamiltonian, 167

Kibble–Zurek mechanism, 174, 442
Kink (see also Soliton), 226, 304, 309,
  312, 314, 317, 349, 430, 431
Korteweg–de Vries soliton, 424
Kubo–Martin–Schwinger condition,
  205, 235, 243, 245, 253

Landau free energy, 290
Landau–Ginzburg equation, 251, 292
Landau–Lifshitz equation, 286
Laplace–Beltrami operator, 264
Legendre transform, 251, 287
Legendre–Fenchel transform, 287–289
Lehmann–Symanzik–Zimmermann
  formula, 57, 140
Limit
  strong, 9, 35, 70
  weak, 9, 35, 70
London penetration depth λ, 437
Lorentz gauge, 120, 124, 150, 219
Lorentz invariance
  breakdown, 390
Low energy
  behavior, 164
  theorem, 126, 163

Macroscopic quantum systems, ix,
  138, 165, 166, 168, 174, 297
Magnon, 114, 118, 139, 143, 328, 332
  anti-ferromagnetic, 286

equation, 142
  ferromagnetic, 286
Matsubara
  formalism, 258
  frequency, 258
  propagator, 265
  summation, 259
Maxwell equations
  classical, 138, 150, 152, 349
Meissner current, 152, 157
Meissner–Ochsenfeld effect, 438
Mermin–Wagner–Coleman theorem,
  135
Moduli space, 474, 480
Monopole (see also Soliton), 138, 226,
  340, 349, 354, 450
  't Hooft–Polyakov, 438
  BPS $SO(3)$, 461
  in GUT, 451
Multi-kink, 485

Neural activity, 396
Neutrino
  mixing, 91, 390, 413
  oscillations, 91, 413
Noether charge, 463
Non-linear $O(3)$ σ model, 455
Noncommutative geometry, 213, 385

Operator
  bounded, 10, 16, 36
  doubling, 207
  energy, 30
  extension, 10, 12
  finite difference, 24
  Hamiltonian, 32, 58, 68
  Hermitian, 4, 304, 344
  momentum, 30, 67, 210
  unbounded, 6, 10, 70
  unitary, 16, 19, 56, 57, 60, 66, 88,
    214, 327
  Weyl, 11
Order, 97
  parameter, viii, 112, 128, 155, 221,
    251, 301, 310, 351
    space, 103, 133, 134, 442, 448

Ordered
  pattern, 99, 137, 158, 166
  state, 128, 298
Oscillator
  damped harmonic, 379
  harmonic, 208, 385, 406
Over-complete set, 19, 39, 49, 266

Particle mixing, 215, 390
Partition function, 240
Path integral, 38, 277, 376
  for coherent states
    fermionic, 279
    generalized, 281
    single mode, 278
Pati–Salam model, 451
Phase transition, 19, 99, 107, 155,
  158, 176, 185, 226, 234, 235, 287,
  298, 354, 361
Phonon, 106, 328
  equation, 332
Pontryagin number, 368, 468
Product
  T-ordered, 236
  T^*-ordered, 237
  T_C-ordered, 237
  $\bar{T}$-ordered, 248
  covariant, 237
  normal
    thermal, 273
Projective space
  complex $P_1(\mathbb{C})$, 447
  real $P_2(\mathbb{R})$, 448

Quantization, 368, 373
Quantum Brownian motion, 214, 374
Quantum coordinate, 302

Real time formalism, 186, 268
Reduction formula, 141, 146
Renormalization, vii, 2, 3, 55, 99,
  114, 140, 146, 187, 239
Representation
  adjoint, 42, 362, 438, 482
  defining, 42
  fundamental, 42, 208, 365, 438

regular, 42
unitary, 16, 41
Rindler
  coordinates, 263
  spacetime, 260
  transformation, 263

S matrix, 144, 148, 351
Schrödinger equation, 188, 189, 334
  Gross–Pitaevskii-type, 329, 336
  non-linear (see also Soliton), 327,
    334
Schur lemma, 43
Schwinger–Dyson equations, 205
  thermal, 252
Self-consistent method, 68
Self-similarity, 26, 30, 421
  self-similarity dimension, 27
Sequence
  exact, 445
  exact homotopy, 445
  fiber homotopy, 445
  short exact, 445
Sine-Gordon (see also Soliton)
  antikink, 430
  equation, 322, 431, 483
  kink, 430
  model, 319
  soliton, 319, 429
  system, 429, 460, 483
Soliton, x, 187, 298, 327, 354, 423,
  466
  N-soliton, 323
  antikink, 317
  coherent state representation, 325
  Davydov, 332
  in Scheibe aggregates, 336
  instanton, 349, 369
  kink, 226, 298, 304, 309, 312, 314,
    317, 431
  monopole, 159, 226, 354, 366
    't Hooft–Polyakov, 340,
      362–364, 438
    Dirac, 364
  Nielsen–Olesen vortex, 340, 434
  non-linear Schrödinger, 327

non-topological, 463
on ferromagnetic chain, 328, 332
ring, 336
sine-Gordon, 235, 298, 319, 429
sphaleron, 349, 365
superfluid vortex, 344
vortex, 138, 157, 174, 235, 297,
    309, 349, 350, 359, 397
Sphaleron (see also Soliton), 159
Spin wave, 118, 286
anti-ferromagnetic, 286
Statistics
Bose–Einstein, 193, 212, 213, 249,
    318
Fermi–Dirac, 249
String
cosmic, 452
GUT $\mathbb{Z}_2$, 452
Strong CP problem, 424, 480
Superconductor
type I, 437
type II, 434
Symmetry
basic, 1, 164
broken phase, 103
BRST, 355
conformal, 369
continuous - group, 99, 129, 139
discrete, 128
dynamical rearrangement of, viii,
    53, 99, 106, 112, 128, 139,
    143, 160, 164, 370
dynamically broken, 291
emergent, 304
explicit breakdown of, 98
generators, 101, 105, 164
global, 112, 119, 143, 145
local gauge, 112, 119, 172, 184, 369
loop-antiloop, 179
ordered phase, 103, 287
permutation, 241
phase, 143, 172
phenomenological, 164
restoration, 100, 128, 169, 224, 354,
    361

spontaneous breakdown, ix, 20, 53,
    97, 104, 118, 137, 155, 164,
    172, 196, 221, 223, 226, 229,
    252, 369, 447
finite volume effects, 126

Target space, 282
Thermo Field Dynamics, 186, 191,
    224, 235, 309, 367, 383
Theta functions, 22, 49
Thin-wall approximation, 464
Thirring model, 235, 323
Tilde-conjugation rule, 198, 270
Time path
closed, 237, 238
Keldysh–Schwinger, 235, 245
Schwinger, 237, 242
Toda lattice, 334
Topological
charge, 435, 455, 457, 458
index, 430
singularities, 138, 155, 156, 173,
    297, 349
vacuum index, 475
Trajectories, 226, 231, 373, 382
chaotic, 232, 417
Transformation
chiral gauge, 124
gauge, 152, 155, 350, 434
large gauge, 476
local gauge, 170, 215
small gauge, 476
Tunneling, 474
Two-slit experiment, 189

Uniaxial nematics, 448
Unitarily inequivalent
Hilbert spaces, 236
representations, 88, 98, 107, 159,
    175, 185, 189, 210, 214, 231,
    379
Unruh effect, 261
Unstable states, 374

Vacuum manifold, 102, 442
Van Hove model, 59, 70

Variational method, 309
Vertex function, 114
  thermal, 249, 251
von Neumann
  lattice, 40, 50
  theorem, viii, 17, 55, 64, 107, 176
Vortex (see also Soliton)
  equation, 360
  Nielsen–Olesen, 340, 434
  superfluid, 344

Ward–Takahashi identities, 113, 120
Wess–Zumino action, 284

Wick
  rotation, 467
  theorem, 223, 271, 273, 313
    thermal, 239, 271
Wightman's function
  thermal, 272
Winding number, 435, 444

Yang–Mills
  vacuum, 474

Zero modes, 340, 370, 424, 480

Printed in the United States
By Bookmasters